Springer Series in Computational Mathematics

2

John R. Rice
Ronald F. Boisvert

Solving Elliptic Problems Using ELLPACK

With 53 Illustrations

Springer-Verlag
New York Berlin Heidelberg Tokyo

John R. Rice

Department of Computer Science, Purdue University
West Lafayette, IN 47907, U.S.A.

Ronald F. Boisvert

Scientific Computing Division, National Bureau of Standards
Gaithersburg, MD 20899, U.S.A.

AMS Subject Classifications: 35J99, 65M99, 65N99

Library of Congress Cataloging in Publication Data
Rice, John R.
Solving elliptic problems using ELLPACK.
(Springer series in computational mathematics ; 2)
Bibliography: p.
Includes index.
1. ELLPACK (Computer system) 2. Differential
equations, Elliptic—Numerical solutions—Data processing.
3. Boundary value problems—Numerical solutions—Data
processing. I. Boisvert, R. F. II. Title. III. Series.
QA377.R53 1984 515.3'53 84-10588

This is to certify that a portion of the book named above was prepared by a United States Government employee as part of his official duties.

Printed and bound by R.R. Donnelley & Sons, Harrisonburg, Virginia.
Printed in the United States of America.

9 8 7 6 5 4 3 2 1

ISBN 0-387-90910-9 Springer-Verlag New York Berlin Heidelberg Tokyo
ISBN 3-540-90910-9 Springer-Verlag Berlin Heidelberg New York Tokyo

PREFACE

ELLPACK is a many faceted system for solving elliptic partial differential equations. It is a forerunner of the very high level, problem solving environments or expert systems that will become common in the next decade. While it is still far removed from the goals of the future, it is also far advanced compared to the Fortran library approach in common current use. Many people will find ELLPACK an easy way to solve simple or moderately complex elliptic problems. Others will be able to solve really hard problems by digging a little deeper into ELLPACK.

ELLPACK is a research tool for the study of numerical methods for solving elliptic problems. Its original purpose was for the evaluation and comparison of numerical software for elliptic problems. Simple examples of this use are given in Chapters 9–11. The general conclusion is that there are many ways to solve most elliptic problems, there are large differences in their efficiency and the most common ways are often less efficient, sometimes dramatically so.

ELLPACK is a research effort in the cooperative creation of a large software system. It contains contributions from over 25 people or sources. It is very large, perhaps it has a total of 80,000 lines of Fortran. A great deal of effort has been put into creating a coherent software structure that can adapt to the widely varying requirements of these people and which allows ELLPACK to grow or shrink gracefully as time and objectives change. The scientific systems programmer can examine our solution to the software engineering problems and perhaps avoid some of the mistakes we made in preliminary versions of ELLPACK.

ELLPACK owes a huge debt to its contributors. These are identified in Section 1.C and specific software contributions are associated with their authors in Chapter 6. A huge dept is also owed to the many funding agencies that supported this work. Our personal efforts were supported by the National Science Foundation, the Department of Energy, Purdue University and the National Bureau of Standards; we thank them. The total cost of developing the software in ELLPACK is of the order of $2 million. Due to the generous funding agencies, ELLPACK is made publicly available for a nominal cost.

Special thanks is due to Connie Sutherlin who prepared many versions of this manuscript, and related reports, on the TROFF system at Purdue University. Her diligence and perseverance are much appreciated. We also thank Calvin Ribbens and Roger Crawfis for valuable assistance in preparing materials for this book. Robert Lynch made many improvements in the exposition, the style and the layout of the book; we are grateful for his very substantial help.

John R. Rice
Purdue University

Ronald F. Boisvert
National Bureau of Standards

November, 1984

CONTENTS

Part 1: THE ELLPACK SYSTEM

Part 2: THE ELLPACK MODULES

Part 3: PERFORMANCE EVALUATION

Part 4: CONTRIBUTOR'S GUIDE

Part 5: SYSTEM PROGRAMMING GUIDE

APPENDICES

PART 1

CHAPTERS 1–5

THE ELLPACK SYSTEM

Part 1 is the user's guide to the ELLPACK system. It describes the structure and objectives of the system, the facilities it has and gives many examples of its application.

Chapter 1

INTRODUCTION

Elliptic partial differential equations (PDEs) are important tools for mathematical modellers in a wide variety of fields. Indeed, many important advances in structural mechanics, atmospheric modelling, nuclear reactor design, electrostatics, and chemical engineering have depended on the ability to solve elliptic equations quickly and accurately. As a result, much research activity in the past 30 years has been directed at improving numerical methods for this class of problems, and the fruits of some of this work may now be found in the growing collection of general-purpose mathematical software for solving elliptic problems [Boisvert and Sweet, 1984]. Our aim in this book is to describe one such package in detail.

ELLPACK is a software system for solving elliptic boundary value problems which includes both a very high-level problem statement language and an extensible library of problem-solving modules. The ELLPACK language is an extension to Fortran that allows users to declare boundary value problems. Powerful executable statements are available which invoke predefined modules for each stage of the solution process. Since a variety of modules representing different methods are available at each stage, ELLPACK has become a flexible tool for both experimenting with numerical methods and solving applications problems.

This chapter presents background material, outlining some key concepts and terminology associated with elliptic boundary value problems and their solution by numerical methods. The presentation here is necessarily very brief. For an elementary introduction to elliptic boundary value problems we recommend standard texts such as [Carrier and Pearson, 1976] or [Zachmanoglou and Thoe, 1976]; more advanced treatments can be found in [Courant and Hilbert, 1953, 1962] and [Mikhlin, 1967]. A number of books provide more detailed descriptions of numerical methods for these problems; for example, see [Forsythe and Wasow, 1960], [Mitchell and Wait, 1977], [Gladwell and Wait, 1979], [Lapidus and Pinder, 1982], and [Birkhoff and Lynch, 1984]. Details of the specific algorithms implemented in the ELLPACK problem-solving modules may be found in Part 2 of this book.

1.A MATHEMATICAL PRELIMINARIES

Elliptic equations describe the equilibrium states of physical systems. Well-posed problems of this type require the determination of a function which satisfies a given partial differential equation (PDE) on some domain R, as well as additional conditions along its boundary ∂R. As a simple example consider the problem of determining the steady-state temperature distribution in a nonhomogeneous isotropic plate. We assume that the plate is thin with insulated plane surfaces so that we may consider the problem to be two-dimensional. The temperature $u(x,y)$ of the plate then satisfies

$$-(k(x,y)u_x)_x - (k(x,y)u_y)_y = 0$$

in the interior of the region R defining the plate, where $k(x,y)$ is the thermal conductivity of the material. We use subscripts to denote partial differentiation; thus, the first term above represents $\partial[k(x,y)\partial u/\partial x]/\partial x$. The three most common types of boundary conditions for this problem are

(a) $u(x,y) = r(x,y)$ on ∂R,

(b) $u_n(x,y) = r(x,y)$ on ∂R,

(c) $p(x,y)u_n(x,y) + q(x,y)u(x,y) = r(x,y)$ on ∂R.

In case (a) the edges of R are held at some fixed temperature, while in case (b) the heat flux across the boundary is specified; if $r = 0$ the edge is insulated. (Here, u_n denotes the derivative of u in the direction of the outward pointing normal to ∂R.) In case (c) the heat flux is a function of the temperature; this occurs when the effect of radiation of heat into the surrounding medium is being modelled. Partial differential equations and boundary conditions of this form arise in a number of different settings. See [Forsythe and Wasow, 1960] and [Birkhoff and Lynch, 1984] for a selection of applications.

A more general equation of this type is

$$\mathrm{L}u = (au_x)_x + (cu_y)_y + fu = g, \tag{1}$$

where a, c, f, and g are functions of x and y and the **ellipticity condition** $ac > 0$ is satisfied in R. Here L denotes the differential **operator** applied to u. Equations written in this way are said to be in **self-adjoint, divergence, or conservation form.** A still more general equation is given by

$$\mathrm{L}u = au_{xx} + 2bu_{xy} + cu_{yy} + du_x + eu_y + fu = g, \tag{2}$$

where a, b, c, d, e, f, and g are functions of x and y. The ellipticity condition $b^2 - ac < 0$ must hold in R for elliptic problems. These are called **second order equations** since they involve at most second order partial derivatives. They are **linear equations** since the coefficient functions a, b, c, d, e, and f and the right side g are independent of u.

Much additional terminology is associated with various special cases of these general equations. If, for instance, $a = c = 1$ and $b = d = e = f = 0$, we have **Poisson's equation**. If, in addition, $g = 0$ the equation is called **Laplace's equation**. When $a = c = 1$, $b = d = e = 0$, and, $f \neq 0$, we have the **Helmholtz equation** (f constant) or the **generalized Helmholtz equation** (f variable). If Lu is the sum of two one-dimensional operators (i.e., a and d are functions of x alone, c and e are functions of y alone, $b = 0$, and f is the sum of a function depending only on x and a function depending only on y), then the equation is called **separable**. Finally, equations with $g = 0$ on R are called **homogeneous**. Specialized numerical methods have been developed for each of these cases.

Two types of domains on which the equation is defined are distinguished. The simplest is the **rectangular domain** $\mathrm{R} = \{(x,y) \mid x_l < x < x_r,\ y_l < y < y_r\}$, and many numerical methods have been developed to take advantage of its special structure. In our terminology **general domain** means a domain R which is bounded and nonrectangular; it may have holes or slits removed from its interior (along which boundary conditions must also be specified), but it must be connected; that is, any two points in the interior of R must be connectable by a polygonal line. Elliptic problems may also be posed on unbounded domains (in these cases boundary conditions "at infinity" are required), but the domains are usually truncated when posed for numerical solution.

The most common type of **boundary condition** specified in elliptic problems takes the general form

$$p(x,y)u_n + q(x,y)u = r(x,y)\,. \tag{3}$$

Again, terminology is associated with various special cases. When $p(x,y)=0$ (solution specified) the condition is known as a **Dirichlet condition**, while $q(x,y)=0$ (flux specified) yields a **Neumann condition**. When $pq \neq 0$ the condition is known as a third kind, elastic, or **Robbins boundary condition**. When $pq=0$ and $p^2+q^2>0$, the condition is called **uncoupled**. If $r=0$ on ∂R then the conditions are termed **homogeneous**. Rectangular domains also admit boundary conditions of **periodic** type,

$$u(x_l,y) = u(x_r,y), \qquad y_l < y < y_r$$

and/or

$$u(x,y_l) = u(x,y_r), \qquad x_l < x < x_r$$

that is, the solution is a periodic function of x, y or both.

Certain terms are also common in describing boundary value problems as a whole (i.e., particular combinations of differential operator, domain, and/or boundary conditions). Among them are

Dirichlet problem
The boundary conditions specify the solution along the entire boundary (i.e., $p=0$ on ∂R).

Neumann problem
The boundary conditions specify the outward-pointing normal derivative along the entire boundary (i.e., $q=0$ on ∂R).

Separable problem
The partial differential operator is separable, the domain is rectangular, and the boundary conditions are either periodic or of the form (3) with p and q constant along each side. Separable problems may be solved very efficiently by discrete analogues of separation of variables. These are the so-called "fast direct" solution techniques.

Self-adjoint problem
The problem is based on a variational principle, that is, it arises from the minimization of an integral which describes the energy of a physical system. A problem is self-adjoint if the operator is self-adjoint and the boundary conditions are "natural" for the corresponding variational equation (see [Forsythe and Wasow, 1960, pp. 159–169]). Equation (2) is self-adjoint if and only if $d=a_x+b_y$ and $e=c_y+b_x$. Equation (1) is always self-adjoint. These problems are inherently symmetric and numerical methods often take advantage of this fact, although it is difficult for methods to do so unless the equation is written in the form (1).

In general, we must have $af \leqslant 0$ on R, $pq \geqslant 0$ on ∂R, and at least one of f and q not identically zero in order to guarantee a unique solution to the boundary value problem defined by (1) or (2) and (3). If a and f are of the same sign, then the homogeneous problem may admit eigensolutions so that the solution to the nonhomogeneous problem is nonunique. The Neumann problem with $f = 0$ has no solution unless g and r satisfy certain consistency conditions [Mikhlin, 1967, Chapter 5], and when they do the solution is unique only up to an additive constant.

A number of important elliptic equations do not take the form of (1) or (2). Among them are

Higher order equations

The simplest equation of this type is the fourth order **biharmonic equation**

$$\mathrm{L}u = u_{xxxx} + 2u_{xxyy} + u_{yyyy} = g,$$

which is important in the modelling of stresses in solid mechanics. Two independent boundary conditions are required to determine u in this case.

Quasilinear equations

These are the simplest types of nonlinear equation. They take the general form of (1) or (2), but have coefficients which are also functions of u, u_x, and u_y. For example, the thermal conductivity in heat conduction problems might be a function of the temperature u. Another quasilinear problem is the classical **Plateau equation**

$$(1 + u_x^2)\, u_{xx} - 2u_x u_y u_{xy} + (1 + u_y^2)\, u_{yy} = 0,$$

whose solutions model the shape of a soap film spanning a given closed loop.

Systems of equations

The biharmonic equation with boundary conditions u and $u_{xx} + u_{yy}$ given may be rewritten as the following system of two second order elliptic equations in two unknown functions

$$u_{xx} + u_{yy} = v, \qquad v_{xx} + v_{yy} = g.$$

More complex systems are the multigroup diffusion equations of nuclear reactor theory [Birkhoff and Lynch, 1984], which are linear, and the equations modelling semiconductors [Fichtner and Rose, 1981], which are quasilinear.

Three-dimensional problems

Boundary value problems involving three space dimensions are also very important if physical systems are to be realistically modelled. To obtain the most general linear equations of this type, one must add a term in u_{zz} to (1), terms in u_{zz}, u_{xz}, u_{yz}, and u_z to (2), and a term in u_z to (3). In this case all coefficients are functions of x, y, and z. They are also functions of u, u_x, u_y, and u_z in the quasilinear case.

The ELLPACK system solves single linear elliptic equations of the form (1) and (2) on general domains in two dimensions and rectangular domains in three dimensions with boundary conditions of the form (3). Periodic boundary conditions are also allowed in the rectangular case. In addition, many problem solving modules are available for special cases of these general problems.

ELLPACK also can be used to solve some quasilinear equations and elliptic systems, although no problem-solving modules solve these directly. However, various facilities are available which allow ELLPACK users to set up simple iterative procedures which use modules designed for linear problems and a single equation. These techniques are described in detail in Chapters 4 and 5.

1.B NUMERICAL METHODS PRELIMINARIES

There are two stages to the solution of elliptic boundary value problems by numerical methods. The first phase, called discretization, requires the replacement of the continuous problem by a discrete one which approximates it. In the case of equations (1) or (2) and (3) one obtains a system of linear algebraic equations. In general, the larger the system the better the approximation. The second phase is the solution of the algebraic system.

DISCRETIZATION METHODS

In the **method of finite differences** one places a rectangular **grid** $G = \{ (x_i, y_j),\ i = 1, \ldots, n, j = 1, \ldots, m \}$ over the domain with the object of determining approximations to the solution at each grid point in $G' = G \cap (\mathrm{R} \cup \partial\mathrm{R})$. To do this we write an algebraic equation for each point in G' which approximates the differential equation locally. The most straightforward approach is to replace the derivatives in the differential equation by simple divided differences.

We illustrate the procedure by considering the case of rectangular R where $G' = G$. In this case the unknowns are $u(x_i, y_j)$, which are denoted by $U_{i,j}$, $i = 1, \ldots, n,\ j = 1, \ldots, m$. In the case of equally spaced grid points the finite difference equation approximating equation (2) at the point (x_i, y_j) is based upon

$$u_{xx} \approx \frac{U_{i+1,j} - 2U_{i,j} + U_{i-1,j}}{h^2}, \qquad u_{yy} \approx \frac{U_{i,j+1} - 2U_{i,j} + U_{i,j-1}}{k^2},$$

$$u_{xy} \approx \frac{U_{i+1,j+1} - U_{i+1,j-1} - U_{i-1,j+1} + U_{i-1,j-1}}{4hk},$$

$$u_x \approx \frac{U_{i+1,j} - U_{i-1,j}}{2h}, \qquad u_y \approx \frac{U_{i,j+1} - U_{i,j-1}}{2k},$$

where h and k are the grid spacings in the x and y directions, respectively. For equation (1) we use

$$(au_x)_x \approx \frac{a_+[U_{i+1,j} - U_{i,j}] - a_-[U_{i,j} - U_{i-1,j}]}{h^2},$$

$$(cu_y)_y \approx \frac{c_+[U_{i,j+1} - U_{i,j}] - c_-[U_{i,j} - U_{i,j-1}]}{k^2},$$

where

$$a_+ = a(x_i + h/2, y_j), \qquad a_- = a(x_i - h/2, y_j),$$

$$c_+ = c(x_i, y_j + k/2), \qquad c_- = c(x_i, y_j - k/2).$$

In the case $b = 0$ these lead to finite difference equations of the form

$$\alpha_0 U_{i,j} + \alpha_1 U_{i+1,j} + \alpha_2 U_{i,j+1} + \alpha_3 U_{i-1,j} + \alpha_4 U_{i,j-1} = g$$

where, in the case of equation (2),

$$\alpha_0 = f - 2\frac{a}{h^2} - 2\frac{c}{k^2},$$

$$\alpha_1 = \frac{a}{h^2} + \frac{d}{2h}, \quad \alpha_2 = \frac{c}{k^2} + \frac{e}{2k}, \quad \alpha_3 = \frac{a}{h^2} - \frac{d}{2h}, \quad \alpha_4 = \frac{c}{k^2} - \frac{e}{2k}.$$

All functions are evaluated at the point (x_i, y_j) in this case. For equation (1) we obtain

$$\alpha_0 = f - \frac{a_+ + a_-}{h^2} - \frac{c_+ + c_-}{k^2},$$

$$\alpha_1 = \frac{a_+}{h^2}, \quad \alpha_2 = \frac{c_+}{k^2}, \quad \alpha_3 = \frac{a_-}{h^2} \quad \alpha_4 = \frac{c_-}{k^2}.$$

Boundary conditions must also be incorporated into the discretization, and Dirichlet and periodic conditions are straightforward to handle. At a point on ∂R where equation (3) with $p \neq 0$ must be satisfied, the simplest approach is to write a finite difference formula for (1) or (2) at this point and then use a finite difference approximation to (3) to eliminate the unknown introduced outside the domain. For example, in the case of a boundary point along the right hand edge (not a corner) one obtains for equation (2)

$$\alpha_0 = f - 2\frac{a}{h^2} - 2\frac{c}{k^2} - \frac{2hq}{p}\left[\frac{a}{h^2} + \frac{d}{2h}\right],$$

$$\alpha_1 = 0, \quad \alpha_2 = \frac{c}{k^2} + \frac{e}{2k}, \quad \alpha_3 = \frac{2a}{h^2}, \quad \alpha_4 = \frac{c}{k^2} - \frac{e}{2k}.$$

where the right side of the equation has also been changed to $-2hr[a/h^2 + d/(2h)]/p$. In the case of equation (1) the corresponding approximation results in

$$\alpha_0 = f - \frac{a_+ + a_-}{h^2} - \frac{c_+ + c_-}{k^2} - \frac{2hq}{p}\left[\frac{a_+}{h^2}\right],$$

$$\alpha_1 = 0, \quad \alpha_2 = \frac{c_+}{k^2}, \quad \alpha_3 = \frac{a_+ + a_-}{h^2}, \quad \alpha_4 = \frac{c_-}{k^2},$$

where the right side is also modified to $g - 2hr[a_+/h^2]/p$.

The finite difference equation at a particular grid point involves at most five unknowns when $b = 0$ (nine otherwise), and hence the formula is known as the **5-point star**. For smooth u and a uniform grid, the error in making these approximations is $O(h^2 + k^2)$, that is, the error goes to zero as fast as $h^2 + k^2$. Thus we say that the difference formulas are **second order accurate**. The difference equations become only slightly more complex for unequally-spaced meshes, and they may easily be generalized to three dimensions (this yields the **7-point star** discretization). Finite difference discretizations with order of accuracy greater than two may be obtained by using the **HODIE method** [Lynch and Rice, 1979] or the techniques of **extrapolation** and **deferred corrections** [Pereyra,1966].

For general domains the chief source of difficulty is the incorporation of boundary conditions. The need to approximate u_n when the outward-pointing normal vector does not correspond to a grid line leads to one-sided approximations that increase the matrix bandwidth. In fact, unless the domain has some very special shape, symmetry will also be destroyed, even for Dirichlet boundary conditions.

The **finite element method** employs an alternative discretization technique. Instead of obtaining values of the solution at a set of points, one selects a set of basis functions $\{\phi_j(x,y), j=1,\ldots,N\}$ and then determines coefficients $c_j, j=1,\ldots,N$ so that the function

$$U(x,y) = \sum_{j=1}^{N} c_j \phi_j(x,y) \tag{4}$$

approximates the solution u as well as possible. (For simplicity we will assume that the boundary conditions imposed on the solution to the differential equation are homogeneous, and that the basis functions ϕ_j satisfy them exactly.) In particular, the coefficients c_j are determined by the requirement that

$$(\mathrm{L}U - g, t_i) = 0, \qquad i=1,\ldots,N, \tag{5}$$

where $(\cdot,\cdot)$ is the inner product defined by $(f,g) = \int_{\mathrm{R}} fg$ and $\{t_i(x,y), i=1,\ldots,N\}$ is a given set of test functions. This is a linear system of algebraic equations which is given more explicitly by

$$\sum_{j=1}^{N} c_j(\mathrm{L}\phi_j, t_i) = (g, t_i), \qquad i=1,\ldots,N. \tag{6}$$

A particular finite element method is obtained by specifying the basis functions and test functions. Intuitively, the ϕ_j are chosen for their ability to approximate u, while the t_i are chosen for their ability to approximate the **residual** $\mathrm{L}U - g$. (The latter observation comes from the fact that (5) requires the residual to be orthogonal to the space spanned by the test functions. If the residual is represented well in this space, then its components orthogonal to it will be small.) In any case, we seek choices of $\{\phi_j\}$ and $\{t_i\}$ so that U can be made arbitrarily close to u by taking N large enough.

The **Galerkin method** or **Rayleigh-Ritz method** uses the choice $t_i = \phi_i$, $i=1,\ldots,N$. It can be shown that with this choice the matrix in (6) is symmetric whenever the operator L is self-adjoint. The integrations in (6) are often done numerically. If Dirac delta functions are used for the t_i, then the method is called **collocation**. Here we choose a set of N points $p_i = (x_i, y_i)$ in R and define the test functions as $t_i(p) = \delta(p - p_i)$. Putting these functions in (5) yields the requirement that the differential equation be satisfied exactly at the collocation points, that is

$$\mathrm{L}U(x_i,y_i) = g(x_i,y_i) \qquad i=1,\ldots,N.$$

A very important feature of finite element methods is the use of **piecewise polynomial basis functions**. To define these we first partition the domain into a set of subdomains called **elements**; rectangles or triangles are the most common choices in two dimensions. A piecewise polynomial defined on this partition is then a function which is a polynomial (usually of low degree) on each element separately. A particular space of piecewise polynomials is selected by the choice of the polynomial degree for each element and the degree of continuity across element boundaries. This gives a finite dimensional linear space, and hence there is a set of basis functions ϕ_j such that any piecewise polynomial in the space may be written in the form (4).

In practice one chooses **basis functions with small support**, that is, basis functions that are identically zero except for a small number of elements. If this is so, the matrix A in (6) becomes very sparse. For Galerkin's method, the matrix entry $A_{ij} = (L\phi_j, \phi_i)$, is nonzero only when the supports of ϕ_i and ϕ_j overlap. For example, consider the **bilinear basis function** defined on a rectangular grid by $\phi(x,y) = \psi_i(x)\psi_j(y)$, where

$$\psi_k(z) = \begin{cases} \dfrac{z - z_{k-1}}{z_k - z_{k-1}}, & z_{k-1} \leqslant z \leqslant z_k, \\ \dfrac{z_{k+1} - z}{z_{k+1} - z_k}, & z_k \leqslant z \leqslant z_{k+1}, \\ 0, & \text{elsewhere}. \end{cases}$$

This element is one at (x_i, y_j) and zero outside the rectangle $x_{i-1} < x < x_{i+1}$, $y_{j-1} < y < y_{j+1}$. It is continuous everywhere, but not differentiable along the lines $x = x_{i-1}, x_i, x_{i+1}$ and $y = y_{j-1}, y_j, y_{j+1}$. (Note that we often may apply Green's theorem (integration by parts) to the integral in (5), thus allowing the use of basis functions with less continuity than is required by the differential operator.) We may associate one such basis function with each grid point and hence there are nine nonzeros per row in the resulting matrix.

In collocation, $A_{i,j}$ is nonzero only when the i-th collocation point is in the support of the j-th basis function. For the same set of basis functions, collocation yields fewer non-zeros per equation than for Galerkin's method and computing them is easier since there are no integrals to do. Due to its simplicity, the class of problems to which collocation is easily applied is greater than for the Galerkin method. However, collocation usually does not give a symmetric matrix even in the self-adjoint case. In addition, one must use basis functions of degree at least two in order for $L\phi_j$ to be nonzero.

Solution Methods

The discretization of an elliptic problem produces a (usually) large sparse system of linear algebraic equations which we denote as $A\mathbf{u} = \mathbf{g}$. We now use lower case bold letters to conform to the usual matrix-vector notation. Methods for solving this system can be classified as either direct or iterative (although there are now various hybrid methods). **Direct methods** produce the exact answer in a finite number of steps (in the absence of rounding error), whereas **iterative methods** produce a sequence of approximations which converge to the solution u only in the limit.

The structure of the matrix A depends upon how we number the unknowns and the order in which we write the equations. In finite differences, both equations and unknowns correspond to grid points, and hence numbering the grid points is equivalent to numbering the equations and unknowns in the same way. In the so-called **natural ordering** we number points in G' from left to right and bottom to top. Similar orderings exist in the finite element case, where numbering the equations and unknowns corresponds to numbering the test functions and basis functions. These lead to matrices with a number of desirable properties:

Sparsity

Most entries in the matrix are zero. In the finite difference discretization above, we have at most nine nonzero coefficients per equation (five when $b = 0$), independent of the number of grid points.

Bandedness

The nonzero entries are confined to a band about the matrix diagonal. The bandwidth is defined to be $\max|i-j|$ for which $A_{i,j} \neq 0$.

Symmetry

$A_{i,j} = A_{j,i}$ for all i and j. In the finite difference and Galerkin examples, a symmetric matrix will always be generated in the case of equation (1).

Definiteness

If $\mathbf{x}^T A \mathbf{x} > 0$ for all nonzero vectors $\mathbf{x}$, then A is said to be **positive definite**. The coefficient matrix A from the finite difference discretization above is negative definite (i.e. $\mathbf{x}^T A \mathbf{x} < 0$ for all non-zero $\mathbf{x}$), but multiplication by -1 makes it positive definite. Galerkin discretizations of (1) also yield positive definite systems. Symmetric positive definite matrices have all eigenvalues real and positive.

A number of very **fast direct methods** have been developed for the special case of separable elliptic problems. These algorithms rely on the particular structure of the matrices generated by certain discretizations (standard finite differences, for example). In the **cyclic reduction** method, half the unknowns are eliminated by taking certain linear combinations of equations, and, since the structure of the matrix is preserved, the process may be reapplied until there is a single equation in one unknown. All other unknowns may then be obtained by back substitution. For an n by n grid problem this requires only $O(n^2 \log_2 n)$ operations. A method of comparable complexity and power is based upon Fourier transforms. In this case one expands g in a finite Fourier series of appropriate type (determined by the boundary conditions) using the **fast Fourier transform** (FFT). Assuming that u has a similar expansion, one can now easily calculate its Fourier coefficients. The FFT is then used to construct u from its trigonometric expansion. This algorithm is applicable to equations of the form

$$\mathrm{L}u = a(x)u_{xx} + d(x)u_x + f(x)u + u_{yy},$$

and

$$\mathrm{L}u = u_{xx} + c(y)u_{yy} + e(y)u_y + f(y)u,$$

but not to the most general separable equation. Finally, if one applies the FFT algorithm after l steps of cyclic reduction one obtains the so-called **FACR(l) method**. Each of these methods requires very little extra storage. See [Swarztrauber, 1977] for details and further references.

Another class of fast direct methods for separable problems is based upon transformation to an initial value problem. This is again a two stage process. One first determines a second set of initial conditions on one side of the domain which will allow one to discard the boundary conditions on the opposite side. Then one simply marches the solution from one side of the domain to the other using initial value techniques. When applied to two-point boundary value problems this technique is known as **shooting**. Unfortunately, the reformulated problem admits an exponential error growth (it is inherently unstable). Thus, the method must be implemented as a sequence of shorter marches; this is known as **multiple shooting**. An implementation of this method, known as the **generalized marching algorithm**, is described in detail in [Bank and Rose, 1977] and [Bank, 1977].

Although each of these fast direct methods is formally restricted to problems where one can perform separation of variables, there are a number of techniques for extending this problem class at the expense of an increase in computing cost. For example, one can solve problems with non-separable equations by solving a sequence of separable problems (i.e. an iterative method with the solution of a separable problem in the innermost loop). See [Gunn, 1965] or [Concus and Golub, 1973]. The dependence upon rectangular domains has been successfully removed in the case of the Helmholtz equation by use of the **capacitance matrix method** [Proskurowski, 1979]. Here, after extending the problem arbitrarily to an embedding rectangle, one can obtain a solution to the original problem at the expense of five to ten fast direct solutions on the rectangle with only a modest increase in storage.

A number of direct methods based upon classical Gauss elimination have been developed for the cases where the fast direct methods are inapplicable. The idea is to take advantage of the sparsity structure of the matrix to reduce the time and storage requirements of a naive application of Gauss elimination. For example, if an N by N matrix is banded one need only store the matrix diagonals lying within the band, thus reducing the work to solve the system from $O(N^3)$ to $O(Nb^2)$ where b is the bandwidth. The storage is reduced from $O(N^2)$ to $O(Nb)$. In the case of standard finite differences on an n by n grid $(N=n^2, b=n)$, one reduces the work from (n^6) to (n^4). This is called **band Gauss elimination**. If the matrix is symmetric, then this can be further reduced by one-half. **Envelope solvers** only store elements from the first nonzero to the last nonzero in each row, thus reducing storage costs further when the matrix band structure is irregular. **Sparse matrix solvers** have even greater potential savings by storing and operating only on nonzero elements. The fact that nonzeros are generated in the course of Gauss elimination is a complicating factor, and sophisticated data management techniques are required to make this algorithm effective. A description of the sparse matrix modules included in ELLPACK can be found in Chapter 6.

An appropriate ordering of the equations and unknowns is crucial to the effectiveness of these methods and thus a number of methods have been developed to reorder the equations generated by discretization. For example, if a band or envelope solver is to be used, then bandwidth or profile reduction is the goal; the **Cuthill-McKee** (or **reverse Cuthill-McKee**) algorithm has this aim [Liu and Sherman, 1976]. If a sparse solver is to be used the goal is an ordering which reduces fill-in (the generation of new nonzeros); the **minimum degree** and **nested dissection** algorithms are of this type [George, 1973]. A nested dissection ordering can lead to sparse Gauss elimination algorithms requiring only (n^3) operations, for instance. When performing Gauss elimination it is also sometimes necessary to effect local reoderings of equations and unknowns to avoid zero (or very small) diagonal elements. This operation, called **pivoting**, is not usually required for systems resulting from the discretization of elliptic problems (it is never necessary for symmetric definite matrices), and hence solvers designed for this problem class often do no pivoting. Occasionally, however, one does encounter problems where pivoting is required.

Although sparse matrix methods make efficient use of the sparsity structure of the matrix, their overhead storage requirements are still substantial. For very large problems one must resort to the use of external storage, which can slow things down considerably. An alternative is the use of **iterative methods**. A general survey of these methods may be found in [Kincaid and Young, 1979] and [Hageman and Young, 1981]. The classical iterative methods for solving $A\mathbf{u} = \mathbf{g}$ are based upon writing the matrix A as

$$A = Q(I - G)$$

where Q, called the **splitting matrix**, is nonsingular. The iteration then takes the form

$$\mathbf{u}^{m+1} = G\mathbf{u}^m + \mathbf{k}, \qquad \mathbf{k} = Q^{-1}\mathbf{g}.$$

Clearly, Q should be chosen so that $Q\mathbf{k} = \mathbf{g}$ is easily solvable. G is called the **iteration matrix**. Iterations of this form converge to the solution of $A\mathbf{u} = \mathbf{g}$ for arbitrary initial guess $\mathbf{u}^0$ if and only if $S(G) < 1$, where $S(G)$ denotes the maximum of the moduli of the eigenvalues of G. $S(G)$ is known as the **spectral radius** of G. In general, the smaller $S(G)$, the faster the convergence. Note that the implementation of these methods normally requires only that one compute matrix-vector products involving parts of A, hence one need only store the nonzero entries of A. Since no "fill-in" occurs, there is little overhead required and these methods are quite storage efficient.

The most common choices for the splitting matrix Q are based upon writing A as $D - L - U$, where D is diagonal and L and U are strictly lower and upper triangular respectively. The resulting iterative methods are:

Method	*Splitting matrix* (Q)
Jacobi	D
Successive overrelaxation (**SOR**)	$(1/\omega)D - L$
Symmetric SOR (**SSOR**)	$[\omega/(2-\omega)][(1/\omega)D-L]D^{-1}[(1/\omega)D-U]$

With the choice $\omega = 1$, SOR is also known as the **Gauss-Seidel method**. When A is symmetric positive definite the SOR and SSOR methods converge if and only if $0 < \omega < 2$. One chooses the parameter ω to make $S(G)$ as small as possible, and convergence much faster than the Jacobi method can be obtained in this way. (In the case of the SSOR method one must also have $S(LU) \leqslant 0.25$ for these schemes to be truly effective.) The theory of these methods is quite well developed for the case of positive definite A [Young, 1971]. If A is indefinite these methods should be used with caution since they are often ineffective in this case.

The convergence of the sequences produced by the Jacobi and SSOR methods, and other symmetrizable methods, may be accelerated using the modified iteration

$$\mathbf{u}^{m+1} = \rho_{m+1}[\gamma_{m+1}(G\mathbf{u}^m + \mathbf{k}) + (1 - \gamma_{m+1})\mathbf{u}^m] + (1 - \rho_{m+1})\mathbf{u}^{m-1},$$

where the scalars ρ_{m+1} and γ_{m-1} may vary at each iteration. The **Chebyshev semi-iterative** and the **conjugate gradient** methods provide prescriptions for the choice of these parameters, each of which is optimal in a certain sense. In the Chebyshev case optimal values for the parameters ω, p, and q are either unknown or impractical to compute directly since this would require an expensive eigenvalue calculation. Various **adaptive procedures** have been developed for estimating them during the course of the iteration [Hageman and Young, 1981].

A rough idea of the relative performance of these iteration methods can be obtained from an analysis of the matrix resulting from the standard finite difference discretization of the Dirichlet problem for the Poisson equation on a square using an n by n grid. The table below gives the number of iterations required to reduce the error by three digits using a variety of methods. Note that **for PDE problems it is unproductive to reduce the error in the solution of the matrix problem below that introduced in the discretization phase.** These estimates are based upon the asymptotic rate of convergence which is known for this problem (see [Hageman and Young, 1981, Chapter 2]). SI denotes **Chebyshev semi-iterative** acceleration. We assume that optimal values of all parameters are used.

Method	*Number of iterations*
Jacobi	$1.4n^2$
Jacobi-SI	$2.2n$
SOR	$1.1n$
SSOR	$2.2n$
SSOR-SI	$2.0n^{1/2}$

The work per iteration is (n^2) in general, but can vary considerably from method to method. For example, an SSOR iteration requires twice the work of an SOR iteration, thus, SSOR's only real advantage is that it can be accelerated. It can be shown that the conjugate gradient acceleration is at least as effective as Chebyshev acceleration in this case.

Finally, we note that the choice of equation and unknown ordering can also effect the performance of iterative methods. As an example consider a rectangular grid resulting from standard finite differences. If one imposes a checkerboard pattern on the grid and numbers all the red points before the black points (this is the **red-black ordering**), then half the unknowns may be eliminated from the Jacobi iteration. After the remaining unknowns are determined, the eliminated ones are obtained by a straightforward computation. These are the so-called **reduced system methods**. Further discussion of these iterative solution techniques may be found in Chapter 7.

One especially promising class of **hybrid methods** are based upon the fact that Gauss-Seidel iteration quickly reduces high frequency components of the error, but not low frequency ones. However, error components that are non-oscillatory with respect to a fine grid are usually oscillatory with respect to a coarser grid. In the **multigrid method** one defines a set of nested grids with the finest one corresponding to the one on which the solution is desired. A fixed number of iterations are performed on the finest grid and then the solution is transferred to the next coarser grid where more iterations are performed, and so on. Direct methods may be applied on the coarsest grid, and then the solution may be transferred back to the finer grids. When performing grid transfers, errors are reintroduced, so that the sequence of transfers from finer to coarser and coarser to finer grids is complex in practice. When properly implemented these methods can be very effective indeed, with operation counts as low as (n^2) [Nicolaides, 1975], [Brandt, 1977].

1.C THE ELLPACK PROJECT

ELLPACK was originally developed as a research tool to evaluate and compare mathematical software for solving elliptic problems. The idea was to create a system where individuals could contribute software modules which either completely solved a class of elliptic problems or performed one of the major steps in a numerical method, e.g. discretization or algebraic equation solution. Modules that partially solved problems would be combined with others to complete the solution. With all the software operating in the same environment one could evaluate the performance of the modules. Several studies of this type have been made (e.g., [Houstis and Rice, 1982], [Rice, 1981]). Part 3 of this book presents some simple examples of performance evaluation.

Considerable effort was put into making ELLPACK easy to use and augment. As a result, the system presented in this book is useful not only for research into the performance of numerical methods and software, but also for education and actual problem solving. Standard elliptic problems of moderate difficulty can be stated and solved in a direct, simple manner. Many more complex problems, including nonlinear, time-dependent, and simultaneous equations can be solved using more advanced ELLPACK facilities.

The ELLPACK project began in 1976, and was coordinated by John R. Rice of Purdue University. The principal participants in the project were

Randolph Bank	University of California at San Diego
Garrett Birkhoff	Harvard University
Ronald Boisvert	National Bureau of Standards
Stanley Eisenstat	Yale University
William Gordon	Drexel University
Elias Houstis	University of Thessaloniki
David Kincaid	University of Texas at Austin
Robert Lynch	Purdue University
Donald Rose	Bell Laboratories
Martin Schultz	Yale University
Andrew Sherman	Exxon Production Research
David Young	University of Texas at Austin

Substantial contributions of software were made by many others including Carl de Boor, John Brophy, Wayne Dyksen, EISPACK, Roger Grimes, Harmut Foerster, LINPACK, William Mitchell, John Nestor, Thomas Oppe, Wlodzimierz Proskurowski, John Respess, Calvin Ribbens, Granville Sewell, Van Snyder, Paul Swarztrauber, Roland Sweet, Linda Thiel, William Ward, and Alan Weiser. ELLPACK was supported by the National Science Foundation, the Department of Energy, and the Office of Naval Research as well as by each of the participants' institutions.

The software described in this book represents the second major revision of the system since its initial release in 1978. The complete system is available from Purdue University; for further information write

The ELLPACK Project
Department of Computer Sciences
Purdue University
West Lafayette, Indiana 47907, USA

1.D BRIEF DESCRIPTION OF ELLPACK

ELLPACK is a programming system for solving elliptic boundary value problems. The problems addressed include linear variable-coefficient elliptic equations of the form

$$au_{xx} + bu_{xy} + cu_{yy} + du_x + eu_y + fu = g,$$

or, in self-adjoint form,

$$(au_x)_x + (cu_y)_y + fu = g,$$

defined on general two-dimensional domains, and their three-dimensional analogues defined on rectangular boxes. For two-dimensional problems, boundary conditions may take the form

$$p_1 u_x + p_2 u_y + qu = r,$$

where p_1, p_2, q, and r are functions of x and y. The three-dimensional case is similar. Periodic boundary conditions are also admitted when the domains are rectangular. In addition, ELLPACK is organized so that it is possible to set up iterations to solve nonlinear problems (i.e. a, b, c, d, e, f, g, p_1, p_2, q, r are also functions of u, u_x, etc.).

ELLPACK users specify the problem they wish to solve in an **ELLPACK program** written in a simple user-oriented **ELLPACK language.** The **ELLPACK preprocessor** processes this program by translating it to a Fortran source program called the **ELLPACK control program** ; this program is then compiled and linked to a precompiled **ELLPACK module library.** Finally, the program is executed, producing a solution to the problem. The process is illustrated in Figure 1.1.

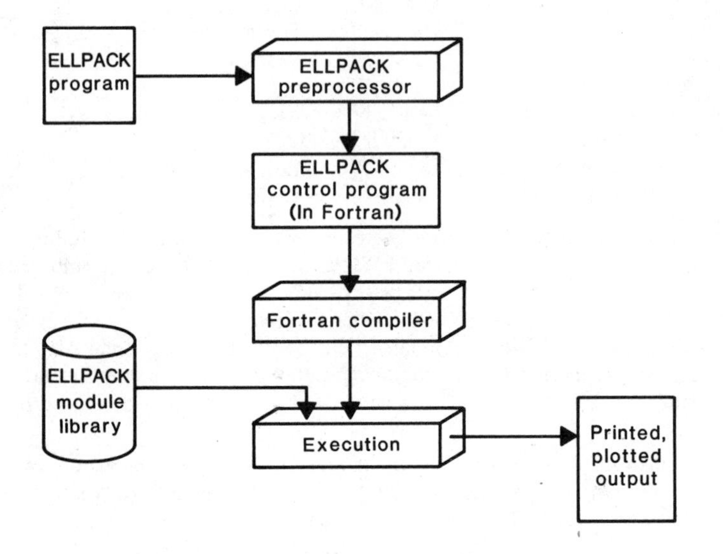

Figure 1.1. Schematic diagram of the processing of an ELLPACK run.

The ELLPACK language is an extension of Fortran, and ordinary Fortran statements can be mixed with ELLPACK statements. Elliptic equations, domains, and boundary conditions can be stated directly in this language, and powerful statements are available to help users solve the problem which they have posed. These statements invoke modules in the ELLPACK module library. There are five basic types of modules:

DISCRETIZATION. The partial differential equation and boundary conditions are approximated by a finite system of linear algebraic equations.

INDEXING. The equations and unknowns of the discrete system are reordered to facilitate solving the system.

SOLUTION. The system of algebraic equations is solved.

TRIPLE. Discretization, indexing and solution are performed as a single step.

OUTPUT. The approximate solution is tabulated or plotted.

One specifies the numerical method to be used by invoking, in turn, a discretization module, an indexing module, and a solution module (or a single triple module) as shown in Figure 1.2. In the ELLPACK language this is done by simply giving their names.

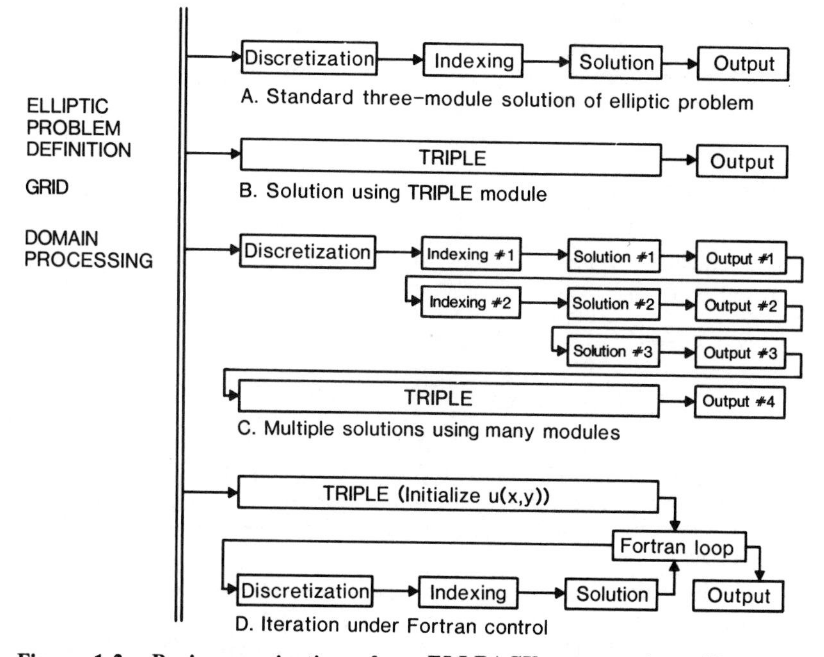

Figure 1.2. Basic organization of an ELLPACK computation. The user specifies the modules to be used and more than one combination may be used on a single ELLPACK run.

A large number of modules are available in ELLPACK for each stage of the computation. For example, discretization by various types of finite difference and finite element methods are possible, as well as solution of algebraic equations by both iterative and direct methods. Detailed descriptions of the available modules are given in Part 2. This easy access to a large repertoire of numerical methods makes ELLPACK useful in comparing solutions obtained by vastly different methods, as well as a "pilot plant" for large scale application problems.

During execution, ELLPACK modules communicate through fixed predefined collections of variables called **interfaces.** This process is illustrated in Figure 1.2 and described in detail in Part 4, where a complete specification of how to add new modules to the ELLPACK system is given.

The final essential ingredient in the solution of an elliptic problem with ELLPACK is the specification of a rectangular grid to cover the domain. When this grid is made finer, the approximations used by discretization modules are more accurate (within the constraints of machine arithmetic), but computer time and memory requirements also increase (since there are more algebraic equations generated). When a nonrectangular domain is specified in ELLPACK, the **domain processor** is invoked. It sets up tables which relate the rectangular grid to the domain in a way useful to discretization and triple modules.

1.E A SIMPLE PROGRAM

We show a very simple elliptic problem and an ELLPACK program which generates an approximate solution. A table of this solution is printed and a contour plot is produced:

ELLIPTIC PROBLEM

Determine a function $u(x,y)$ satisfying the partial differential equation

$$u_{xx} + u_{yy} + 3u_x - 4u = \exp(x+y)\sin(\pi x),$$

in the rectangle $0 < x < 1$, $-1 < y < 2$ and satisfying the boundary conditions

$$\begin{array}{lll} u = 0, & x = 0, & -1 < y < 2, \\ u = \sin(\pi x) - \frac{x}{2}, & y = -1. & 0 < x < 1, \\ u = \frac{y}{2}, & x = 1, & -1 < y < 2, \\ u = x, & y = 2 & 0 < x < 1, \end{array}$$

The ordinary finite difference approximation is used to discretize the problem at points of a 6 by 6 grid i.e., grid with spacing 1/5 in the x direction and 3/5 in the y direction. The resulting linear system is solved with ordinary Gauss elimination for band matrices.

This program consists of several **segments** whose names (EQUATION, BOUNDARY, and so on) begin in column 1 of a line; the rest is written in free format (excluding column 1). The dollar sign is a separator to allow more than one item on one line in a segment. Parts of the program include Fortran expressions (+3., EXP(X+Y)*SIN(PI*X), etc.) which must follow the rules of Fortran. Lines beginning with * are comments.

```
*         ***************************************************
*         *                                                 *
*         *                                                 *
*         *  EXAMPLE ELLPACK PROGRAM 1.E1                   *
*         *                                                 *
*         *                                                 *
*         ***************************************************

OPTIONS.          TIME  $  MEMORY

EQUATION.         UXX + UYY + 3.0*UX - 4.0*U  =  EXP(X+Y)*SIN(PI*X)

BOUNDARY.         U = 0.0                  ON  X = 0.0
                  U = SIN(PI*X) - X/2.0    ON  Y =-1.0
                  U = Y/2.0                ON  X = 1.0
                  U = X                    ON  Y = 2.0

GRID.             6 X POINTS
                  6 Y POINTS

DISCRETIZATION. 5 POINT STAR
SOLUTION.         LINPACK BAND

OUTPUT.           TABLE (U)
                  PLOT (U)
END.
```

Output of ELLPACK Run:

```
--------------------
 PREPROCESSOR OUTPUT
--------------------

    2.25 SECONDS TO INITIALIZE PREPREPROCESSOR
    1.20 SECONDS TO PARSE ELLPACK PROGRAM
    3.60 SECONDS TO CREATE ELLPACK CONTROL PROGRAM
    7.05 SECONDS TOTAL

         ELLPACK VERSION --- APRIL 5, 1984

    APPROXIMATE MEMORY REQUIREMENTS (IN WORDS)

    PROBLEM DEFINITION        95
    DISCRETIZATION           233
    INDEXING                  35
    WORKSPACE                675
    OTHER                     36
    TOTAL MEMORY            1074

----------------------
 DISCRETIZATION MODULE
----------------------

    5 - P O I N T   S T A R

    DOMAIN                               RECTANGLE
    DISCRETIZATION                         UNIFORM
    NUMBER OF EQUATIONS                         16
    MAX NO. OF UNKNOWNS PER EQ.                  5
    MATRIX IS                        NON-SYMMETRIC

    EXECUTION SUCCESSFUL
```

```
 ----------------
 SOLUTION MODULE
 ----------------

   L I N P A C K   B A N D

   NUMBER OF EQUATIONS        16
   LOWER BANDWIDTH             4
   UPPER BANDWIDTH             4
   REQUIRED WORKSPACE        224

   EXECUTION SUCCESSFUL

 ---------------
 ELLPACK OUTPUT
 ---------------

          ++++++++++++++++++++++++++++++++++++++++++++++
          +                                            +
          +    TABLE OF U       ON   6 X   6  GRID     +
          +                                            +
          ++++++++++++++++++++++++++++++++++++++++++++++

    X-ABSCISSAE ARE
    ---------------
    0.000000E+00    2.000000E-01    4.000000E-01    6.000000E-01
    8.000000E-01    1.000000E+00

       Y = 2.000000E+00
       ----------------
    0.000000E+00    2.000000E-01    4.000000E-01    6.000001E-01
    8.000000E-01    1.000000E+00

       Y = 1.400000E+00
       ----------------
    0.000000E+00   -6.890895E-02   -4.787898E-02    9.788278E-02
    3.680595E-01    7.000000E-01

       Y = 8.000001E-01
       ----------------
    7.450581E-09   -6.914875E-02   -6.634721E-02    2.381689E-02
    1.942513E-01    4.000000E-01

       Y = 2.000000E-01
       ----------------
    0.000000E+00   -6.218192E-02   -8.088307E-02   -5.358129E-02
    1.471775E-02    1.000000E-01

       Y =-4.000000E-01
       ----------------
    0.000000E+00   -2.025655E-03   -2.528100E-02   -7.325266E-02
   -1.359451E-01   -2.000000E-01

       Y =-1.000000E+00
       ----------------
    0.000000E+00    4.877852E-01    7.510565E-01    6.510564E-01
    1.877854E-01   -5.000000E-01

 ---------------
 ELLPACK OUTPUT
 ---------------

    CONTOUR PLOT OF     U
    GRID               20   BY    20
    EXECUTION SUCCESSFUL
```

```
---------------
 ELLPACK OUTPUT
---------------

          +++++++++++++++++++++++++
          +                       +
          +    EXECUTION  TIMES   +
          +                       +
          +++++++++++++++++++++++++

            SECONDS       MODULE NAME

               0.10       5-POINT STAR
               0.05       LINPACK BAND SETUP
               0.03       LINPACK BAND
               0.32       TABLE
               6.22       PLOT
               6.77       TOTAL TIME
```

This example is the simplest case of an ELLPACK program: one defines the elliptic problem in the EQUATION and BOUNDARY segments, OPTIONS are chosen, a rectangular grid is defined in the GRID segment, the solution method is specified in the DISCRETIZATION and SOLUTION segments, and the desired output is specified in the OUTPUT segment. Every ELLPACK program ends with END.

The ELLPACK preprocessor lists the program with an identifying heading. It also prints the memory estimates in words as requested in the OPTIONS segment along with its execution time. Each ELLPACK module prints a simple summary message. The OUTPUT segment contains two requests, one is a table of the solution at the grid points and the other is a contour plot produced by some graphics device, see Figure 1.3. The graphics connected to ELLPACK will vary from installation to installation.

1.F ORGANIZATION OF THE BOOK

Part 1 (Chapters 1–5) of this book is the **User's Guide.** The basic features of the ELLPACK language are presented in the next chapter, then three more examples are presented in detail in Chapter 3 (the first solves the same problem by two different methods, another illustrates non-rectangular geometry and the third shows how Fortran can be interspersed with ELLPACK statements.) Chapters 4 and 5 describe and illustrate more advanced features of ELLPACK. There are a number of examples of advanced applications in these two chapters.

Part 2 (Chapters 6–7) contains summary descriptions of the more than 50 **ELLPACK modules** available and an overview of the **ITPACK** software included in ELLPACK for the iterative solution of large sparse linear algebraic systems.

Part 3 (Chapters 8–11) presents a basic **performance evaluation** of many of the ELLPACK modules. The objective is to give the reader some feel for the properties of various methods (software modules) and not to present a complete scientific evaluation.

Part 4 (Chapters 12–16) is a **Contributor's Guide;** it provides the information needed to prepare a new module for the ELLPACK system. The ELLPACK system is designed so that new modules can be easily added (and corresponding

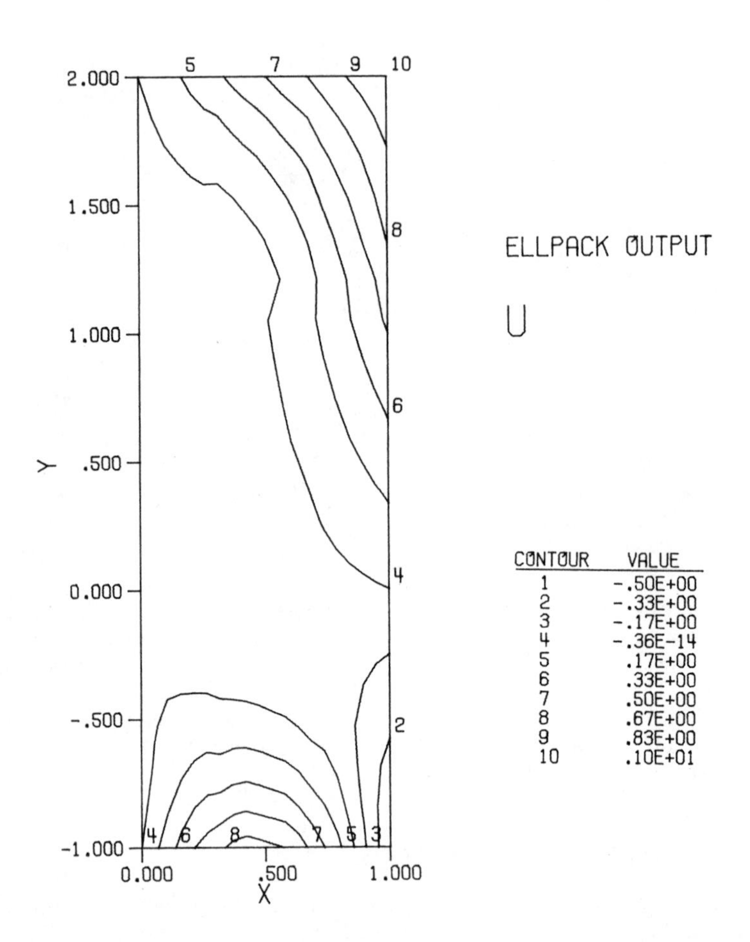

Figure 1.3. The contour plot produced by PLOT(U) in the example ELLPACK 1.E1 program. This plot is made with an electrostatic printer.

additions made to the language). There is also useful information here for those who wish to attempt advanced ELLPACK applications; otherwise this and the following Part 5 are not essential to the use of ELLPACK.

Part 5 (Chapters 17 and 18) is a **Systems Programming Guide:** it provides detailed information on how to install ELLPACK and to make modifications to it. The basic ELLPACK system can be installed without much difficulty; one needs to know how to manipulate files and create a library from a set of Fortran programs. Tailoring the ELLPACK system is more complicated and, while it does not require the specific expertise of a systems programmer, it is helpful if one has had experience with complex software systems.

The **Appendices** contain reference material; there are brief summaries of the **PG system** and the **TOOLPACK template processor** which are used to create the ELLPACK preprocessor. There is also the **PDE population,** a set of more than 60 linear elliptic partial differential equations on two dimensional rectangular domains. These can be used as a problem population for a systematic performance evaluation.

1.G REFERENCES

Bank, R. E. [1977], "Marching algorithms for elliptic boundary value problems, II: the variable coefficient case", SIAM J. Numer. Analy. 14, pp. 950–970.

Bank, R. E. and D. J. Rose [1977], "Marching algorithms for elliptic boundary value problems, I: the constant coefficient case", SIAM J. Numer. Analy. 14, pp. 792–829.

Birkhoff, G. and R. E. Lynch [1984], *Numerical Solution of Ellpitic Problems*, SIAM Publications, Philadelphia.

Boisvert, R. F. and R. A. Sweet [1983], "Mathematical software for elliptic boundary value problems", in *Sources and Development of Mathematical Software* (W. Cowell, ed.), Prentice-Hall, Englewood Cliffs, NJ.

Brandt, A. [1977], "Multi-level adaptive solutions to boundary-value problems", Math. Comp. 31, pp. 333–390.

Carrier, G. F. and C. E. Pearson [1976], *Partial Differential Equations — Theory and Technique*, Academic Press, New York.

Concus, P. and G. H. Golub [1973], "Use of fast direct methods for the efficient numerical solution of non-separable elliptic equations", SIAM J. Numer. Analy. 10, pp. 1103–1120.

Courant, R. and D. Hilbert [1953], *Methods of Mathematical Physics*, vol. 1, Interscience, New York.

Courant, R. and D. Hilbert [1962], *Methods of Mathematical Physics*, vol. 2, Interscience, New York.

Fichter, W. and D. J. Rose [1981], "On the numerical solution of nonlinear elliptic pdes arising from semiconductor device modelling", in *Elliptic Problem Solvers*, (M. Schultz, ed.), Academic Press, New York, pp. 277–284.

Forsythe, G. E. and W. R. Wasow [1960], *Finite Difference Methods for Partial Differential Equations*, John Wiley & Sons, New York.

George, J. A. [1973], "Nested dissection of a regular finite element mesh", SIAM J. Numer. Analy., pp. 345–363.

Gladwell, I. and R. Wait (eds.) [1979], *A Survey of Numerical Methods for Partial Differential Equations*, Oxford University Press, Oxford.

Gunn, J. E. [1965], "The solution of difference equations by semi-explicit iterative techniques", SIAM J. Numer. Analy. 2, pp. 24–55.

Hageman, L. A. and D. M. Young [1981], *Applied Iterative Methods*, Academic Press, New York.

Houstis, E. N. and J. R. Rice [1982], "High order methods for elliptic partial differential equations with singularities", Internat. J. Numer. Meth. Engin. 18, pp. 737–754.

Kincaid, D. R. and D. M. Young [1979], "Survey of iterative methods", in *Encyclopedia of Computer Science and Technology*, vol. 13 (J. Belzer, A. Holzman, and A. Kent, eds.), Marcel Dekker, New York, pp. 354–391.

Lapidus, L. and G. F. Pinder [1982], *The Numerical Solution of Partial Differential Equations in Science and Engineering*, John Wiley & Sons, New York.

Liu, W. H. and A. H. Sherman [1976], "A comparative analysis of the Cuthill-McKee and reverse Cuthill-McKee ordering algorithms for sparse matrices", SIAM J. Numer. Analy. 13, pp. 198–213.

Lynch, R. E., and J. R. Rice [1979], "High accurate finite difference approximations to solutions of elliptic partial differential equations", Proc. Nat. Acad. Sci. (USA) 75, pp. 2541–2544.

Mikhlin, S. G. (ed.) [1967], *Linear Equations of Mathematical Physics*, Holt, Rinehart and Winston, New York.

Mitchell, A. R. and R. Wait [1977], *The Finite Element Method in Partial Differential Equations*, John Wiley & Sons, New York.

Nicolaides, R. A. [1975], "On multiple grid and related techniques for solving discrete elliptic systems", J. Comput. Phys. 19, pp. 418–531.

Pereyra, V. [1966], "On improving an approximate solution of a functional equation by deferred corrections", Numer. Math. 8, pp. 376–391.

Proskurowski, W. [1979], "Numerical solution of Helmholtz's equation by implicit capacitance methods", ACM Trans. Math. Softw. 5, pp. 36–59.

Rice, J. R. [1981], "On the effectiveness of iteration for the Galerkin method equations", in *Computer Methods for Partial Differential Equations* IV (R. Vichnevetsky and R. S. Stepleman, eds.), IMACS, New Brunswick, N. J., pp. 68–73.

Rice, J. R. [1982a], "Software for scientific computation", in *Computer Aided Design of Technological Systems* (G. Leininger, ed.), IFAC, Pergamon.

Swarztrauber, P. N. [1977], "The methods of cyclic reduction, Fourier analysis and the FACR algorithm for the discrete solution of Poisson's equation on a rectangle", SIAM Rev. 19, pp. 490–501.

Young, D. M. [1971], *Iterative Solution of Large Linear Systems*, Academic Press, New York.

Zachmanoglou, E. C. and D. W. Thoe [1976], *Introduction to Partial Differential Equations with Applications*, Waverly Press, Baltimore.

Chapter 2

THE ELLPACK LANGUAGE

2.A ORGANIZATION OF AN ELLPACK PROGRAM

An ELLPACK program should be interpreted like the main program of a Fortran job. The basic blocks of statements in an ELLPACK program are **segments.** The segments that define the elliptic problem and options are like declarations: they must come first and they are not executed. The other segments (except END) are executed and the flow of the computation is controlled, as in Fortran, by placing them in the proper sequence. Ordinary Fortran statements may be interspersed among the executable segments to provide DO-loops, IFs, etc. There is also a facility to specify Fortran subprograms.

A brief summary of the segments is given in groups.

Group 1 Segments define the elliptic problem and should be viewed as declarations in the program. They must appear before any from Group 2; EQUATION and BOUNDARY must appear exactly once.

EQUATION.

Specifies the partial differential equation.

BOUNDARY.

Specifies the domain and boundary conditions. It must follow the EQUATION segment.

HOLE.

Defines a hole in the domain and associated boundary conditions. This segment can appear more than once if several holes are present. It must follow the BOUNDARY segment.

ARC.

Defines an interface or slit in the domain on which additional conditions are prescribed. Its use is governed by the same rules as the HOLE segment.

Group 2 Segments specify the executable ELLPACK modules and may appear more than once. These should be viewed as specifying the numerical method to solve the elliptic problem; one might think of them as "calling" the ELLPACK modules. A specific ordering is usually required; e.g., GRID, DISCRETIZATION, INDEXING, SOLUTION, OUTPUT or GRID, TRIPLE, OUTPUT.

GRID.

Specifies a set of vertical and horizontal grid lines. GRID can appear with a variable size to change the grid provided that the maximum grid size is declared in an OPTION segment before the first GRID segment.

DISCRETIZATION.

Specifies a module to create a discrete approximation to the elliptic problem; this generates a system of linear algebraic equations. (This is the optional first phase of an ELLPACK solution algorithm.)

INDEXING.

Specifies a module to reorder the linear equations and the unknowns. (This is the optional second phase of an ELLPACK solution algorithm.)

SOLUTION.

Specifies a module to solve the linear equations. (This is the final phase of an ELLPACK solution algorithm.)

TRIPLE.

Specifies a combination method which includes discretization, indexing and solution all in one module.

PROCEDURE.

Specifies various other optional actions in solving or analyzing the problem.

OUTPUT.

Specifies ELLPACK-generated output (printed and graphical).

Group 3 Segments may appear anywhere in the program and as many times as desired.

*

Specifies a comment.

DECLARATIONS.

Provides Fortran declarations for the user-provided executable Fortran statements.

GLOBAL.

Provides Fortran declarations (primarily COMMON blocks) that are placed within all Fortran programs generated by ELLPACK to define the elliptic problem.

OPTIONS.

Specifies which of various options are desired. Should be put near the top of the program.

FORTRAN.

Specifies that the statements which follow are user-supplied executable Fortran statements. If this segment appears in the declaration part of an ELLPACK program (among Group 1 segments) the Fortran statements are executed as though they appeared at the end of the Group 1 segments.

(blank line)

Allowed anywhere

Group 4 Segments specify various information and appear after all the other segments.

SUBPROGRAMS.

Specifies that a set of complete Fortran subprograms (FUNCTION or SUBROUTINE) follows. This segment must appear just before the ELLPACK END segment.

END.

Specifies the end of the ELLPACK program.

Two or more letters of the beginning of a segment name form an acceptable abbreviation. **Blanks are not allowed within segment names.** All segment names and their abbreviations must end with a period. Each EQUATION, DISCRETIZATION, INDEXING, SOLUTION, TRIPLE and PROCEDURE segment must be on a single line. **Line continuations** are made by putting an ampersand (&) as the last character at the end of the line (this may be done several times if necessary). No segment can be longer than 1000 characters. The OPTIONS, OUTPUT, BOUNDARY, GRID, HOLE, and ARC segments may use several lines and the separator $ may be used to place several parts of these segments on one line. If these segments are broken in the middle of a word or expression, then the continuation convention (ampersand at end of line) must be used. The segments FORTRAN, SUBPROGRAMS, DECLARATIONS, and GLOBAL start with the segment name on a separate line followed by lines of Fortran code.

The independent variables are denoted by X, Y, and Z (X and Y for two-dimensional problems). The dependent variable is denoted by U, its first derivatives u_x, u_y and u_z by UX, UY, and UZ, and its second derivatives by UXX, UYY, UXY, and so on. These are reserved names in ELLPACK and Fortran variables with these names cannot be used safely. Once the PDE is solved the ELLPACK functions U(X,Y), UX(X,Y), etc. (U(X,Y,Z), UX(X,Y,Z), etc. in three dimensions) become defined and may be used as ordinary Fortran functions. The complete set of **reserved names in ELLPACK** is

X, Y, Z, U, UX, UXX, UY, UYY, UYX, UXY, UZ, UZZ, UZX, UXZ, UZY, UYZ, TRUE, ERROR, RESIDU, ON, FOR, TO, LINE, PI, and **any six character name starting with C, I, L, Q or R followed by a digit**.

The words ON, FOR, TO and LINE must actually only be avoided in Fortran functions of the BOUNDARY segment. The variable PI is set to the mathematical constant π and can be used anywhere in the ELLPACK program except in the SUBPROGRAMS segment. The six character names are internal Fortran variables for ELLPACK; their use would create a name conflict. The meanings of the initial characters are:

C	Common block	I	Integer variable or function
L	Logical variable or function	Q	Subroutines
R	Real (or double precision) variable or function		

2.B SEGMENTS WHICH DEFINE THE PROBLEM

In this section we describe the rules (syntax) for defining the PDE problem.

EQUATION. SEGMENT

The EQUATION segment specifies the partial differential equation to be solved. An equation can be specified in either of two forms. The most **general two-dimensional form** for the EQUATION segment is

± ⟨expression⟩*UXX ± ⟨expression⟩*UXY ± ⟨expression⟩*UYY &
± ⟨expression⟩*UX ± ⟨expression⟩*UY ± ⟨expression⟩*U &
= ⟨expression⟩

where ⟨expression⟩ represents any legal Fortran real (or double precision) arithmetic expression involving the variables X and Y. External functions appearing in these expressions are defined in the SUBPROGRAM segment. If an ⟨expression⟩ on the left of the equal sign is identically 1, then the 1.0* may be dropped; if ⟨expression⟩ is identically zero the entire term may be omitted. For three-dimensional problems terms involving UZZ, UXZ, UYZ, and UZ are added, and each coefficient expression may also involve Z. If the equation requires more than one line, the continuation character (&) must be used at the end of each line to be continued.

Some examples of the EQUATION segment are given below:

```
*       LAPLACE'S EQUATION IN 3-D
EQUATION.  UXX + UYY + UZZ = 0.
*       AN EQUATION WITH CONSTANT COEFFICIENTS
EQUATION.   -4.*UXX + .377*UX - 3.*PI*UYY + 3.E+4*UY            &
                                = SIN(X+COS(X*Y))
* THE COEFFICIENTS OF UYY AND U ARE GIVEN AS FORTRAN FUNCTIONS.
* THESE ARE SUPPLIED BY THE USER IN THE SUBPROGRAMS SEGMENT.
EQUATION.   (X**2 + Y**2 + 16.)*UXX + VALUYY(X,Y)*UYY           &
        -2.234E-3*ATAN2(Y,X)*UX + 1.4*UY - VALU(X,Y)*U = 0.
```

The second type of equation recognized by ELLPACK is the **self-adjoint equation** form:

± ((⟨expression⟩*UX)X ± ((⟨expression⟩*UY)Y ± ⟨expression⟩*U &
= ⟨expression⟩

where ⟨expression⟩ follows the same rules as above. The notation ((⟨expression⟩*UX)X denotes $\frac{\partial}{\partial x}(\langle\text{expression}\rangle \frac{\partial u}{\partial x})$; thus, for example, the ELLPACK equation

```
(P(X,Y)*UX)X + (Q(X,Y)*UY)Y + R(X,Y)*U = G(X,Y)
```

represents the mathematical equation

$$(p(x,y)u_x)_x + (q(x,y)u_y)_y + r(x,y)u = g(x,y).$$

A term of the form ((⟨expression⟩*UZ)Z is added in the three-dimensional case.

There is an alternate way to indicate a self-adjoint equation by using the OPTIONS segment as follows:

```
*            SELF-ADJOINT, ALTERNATE FORM
*
OPTION.      SELF-ADJOINT = .TRUE.
*
EQUA.   P(X,Y)*UXX + Q(X,Y)*UYY + R(X,Y)*U = G(X,Y)
```

Terms involving UX and UY are ignored if this form of the equation is used.

The ELLPACK preprocessor does not make an exhaustive evaluation to determine whether the equation is actually elliptic. Some simple checks are made, however. For example, it requires that terms in UXX and UYY always appear (implicitly, in the self-adjoint case). In addition, ELLPACK users should recognize that individual problem solving modules may restrict the problems to which they apply. Thus, some modules may require a self-adjoint equation, some a Helmholtz equation, and so on.

BOUNDARY. SEGMENT

The BOUNDARY segment specifies the boundary of the domain R and the boundary conditions on it. We first describe how general two-dimensional domains are specified in ELLPACK; the special facilities for the simpler cases of rectangular two- and three-dimensional domains are described after that. The boundary is broken up into a series of **pieces** which must join together in sequence. A condition and piece are specified by

⟨condition⟩ ON ⟨piece⟩

where ⟨condition⟩ is one of the following:

PERIODIC

or

± ⟨expression⟩*UX ± ⟨expression⟩*UY ± ⟨expression⟩*U = ⟨expression⟩

where ⟨expression⟩ is a legal Fortran expression. The three terms on the left can be in any order, and any term may be omitted if its coefficient expression is zero. If the ⟨condition⟩ preceeding ON is omitted, then the preceding ⟨condition⟩ is used as the default.

Periodic boundary conditions may only be applied in the case of rectangular domains. If PERIODIC is specified on one side, then it must also be specified on the opposite side. Any of the other usual types of boundary conditions can be specified using the second form. For instance, a Dirichlet condition is specified as

U = F(X,Y)

and a Neumann condition ($u_n(x,y) = f(x,y)$, mathematically) as

A(X,Y)*UX + B(X,Y)*UY = F(X,Y),

where (X,Y) is a point on the boundary and (A(X,Y), B(X,Y)) is the unit vector normal to the boundary (pointing outward). The latter reduces to ± UX = F(Y) or ± UY = F(X) for rectangular domains.

A general two-dimensional domain is specified as a sequence of parametrized pieces. The general form of ⟨piece⟩ is

X = ⟨expression⟩, Y = ⟨expression⟩ FOR ⟨parameter⟩ = ⟨a⟩ TO ⟨b⟩

where

⟨parameter⟩	is a real Fortran variable that parametrizes the piece
⟨expression⟩	is a Fortran expression in the parameter
⟨a⟩, ⟨b⟩	are Fortran expressions that evaluate to constants which determine the initial and final value of the parameter.

The pieces are assumed to be given in counterclockwise order (this may be overridden by putting CLOCKWISE = .TRUE. in an OPTION segment). Each piece starts on a new line unless the $ separator is used. The parameter must increase from ⟨a⟩ to ⟨b⟩. **It is essential that the parametrization be of ordinary size and not vary erratically along the boundary.** The continuity of joining the boundary pieces is checked and the joining must be done accurately. Two simple examples of nonrectangular boundary and boundary condition specification follow:

```
*                    CIRCULAR  DISK WITH CENTER (1,1)
*
BOUND. U = 0.0 ON X = 1.-COS(PI*THETA), Y = 1.-SIN(PI*THETA)   &
                                        FOR THETA = 0. TO 2.
*           QUARTER ANNULUS
BOUNDARY.
*
 U=100.          ON X=SIN(PI*T),  Y=COS(PI*T)           FOR T=0. TO 0.5
 U=100.*(2.-X)   ON X=R,     Y=0.                       FOR R=1. TO 2.0
 U=0.0           ON X=2.*COS(PI*T),Y=2.*SIN(PI*T)       FOR T=0. TO 0.5
 U=100.*(2.-Y)   ON X=0.0,  Y=2.-R                      FOR R=0. TO 1.0
```

The reserved words ON, FOR, LINE and TO cannot have blanks in them and **must have blanks on both sides of them.**

There is a special simple form for **straight line pieces** of the boundary. In this case ⟨piece⟩ appears as:

LINE ⟨x constant⟩, ⟨y constant⟩ TO ⟨x constant⟩, ⟨y constant⟩

where (⟨x constant⟩, ⟨y constant⟩) are the coordinates of the end points of the piece; they may be any Fortran expressions that evaluate to a constant. Straight line sides may be connected by the following **multiple line form**:

line sides may be connected by the following **multiple line form:**

```
⟨condition⟩   ON LINE   ⟨x1⟩, ⟨y1⟩   TO   ⟨x2⟩, ⟨y2⟩
⟨condition⟩                          TO   ⟨x3⟩, ⟨y3⟩
    ...                                       ...
⟨condition⟩                          TO   ⟨xk⟩, ⟨yk⟩
```

The boundary condition ⟨condition⟩ may be omitted if it is the same as for the preceding pieces (both for straight line pieces and parameterized pieces). Several groups of TO ⟨x⟩, ⟨y⟩ can be placed on one line as long as the same boundary condition holds, as for example

U = 1.0 ON LINE ⟨x1⟩,⟨y1⟩ TO ⟨x2⟩,⟨y2⟩ TO ··· TO ⟨xk⟩,⟨yk⟩

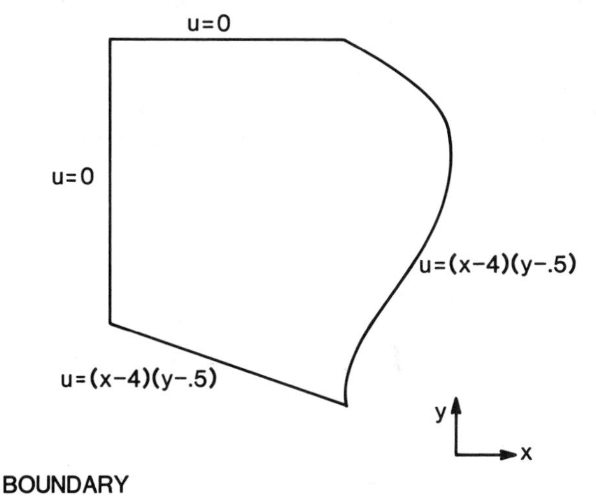


```
BOUNDARY
    U=0.0 ON LINE 4.,4. TO 1.,4.
                        TO 1.,0.5
    U=(X-4)*(Y-0.5)     TO 4.,-0.5
                        ON X=4. + 0.1*P*(P-4.5)**2, &
                           Y=-0.5 + P FOR P=0. TO 4.5
```

Figure 2.1. A nonrectangular domain with its parametrization and boundary conditions given.

The complex example in Figure 2.1 of a general domain is specified as follows:

```
*                 FOUR SIDED, NONRECTANGULAR DOMAIN
BOUNDARY.
*
      U = 0.0 ON LINE 4.,4.  TO 1., 4.
                             TO 1., 0.5
      U = (X-4.)*(Y-.5)      TO 4.,-0.5
                             ON X = 4.+.1*P*(P-4.5)**2   &
                                Y =-.5+P FOR P=0. TO 4.5
```

This example shows how omitting the boundary condition specifies it to be the previous one and how the LINE specifications continues from piece to piece.

A second complicated example follows:

```
*           SIX SIDED REGION WITH 3 STRAIGHT SIDES
*
BOUNDARY.
*
      U = 0.0              ON X=-T, Y=(T-1.)**2      FOR T = 1. TO 2.
                           ON X=(P-1)**2-2, Y=P      FOR P = 1. TO 2.
      U = (2.*X-Y)**2      ON LINE -1.,2. TO .5,2.   TO 1.,1. TO 0.,0.
      PI*X*COS(PI*X**2)*UX+UY-3.0*U = .5                             &
             ON X=-SQRT(PHI), Y=SIN(PI*PHI)/5. FOR PHI = 0.0 TO 1.0
```

There is a special abbreviated form for **rectangular domains** (in two or three dimensions) where ⟨piece⟩ is of the form

⟨variable⟩ = ⟨constant⟩

Here ⟨variable⟩ is one of X, Y, Z and ⟨constant⟩ is a Fortran expression that evaluates to a constant. Two examples of defining a two-dimensional rectangular domain and its boundary conditions follow:

```
**          ABBREVIATED BOUNDARY FORM FOR A RECTANGLE
BOUND.
        U = 1.0                         ON X = 0.0
        U + X*UX = (X+Y)*EXP(Y)         ON X = 1.0
        PERIODIC                        ON Y = 0.0
                                        ON Y = 1.0
*
*       BOUNDARY CONDITION CARRIED FORWARD FROM PIECE TO PIECE
*
BO.     U = 0.0                         ON Y = 0.0
                                        ON Y = PI/2.
                                        ON X = 0.0
        U = EXP(1.0)*SIN(2.*PI*Y)       ON X = EXP(1.0)
```

The preceding example can be written in a more compact form using the $ separator as follows

```
BOUNDARY.
   U  = 0.0 ON Y = 0.0 $ ON Y = PI/2. $ ON X= 0.0
   U = EXP(1.0)*SIN(2.*PI*Y) ON X = EXP(1.0)
```

The extension of this notation to three-dimensional boxes is straightforward; six (rather than four) sides and conditions are required and boundary conditions can include U, UX, UY, and UZ terms.

It is important to use the abbreviated form of the BOUNDARY segment whenever the domain is rectangular. If this is not done, the domain will be considered nonrectangular by ELLPACK and modules restricted to rectangular domains will not work.

2.C THE GRID SEGMENT

The GRID segment defines the rectangular grid placed over the problem domain. The numerical methods use this grid and thus GRID must appear before any particular ELLPACK problem solving module. The general form of the segment is a set of terms:

⟨n⟩ ⟨variable⟩ POINTS ⟨ point list⟩

where

⟨n⟩	=	number of points; must be a constant unless MAX X POINTS and/or MAX Y POINTS is set in an OPTION segment,
⟨variable⟩	=	variable involved (one of X, Y or Z),
⟨point list⟩	=	list of grid coordinates in increasing order.

Each term must be on a separate line or separated by a $. For two dimensional domains there must be one set of points specified for X and another for Y. In three dimensions there must also be a specification for Z. If the following grid is specified

$$n_1 \text{ X POINTS } x_1, x_2, \ldots, x_{n_1}$$

$$n_2 \text{ Y POINTS } y_1, y_2, \ldots, y_{n_2}$$

then the rectangular grid is made up of the lines

$$x = x_1, \quad x = x_2, \quad \cdots, \quad x = x_{n_1},$$

$$y = y_1, \quad y = y_2, \quad \cdots, \quad y = y_{n_2}.$$

See Figure 2.2 for an example 4 by 5 grid (4 X POINTS and 5 Y POINTS).

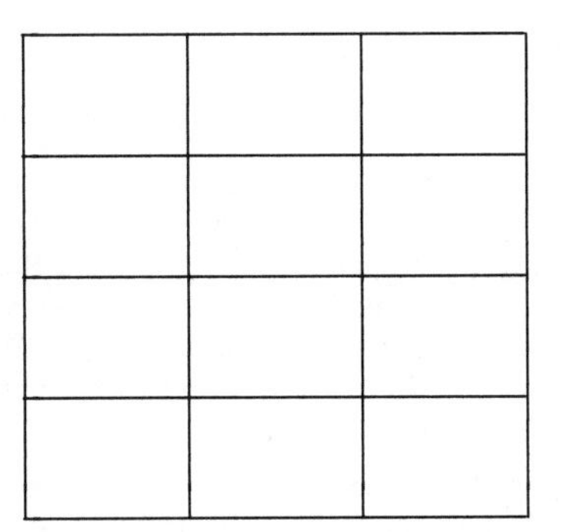

Figure 2.2. The rectangular grid defined by 4 X POINTS $ 5 Y POINTS.

For **uniformly spaced grids** on non-rectangular domains, ⟨point list⟩ may take the form

⟨a⟩ TO ⟨b⟩

where

⟨a⟩ is the initial value of the grid variable,
⟨b⟩ is the final value of the grid variable.

In this case the points used to discretize the variable are

$$p_i = (i-1)\frac{(b-a)}{(n-1)} + a, \quad i = 1, 2, \ldots, n$$

For **rectangular domains** ⟨a⟩ TO ⟨b⟩ is not used and the initial and final values of the variable correspond to the sides of the rectangle.

Some possible combinations are illustrated by the following examples:

```
*
*                    GENERAL NONUNIFORM CASE
*
GRID.      7 X POINTS  -1.0, -.8, -.5, 0.0, .5, .8, 1.0
           5 Y POINTS   0.0,  .2,  .5,  .8, 1.0
*
*                    UNIFORM GRID - FOR NONRECTANGULAR DOMAINS
*
GRID.      7 X POINTS -1.0 TO 1.0  $  4 YPOINTS  0.0 TO 1.0
```

```
*
*                    UNIFORM GRID - ON A RECTANGLE
*                    FORM VALID ONLY FOR RECTANGULAR DOMAINS
*
GRID.     7 X-POINTS  $  4 Y-POINTS
*
*                    MIXED CASE FOR 3-D, - ALWAYS RECTANGULAR
*
GRID.     7 X-POINTS  $  4 YPOINTS
          6 Z POINTS -1.0, -.7, -.25, .25, .7, 1.0
```

The purpose of the GRID segment is to setup a discrete domain for the computation. GRID is executable, and hence one can have several grids in a single ELLPACK program. When GRID is executed two things happen:

1. A rectangular grid is defined.

2. The relationship between the grid and the domain is determined.

For rectangular domains the second step is not required; the domain boundary simply corresponds to the extreme grid lines. For nonrectangular domains the situation is more complex; problem solving modules need to know the status of each grid point (e.g., inside or outside the domain), as well as information about each intersection of the boundary with grid lines. The **ELLPACK domain processor** provides this information based on the user-supplied specifications from the BOUNDARY, HOLE, ARC, and GRID segments.

The following ELLPACK program segment shows how one can solve the same problem using several grids. This is useful in determining the convergence properties of the numerical methods. An ordinary Fortran DO loop is used to solve the problem with grids of size 9×5, 17×9, and 33×17. (Use of the FORTRAN segment is described further in Section 2.E.) Note that one must declare the maximum number of grid lines in the OPTIONS segment (see Section 2.F).

```
EQUATION.    *** EQUATION DEFINED HERE ***
BOUNDARY.    *** RECTANGULAR DOMAIN DEFINED HERE ***
*
OPTION.     MAX X POINTS = 33
OPTION.     MAX Y POINTS = 17
FORTRAN.
C           LOOP OVER THREE GRIDS
            DO 100 K=1,3
               NX = 1 + 2**(K+2)
               NY = 1 + 2**(K+1)
*
GRID.          NX X POINTS
               NY Y POINTS
*
      *** ELLPACK PROGRAM CONTINUES ***
```

Since a new discrete domain is constructed each time the GRID segment is executed, one can even change the domain between successive executions of GRID. To do this one makes the parameterization of the boundary depend upon user-supplied functions which, in turn, depend upon a parameter which changes before each invocation of GRID. Note that one cannot change the number of boundary pieces in this way, only their definitions. In the following example, the functions XP and YP define an ellipse whose eccentricity depends upon the parameter ECC which is passed through a Fortran COMMON block. The problem is solved four times for four domains characterized by different values of ECC.

```
DECLARE.
            COMMON / PARAM / ECC
EQUATION.       *** EQUATION DEFINED HERE ***
BOUNDARY.       U = 0.0 ON X = XP(T), Y = YP(T) FOR T = 0.0 TO 2.0*PI
*
FORTRAN.
C           SOLVE FOR FOUR DIFFERENT ELLIPSES
            DO 100 K=1,4
               ECC = 1.0 + (K-1)/3.0
*
GRID.          21 X POINTS     0.0 TO 4.0
               11 Y POINTS     0.0 TO 2.0
*
            *** ELLPACK PROGRAM CONTINUES ***
*
SUBPROGRAMS.
              FUNCTION XP(T)
              COMMON / PARAM / ECC
              XP = ECC*(1.0 + COS(T))
              RETURN
              END
              FUNCTION YP(T)
              YP = 1.0 + SIN(T)
              RETURN
              END
END.
```

Note that the GRID segment here must be executed inside the Fortran DO-loop even though the position of the grid lines is the same each time.

2.D SEGMENTS WHICH SPECIFY THE NUMERICAL METHODS

We next describe the four segments which specify ELLPACK library modules to be used in solving the elliptic problem. These are DISCRETIZATION, INDEXING, SOLUTION, and TRIPLE. Most ELLPACK programs have three segments present (DISCRETIZATION, INDEXING, and SOLUTION) corresponding to the three steps in approximately solving the problem (see Figure 1.2). INDEXING is optional; the default is AS IS. A TRIPLE segment incorporates all three of these steps.

ELLPACK modules communicate among one another by using fixed sets of variables which define interfaces. Each DISCRETIZATION module produces a system of linear algebraic equations in a standard format for the discretization interface. INDEXING modules produce a set of permutation vectors for the indexing interface, and SOLUION and TRIPLE modules produce a solution vector for the solution interface. It is important to note that ELLPACK interfaces of a given type are unchanged by the execution of modules of other types. Thus, as in the last example above, several SOLUTION segments or cmbinations of INDEXING and SOLUTION segments may follow a single DISCRETIZATION segment. This feature makes ELLPACK useful for studying the effectiveness of various PDE solution methods.

A summary description of each ELLPACK module is given in Chapter 6. References are given there for further information about the numerical methods used. Note that **each module has restrictions** on its use (such as a self-adjoint equation or a rectangular domain). One should read the descriptions before using the modules.

It is also important to note that **not all combinations of modules are legal.** ELLPACK users should understand the basic premise of each module so they can determine whether a combination of modules is legal. Some illegal combinations are fairly obvious such as using a symmetric linear equation solver with a DISCRETIZATION module that does not yield a symmetric linear system. Similarly, an INDEXING module which tries to minimize matrix bandwidth is not likely to help with a module for solving the equations iteratively. A table at the beginning of Chapter 6 also gives some guidance as to which combinations are legal. The ELLPACK preprocessor does not check the legality of module combinations.

The following is a brief summary of several modules which are available in each of the DISCRETIZATION, INDEXING, SOLUTION, and TRIPLE segments.

DISCRETIZATION. SEGMENT

This segment names a module to be used to form a linear system of equations. The content of this segment is a single module name and the list of available modules is expandable. The basic set in ELLPACK consists of

5-POINT STAR
: Ordinary second order divided central differences (restricted to two dimensional domains).

7-POINT STAR
: Ordinary second order divided central differences (restricted to three-dimensional rectangular domains with self-adjoint operator and Dirichlet boundary conditions).

SPLINE GALERKIN
: Galerkin method with piecewise polynomials of general degree and smoothness (restricted to self-adjoint problems on two dimensional rectangular domains).

HERMITE COLLOCATION
: Collocation method with bicubic Hermite piecewise polynomials (restricted to rectangular domains in two dimensions).

HODIE
: Higher order finite differences (restricted to rectangular domains in two dimensions and boxes in three dimensions with Dirichlet boundary conditions).

COLLOCATION
: Collocation method with Hermite bicubics on nonrectangular domains (restricted to two dimensions).

The complete set of modules supplied with the ELLPACK system contains about 10 more discretization modules. A list of these modules is given at the start of Chapter 6.

INDEXING. SEGMENT

These modules take the linear system produced by DISCRETIZATION and reorganize it by renumbering the equations and/or unknowns. For example, one may wish to have the nested dissection ordering of the equations before using Gauss elimination to solve them. The basic set in ELLPACK consists of:

NESTED DISSECTION
: Computes the nested dissection ordering of the equations. Used with sparse Gauss elimination.

AS IS
: The ordering is that of the generation of the equations and unknowns by the discretization modules. If no indexing module is specified, then AS IS is used as the default.

RED-BLACK
: The variables and unknowns are numbered as on a checkerboard, all "red" points before the "black" points. This is used with REDUCED SYSTEM and SOR iteration primarily.

MINIMUM DEGREE
: Computes a mimimal degree ordering of the equations. This is used with sparse Gauss elimination.

There are several other INDEXING modules supplied with the complete ELLPACK system described in Chapter 6.

SOLUTION. SEGMENT

These modules solve the linear system of equations. This step may also involve reformatting the equations. For example, modules for solving banded systems require the equations to be in a certain band matrix format before the Gauss elimination is done, so they do the reformatting as well as the solution of the equations. The basic set in ELLPACK is

BAND GE
: Gauss elimination with scaled partial pivoting for a general band matrix.

LINPACK SPD BAND
: Cholesky elimination for a symmetric positive definite band matrix.

JACOBI CG
: Jacobi iteration with conjugate gradient acceleration.

REDUCED SYSTEM CG
: Reduced system iteration with conjugate gradient acceleration (requires the RED-BLACK indexing).

SOR
: SOR iteration.

SPARSE LU UNCOMPRESSED
: General sparse matrix Gauss elimination.

There are several other solution modules described in Chapter 6 and supplied with the complete ELLPACK system.

TRIPLE. SEGMENT

These modules combine the functions of discretization, indexing and solution of the resulting linear system. The one included in the basic set of ELLPACK modules is FFT 9-POINT which solves the Helmholtz equation on two dimensional rectangles with second, fourth, or sixth (the last for Poisson problems only) order finite differences using fast Fourier transform techniques. There are several other triple modules described in Chapter 6 and supplied with the complete ELLPACK system.

Module names (unlike segment names and ELLPACK reserved words) may have dashes and blanks in them. All dashes and blanks are removed before the word is analyzed by the preprocessor. The following are equivalent:

```
5 POINT STAR 5-POINT STAR 5 POINT-STAR 5-POI NT STAR 5POINTSTAR
```

Many modules accept parameters which are placed in parentheses following the module name. These parameters may be specified in any order and default values are provided; setting a parameter value for one use of a module does not affect the default value for later use. The **module parameters** are specified as ⟨parameter⟩ = ⟨value⟩, where ⟨parameter⟩ is an actual variable in the Fortran program generated by ELLPACK, so one must not use the same name for something else. This is true even if the default parameter value is used. The ⟨value⟩ is an actual value; it may be a constant or Fortran variable. Two simple examples of the use of parameters follow:

```
SOL.   SOR(OMEGA = 1.8, ITMAX=100, IADAPT=0)
SOL.   SOR(ZETA = TOLER, OMEGA = 1.88)
```

which set the relexation parameter OMEGA, etc. different from the default values for the SOR iteration.

Examples of method specifications follow:

```
*              ORDINARY FINITE DIFFERENCES AND GAUSS ELIMINATION

DISCRETIZATION.  5-POINT STAR
SOLUTION.        BAND GE

*                A FINITE ELEMENT METHOD WITH ITERATION

DISCRETIZATION.  SPLINE GALERKIN(DEGREE=3,NDERV=2)
SOLUTION.        SOR

*                A SINGLE MODULE FOR THE PROBLEM

TRIPLE.          FFT 9-POINT

*                SOLVE THE SAME PROBLEM BY A GAUSS ELIMINATION
*                SPARSE MATRIX METHOD AND TWO ITERATIVE METHODS
*                THE OUTPUT BETWEEN SOLUTIONS IS OMITTED

DIS.             5-POINT STAR
INDEX.           MINIMUM DEGREE
SOLUTION.        SPARSE LU UNCOMPRESSED
SOL.             JACOBI CG(ITMAX=200, ZETA=1.E-4)
INDEX.           RED-BLACK
SO.              REDUCED SYSTEM CG(ITMAX=200,ZETA=1.E-4,IADAPT=1)
```

2.E FORTRAN AND PROGRAM CONTROL

Fortran has two distinct uses in ELLPACK. The first and simplest is to define various functions that appear in the problem. Simple expressions like SIN(X+2.5*Y) can be inserted wherever needed, but more complex functions may need several Fortran statements. These functions can be defined as ordinary Fortran function subprograms and appended to the ELLPACK program in the SUBPROGRAM segment just before the END segment. They can then be used to define coefficients in the EQUATION or BOUNDARY segments just like built-in Fortran functions.

The second use of Fortran is to allow special calculations to be done. These might range from something simple like printing a heading and a few key parameters to complex auxiliary computations that require the full range of Fortran facilities. One might do computations that interact with the ELLPACK modules to solve nonlinear or other special problems. The more complicated uses are presented and illustrated in Chapters 4 and 5.

One of the primary uses of Fortran in ELLPACK is to post-process the computed solution in some way. The computed solution is accessible through several Fortran functions which become defined after a solution is computed, i.e. after a SOLUTION or TRIPLE segment executes. For two dimensional problems the functions U(X,Y), UX(X,Y), UY(X,Y), UXX(X,Y), UXY(X,Y), and UYY(X,Y) may be used to evaluate $u(x,y)$, $u_x(x,y)$, $u_y(x,y)$, $u_{xx}(x,y)$, and $u_{yy}(x,y)$, respectively, for any (x,y) in the domain. In three dimensions these become functions of three variables (X,Y,Z), and the functions UZ, UXZ, UYZ, and UZZ are added.

There are three segments for Fortran use.

FORTRAN. SEGMENT

The FORTRAN segment indicates lines of executable Fortran code to be inserted into the control program generated by ELLPACK. The statements are inserted exactly where they appear with respect to the order of module and output execution. FORTRAN segments may be used to control the order of module execution. **The ELLPACK system uses statement labels starting at 20000, so such labels must be avoided in the users' program.** The use of this segment is illustrated as follows:

```
*                PRINT A HEADING
FORTRAN.
C       Q1DATE IS AN ELLPACK UTILITY TO PROVIDE THE DATE
        CALL Q1DATE(IMO,IDAY,IYR)
        WRITE(6,20) IMO, IDAY, IYR
   20   FORMAT(///20X, 'WING LIFT CALCULATION', 5X, I2,
     A  2('-',I2)/ 20X, 'USING FINE GRID AND CUBIC SPLINES'///)

*
*       COMPUTE SOME PROPERTIES OF THE SOLUTION.
*       AFTER THE PROBLEM IS SOLVED THE FUNCTIONS U, UX AND UY
*       BECOME DEFINED FROM THE APPROXIMATE SOLUTION.
FORTRAN.
        DMAX = 0.0
        USUM = 0.0
        DO 10 I = 1,10
```

```
            YG = (I-1)*.2
            DO 10 J = 1,10
               XG = (I-1)*.1
               DMAX = AMAX1(DMAX, UX(XG,YG)**2 + UY(XG,YG)**2)
   10          USUM = DSUM + ABS(SQRT(U(XG,YG))
         USUM = USUM*.02
         PRINT 20, DMAX,USUM
   20    FORMAT(//'DMAX =', F10.4, 10X, 'SIZE SQRT(U)' = F10.4)
```

DECLARATION. SEGMENT

The DECLARATION segment indicates lines of Fortran declaration statements to be placed at the beginning of the ELLPACK control program. For example:

```
DECLARATIONS.
      INTEGER DIGIT, COUNTS(10)
      REAL MAXU4, MINU4, LOADS(20)
```

The DECLARATION segment is primarily used to declare variables used in FORTRAN segments. Note that these declarations do *NOT* affect variables used in the EQUATION and BOUNDARY segments, since those are buried inside various functions created by the ELLPACK preprocessor. To declare such variables the GLOBAL segment must be used (see Chapter 4). Fortran comments may not appear in this segment.

There is also a real **work space array** R1WORK always available for use. This array is used for temporary storage by modules, but it may also be used for scratch storage in Fortran segments. Note that the contents of R1WORK may be altered by any ELLPACK module. Its size (given by the Fortran variable I1KWRK) is usually fairly large and it can be made larger by using the OPTIONS segment. R1WORK resides in blank COMMON so no blank COMMON declarations can appear in this segment.

SUBPROGRAMS. SEGMENT

The SUBPROGRAMS segment defines complete Fortran functions or subroutines. For example, the user can define A(X,Y) or A(X,Y,Z) to be the coefficient of UXX in the PDE. This segment must be **at the end of the ELLPACK program,** just before the END segment.

Lines with a $ in column 1 are handled specially in SUBPROGRAMS segments; the $ is stripped off and the line is copied, shifted one character to the left. The $ is for those Fortran systems (mercifully rare) that require control cards for each Fortran subprogram.

2.F OUTPUT AND OPTIONS SEGMENTS

OPTIONS. SEGMENT

This segment sets various switches in the ELLPACK system. The OPTIONS segment must be near the start of the program, similar to a declaration. Options are not dynamic and, if one is given more than once, the last appearance is used. Some options may be changed during execution by setting internal ELLPACK Fortran variables in FORTRAN segments. If this can be done, the Fortran variables are listed with the description of the option. Several other options are described in Chapter 4.

CLOCKWISE = k	Select orientation of boundary parametrization.
k = .TRUE.	Clockwise orientation
k = .FALSE.	Counter clockwise orientation (default)
L1CLKW	Fortran variable for CLOCKWISE
INTERPOLATION = k	Select the method of interpolation to define U(X,Y), etc. off the grid (for finite difference methods only).
k = QUADRATICS	Local quadratic polynomials (default)
k = SPLINES	Use B-splines of degree appropriate for the order of the discretization module. This interpolation is appropriate primarily for rectangular domains; for general domains the solution is first extended to grid points near the domain boundary by low order extrapolation and then set to zero elsewhere. See [deBoor, 1978] for a description of B-splines; the interpolation routines were adapted from deBoor's PPPACK software.
LEVEL = k	Set levels (0-5) for amount of printed output in ELLPACK run.
k = 0	Requests no output from modules except fatal error messages
k = 1	Request minimal output (default)
k = 2	Requests reasonable summary of what happened
k = 3, 4, 5	More and more intermediate output, primarily useful for debugging
I1LEVL	Fortran variable for LEVEL
MAX X POINTS = NX MAX Y POINTS = NY MAX Z POINTS = NZ	Set maximum values for the GRID statement. This is required whenever a variable is used to give the number of points in the GRID segment.
MEMORY	Requests a printed estimate of the memory used in the ELLPACK run with some breakdown.
NO EXECUTION	Run the preprocessor, but no the ELLPACK program.
PAGE = k	Select type of pagination for module output.
k = 0	No page advances.
k = 1	New page before DISCRETIZATION, TRIPLE, TABLE, or SUMMARY (default).
k = 2	New page before every module and OUTPUT segment.
I1PAGE	Fortran variable for PAGE.
SELF-ADJOINT = k	Set the switch for self adjoint form of the PDE. k may be .TRUE. or .FALSE. See Section 2.B.
L1SELF	Fortran logical variable for SELF-ADJOINT.

TIME — Requests a printed summary of the execution times of each module.

  L1TIME — Fortran variable for TIME (can only turn TIME off).

MAX WORKSPACE = k — Limit the automatic workspace estimate and set the workspace array R1WORK to have dimension at most I1KWRK = k.

MIN WORKSPACE = k — Set workspace array R1WORK to have dimension at least I1KWRK = k.

The INTERPOLATION option specifies how the functions U(X,Y), UX(X,Y), etc. are defined for some, mainly finite difference, modules. If a discretization produces approximate values only on a grid of points, then an interpolation algorithm is used to provide values off the grid. Only one interpolation algorithm can be used in an ELLPACK run. The choice INTERPOLATION = SPLINES usually involves a substantial computation, but is more appropriate for use with higher order accurate finite difference discretizations.

ELLPACK modules use the blank COMMON array R1WORK as a workspace for scratch storage while they execute. Modules provide an estimate for the size of this array to the preprocessor, which uses the largest such dimension required by any single module for the run. Occasionally these estimates are too high (in which case the program is storage-inefficient) or too low (in which case the run fails). The options MAX-WORKSPACE and MIN-WORKSPACE can be used to remedy these problems. User-defined Fortran code (in FORTRAN segments) can also use this array for scratch storage, and the MIN-WORKSPACE option can be used to indicate user requirements in this case. Note that the contents of R1WORK are changed whenever an ELLPACK module executes.

Some OPTIONS segment examples follow:

```
*
*      REQUEST OUTPUT OF BASIC PERFORMANCE STATISTICS
*
OPTIONS.   TIME $ MEMORY
*
*      ENLARGE WORKSPACE, REQUEST MORE DETAILED PRINTED OUTPUT
*
OPT.          MIN-WORKSPACE=7500 $ LEVEL=2
*
*      ENLARGE WORKSPACE, SUPPRESS PAGINATION
*
OPT.          MIN-WORKSPACE=12000 $ PAGE=0
```

OUTPUT. Segment

This segment produces various kinds of output from the computation. The requests are of the forms:

⟨type⟩ or ⟨type⟩(⟨function⟩) or ⟨type⟩(⟨function⟩,⟨grid⟩)

where ⟨type⟩ is a keyword, ⟨function⟩ is a function name and ⟨grid⟩ defines a grid. The default ⟨grid⟩ is the one defined in the GRID segment. A uniform grid within the standard grid is defined by NX,NY or NX,NY,NZ where NX, NY and

NZ must be actual integer constants, the number of grid lines for each of the X,Y,Z variables respectively. The types are:

MAX(f)
MAX(f,grid) — Print maximum value, simple least squares value (root mean square or L_2 norm) and average absolute value (L_1 norm) of f, all based on the grid. These values are, respectively,

$$\max_{i,j} |f(x_i,y_j)|$$

$$\left[\frac{1}{N}\sum_{i,j} f^2(x_i,y_j)\right]^{1/2}$$

$$\frac{1}{N}\sum_{i,j} |f(x_i,y_j)|$$

where N is the total number of points from the grid that are in the domain or on its boundary. The maximum and summations are taken over these points.

RMS(f)
RMS(f,grid) — Same as MAX. (RMS is for root mean square).

NORM(f)
NORM(f,grid) — Same as MAX.

PLOT(f)
PLOT(f,grid) — Produces a contour plot of a function (f) of two variables. Here the grid size determines the smoothness of the contour lines, the default grid is 20 by 20. The finer the grid the smoother the contour lines, and the more expensive the computation. Sample plots are shown in Figure 17.1.

PLOT DOMAIN — Display the domain with grid lines. An example is given in Figure 17.1.

TABLE(f)
TABLE(f,grid) — Print table of function f at grid points.

SUMMARY(f)
SUMMARY(f,grid) — Equivalent to MAX(f) $ TABLE(f).

The function f may be one of the following standard ELLPACK functions or any user named function of two real variables (or three variables in three dimensions):

U, UX, UY, UZ,
UXX, UYY, UZZ,
UXY, UXZ, UYZ — The solution function and its derivatives (defined after a SOLUTION or TRIPLE segment has been executed).

TRUE — The known solution of the problem. Defined by the user in the SUBPROGRAM segment as

REAL FUNCTION TRUE(X,Y)

or

REAL FUNCTION TRUE(X,Y,Z).

ERROR — The error in the computed solution. The function TRUE must be provided; otherwise TRUE=0 is used.

RESIDU — If L$u = f$ represents the partial differential equation, then the residual is L$U - f$ where U is the computed solution. If the equation is given in self-adjoint form

$$(p(x,y)\, u_x)_x + (q(x,y)\, u_y)_y + r(x,y)\, u = f(x,y)$$

then the user must supply the Fortran functions

REAL FUNCTION CDXU(X,Y)

and

REAL FUNCTION CDYU(X,Y)

which return the values of $\partial p/\partial x$ and $\partial q/\partial y$ respectively. In three dimensions the functions CDXU, CDYU, CDZU with arguments X,Y,Z are required. Replace the REAL declaration by DOUBLE PRECISION for these functions when using a double precision version of ELLPACK.

The following illustrates the use of the OUTPUT segment.

```
*   CHECK HOW GOOD A SOLUTION IS (TRUE SOLUTION KNOWN)
*
OUTPUT.     MAX(ERROR) $ PLOT(ERROR) $ MAX(RESIDU)
*
*   QUICK LOOK AT RESULTS
*
OUTPUT.     SUMMARY(U) $ PLOT(U)
*
*   TABULATE FUNCTIONS RELATED TO SOLUTION AND THE USER-DEFINED
*   FUNCTION FUNK
*
OUT.        TABLE (U) $ TABLE(UXX) $ TABLE(FUNK)
*
*   TABULATE ON A GRID DIFFERENT THAN IN DISCRETIZATION
*
OUT.        TABLE(U,12,12)  $  TABLE(ERROR,6,6)
*
SUBPROGRAMS.
        REAL FUNK(X,Y)
        FUNK = UX(X,Y)**2 + UY(X,Y)**2
        RETURN
        END
```

2.G DEBUGGING ELLPACK PROGRAMS

Program bugs are of two types: syntactic (something is wrong with the language input) and operational (something is wrong with the program operation).

SYNTAX ERRORS

Errors in syntax are further subdivided into four groups:

(a) **ELLPACK language errors.** The ELLPACK preprocessor provides diagnostics whenever there is an ELLPACK language error. Sometimes the messages are not as clear as one would hope, especially when there are several errors in one segment or when an error makes a Fortran statement appear to be an

ELLPACK statement. The latter occurs, for example, when a Fortran statement other than a comment starts in column 1 or when the segment name FORTRAN fails to start in column 1. Fortunately, the ELLPACK language has a rather simple structure so that ELLPACK language errors rarely are difficult to locate.

(b) **Fortran expression errors in ELLPACK segments.** The ELLPACK preprocessor does not check Fortran expressions; they are simply passed on to the Fortran compiler for the second stage of the language translation. The information one receives from the Fortran compiler depends entirely on how ELLPACK is installed locally; thus no general statement can be made about these errors. ELLPACK should be installed so that it is possible to obtain a copy of the ELLPACK control program by setting a switch. This Fortran program will contain the erroneous expression along with the local Fortran compiler's diagnostics and this should be sufficient to identify the erroneous expression.

There is, in addition, one subtle type of Fortran expression error, the **accidental keyword misuse error.** This error is most likely to occur in the BOUNDARY segment and we give some examples in this case. Suppose the correct ELLPACK program is to contain the following two boundary condition specifications:

```
U = A(X,Y) * BOND(Y) ON X=3.
U = 0.0 ON X=F(T,A(2,2)), Y=A(1,1)*FORT(T)   FOR T=0. TO 3.
```

Typing errors or editing can result in the following variations of these statements.

```
U = A(X,Y)*B ON D(Y)   ON  X=3.
U = 0.0 ON X = F(T,A 2,2)), Y = A(1,1)*FORT(T)    FOR T=0. TO 3.
U = 0.0 ON X = F(T,A(2,2) , Y = A(1,1)*FORT(T)    FOR T=0. TO 3.
U = 0.0 ON X = F(T,A(2,2)), Y = A(1,1)* FOR T(T)  FOR T=0. TO 3.
```

In the first variation, the middle two letters ON of BOND are taken to be the keyword ON. In the last variation, the first three letters FOR of FORT are taken to be the keyword FOR. In the middle two variations a missing parenthesis makes a comma in F(T,A(2,2)) appear to be the comma that separates x = ⟨ expression ⟩ from the y = ⟨ expression ⟩ portion of the statement. Such errors lead to fatal Fortran errors and may cause the ELLPACK preprocessor to get lost also.

(c) **Errors in FORTRAN segments.** The Fortran code from FORTRAN, GLOBAL and DECLARATION segments are copied directly to the ELLPACK control program. Errors in these Fortran statements show up only when that program is compiled. The diagnostics from the local Fortran compiler should be obtained and these errors then should be easily identified. Note that Fortran comments are not allowed in DECLARATION and GLOBAL segments.

(d) **Errors in Fortran subprograms.** All the programs of the SUBPROGRAMS segment are appended to the end of the ELLPACK control program. Any errors in them are detected by the Fortran compiler and, again, one should obtain a copy of the control program with the diagnostic messages. Note that one cannot use the ELLPACK preprocessor variables (introduced later in Section 4.C) in this segment because the ELLPACK preprocessor completely ignores these lines of code.

OPERATIONAL ERRORS

Errors in syntax are a nuissance, but rarely much more. Operational errors are usually much more troublesome and have, at least, the following causes:

(a) **ELLPACK system error**. The ELLPACK system does not work the way it is supposed to.

(b) **ELLPACK module error**. An ELLPACK module does not work the way it is supposed to.

(c) **Module sequencing error**. The ELLPACK program uses a combination of modules which do not interface properly.

(d) **Module use error**. An ELLPACK module is applied to a problem which it is not intended to solve.

(e) **Normal program bug**. The ELLPACK program has an erroneous formulaton of the elliptic problem or the method to solve it.

We discuss three approaches to locating such errors.

1. **Obtain intermediate output.** The LEVEL option may be used to request output from the ELLPACK modules. The higher values of LEVEL (e.g. 4 or 5) are intended for debugging these modules and give information which most users will find difficult to interpret. However, LEVEL=2 should produce a reasonable amount of information, some of which is understandable without detailed knowledge of the modules. For example, if the discretization module prints

```
ELEMENTS OF THE DISCRETIZATION MATRIX
```

followed by all zeros, it is clear that there is an error in EQUATION or the discretization module has failed. **Caution:** Use a coarse grid when using a high LEVEL of output or you might receive many pages of output giving every element of large matrices and vectors.

It might also be useful to use the TABLE facilities of the OUTPUT segment presented in Section 4.B and the PROCEDURE DISPLAY MATRIX PATTERN presented in Section 4.A.

2. **Define an elliptic problem where the method is exact.** If the true solution of the PDE is a simple polynomial (such as $1+x+y$ or x^2+y^2), most discretizations should be exact and thus you should obtain an error the size of round-off. If a program is not working, make a copy of it and modify the right side of the PDE and boundary conditions to make the true solution such a simple polynomial. If the program still does not work, try changing the ELLPACK modules used. If 5-POINT-STAR and HERMITE COLLOCATION used with BAND GE both fail to give the true solution within round-off error there is almost surely an error in the formulation of the problem. Note that a coarse (say, 4 by 4) grid can be used for this testing.

3. **Define an elliptic problem with a known true solution.** If one is unsure of the correctness of a numerical solution obtained, one can frequently define an almost identical problem whose true solution is known and which behaves somewhat like the original problem. The method used can then be tested on this related problem to estimate the effectiveness of the methods being used. This kind

of checking is particularly important for solving complex problems by iteration and other techniques as described in Chapter 5.

Program 5.E1 illustrates a general technique to make any PDE with any boundary conditions have a given function as true solution merely by changing the right sides of the PDE and boundary conditions. It might be rather tedious to construct a function similar to the solution of the original problem.

16.H REFERENCES

de Boor, C. [1978], *A Practical Guide to Splines*, Springer-Verlag, New York.

Chapter 3

EXAMPLES

This chapter gives example ELLPACK programs with output to illustrate the use of the facilities of Chapter 2. The first example is a revision of the example of Chapter 1; the domain has been made non rectangular, and a normal derivative boundary condition is used on one piece. The second example is a completely general equation with mixed boundary conditions on a rectangular domain. The third example shows how ELLPACK and Fortran interact. A problem is discretized and then solved several times with an iterative method; each time the convergence test is changed and the purpose is to examine the effect on accuracy achieved and execution time.

3.A EXAMPLE OF CHAPTER 1 REVISED

The first example shows a simple case of a general domain, a quadrilateral. The quadrilateral could be specified completely by the LINE facility, but actual parametrized pieces are given to illustrate their use. Note how a rectangular grid is placed over the domain. Ordinary finite differences are used along with band Gauss elimination. Once the problem is discretized, several indexing or solution modules may be applied.

```
*       ***********************************************************
*       *                                                         *
*       *  EXAMPLE ELLPACK PROGRAM 3.A1                           *
*       *                                                         *
*       *     THIS IS THE SAME EQUATION AS THE EXAMPLE IN         *
*       *     CHAPTER 1.  THE BOUNDARY CONDITIONS ARE CHANGED     *
*       *     AND THE DOMAIN IS NO LONGER RECTANGULAR.            *
*       *                                                         *
*       ***********************************************************
OPTION.    CLOCKWISE = .TRUE.
EQUATION.  UXX  +  UYY  +  3*UX  -  4*U  =  EXP(X+Y)*SIN(PI*X)
BOUNDARY.  UX = 0.         ON  X = 0., Y = T    FOR  T = -1. TO 2.
           U  = X**2       ON  X = R,  Y = 2.   FOR  R =  0. TO 1.
           U  = 1. + Y/2.  ON  LINE 1.,2.  TO 1., 0.
           U  = X*Y/2.     ON  X = 1.-S, Y = -S FOR  S =  0. TO 1.
GRID.      6  X POINTS    0. TO 1.
           6  Y POINTS   -1. TO 2.
DISCRET.   5 POINT STAR
SOLUTION.  BAND GE
OUTPUT.    TABLE(U)  $  PLOT(U)
END.
```

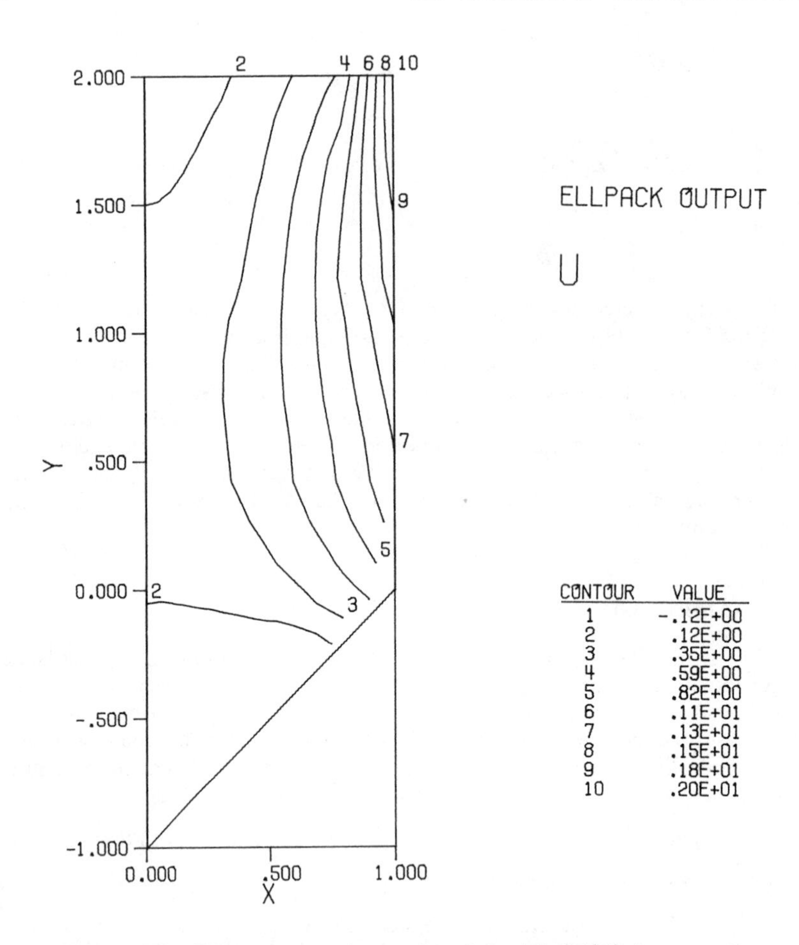

Figure 3.1. The contour plot produced by PLOT(U) in example ELLPACK program 3.A1.

Output of ELLPACK Run (The contour plot is shown in Figure 3.1)

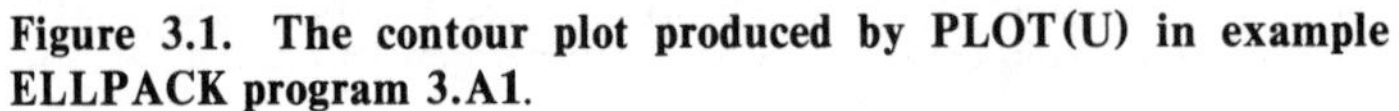

PREPROCESSOR OUTPUT

```
2.23 SECONDS TO INITIALIZE PREPREPROCESSOR
1.27 SECONDS TO PARSE ELLPACK PROGRAM
3.92 SECONDS TO CREATE ELLPACK CONTROL PROGRAM
7.42 SECONDS TOTAL
     ELLPACK VERSION ---- APRIL 5, 1984
```

```
------------------
DOMAIN PROCESSOR
------------------

   BOUNDARY POINTS FOUND          19
   BOUNDARY PIECES FOUND           4
   GRID SIZE                6 BY   6
   EXECUTION SUCESSFUL

-----------------------
DISCRETIZATION MODULE
-----------------------

   5 - P O I N T   S T A R

   DOMAIN                      NONRECTANGULAR
   UNIFORM GRID                         6 X   6
   NUMBER OF EQUATIONS                       19
   MAX NO. OF UNKNOWNS PER EQ.                5
   EXECUTION SUCCESSFUL

-----------------
SOLUTION MODULE
-----------------

  B A N D   G E

  NUMBER OF EQUATIONS        19
  LOWER BANDWIDTH            10
  UPPER BANDWIDTH            10
  REQUIRED WORKSPACE        608

  EXECUTION SUCCESSFUL

----------------
ELLPACK OUTPUT
----------------

          +++++++++++++++++++++++++++++++++++++++++++++
          +                                           +
          +    TABLE OF U      ON    6 X    6  GRID   +
          +                                           +
          +++++++++++++++++++++++++++++++++++++++++++++

   X-ABSCISSAE ARE
   ---------------
   0.000000E+00     2.000000E-01     4.000000E-01     6.000000E-01
   8.000000E-01     1.000000E+00

      Y = 2.000000E+00
      ----------------
   0.000000E+00     4.000001E-02     1.600000E-01     3.600000E-01
   6.400000E-01     2.000000E+00

      Y = 1.400000E+00
      ----------------
   1.379495E-01     1.847529E-01     3.251631E-01     6.239269E-01
   1.098070E+00     1.700000E+00

      Y = 8.000001E-01
      ----------------
   2.546195E-01     2.956870E-01     4.188895E-01     6.509129E-01
   9.915969E-01     1.400000E+00

      Y = 2.000000E-01
      ----------------
   1.951296E-01     2.248628E-01     3.140624E-01     4.793794E-01
   7.290915E-01     1.100000E+00

      Y =-4.000000E-01
      ----------------
  -5.476528E-02    -6.280090E-02    -8.690773E-02    -1.200000E-01
   0.000000E+00     0.000000E+00
```

```
        Y =-1.000000E+00
        ----------------
   -5.607122E-02      0.000000E+00      0.000000E+00      0.000000E+00
    0.000000E+00      0.000000E+00

----------------
 ELLPACK OUTPUT
----------------

    CONTOUR PLOT OF      U
    GRID                20   BY    20
    EXECUTION SUCCESSFUL
```

3.B GENERAL EQUATION WITH MIXED BOUNDARY CONDITIONS

The second example shows a comparison of two different methods in solving a problem with mixed boundary conditions. Ordinary finite differences (5 POINT STAR) with Gauss elimination (BAND GE) gives an error of 1.5 percent while collocation (HERMITE COLLOCATION) with Gauss elimination gives an error of 0.0055 percent. The execution times on a VAX 11/780 are 0.18 sec. and 2.45 sec., both relatively small.

```
*       ***********************************************************
*       *                                                         *
*       *  EXAMPLE ELLPACK PROGRAM 3.B1                           *
*       *                                                         *
*       *       THIS PROBLEM HAS MIXED BOUNDARY CONDITONS.        *
*       *       THE PROGRAM COMPARES TWO DISTINCT METHODS FOR     *
*       *       SOLVING THE SAME PROBLEM.                         *
*       *                                                         *
*       *                                                         *
*       ***********************************************************
OPTIONS.        LEVEL=1 $ TIME
EQUATION.       UXX + (1.0+Y**2)*UYY - UX - (1.0+Y**2)*UY = F(X,Y)
BOUNDARY.        - U + UX = 0.             ON  X=1.
                   U      = TRUE(X,Y)      ON  Y=0.
                   U + UX = 2.0*EXP(Y)     ON  X=0.
                   U      = TRUE(X,Y)      ON  Y=1.
GRID.           4 X POINTS  $  5 Y POINTS
OUT.            MAX(TRUE)
DIS.            5 POINT STAR
SOL.            BAND GE
OUT.            TABLE(U) $ MAX(ERROR,7,9)
DIS.            HERMITE COLLOCATION
SOL.            BAND GE
OUT.            TABLE(U) $ MAX(ERROR,7,9)
SUBPROGRAMS.
      FUNCTION TRUE(X,Y)
C       THE STANDARD ELLPACK TRUE SOLUTION FUNCTION (IF KNOWN)
        TRUE = EXP(X+Y) + ((X*(X-1.0))**2)*ALOG(1.0+Y**2)
        RETURN
      END
      FUNCTION F(X,Y)
C       CONSTRUCT F SO TRUE IS AS GIVEN
        F = ALOG(1.0+Y**2)*(2.0+X*(-14.0+X*(18.0-4.0*X)))
     A  + 2.0*((X*(X-1.0))**2)*(1.0-Y-2.0*Y**2/(1.0+Y**2))
        RETURN
      END
END.
```

Output of ELLPACK Run:

```
----------------------
 PREPROCESSOR OUTPUT
----------------------

     2.07 SECONDS TO INITIALIZE PREPREPROCESSOR
     1.52 SECONDS TO PARSE ELLPACK PROGRAM
     4.30 SECONDS TO CREATE ELLPACK CONTROL PROGRAM
     7.88 SECONDS TOTAL

          ELLPACK VERSION --- APRIL 5, 1984
-----------------
 ELLPACK OUTPUT
-----------------

     +++++++++++++++++++++++++++++++++++++++++++++++++++++++++++++
     +                                                           +
     +  MAX( ABS(TRUE  ) ) ON   4 X   5 GRID = 7.3890557E+00     +
     +                                                           +
     +  L1 NORM( TRUE  )   ON   4 X   5 GRID = 3.1028562E+00     +
     +                                                           +
     +  L2 NORM( TRUE  )   ON   4 X   5 GRID = 3.4975860E+00     +
     +                                                           +
     +++++++++++++++++++++++++++++++++++++++++++++++++++++++++++++
------------------------
 DISCRETIZATION MODULE
------------------------

     5 - P O I N T   S T A R

     DOMAIN                                  RECTANGLE
     DISCRETIZATION                            UNIFORM
     NUMBER OF EQUATIONS                            12
     MAX NO. OF UNKNOWNS PER EQ.                     5
     MATRIX IS                           NON-SYMMETRIC

     EXECUTION SUCCESSFUL

-----------------
 SOLUTION MODULE
-----------------

    B A N D   G E

    NUMBER OF EQUATIONS         12
    LOWER BANDWIDTH              4
    UPPER BANDWIDTH              4
    REQUIRED WORKSPACE         168

    EXECUTION SUCCESSFUL

-----------------
 ELLPACK OUTPUT
-----------------

            ++++++++++++++++++++++++++++++++++++++++++++++++
            +                                              +
            +    TABLE OF U      ON   4 X   5  GRID        +
            +                                              +
            ++++++++++++++++++++++++++++++++++++++++++++++++

    X-ABSCISSAE ARE
    ---------------

    0.000000E+00    3.333333E-01    6.666667E-01    1.000000E+00
       Y = 1.000000E+00
       ----------------
    2.718282E+00    3.827898E+00    5.328720E+00    7.389056E+00
```

```
   Y = 7.500000E-01
   ----------------
2.109272E+00    2.973749E+00    4.135154E+00    5.714677E+00

   Y = 5.000000E-01
   ----------------
1.642888E+00    2.307862E+00    3.209331E+00    4.439072E+00

   Y = 2.500000E-01
   ----------------
1.282753E+00    1.792140E+00    2.495316E+00    3.464102E+00

   Y = 0.000000E+00
   ----------------
1.000000E+00    1.395612E+00    1.947734E+00    2.718282E+00
```

```
----------------
ELLPACK OUTPUT
----------------

 ++++++++++++++++++++++++++++++++++++++++++++++++++++++++++++
 +                                                          +
 +  MAX( ABS(ERROR ) ) ON   7 X   9 GRID = 4.2617321E-02    +
 +                                                          +
 +  L1 NORM( ERROR )    ON   7 X   9 GRID = 1.0838118E-02    +
 +                                                          +
 +  L2 NORM( ERROR )    ON   7 X   9 GRID = 1.5031885E-02    +
 +                                                          +
 ++++++++++++++++++++++++++++++++++++++++++++++++++++++++++++
```

```
----------------------
DISCRETIZATION MODULE
----------------------

 H E R M I T E   C O L L O C A T I O N

 CASE                                     NONHOMGENEOUS
 DOMAIN                                       RECTANGLE
 X INTERVAL                   0.    E+00, 0.100E+01
 Y INTERVAL                   0.    E+00, 0.100E+01
 BOUND. COLLOC. PTS. PARAMS.              0.    E+00
                                          0.    E+00
 GRID                                       4 X    5
 HX                                       0.333E+00
 HY                                       0.250E+00
 OUTPUT LEVEL                                     1
 NUMBER OF EQUATIONS                             80
 MAX NO. OF UNKNOWNS PER EQ.                     16
 EQUATIONS ARE NOT SYMETRIC

 EXECUTION SUCCESSFUL
```

```
----------------
SOLUTION MODULE
----------------

 B A N D   G E

 NUMBER OF EQUATIONS          80
 LOWER BANDWIDTH              27
 UPPER BANDWIDTH              27
 REQUIRED WORKSPACE         6640

 EXECUTION SUCCESSFUL
```

```
----------------
ELLPACK OUTPUT
----------------
```

```
          +++++++++++++++++++++++++++++++++++++++++
          +                                       +
          +   TABLE OF U      ON   4 X   5 GRID   +
          +                                       +
          +++++++++++++++++++++++++++++++++++++++++
 X-ABSCISSAE ARE
 ---------------
 0.000000E+00   3.333333E-01   6.666667E-01   1.000000E+00
    Y = 1.000000E+00
    ----------------
 2.718580E+00   3.828192E+00   5.329034E+00   7.389400E+00
    Y = 7.500000E-01
    ----------------
 2.117355E+00   2.976796E+00   4.145617E+00   5.754802E+00
    Y = 5.000000E-01
    ----------------
 1.648983E+00   2.312166E+00   3.222430E+00   4.481803E+00
    Y = 2.500000E-01
    ----------------
 1.284152E+00   1.795079E+00   2.504001E+00   3.490397E+00
    Y = 0.000000E+00
    ----------------
 1.000000E+00   1.395631E+00   1.947759E+00   2.718282E+00
---------------
ELLPACK OUTPUT
---------------

   +++++++++++++++++++++++++++++++++++++++++++++++++++++
   +                                                   +
   + MAX( ABS(ERROR ) ) ON   7 X  9 GRID = 4.0960312E-04 +
   +                                                   +
   + L1 NORM( ERROR )   ON   7 X  9 GRID = 1.5688321E-04 +
   +                                                   +
   + L2 NORM( ERROR )   ON   7 X  9 GRID = 1.9722113E-04 +
   +                                                   +
   +++++++++++++++++++++++++++++++++++++++++++++++++++++

---------------
ELLPACK OUTPUT
---------------

       ++++++++++++++++++++++++
       +                      +
       +   EXECUTION  TIMES   +
       +                      +
       ++++++++++++++++++++++++
        SECONDS      MODULE NAME
           0.13      MAX
           0.12      5-POINT STAR
           0.03      BAND GE SETUP
           0.03      BAND GE
           0.28      TABLE
           0.17      MAX
           0.25      HERMITE COLLOCATION
           0.22      BAND GE SETUP
           1.98      BAND GE
           0.35      TABLE
           0.45      MAX
           4.10      TOTAL TIME
```

3.C HOW FORTRAN AND ELLPACK INTERACT

The third example is somewhat more complicated. The solution module JACOBI CG has a parameter ZETA to control iteration termination; iteration on the linear system is stopped when the estimated error is less than ZETA. The object here is to test the effect of changing ZETA; thus, the problem is solved three times, using values of 10^{-3}, 10^{-4} and 10^{-5}.

An important feature here is that parameters of the module JACOBI CG are changed each time it is invoked. Caution must be used as this does not always work; some parameters (e.g. those which affect array sizes) affect the program at preprocessing time rather than at execution time. For example, SPARSE LDLT (NSP=NWORK) will fail because a numerical value for NSP is required by the preprocessor and the value of the Fortran variable NWORK is not known until execution time. Another feature of this example is the use of self-adjoint form for the PDE.

The results show that the stopping criterion has a substantial effect on the number of iterations. The results are summarized as follows:

ZETA	=	10^{-3}	10^{-4}	10^{-5}
Number of iterations	=	30	46	50

The maximum error in solving the elliptic problem is unaffected by these changes as it is due to the discretization error and not to the error in solving the linear system. Even with ZETA = 10^{-3} the error in solving the linear system is less than the error in discretizing the elliptic problem (which has maximum absolute value of 0.08 in this case); thus no improvement is made by taking more iterations.

```
*       ***********************************************************
*       *                                                         *
*       *   EXAMPLE ELLPACK PROGRAM 3.C1                          *
*       *                                                         *
*       *        SELF-ADJOINT PROBLEM SOLVED BY FINITE DIFFERENCES *
*       *        AND ITERATION. THE PROGRAM TESTS THE EFFECT      *
*       *        OF CHANGING THE STOPPING CRITERION.              *
*       *                                                         *
*       ***********************************************************
EQUATION.     ( W(X,Y)*UX )X  +  ( W(X,Y)*UY )Y  =  F(X,Y)
BOUNDARY.     UX = 0.0   ON   X = 0.5
              U  = 0.0   ON   X = 1.0
              UY = 0.0   ON   Y = 0.5
              U  = 0.0   ON   Y = 1.0
GRID.         17 X POINTS
              17 Y POINTS
OUT.          PLOT(TRUE)  $  MAX(TRUE)
DISCRET.      5 POINT STAR
FORTRAN.
C                PRINT NUMBER OF ITERATIONS FOR JACOBI CG WITH
C                STOPPING CRITERION ZETA=1/10.**N FOR N=3,4,5
           DO 100 NZETA = 3 , 5
              PRINT 10, 1./10.**NZETA
     10       FORMAT(/5X,'* * ZETA =',E10.3,' * *')
SOL.          JACOBI CG (ITMAX = 50, ZETA = 1./10.**NZETA)
OUT.          MAX(ERROR)
```

```
FORTRAN.
  100 CONTINUE

SUBPROGRAMS.
      FUNCTION W(X,Y)
       COMMON /CONCOM/ PI
C      ELLPACKS PI IS NOT AVAILABLE IN SUBPROGRAMS
       DATA PI/3.14159265358979/
       W = ((PI*COS(PI*X)*SIN(PI*Y))**2 +
     A    (PI*SIN(PI*X)*COS(PI*Y))**2)**0.15
       RETURN
      END
      FUNCTION TRUE(X,Y)
       COMMON /CONCOM/ PI
       TRUE = SIN(PI*X)*SIN(PI*Y)
       RETURN
      END
      FUNCTION F(X,Y)
       COMMON /CONCOM/ PI
C      CONSTRUCT F SO TRUE IS AS GIVEN
       PI2 = PI * PI
       SINPIX = SIN(PI*X)
       SINPIY = SIN(PI*Y)
       COSPIX = COS(PI*X)
       COSPIY = COS(PI*Y)
       TU = SINPIX*SINPIY
       TUX = PI*COSPIX*SINPIY
       TUXX = -PI2*TU
       TUY = PI*SINPIX*COSPIY
       TUYY = -PI2*TU
       F = W(X,Y)*(TUXX + TUYY) + CDXU(X,Y)*TUX + CDYU(X,Y)*TUY
       RETURN
      END
END.
```

Output of ELLPACK Run (Abbreviated: **** indicates deleted lines.)
(The contour plot is shown in Figure 3.2.)

```
 ----------------------
  PREPROCESSOR OUTPUT
 ----------------------

    2.23 SECONDS TO INITIALIZE PREPREPROCESSOR
    1.35 SECONDS TO PARSE ELLPACK PROGRAM
    3.92 SECONDS TO CREATE ELLPACK CONTROL PROGRAM
    7.50 SECONDS TOTAL

         ELLPACK VERSION --- APRIL 5, 1984

 ----------------
  ELLPACK OUTPUT
 ----------------

    CONTOUR PLOT OF      TRUE
    GRID                20  BY   20
    EXECUTION SUCCESSFUL

 ----------------
  ELLPACK OUTPUT
 ----------------

    ++++++++++++++++++++++++++++++++++++++++++++++++++++++++++++++
    +                                                            +
    +  MAX( ABS(TRUE   ) ) ON  17 X  17 GRID = 9.9999988E-01     +
                                  ****
```

```
 ------------------------
 DISCRETIZATION MODULE
 ------------------------

    5 - P O I N T   S T A R

    DOMAIN                              RECTANGLE
    DISCRETIZATION                        UNIFORM
    NUMBER OF EQUATIONS                       256
    MAX NO. OF UNKNOWNS PER EQ.                 5
    MATRIX IS                           SYMMETRIC

    EXECUTION SUCCESSFUL

    * * ZETA = 0.100E-02 * *
 -----------------
 SOLUTION MODULE
 -----------------

I T P A C K     J A C O B I   C G
JACOBI CG  HAS CONVERGED IN     30 ITERATIONS
EXECUTION SUCCESSFUL
 ----------------
 ELLPACK OUTPUT
 ----------------

    ++++++++++++++++++++++++++++++++++++++++++++++++++++++++++++
    +                                                          +
    +  MAX( ABS(ERROR ) ) ON  17 X  17 GRID = 1.1345798E-01   +
                                   ****

    * * ZETA = 0.100E-03 * *
 -----------------
 SOLUTION MODULE
 -----------------

I T P A C K     J A C O B I   C G
JACOBI CG  HAS CONVERGED IN     38 ITERATIONS
EXECUTION SUCCESSFUL
 ----------------
 ELLPACK OUTPUT
 ----------------

    ++++++++++++++++++++++++++++++++++++++++++++++++++++++++++++
    +                                                          +
    +  MAX( ABS(ERROR ) ) ON  17 X  17 GRID = 1.1345553E-01   +
                                   ****

    * * ZETA = 0.100E-04 * *
 -----------------
 SOLUTION MODULE
 -----------------

I T P A C K     J A C O B I   C G
*** W A R N I N G ***********
    IN ITPACK ROUTINE JACOBI CG
    RPARM(1) = 0.100E-04 (ZETA)
    A VALUE THIS SMALL MAY HINDER CONVERGENCE
    SINCE MACHINE PRECISION SRELPR = 0.596E-07
    ZETA RESET TO  0.298E-04
*** W A R N I N G ***********
    IN ITPACK ROUTINE Q5I9ZB
    F(A) AND F(B) HAVE SAME SIGN
JACOBI CG  HAS CONVERGED IN     49 ITERATIONS
EXECUTION SUCCESSFUL
```

```
-------------------
 ELLPACK OUTPUT
-------------------

++++++++++++++++++++++++++++++++++++++++++++++++++++++++++++++
+                                                            +
+  MAX( ABS(ERROR ) ) ON  17 X  17 GRID = 1.1341274E-01     +
                           ****
```

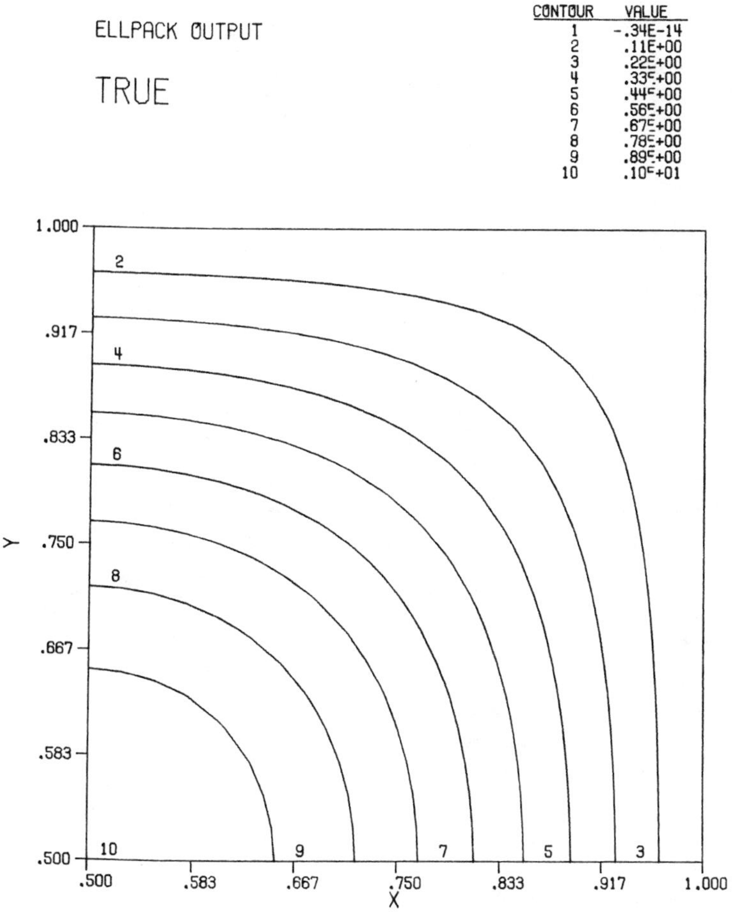

Figure 3.2. The contour plot produced by PLOT(TRUE) in Example 3.C1.

Chapter 4

ADVANCED ELLPACK FEATURES

This chapter presents features of ELLPACK which give one more control over the problem solving process. These features provide:

1. the capability to handle more general problems,
2. means to construct iterative methods (e.g., for nonlinear problems),
3. the ability to reduce computer resource usage,
4. means to study the methods in ELLPACK,
5. more convenient programming in certain applications.

The use of these features to solve more complex problems is illustrated in the final section of this chapter. Even more complicated examples are presented in Chapter 5.

The additional ELLPACK language features are:

OPTIONS.	To control storage To set problem characteristics
HOLE, ARC.	To handle more complex domains
GLOBAL.	To parametrize the PDE and boundary conditions
PROCEDURE.	To display the pattern of nonzeros in the matrix To initialize unknowns for iteration methods To compute eigenvalues of the discretization matrix To assist the domain processor with difficult regions To handle problems with nonunique solutions To list the ELLPACK modules To transform the PDE by subtracting a known function
TRIPLE.	To initialize the solution function U
OUTPUT.	To tabulate variables defining the elliptic problem: discrete equations, unknowns and indexes, domain and boundary

In addition, there is a section describing how one can access internal ELLPACK variables, including "preprocessor" or "template" variables.

4.A ADDITIONAL SEGMENTS

There are four new segments in ELLPACK described here.

HOLE. SEGMENT

This segment defines a hole to be removed from the domain of the problem. Its form is exactly like BOUNDARY, except that the name HOLE is used. **HOLE segments must appear after the BOUNDARY segment,** and several HOLE segments may appear. The boundary of the hole must be parametrized in the **opposite** direction of that of the domain boundary. Thus if CLOCKWISE = .TRUE. appears as an option (specifying the domain boundary is defined clockwise) then the boundaries of the holes must be specified counterclockwise.

The **grid must be fine enough** so that at least one interior grid point lies on any grid line between the boundary of a hole and the boundary of the domain. **The short notation for rectangular domains cannot be used if there are holes in the rectangle.** The reason is that short cuts are taken for rectangles in the ELLPACK preprocessor which leave it unprepared for a HOLE segment.

ARC. SEGMENT

This segment defines an arc or curved slit to be removed from the domain of the problem as well as side (boundary) conditions that apply on it. Its form is exactly like BOUNDARY, and the same restrictions apply to ARC that apply to HOLE. Note that a single boundary condition is given on the arc. If "two-sided" boundary conditions are needed, then long, narrow holes must be specified. See Section 5.A for examples which illustrate the technique. Arcs cannot divide the domain into two or more disjoint parts.

GLOBAL. SEGMENT

This segment puts user-specified Fortran declaration statements in the ELLPACK control program as well as in all the Fortran subprograms generated by the ELLPACK preprocessor (those that define the PDE, the domain, and the boundary conditions). **It does not affect the ELLPACK library subprograms (modules).** Specifically, the internal ELLPACK subprograms affected are

Q1PCOE The PDE coefficients subroutine

R1PRHS The PDE right hand side function

Q1BCOE The boundary condition coefficients subroutine

R1BRHS The boundary condition right hand side function

Q1BDRY The boundary coordinates subroutine

The GLOBAL facility allows one to parametrize the elliptic problem and provide control of these parameters at the ELLPACK program level. To do this, one simply includes Fortran COMMON blocks in the GLOBAL segment. If the segment

```
GLOBAL.
        COMMON/PARAM/A,K
```

is included, then the generated right side function R1PRHS in the ELLPACK control program is

```
      REAL FUNCTION R1PRHS(X,Y)
      COMMON/PARAM/A,K
      R1RPRHS = .LT. EXPRESSION FROM EQUATION SEGMENT .GT.
      RETURN
      END
```

Consider the following ELLPACK program fragment (with many lines omitted)

```
EQ.        UXX + UYY - K * U = A * (1. + X)/(A + X * Y)
                    .
                    .
                    .
GLOBAL.
           COMMON/PARAM/ A,K
           REAL K
                    .
                    .
                    .
FORTRAN.
           DO 10 I = 1, 8
              A = 2.0 * I - 1.0
              K = I

TRIPLE.       FFT 9-POINT(IORDER=4)
OUT.          PLOT(U) $ SUMMARY(U)

FORTRAN.
      10   CONTINUE
                    .
                    .
```

This problem is solved 8 times with 8 different values of the parameters A and K.

PROCEDURE. SEGMENT

The PROCEDURE segment provides facilities that are useful in solving or analyzing an elliptic problem but that are not one of the standard steps in solving the problem. The ELLPACK system is designed to allow one easily to add procedures for particular applications or special situations. The form of the PROCEDURE segment is the same as DISCRETIZATION.

There are several PROCEDURE facilities in the ELLPACK system. Detailed descriptions are given in Chapter 6 for each of these; some examples are given below.

DISPLAY MATRIX PATTERN

Provides a printout of the pattern of nonzero elements in the matrix of the discretization.

EIGENVALUES

Computes eigenvalues of the discretization matrix.

SET UNKNOWNS FOR 5 POINT STAR

Uses the values of a user-specified Fortran function to initialize the solution vector of the discretization matrix. This is useful when using an iterative method for solving the linear system. For examples of this application see Section 4.B.

NONUNIQUE (X = ⟨value⟩, Y = ⟨value⟩, U = ⟨value⟩)

Provides a value of u at a point (x,y) to make the solution of a problem unique whose solution is not otherwise unique; the Neumann problem is an example. Some modules do not use this information or require that the point have a special (x,y) location. This procedure also has a Z = ⟨value⟩ in three dimensions.

DOMAIN FILL

Provides additional information to the ELLPACK domain processor to allow it to process domains with narrow subregions or sharp corners.

REMOVE

Automatically transforms the PDE and boundary conditions by subtracting a known function from the solution. Some examples of the use of these procedures are given in Section 4.B.

LIST MODULES

Lists all the modules in the ELLPACK system being used along with the names of their parameters.

4.B ADDITIONAL FEATURES OF BASIC SEGMENTS

There are several additional facilities useful for complex ELLPACK applications. These are described in this section.

INITIALIZATION OF THE SOLUTION FUNCTION U(X,Y) BY TRIPLE SEGMENTS.

Iteration methods may be constructed in ELLPACK to solve nonlinear problems or time dependent problems. Each of these processes must be initialized; the ELLPACK facilities to do this are in a TRIPLE segment because initialization has the same effect as completely solving the elliptic problem. Here, however, we do not expect to obtain much accuracy. The three triple modules for this purpose are:

SET U BY BLENDING

Use blending function interpolation to define U(X,Y) as a smooth function which exactly matches the elliptic problem's boundary conditions. Applicable only for two-dimensional rectangular domains and boundary conditions with constant coefficients. This procedure provides a good method for generating an initial guess of the unknowns for iterative method solution modules; the technique is shown below.

SET U BY BICUBICS

Use Hermite bicubics to define U(X,Y) which interpolates the elliptic problem's boundary conditions at the boundary grid points plus two points in between each pair of boundary grid points plus at the corners. Applicable only for two dimensional rectangular domains and uncoupled boundary conditions (only U, UX or UY specified at any point). The interpolant U(X,Y) is identically zero in the middle of the rectangle (i.e. except in elements adjacent to the boundary). This procedure is primarily for boundary layer problems where the interpolant provides an approximation to the differences between the "smooth" solution in the interior and the actual solution.

SET (U = ⟨fname⟩)
The values of the real Fortran function

FUNCTION ⟨fname⟩ (X,Y)

are tabulated at the grid points and then extended by interpolation to define U(X,Y), etc., everywhere. A third argument Z is added in three dimensions.

In each case these TRIPLE segments create all the standard ELLPACK functions U, UX, UY, UXX, UXY and UYY.

Initialization of the Solution Vector by PROCEDURE Segments

Several ELLPACK solution modules are available which use iterative methods to solve the linear system of algebraic equations generated by discretization modules. Several of these are described in detail in Chapter 7. Such methods begin with an initial approximation to the solution vector and produce successively more accurate approximations until some convergence criterion is satisfied. The default initial vector in ELLPACK is the zero vector. A considerable reduction in the number of iterations is possible if a more accurate initial vector is used, and several ELLPACK procedures are available to aid users in generating this initial approximation. These are the SET UNKNOWNS procedures. Since the ordering of the solution vector depends upon the discretization module used, separate procedures are used to initialize the solution vector for different discretizations. For example,

```
SET UNKNOWNS FOR 5-POINT STAR (U = ⟨FUNCT⟩ )
```

uses the values of the user-defined Fortran function ⟨FUNCT⟩ (X,Y) to intitialize the solution vector. A typical use of this procedure is

```
DIS.   5-POINT STAR
PRO.   SET UNKNOWNS FOR 5-POINT STAR (U=GUESS)
SOL.   SOR
```

where GUESS(X,Y) is defined in the SUBPROGRAMS segment. Note that the PROCEDURE must be placed after the DISCRETIZATION, since it uses the ordering of unknowns generated by the discretization. SET UNKNOWNS procedures are available for several other discretizations; consult Chapter 6 for a complete list.

For problems on rectangular domains one can also automatically generate an initial solution vector using the SET U triples, as the following example shows.

```
TRI.   SET U BY BLENDING
DIS.   5-POINT STAR
PRO.   SET UNKNOWNS FOR 5-POINT STAR (U=U)
SOL.   SOR
```

Here the TRIPLE segment defines the function U(X,Y) to be a smooth function which exactly satisfies the boundary conditions. The DISCRETIZATION segment leaves U unchanged (only the execution of a TRIPLE or SOLUTION causes U to be redefined), and hence the PROCEDURE segment uses the blending function interpolant to initialize the solution vector.

PROBLEM TRANSFORMATION BY PROCEDURE SEGMENTS

Several procedure modules are available which transform the declared PDE problem so that the solution is $v(x,y) = u(x,y) - r(x,y)$, where u is the original unknown function and r is known. If the original equation is represented by $\mathrm{L}u = f$ with boundary conditions $\mathrm{B}u = g$, then the transformed problem is $\mathrm{L}v = f - \mathrm{L}r$ with boundary conditions $\mathrm{B}v = g - \mathrm{B}r$. By subtracting appropriate functions r one can often significantly simplify the problem. For instance one might hope to

* remove a known singularity in the solution,
* replace the boundary conditions with homogeneous ones, or
* remove a known boundary layer in the solution.

The problem transformation is done internally and automatically by ELLPACK when one of several REMOVE procedures is executed. There are three separate REMOVE modules.

REMOVE (r = ⟨fname⟩)

Transform the problem using a user-defined function r whose name is specified. The function r should be defined in the SUBPROGRAMS segment as

FUNCTION ⟨fname⟩ (X,Y)

or

FUNCTION ⟨fname⟩ (X,Y,Z)

If periodic boundary conditions are specified, then this function must have the same periodicity. The problem transformation requires that second (and possibly first) partial derivatives of r be known. These are computed numerically, and hence the problem transformation will not be "exact". This module may be useful in removing a known singularity r from the solution. The transformed problem may often be solved with greater accuracy in this case.

REMOVE BLENDED BC

Transforms the problem to one with homogeneous boundary conditions. A function r which exactly satisfies the boundary conditions is computed internally using blending function interpolation (this is the same function as that computed by the triple module SET U BY BLENDING). This module is restricted to problems on rectangular domains with constant coefficient Robbins type boundary conditions. The boundary conditions need not match at the corners of the domain. The derivatives of r are computed partially numerically. The homogenization of boundary conditions is useful since some discretization modules are considerably more efficient in this case. For example, the linear system generated by the discretization module HERMITE COLLOCATION on an n-by-m grid is normally of size $4nm$ with bandwidth $4m+7$, but this is reduced to size $4(n-1)(m-1)$ and bandwidth $2m+3$ in the homogeneous case by use of the module INTERIOR COLLOCATION.

REMOVE BICUBIC BC

Transforms the problem using a Hermite bicubic piecewise polynomial r which exactly satisfies the boundary conditions at each boundary grid point plus two points in between each pair of boundary grid points. Furthermore, it is identically zero on all grid squares which are not adjacent to the boundary. This module is restricted to problems on rectangular domains with uncoupled boundary conditions (i.e., either U, UX, or UY specified). The derivatives of r are computed exactly. This function r may be useful in removing boundary layers from the solution.

These PROCEDURE modules must be executed before the first DISCRETIZATION or TRIPLE module in the ELLPACK program. After the transformed problem is solved, the function U(X,Y) returns the computed value of the original unknown function u. A sample ELLPACK program which illustrates the use of REMOVE BLENDED BC to homogenize boundary conditions follows:

```
*  USE REMOVE PROCEDURE TO HOMOGENIZE BOUNDARY CONDITIONS
*
EQUATION.   UXX + UYY - X  Y * UY = 1.0
BOUNDARY.     U = 0.   ON X=0.
              U = 1.   ON Y=0.
             UX = Y    ON X=1.
             UY = X    ON Y=1.
PROCEDURE.   REMOVE BLENDED BC
DISCRETI.    INTERIOR COLLOCATION
INDEXING.    INTERIOR COLLORDER
SOLUTION.    BAND GE
OUTPUT.     PLOT(U)
END.
```

TABULATE INTERFACE VARIABLES BY THE OUTPUT SEGMENT.

Certain tables of internal ELLPACK interface variables can be printed. These can be useful for complicated problems where one has to interact with the internal data structures of ELLPACK. The additional OUTPUT statements are given below with a brief description of the resulting output.

TABLE PROBLEM

List current values of ELLPACK variables which define the problem.

TABLE BOUNDARY

List arrays produced by the domain processor for nonrectangular domains.

TABLE DOMAIN

List arrays that define a nonrectangular domain's relation to the rectangular grid

TABLE EQUATIONS

List linear equations produced by discretization modules

TABLE INDEXES

List ELLPACK indexing arrays

TABLE UNKNOWNS
List unknowns of the linear system

To understand the information provided by these statements, users must be familiar with the ELLPACK interfaces defined in Chapter 13.

Control of Dimensions by the OPTIONS Segment

The ELLPACK system creates a Fortran program with dimensions declared for all variables. Sometimes ELLPACK creates arrays for a particular problem which are not used or which are declared larger than needed. The dimension of each such array is controlled by a variable in the ELLPACK preprocessor and these can be set in the OPTIONS segment. Effective use of this facility requires one to become familiar with the ELLPACK control program and the ELLPACK interfaces. Each of the ELLPACK arrays is in its own COMMON block which is named the same except the first letter is C. **The workspace array R1WORK is in blank COMMON.**

Simple examples of this facility are the options MAX WORKSPACE = k, MIN WORKSPACE = k already introduced in Chapter 2.

Additional options for setting dimension declarations are

EQUATIONS = k
Set the number of equations to k. Each Fortran array in the ELLPACK control program with dimension equal to the number of equations is row dimensioned at k.

COEFFICIENTS = k
Set the maximum number of nonzero coefficients per equation to k.

INDEXING = k
Set the size of the indexing array I1ENDX to k.

UNKNOWNS = k
Set the number of unknowns (sizes of the arrays R1UNKN and I1UNDX) to k.

BOUNDARY POINTS = k
Set the number of boundary-grid intersection points to k.

BANDWIDTH = k
Set the bandwidth of the arrays used by the band matrix solvers to k.

More detailed control is possible using specific variables associated with individual arrays. For example, the dimension of the array R1UNKN of unknowns is I0MUNK and the statement

```
OPTION.     I0MUNK=388
```

sets this dimension to 388 independently of what the normal size of this array is. Table 4.1 gives a sample set of the more important array names, their dimension variables and a brief description of the array. Note that most of these variables have third character M; the second character is 0 to indicate that this variable belongs to the ELLPACK preprocessor.

CONTROL OF PROBLEM CHARACTERISTICS BY THE OPTIONS OR FORTRAN SEGMENTS

The ELLPACK preprocessor automatically examines the PDE and sets values of the following Fortran logical variables in the control program.

L1LAPL   Laplace's equation
L1CSTC   Constant coefficients
L1POIS   Poisson equation
L1HMEQ   Homogeneous PDE

These automatic settings can be reset by assigning new values to these variables in an OPTIONS or FORTRAN segment. Putting the assignment in a Fortran segment allows the setting to vary during a single run. Thus, for example, setting

```
OPTION.   L1CSTC = .FALSE.
```

would have the PDE

```
EQUATION.     3*UXX + 2*UYY - 7*U = 0
```

classified as having variable coefficients and thus a module which does something special for constant coefficients would not do that in this run. By doing this one can evaluate the benefit of a module's special action for constant coefficients.

Table 4.1. Variables for Control of Selected Array Dimensions in ELLPACK

Array Name	*Preprocessor Dimension Control*	*Description*
R1COEF	I0MNEQ, I0MNCO	These define the linear system
I1IDCO	I0MNEQ, I0MNCO	
R1BBBB	I1MEQ	
I1IDCO	IOIDCO, IOIDC2	Provides control of individual arrays of the linear system
R1BBBB	IOBBBB	
I1ENDX	I0MEND	Equation reordering permutation vector Unknown reordering permutation vector The unknown variables of the linear system
I1UNDX	I0MUND	
R1UNKN	I0MUNK	
R1XBND	I0MBPT	These provide information about boundaries of general domains
R1YBND	I0MBPT	
R1BPAR	I0MBPT	
I1PECE	I0MBPT	
I1BPTY	I0MBPT	
I1BGRD	I0MBPT	
I1BNGH	I0MBPT	
R1WORK	I0KWRK	Module workspace

Warning: These options must be used with caution because:

1. There is no guarantee that a module acts upon these variables,
2. A module can do its own analysis of the problem, and hence there might be no effect, and

3. Since properties of the PDE are known to the preprocessor, they sometimes determine the dimensions of arrays or even the selection of subprograms loaded.

In summary, these options should be tried on an experimental basis. Although there are many situations where they are convenient, there are some where they cause the ELLPACK run to fail (perhaps in a mysterious way).

4.C ACCESS TO PREPROCESSOR VARIABLES

In some applications of ELLPACK one needs to refer to values which the preprocessor computes and which are inconvenient (or impossible) to compute while writing the ELLPACK program. The simplest instance of this is the dimension of an array, say the workspace R1WORK or the unknowns R1UNKN. If one wants to create a new array of the same or related size, one must know the size to dimension the new array. There is a mechanism which allows access to certain variables of this type, called **template variables**, in ELLPACK. Template variables may appear in any Fortran statement in an ELLPACK program (except in the SUBPROGRAM segment). In the above cases, the dimensions of the R1WORK and R1UNKN arrays are $I1MRWK and $I1KWRK; these are exactly the same names as listed in Table 4.1 preceeded by a $. Thus, the ELLPACK program fragment

```
DECLARATIONS.
        REAL  COPYU($I1MUNK),  WORK2($I1KWRK,2)
GLOBAL.
        COMMON/PASSER/UNKOLD($I1MUNK)
```

is read by the preprocessor and correct numerical values substituted for template variables that appear. This substitution is made when the ELLPACK control program is generated so that it will compile correctly.

Table 4.2 gives the more useful of these variables, a few others may be identified from the tables in Chapter 13.

Table 4.2. Available Template Variables

Name	*Description*
$I1KBAN	Estimated half-band width of discretization matrix
$I1KWRK	Dimension of R1WORK, workspace
$I1MEND	Dimension of I1ENDX, indexing vector
$I1MUND	Dimension of I1UNDX, indexing vector
$I1MNCO	Column dimension of R1COEF, coefficient matrix
$I1MUNK	Dimension of I1UNKN, maximum number of unknowns
$I1MNEQ	Row dimension of R1COEF, coefficient matrix
$I1NBND	Number of boundary pieces
$I1MBPT	Maximum number of boundary points
$I1NGRX	Dimension of R1NGRX, *x*-points grid vector
$I1NGRY	Dimension of R1NGRY, *y*-points grid vector
$I1NGRZ	Dimension of R1NGRZ, *z*-points grid vector

If the variables are used in a context where they are not followed by a blank or special character (which is unlikely), the six characters may be enclosed in parentheses; that is, $I1KBAN and $(I1KBAN) are treated the same. Thus the ELLPACK statements:

```
FORTRAN.
      WRITE(I1OUTP,20) ((R1COEF(I,J), J=1. I1MNCO), I=1, I1MNEQ)
   20 FORMAT(/ / 'THE $(I1MNEQ)X$(I1MNCO) COEFFICIENT ARRAY' /
      A            ( $(I1MNCO)F10.5))
```

causes the following Fortran code to be generated in the ELLPACK Control Program when I1MNEQ=50 and I1MNCO=5:

```
      WRITE(I1OUTP, 20) ((R1COEF(I,J), J=1, I1MNCO), I=1, MNEQ)
   20 FORMAT (/ / 'THE 50X5 COEFFICIENT ARRAY' /
      A            (5F10.5))
```

The Fortran 77 PARAMETER statement can be used with these variables to create dimensions for arrays of related sizes. For example, the statements:

```
DECLARATIONS.
      PARAMETER (NSIDE = ($I1NGRX-2) * ($I1NGRY-2))
      PARAMETER (NEDGE = 2*($I1NGRX + $I1NGRY-1))
      REAL INTER(NSIDE,3), RECTB(NEDGE), UINTER(NSIDE)
      INTEGER IDEDGE(NEDGE), KTYPE(NSIDE)
```

give numerical values to NSIDE and NEDGE when the ELLPACK control program is generated. Thus, one has five arrays whose dimensions are related to the number of interior and edge points of the rectangular grid.

4.D ADVANCED ELLPACK EXAMPLES

This section presents four example problems solved with ELLPACK. The first illustrates how to make a parameter study (vary physical parameters) for an application to the solidification of alloys. This is an actual application of ELLPACK to a real world problem. The second example shows how to use ELLPACK procedures to analyze numerical methods. While this example is artificially simplified here, these procedures can be very useful in practice. The third example illustrates how to solve nonlinear problems using Picard iteration. Finally, there is an example solving a problem on an elliptical domain with an elliptical hole in it.

EXAMPLE 4.D1 PARAMETER STUDY FOR ALLOY SOLIDIFICATION

The following elliptic boundary value problem is of interest in the study of the solidification of metallic alloys:

$$\nabla^2 u - (\beta/2)^2 u = 0$$

$$u = 0 \qquad \text{on} \quad y = y^*$$

$$u_n = 0 \qquad \text{on} \quad x+0 \text{ and } x+1/2$$

$$-w'(x)u_x + u_y = -\beta \sinh(\beta y/2) \qquad \text{on} \quad y = w(x)$$

The domain represents a liquid alloy ahead of a solidification front $w(x) = \delta \cos(2\pi x)$ moving at constant velocity V in the y direction (see Figure 4.1). The coordinate system is taken to be moving at this velocity and the system is assumed at steady state. The function u is then related to the concentration of solute in the liquid according to the formula

$$c(x,y) = c_0(y) + u(x,y) \exp(-\beta y/2)$$

where $c_0(y) = 1 + \exp(-\beta y)$ is the concentration for an unperturbed solid-liquid interface ($\delta = 0$). The sides of the container are at $x = 1/2$ and $x = -1/2$, but the domain is truncated at $x = 0$ due to symmetry. Also, the boundary condition along the topmost edge is actually $u \rightarrow 0$ as $y \rightarrow \infty$, but the domain is truncated to some finite value y^* as shown in Figure 4.1. For more information see [Coriell, et al., 1981].

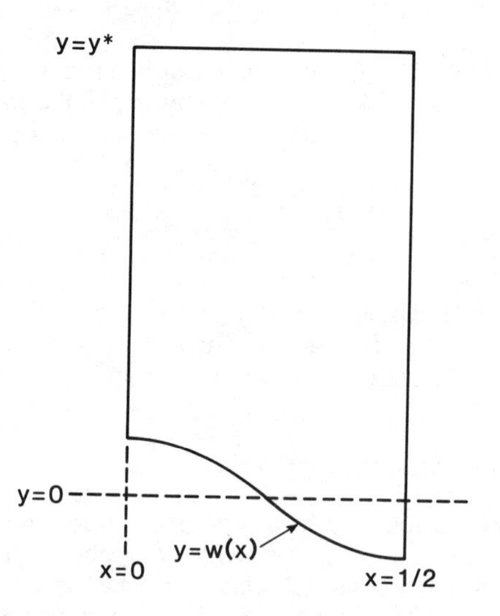

Figure 4.1. Typical domain for Example 4.D1.

We show how to perform parameter studies with respect to β, δ, and y^*. The parameter δ is the amplitude of the solid-liquid interface, and $\beta = VL/D$ where V is the solidification velocity, L is the actual half-width of the container, and D is the diffusivity of the solute in the liquid. In each case we wish to determine the solute distribution along the solid-liquid interface.

The following ELLPACK program solves this problem. Note the use of the GLOBAL segment to parametrize the program. The parameter values for the run are read and set in the Fortran segment. The parameter variables are then used in the EQUATION and BOUNDARY segments, as well as in the user-defined subprograms. Once the PDE is solved, the solute concentration function c is plotted and its value along the curve $w(x)$ is tabulated.

```
*       **********************************************************
*       *                                                        *
*       *  EXAMPLE ELLPACK PROGRAM 4.D1                          *
*       *                                                        *
*       *  REMARKS                                               *
*       *    MODEL OF SOLUTE SEGREGATION DURING UNIDIRECTIONAL   *
*       *    SOLIDIFICATION OF A BINARY ALLOY WITH A CURVED      *
*       *    SOLID-LIQUID INTERFACE                              *
*       *                                                        *
*       **********************************************************
GLOBAL.
      COMMON / PARAMS / BETA,BOV2,BOV2SQ,YINF,DELTA,TWOPI,TWOPID
OPTION.        CLOCKWISE = .TRUE.
EQUATION.      UXX + UYY - BOV2SQ*U = 0.0
BOUNDARY.  UX = 0.0  ON X=0.0,  Y=T        FOR T=W(0.0) TO YINF
         U = 0.0  ON X=T,      Y=YINF    FOR T=0.0 TO 0.5
        UX = 0.0  ON X=0.5,    Y=YINF-T  FOR T=0.0 TO YINF-W(0.5)
            -DW(X)          *UX + UY = -BETA*SINH(BOV2*Y)       &
                        ON X=0.5 - T,Y=W(0.5-T)  FOR T=0.0 TO 0.5
FORTRAN.
C                   SET PROBLEM PARAMETERS
      YINF  = 0.5
      BETA  = 2.0
      DELTA =0.2
      WRITE(6,2000) YINF,BETA,DELTA
 2000 FORMAT('1 PARAMETERS FOR THIS RUN ARE ' /
     A        ' ---------------------------- ' //
     B        5X,'YINF  = ',1PE12.5 / 5X,'BETA  = ',1PE12.5 /
     C        5X,'DELTA = ',1PE12.5 /)
      BOV2 = 0.5*BETA
      BOV2SQ = BOV2*BOV2
      TWOPI = 2.0*PI
      TWOPID = TWOPI*DELTA
GRID.          11 X POINTS     0.0 TO 0.5
               16 Y POINTS     W(0.5) TO YINF
*              SOLVE PDE PROBLEM FOR FUNCTION U
DIS.           5-POINT STAR
SOLUTION.      LINPACK BAND
*              PLOT CONTOURS OF SOLUTE CONCENTRATION
OUTPUT.        PLOT(C)
FORTRAN.
C                TABULATE CONCENTRATION ALONG INTERFACE
      NPTS = 21
      WRITE(6,2001)
 2001 FORMAT('      TABLE OF CONCENTRATION ALONG INTERFACE' /
     A       '      --------------------------------------' //
     B       '              X                C(X,W(X))        '/)
      DX = 0.5/FLOAT(NPTS-1)
      DO 100 I = 1,NPTS
         X = FLOAT(I-1)*DX
         CVAL = C(X,W(X))
         WRITE(6,2002) X,CVAL
  100 CONTINUE
 2002 FORMAT(1X,1P,2E18.6)
SUBPROGRAMS.
      REAL FUNCTION W(X)
C                SHAPE OF THE SOLID-LIQUID INTERFACE
      COMMON / PARAMS / BETA,BOV2,BOV2SQ,YINF,DELTA,TWOPI,TWOPID
      W = DELTA*COS(TWOPI*X)
      RETURN
      END
      REAL FUNCTION DW(X)
C                  DERIVATIVE OF SOLID-LIQUID INTERFACE SHAPE
```

```
      COMMON / PARAMS / BETA,BOV2,BOV2SQ,YINF,DELTA,TWOPI,TWOPID
      DW = -TWOPID*SIN(TWOPI*X)
      RETURN
      END
      REAL FUNCTION C(X,Y)
C              COMPUTE CONCENTRATION FROM PDE SOLUTION
      COMMON / PARAMS / BETA,BOV2,BOV2SQ,YINF,DELTA,TWOPI,TWOPID
      C = 1.0 + EXP(-BETA*Y) + U(X,Y)*EXP(-BOV2*Y)
      RETURN
      END
END.
```

Example 4.D2 Use of PROCEDURES to Analyze a Method

One of the objectives of ELLPACK is to assist in the analysis of numerical methods for PDEs. Two ELLPACK tools for this purpose are the procedures DISPLAY MATRIX PATTERN and EIGENVALUES. The first is useful for analyzing various techniques and data structures for Gauss elimination methods, and the second is useful in analyzing the applicability of iterative methods. The following example shows the results of applying these tools to the 5 POINT STAR discretization with three indexings: AS IS, RED BLACK and NESTED DISSECTION.

```
*       ***********************************************************
*       *                                                         *
*       *  EXAMPLE ELLPACK PROGRAM 4.D2                           *
*       *                                                         *
*       *     STUDY OF ZERO PATTERNS AND EIGENVALUES OF           *
*       *     MATRICES OBTAINED FROM 5 POINT STAR WITH            *
*       *     DIFFERENT ORDERINGS OF THE EQUATIONS.               *
*       *                                                         *
*       ***********************************************************
*
EQUATION.   - UXX - UYY = 0.
BOUND.        U = 0. ON X = 0.0
                     ON X = 1.0
                     ON Y = 0.0
                     ON Y = 1.0
GRID.         8 X POINTS $ 8 Y POINTS
OPTION.       TIME
DIS.          5 POINT STAR
INDEX.        AS IS
PROC.         DISPLAY MATRIX PATTERN (MATNBR=1,MATNBC=1,MATBLK=6)
FORT.
        H = 1./7.
        WRITE(I1OUTP,1000)
        DO 10 M = 1,6
          DO 10 N = 1,6
            E = (4.-2.*COS(PI*H*M)-2.*COS(PI*H*N)) / (PI**2*H*H)
            WRITE(I1OUTP,1010) M, N, E
   10   CONTINUE
 1000   FORMAT(//4X,1HM,3X,1HN,5X,17HEXACT EIGENVALUES   /)
 1010   FORMAT(1X,2I4,4X,E15.6)
PROC.           EIGENVALUES (SCALE = 1./PI**2)
INDEX.        RED BLACK
PROC.         DISPLAY MATRIX PATTERN (MATNBR=1,MATNBC=1,MATBLK=18)
INDEX.        NESTED DISSECTION (NDTYPE=5)
PROC.         DISPLAY MATRIX PATTERN (MATNBR=1,MATNBC=1,MATZER=1H)
END.
```

Output of ELLPACK Run:

```
 ----------------------
  PREPROCESSOR OUTPUT
 ----------------------

     2.30 SECONDS TO INITIALIZE PREPREPROCESSOR
     1.65 SECONDS TO PARSE ELLPACK PROGRAM
     4.30 SECONDS TO CREATE ELLPACK CONTROL PROGRAM
     8.25 SECONDS TOTAL

          ELLPACK VERSION --- APRIL 5, 1984

 ------------------------
  DISCRETIZATION MODULE
 ------------------------

     5 - P O I N T   S T A R

     DOMAIN                                  RECTANGLE
     DISCRETIZATION                            UNIFORM
     NUMBER OF EQUATIONS                            36
     MAX NO. OF UNKNOWNS PER EQ.                     5
     MATRIX IS                               SYMMETRIC

     EXECUTION SUCCESSFUL

 -------------------
  INDEXING MODULE
 -------------------

     A S   I S

     EQUATIONS INDEXED          36
     UNKNOWNS INDEXED           36

     EXECUTION SUCCESSFUL

 --------------------
  PROCEDURE MODULE
 --------------------

    D I S P L A Y   M A T R I X   P A T T E R N

                 1            2            3
      123456 789012 345678 901234 567890 123456

    1 DX.... X..... ...... ...... ...... ......
    2 XDX... .X.... ...... ...... ...... ......
    3 .XDX.. ..X... ...... ...... ...... ......
    4 ..XDX. ...X.. ...... ...... ...... ......
    5 ...XDX ....X. ...... ...... ...... ......
    6 ....XD .....X ...... ...... ...... ......

    7 X..... DX.... X..... ...... ...... ......
    8 .X.... XDX... .X.... ...... ...... ......
    9 ..X... .XDX.. ..X... ...... ...... ......
   10 ...X.. ..XDX. ...X.. ...... ...... ......
   11 ....X. ...XDX ....X. ...... ...... ......
   12 .....X ....XD .....X ...... ...... ......

   13 ...... X..... DX.... X..... ...... ......
   14 ...... .X.... XDX... .X.... ...... ......
   15 ...... ..X... .XDX.. ..X... ...... ......
   16 ...... ...X.. ..XDX. ...X.. ...... ......
   17 ...... ....X. ...XDX ....X. ...... ......
   18 ...... .....X ....XD .....X ...... ......

   19 ...... ...... X..... DX.... X..... ......
   20 ...... ...... .X.... XDX... .X.... ......
   21 ...... ...... ..X... .XDX.. ..X... ......
   22 ...... ...... ...X.. ..XDX. ...X.. ......
   23 ...... ...... ....X. ...XDX ....X. ......
   24 ...... ...... .....X ....XD .....X ......
```

```
25  ...... ...... ...... X..... DX.... X.....
26  ...... ...... ...... .X.... XDX... .X....
27  ...... ...... ...... ..X... .XDX.. ..X...
28  ...... ...... ...... ...X.. ..XDX. ...X..
29  ...... ...... ...... ....X. ...XDX ....X.
30  ...... ...... ...... .....X ....XD .....X

31  ...... ...... ...... ...... X..... DX....
32  ...... ...... ...... ...... .X.... XDX...
33  ...... ...... ...... ...... ..X... .XDX..
34  ...... ...... ...... ...... ...X.. ..XDX.
35  ...... ...... ...... ...... ....X. ...XDX
36  ...... ...... ...... ...... .....X ....XD
 EXECUTION SUCCESSFUL
 M    N       EXACT EIGENVALUES

 1    1        0.196666E+01
 1    2        0.472188E+01
 1    3        0.870329E+01
 1    4        0.131223E+02
 1    5        0.171037E+02
 1    6        0.198590E+02
 2    1        0.472188E+01
 2    2        0.747710E+01
 2    3        0.114585E+02
 2    4        0.158775E+02
 2    5        0.198590E+02
 2    6        0.226142E+02
 3    1        0.870329E+01
 3    2        0.114585E+02
 3    3        0.154399E+02
 3    4        0.198590E+02
 3    5        0.238404E+02
 3    6        0.265956E+02
 4    1        0.131223E+02
 4    2        0.158775E+02
 4    3        0.198590E+02
 4    4        0.242780E+02
 4    5        0.282594E+02
 4    6        0.310146E+02
 5    1        0.171037E+02
 5    2        0.198590E+02
 5    3        0.238404E+02
 5    4        0.282594E+02
 5    5        0.322408E+02
 5    6        0.349960E+02
 6    1        0.198590E+02
 6    2        0.226142E+02
 6    3        0.265956E+02
 6    4        0.310146E+02
 6    5        0.349960E+02
 6    6        0.377513E+02

-------------------
PROCEDURE MODULE
-------------------

   E I G E N V A L U E S

   NUMBER OF EQUATIONS                 36
   REQUIRED WORKSPACE                1440
   SCALE FACTOR              1.013212E-01

N            EIGENVALUE(N)                MAGNITUDE       ANGLE/PI

1   7.704352E-01 + 0.000000E+00 I    7.704352E-01    0.000000E+00
2   7.142065E-01 + 0.000000E+00 I    7.142065E-01    0.000000E+00
3   7.142062E-01 + 0.000000E+00 I    7.142062E-01    0.000000E+00
```

```
 4   6.579775E-01 + 0.000000E+00 I   6.579775E-01   0.000000E+00
 5   6.329530E-01 + 0.000000E+00 I   6.329530E-01   0.000000E+00
 6   6.329528E-01 + 0.000000E+00 I   6.329528E-01   0.000000E+00
 7   5.767238E-01 + 0.000000E+00 I   5.767238E-01   0.000000E+00
 8   5.767236E-01 + 0.000000E+00 I   5.767236E-01   0.000000E+00
 9   5.427681E-01 + 0.000000E+00 I   5.427681E-01   0.000000E+00
10   5.427680E-01 + 0.000000E+00 I   5.427680E-01   0.000000E+00
11   4.954705E-01 + 0.000000E+00 I   4.954705E-01   0.000000E+00
12   4.865394E-01 + 0.000000E+00 I   4.865394E-01   0.000000E+00
13   4.865386E-01 + 0.000000E+00 I   4.865386E-01   0.000000E+00
14   4.615150E-01 + 0.000000E+00 I   4.615150E-01   0.000000E+00
15   4.615142E-01 + 0.000000E+00 I   4.615142E-01   0.000000E+00
16   4.052857E-01 + 0.000000E+00 I   4.052857E-01   0.000000E+00
17   4.052853E-01 + 0.000000E+00 I   4.052853E-01   0.000000E+00
18   4.052848E-01 + 0.000000E+00 I   4.052848E-01   0.000000E+00
19   4.052848E-01 + 0.000000E+00 I   4.052848E-01   0.000000E+00
20   4.052848E-01 + 0.000000E+00 I   4.052848E-01   0.000000E+00
21   4.052848E-01 + 0.000000E+00 I   4.052848E-01   0.000000E+00
22   3.490568E-01 + 0.000000E+00 I   3.490568E-01   0.000000E+00
23   3.490561E-01 + 0.000000E+00 I   3.490561E-01   0.000000E+00
24   3.240324E-01 + 0.000000E+00 I   3.240324E-01   0.000000E+00
25   3.240318E-01 + 0.000000E+00 I   3.240318E-01   0.000000E+00
26   3.151011E-01 + 0.000000E+00 I   3.151011E-01   0.000000E+00
27   2.678032E-01 + 0.000000E+00 I   2.678032E-01   0.000000E+00
28   2.678027E-01 + 0.000000E+00 I   2.678027E-01   0.000000E+00
29   2.338479E-01 + 0.000000E+00 I   2.338479E-01   0.000000E+00
30   2.338477E-01 + 0.000000E+00 I   2.338477E-01   0.000000E+00
31   1.776189E-01 + 0.000000E+00 I   1.776189E-01   0.000000E+00
32   1.776188E-01 + 0.000000E+00 I   1.776188E-01   0.000000E+00
33   1.525946E-01 + 0.000000E+00 I   1.525946E-01   0.000000E+00
34   9.636535E-02 + 0.000000E+00 I   9.636535E-02   0.000000E+00
35   9.636513E-02 + 0.000000E+00 I   9.636513E-02   0.000000E+00
36   4.013628E-02 + 0.000000E+00 I   4.013628E-02   0.000000E+00
```

```
-----------------
 INDEXING MODULE
-----------------

    R E D - B L A C K

------------------
 PROCEDURE MODULE
------------------

   D I S P L A Y   M A T R I X   P A T T E R N

                1                  2         3
      123456789012345678 901234567890123456

   1  D................. ..............X..X
   2  .D................ .............X..XX
   3  ..D............... ............X..XX.
   4  ...D.............. ...........X.XX..X
   5  ....D............. ..........X.XX..X.
   6  .....D............ .........X..X..X..
   7  ......D........... ........X..X..X...
   8  .......D.......... .......X..XX.X....
   9  ........D......... ......X..XX.X.....
  10  .........D........ .....X.XX..X......
  11  ..........D....... ....X.XX..X.......
  12  ...........D...... ...X..X..X........
  13  ............D..... ..X..X..X.........
  14  .............D.... .X..XX.X..........
  15  ..............D... X..XX.X...........
  16  ...............D.. .XX..X............
  17  ................D. XX..X.............
  18  .................D X..X..............

  19  ..............X.XX D.................
```

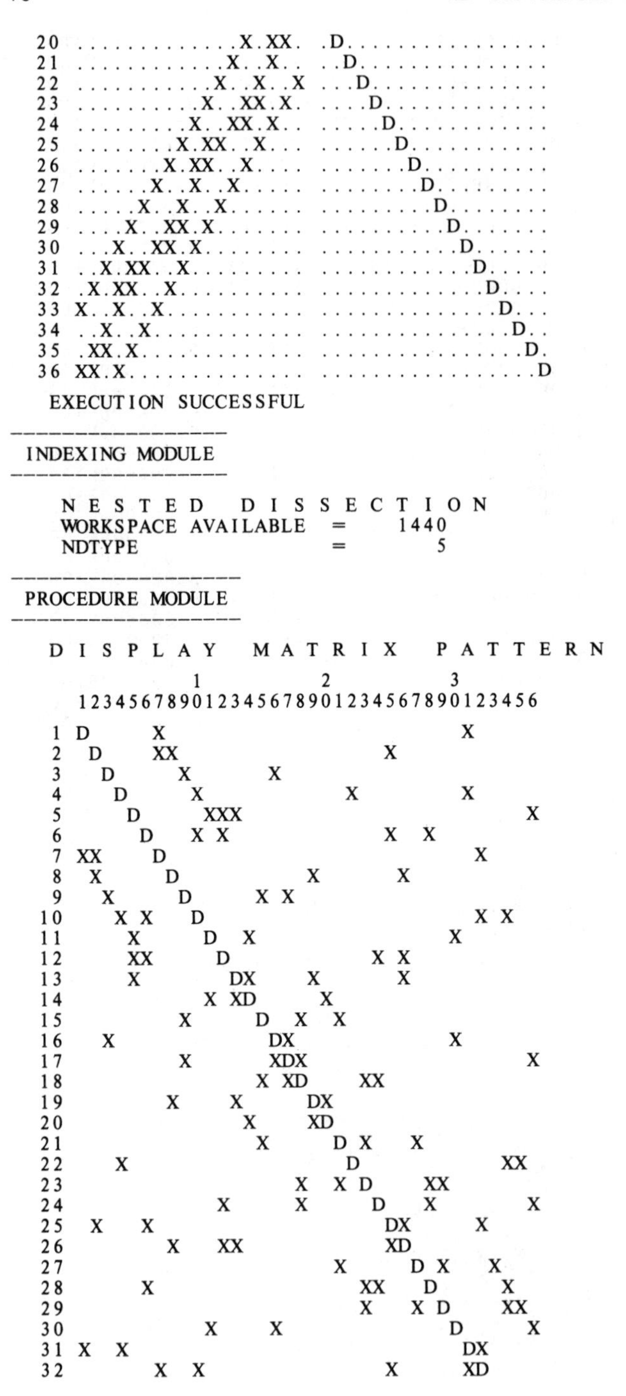

```
20 .............X.XX. .D................
21 ............X..X.. ..D...............
22 ...........X..X..X ...D..............
23 ..........X..XX.X. ....D.............
24 .........X..XX.X.. .....D............
25 ........X.XX..X... ......D...........
26 .......X.XX..X.... .......D..........
27 ......X..X..X..... ........D.........
28 .....X..X..X...... .........D........
29 ....X..XX.X....... ..........D.......
30 ...X..XX.X........ ...........D......
31 ..X.XX..X......... ............D.....
32 .X.XX..X.......... .............D....
33 X..X..X........... ..............D...
34 ..X..X............ ...............D..
35 .XX.X............. ................D.
36 XX.X.............. .................D
  EXECUTION SUCCESSFUL

------------------
 INDEXING MODULE
------------------

    N E S T E D   D I S S E C T I O N
    WORKSPACE AVAILABLE    =     1440
    NDTYPE                 =        5

-------------------
 PROCEDURE MODULE
-------------------

  D I S P L A Y   M A T R I X   P A T T E R N

            1         2         3
   123456789012345678901234567890123456

 1 D     X                       X
 2  D    XX                X
 3   D     X      X
 4    D     X           X        X
 5     D     XXX                      X
 6      D   X X            X  X
 7 XX    D                        X
 8  X     D          X      X
 9   X     D     X X
10    X X   D                     X X
11     X     D  X               X
12     XX     D           X X
13     X       DX    X      X
14           X XD     X
15         X     D  X  X
16   X            DX            X
17         X      XDX                 X
18               X XD    XX
19        X    X     DX
20              X    XD
21               X     D X   X
22    X                 D           XX
23                  X  X D    XX
24            X     X     D   X       X
25  X   X                  DX     X
26        X   XX           XD
27                     X     D X   X
28      X                XX   D     X
29                       X   X D    XX
30           X    X             D     X
31 X  X                          DX
32       X  X              X     XD
```

```
 33                             X     D X
 34          X           X       XX     D
 35                      X        X   X D
 36     X          X        X      X      D
  EXECUTION SUCCESSFUL

---------------
 ELLPACK OUTPUT
---------------

          +++++++++++++++++++++++++
          +                       +
          +    EXECUTION  TIMES   +
          +                       +
          +++++++++++++++++++++++++

            SECONDS       MODULE NAME

               0.15       5-POINT STAR
               0.05       AS IS
               0.77       DISPLAY MATRIX PATTERN
               2.67       EIGENVALUES
               0.03       RED-BLACK
               0.63       DISPLAY MATRIX PATTERN
               0.13       NESTED DISSECTION
               0.57       DISPLAY MATRIX PATTERN
               5.35       TOTAL TIME
```

Example 4.D3 Nonlinear PDE Solution by Picard Iteration

An ELLPACK program is shown below for the equation

$$u_{xx} + u_{yy} = u^2 (x^2 + y^2) e^{-xy}$$

on the unit square with boundary values so the true solution is e^{xy}. The method of Picard (fixed point iteration) is used to solve this nonlinear problem through a sequence of linear ones [Ortega and Rheinboldt, 1970]. HODIE higher order finite differences (see Chapter 6) and Gauss elimination are used, and the solution is obtained quickly. See Section 5.B for another example of using Newton iteration with ELLPACK for a nonlinear problem.

```
*      ***********************************************************
*      *                                                         *
*      *   EXAMPLE ELLPACK PROGRAM 4.D3                          *
*      *                                                         *
*      *       NONLINEAR POISSON PROBLEM WITH U**2 ON RIGHT      *
*      *       SIDE.  FIXED POINT ITERATION IS USED.             *
*      *                                                         *
*      *                                                         *
*      ***********************************************************
OPT.           MEMORY
DECLARATION.
      REAL UOLD(5,5)
EQU.           UXX + UYY = (X**2 + Y**2) * EXP(-X*Y) * U(X,Y)**2
BOUND.         U = 1.0      ON  X = 0.0
                            ON  Y = 0.0
               U = EXP(Y)   ON  X = 1.0
               U = EXP(X)   ON  Y = 1.0
GRID.          7  X POINTS  $  7  Y POINTS
*              INITIALIZE PICARD ITERATION
```

```
TRIPLE.         SET U BY BLENDING
FORTRAN.
C               INITIALIZE UOLD ARRAY FOR CONVERGENCE TEST
      DO 10 I = 1, 5
         DO 10 J = 1,5
   10       UOLD(I,J) = 0.
C
C     ITERATION TO SOLVE THE PDE
      DO 50 ITER = 1,6
DIS.         HODIE HELMHOLTZ(METHOD=3)
SOL.         BAND GE
FORTRAN.
C            COMPUTE MAX CHANGE OF U AND ERROR ON THE GRID POINTS
             DIFMAX = 0.0
             ERRMAX = 0.0
             DO 20 I = 1,5
                X = I/6.
                DO 20 J = 1,5
                   Y = J/6.
                   DIFMAX = AMAX1(DIFMAX,ABS(U(X,Y)-UOLD(I,J)))
                   UOLD(I,J) = U(X,Y)
                   ERRMAX = AMAX1(ERRMAX,ABS(UOLD(I,J)-TRUE(X,Y)))
   20        CONTINUE
             WRITE(6,30) ITER
   30        FORMAT(//8X,10(1H*),11HITERATION = ,I3,2X,20(1H*)/)
             WRITE(6,40) DIFMAX,ERRMAX
   40        FORMAT(8X,25HMAX CHANGE IN UNKNOWNS = , E16.5/
     A              8X,25HMAX ERROR               = , E16.5)
             I1LEVL = 0
   50 CONTINUE
SUB.
      FUNCTION TRUE(X,Y)
         TRUE = EXP(X*Y)
         RETURN
      END
END.
```

Output of ELLPACK Run:

```
 ----------------------
  PREPROCESSOR OUTPUT
 ----------------------

     2.02 SECONDS TO INITIALIZE PREPREPROCESSOR
     1.48 SECONDS TO PARSE ELLPACK PROGRAM
     3.58 SECONDS TO CREATE ELLPACK CONTROL PROGRAM
     7.08 SECONDS TOTAL

          ELLPACK VERSION --- APRIL 5, 1984

     APPROXIMATE MEMORY REQUIREMENTS (IN WORDS)

     PROBLEM DEFINITION          97
     DISCRETIZATION             554
     INDEXING                    53
     WORKSPACE                  500
     OTHER                       45
     TOTAL MEMORY              1249

 ----------------
  TRIPLE MODULE
 ----------------

   S E T   U   B Y   B L E N D I N G

   EXECUTION SUCCESSFUL
```

```
------------------------
 DISCRETIZATION MODULE
------------------------

    H O D I E - H E L M H O L T Z

    DOMAIN                                    RECTANGLE
    GRID                          UNIFORM,     7 X   7
    HX                                        1.667E-01
    HY                                        1.667E-01
    SIZE OF WORKSPACE                                44
    DISCRETIZATION ORDER                              4
    DISCRETIZATION METHOD                             3
    MATRIX IS                                 SYMMETRIC

    NUMBER OF EQUATIONS                              25
    MAX NO. OF UNKNOWNS PER EQ.                       9
    EXECUTION SUCCESSFUL

-----------------
 SOLUTION MODULE
-----------------

   B A N D   G E

   NUMBER OF EQUATIONS           25
   LOWER BANDWIDTH                6
   UPPER BANDWIDTH                6
   REQUIRED WORKSPACE           500

   EXECUTION SUCCESSFUL

       **********ITERATION =  1   ********************

       MAX CHANGE IN UNKNOWNS =       0.20045E+01
       MAX ERROR              =       0.41002E-02

       **********ITERATION =  2   ********************

       MAX CHANGE IN UNKNOWNS =       0.43652E-02
       MAX ERROR              =       0.26500E-03

       **********ITERATION =  3   ********************

       MAX CHANGE IN UNKNOWNS =       0.28872E-03
       MAX ERROR              =       0.23723E-04

       **********ITERATION =  4   ********************

       MAX CHANGE IN UNKNOWNS =       0.18835E-04
       MAX ERROR              =       0.48876E-05

       **********ITERATION =  5   ********************

       MAX CHANGE IN UNKNOWNS =       0.11921E-05
       MAX ERROR              =       0.60797E-05

       **********ITERATION =  6   ********************

       MAX CHANGE IN UNKNOWNS =       0.23842E-06
       MAX ERROR              =       0.59605E-05
```

Discussion

We discuss three of the more interesting points about this program.

1. **Treatment of the nonlinearity.** The **fixed point iteration** method is used to handle the nonlinearity. This method is very simple to implement in ELLPACK; in this case it corresponds to the iteration

$$u_{xx}^{(n+1)} + u_{yy}^{(n+1)} = (u^{(n)})^2(x^2+y^2)\, e^{-xy}, \quad n = 0, 1, 2, \ldots$$

where one has a linear problem to obtain $u^{(n+1)}$ from $u^{(n)}$. One could also try the iteration

$$u_{xx}^{(n+1)} + u_{yy}^{(n+1)} - u^{(n)}(x^2+y^2)\, e^{-xy}\, u^{(n+1)} = 0$$

in much the same way. In general, one splits the nonlinear equation $F(u) = 0$ into the form $L(u)u = g(u)$, and then uses the iteration

$$L^{(u^{(m)})} u^{(m+1)} = g(u^{(m)}).$$

The strength of fixed point iteration is its simplicity. Its weakness is that one cannot predict whether it will work and, if it does work, how well it will work. In this example it works well. When one formulation of the iteration fails, others should be attempted.

Note that the ELLPACK function U(X,Y) is used directly in the right side of the PDE. This facility can be used for more complex PDEs, for example

```
EQ. 7.*UXX + (1.+U(X,Y)**2)*UYY + SIN(UX(X,Y))*UX - UY(X,Y)*U = 0
```

for the equation

$$7u_{xx} + (1+u^2)\, u_{yy} + \sin(u_x)\, u_x - u_y\, u = 0\,.$$

The discretization module uses the current definition of U(X,Y), UX(X,Y), etc. in forming the linear system for the problem. This means that U(X,Y), etc. must be initialized.

There is a **word of caution about nonlinearities.** This technique works in most situations, but there are cases where it fails because the discretization module may change something about how the solution U(X,Y) is computed during its execution. For example, suppose that one changes the grid size between two uses of 5 POINT STAR. After the first use, the unknowns are stored in a table corresponding to the first grid. Once the grid is changed this table no longer corresponds to the existing grid. When 5 POINT STAR is used again, it sets variables to determine how to evaluate the solution on the new grid, and U(X,Y) becomes improperly defined. If one suspects there may be a problem like this, the information about U(X,Y), etc., should be moved to user defined arrays and then used from there. An example of this is given in Chapter 5.

2. **Initialization of the iteration.** The unknown function U(X,Y) is initialized directly by the ELLPACK triple SET U BY BLENDING. Once this triple is executed then U(X,Y), UX(X,Y), etc. are defined and, in fact, the blending function method frequently produces a very good initial approximation. If one knows a good approximation to U(X,Y), say START, (X,Y) then one can use the ELLPACK triple SET(U = START). Both of these triples define U just as though a numerical method were used to approximate u with a more standard set of ELLPACK modules.

3. **Testing for convergence.** The convergence test used in this example is based upon the maximum absolute difference in the values of the solution function U at grid points from one iteration to the next. A more appropriate test is based on the change in the solution vector of the discretization linear system. For finite difference modules these are the same, but not so for discretizations of finite element type. To perform such a test, one declares the array UOLD as UOLD($I1MNEQ) and replaces the DO 20 loop by the following Fortran statements

```
C     I1NEQN = LENGTH OF SOLUTION VECTOR
C     R1UNKN = SOLUTION VECTOR
C     UOLD   = SOLUTION VECTOR OF LAST ITERATION,
               DECLARED AS UOLD (I1MNEQ)
C
      DIFMAX = 0.0
      DO 20 J=1,I1NEQN
         DIFMAX=AMAX1(DIFMAX,ABS(R1UNKN(J)-UOLD(J)))
         UOLD(J)=R1UNKN(J)
   20 CONTINUE
      WRITE(6,30)DIFMAX
   30 FORMAT(8X,'MAX CHANGE IN SOLUTION=',E16.5)
```

The vector UOLD should be initialized to 0.0 before the first iteration. I1NEQN and R1UNKN are internal ELLPACK variables. See Chapter 13 for further details.

Example 4.D4 Nonrectangular Domain with a Hole

This example illustrates the use of the HOLE segment and nonrectangular domains.

```
*       ***********************************************************
*       *                                                         *
*       *  EXAMPLE ELLPACK PROGRAM 4.D4                           *
*       *                                                         *
*       *       THIS PROGRAM USES THE HOLE FEATURE IN ELLPACK.    *
*       *       THE REGION IS BETWEEN TWO CONFOCAL ELLIPSES.      *
*       *                                                         *
*       ***********************************************************
OP.     TIME $ MEMORY $ CLOCKWISE=.TRUE.
EQ.     UXX  +  UYY  =  0.0
BOUND.  U = 0.   ON  X = COSH(3.0)*SIN(T), Y =  SINH(3.0)*COS(T) &
                                           FOR  T = 0.0 TO 2*PI
HOLE.   U = 1.   ON  X = COSH(2.3)*SIN(T), Y = -SINH(2.3)*COS(T) &
                                           FOR  T = 0.0 TO 2*PI
GRID    17 X POINTS,   -COSH(3.0) TO COSH(3.0)
        17 Y POINTS,   -SINH(3.0) TO SINH(3.0)
DIS.    5 POINT STAR
SOL.    BAND GE
OUT.    PLOT-DOMAIN
        MAX(TRUE) $ MAX(U) $ MAX(ERROR)
SUBPROGRAMS.
      FUNCTION TRUE(X,Y)
         R1   = SQRT((X-1.0)**2+Y**2)
         R2   = SQRT((X+1.0)**2+Y**2)
         U    = ACOSH(0.5*(R1+R2))
         TRUE = (3.0-U)/(3.0-2.3)
         RETURN
      END
      FUNCTION ACOSH(X)
         ACOSH = ALOG(X+SQRT(X**2-1.0))
         RETURN
      END
END.
```

Output of ELLPACK Run (The contour plot is shown in Figure 4.2)

```
----------------------
 PREPROCESSOR OUTPUT
----------------------
```

```
    2.20 SECONDS TO INITIALIZE PREPREPROCESSOR
    1.25 SECONDS TO PARSE ELLPACK PROGRAM
    3.82 SECONDS TO CREATE ELLPACK CONTROL PROGRAM
    7.27 SECONDS TOTAL

         ELLPACK VERSION --- APRIL 5, 1984

    APPROXIMATE MEMORY REQUIREMENTS (IN WORDS)

    PROBLEM DEFINITION        1471
    DISCRETIZATION            4438
    INDEXING                   521
    WORKSPACE                12173
    OTHER                      279
    TOTAL MEMORY             18882

-----------------
DOMAIN PROCESSOR
-----------------

    BOUNDARY POINTS FOUND              60
    BOUNDARY PIECES FOUND               1
    GRID SIZE                  17 BY   17
    EXECUTION SUCESSFUL

------------------
DOMAIN  PROCESSOR
------------------

     H O L E   P R O C E S S O R

    BOUNDARY POINTS FOUND              32
    BOUNDARY PIECES FOUND               1
    GRID SIZE                  17 BY   17
    EXECUTION SUCESSFUL

----------------------
DISCRETIZATION MODULE
----------------------

    5 - P O I N T   S T A R

    DOMAIN                          NONRECTANGULAR
    UNIFORM GRID                          17 X  17
    NUMBER OF EQUATIONS                        146
    MAX NO. OF UNKNOWNS PER EQ.                  5
    EXECUTION SUCCESSFUL

----------------
SOLUTION MODULE
----------------

   B A N D   G E

   NUMBER OF EQUATIONS        146
   LOWER BANDWIDTH             14
   UPPER BANDWIDTH             14
   REQUIRED WORKSPACE        6424

   EXECUTION SUCCESSFUL

---------------
ELLPACK OUTPUT
---------------

    P L O T   D O M A I N
    EXECUTION SUCCESSFUL

---------------
ELLPACK OUTPUT
---------------
```

```
++++++++++++++++++++++++++++++++++++++++++++++++++++++++++++++++
+                                                              +
+  MAX( ABS(TRUE   ) ) ON  17 X  17 GRID = 9.7972941E-01  +
+                                                              +
+  L1 NORM( TRUE   )   ON  17 X  17 GRID = 4.0751064E-01  +
+                                                              +
+  L2 NORM( TRUE   )   ON  17 X  17 GRID = 4.9531445E-01  +
+                                                              +
++++++++++++++++++++++++++++++++++++++++++++++++++++++++++++++++
```

```
----------------
ELLPACK OUTPUT
----------------
```

```
++++++++++++++++++++++++++++++++++++++++++++++++++++++++++++++++
+                                                              +
+  MAX( ABS(U      ) ) ON  17 X  17 GRID = 9.7954005E-01  +
+                                                              +
+  L1 NORM( U      )   ON  17 X  17 GRID = 4.0747702E-01  +
+                                                              +
+  L2 NORM( U      )   ON  17 X  17 GRID = 4.9526441E-01  +
+                                                              +
++++++++++++++++++++++++++++++++++++++++++++++++++++++++++++++++
```

```
----------------
ELLPACK OUTPUT
----------------
```

```
++++++++++++++++++++++++++++++++++++++++++++++++++++++++++++++++
+                                                              +
+  MAX( ABS(ERROR ) ) ON  17 X  17 GRID = 1.4755726E-03  +
+                                                              +
+  L1 NORM( ERROR )   ON  17 X  17 GRID = 5.9809396E-04  +
+                                                              +
+  L2 NORM( ERROR )   ON  17 X  17 GRID = 7.1683980E-04  +
+                                                              +
++++++++++++++++++++++++++++++++++++++++++++++++++++++++++++++++
```

```
----------------
ELLPACK OUTPUT
----------------
```

```
+++++++++++++++++++++++++
+                       +
+    EXECUTION  TIMES   +
+                       +
+++++++++++++++++++++++++

  SECONDS      MODULE NAME

     1.63      DOMAIN
     0.95      HOLE
     0.32      5-POINT STAR
     0.18      BAND GE SETUP
     0.95      BAND GE
     0.22      PLOT-DOMAIN
     0.18      MAX
     0.25      MAX
     0.33      MAX
     5.12      TOTAL TIME
```

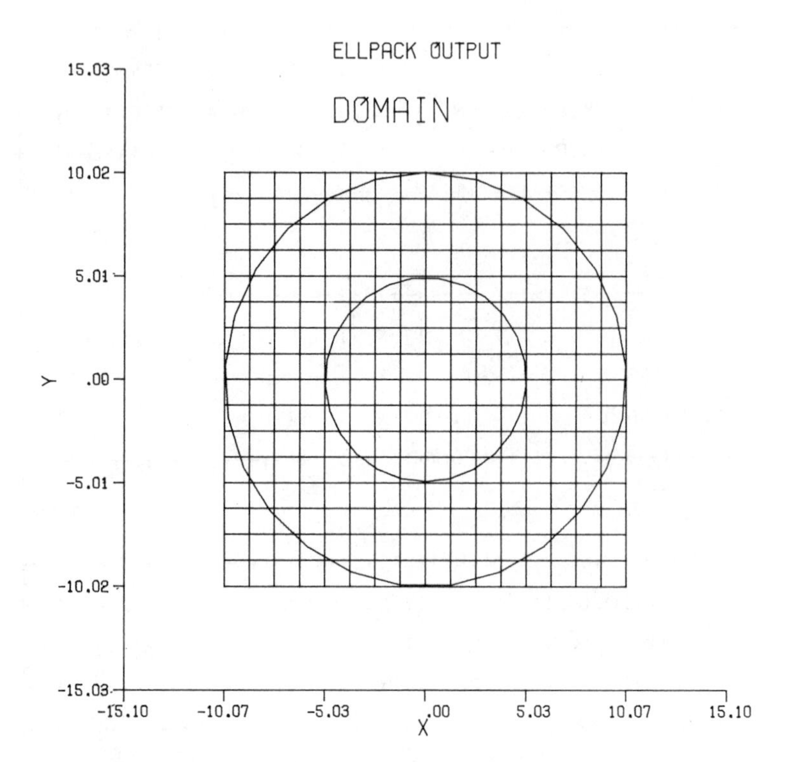

Figure 4.2. The plot produced by PLOT DOMAIN in the example program 4.D4. The domain is the anulus between the two almost circular, confocal ellipses.

4.E REFERENCES

Coriell, S.R., R.F. Boisvert, R.G. Rehm and R.F. Sekerka [1981], "Lateral solute segregation during undirectional solidification of a binary alloy with a curved solid-liquid interface II — Large departures from planarity", J. Crystal Growth 54, pp. 167–175.

Ortega, J.M. and W.C. Rheinboldt [1970], *Iterative Solution of Nonlinear Equations in Several Variables*, Academic Press, New York.

Chapter 5

EXTENDING ELLPACK TO NONSTANDARD PROBLEMS

ELLPACK can be used to solve or study problems where its "automatic" problem solving capabilities do not apply. To use ELLPACK in this way requires that one have an understanding of both the ELLPACK system and the numerical methods involved. Even more complicated applications depend on knowing some of the details of the implementations in the ELLPACK modules. We illustrate this use of ELLPACK with five examples.

1. A problem with a double-valued boundary condition along a slit.
Such problems arise, for example, when thin conducting plates are placed in electrical fields. Many problems of this type can be solved using ELLPACK once one learns the technique.

2. Diffusion problem with interior interface conditions.
Diffusion in a heterogeneous medium introduces interfaces with a derivative jump conditions. The 5-POINT STAR equations are modified along the material interface to satisfy this condition.

3. Three nonlinear problems.
Newton iteration methods for handling nonlinearities can be implemented easily in ELLPACK. The examples given show very rapid convergence and two are real world applications.

4. A time dependent problem.
Consider the parabolic problem

$$u_t = \mathrm{L}u$$

where L is a linear elliptic operator in two or three variables. Many numerical methods for such problems are implicit so they can take large time steps. At each time step these methods solve an elliptic boundary value problem. Several of these methods can be concisely formulated to use ELLPACK's elliptic problem solving capabilities. We illustrate this technique for the PDE

$$u_t = 4y^2 u_{xx} + u_{yy} + (2 + \tan(x + y + t))\, u_y + u + f(x,y,t).$$

5. A set of problems with two simultaneous elliptic equations.
This set involves pairs of simultaneous, nonlinear elliptic equations. A particular iteration is defined which might loosely be called a Gauss-Seidel iteration; this can be viewed as an extension of Picard's method.

5.A SPECIAL INTERIOR BOUNDARY CONDITIONS

ELLPACK has the capability to handle auxiliary conditions along curves inside the domain. These might be conditions on slits such as occurs when a thin conducting plate is inserted in an electric field. The ARC segment allows one to specify single valued boundary conditions in a straight forward way. See Figure 5.1 for the domain used. This example also illustrates the various output that one can obtain from ELLPACK. For example, TABLE-DOMAIN, and TABLE-BOUNDARY are primarily useful for those who want to know how ELLPACK works internally.

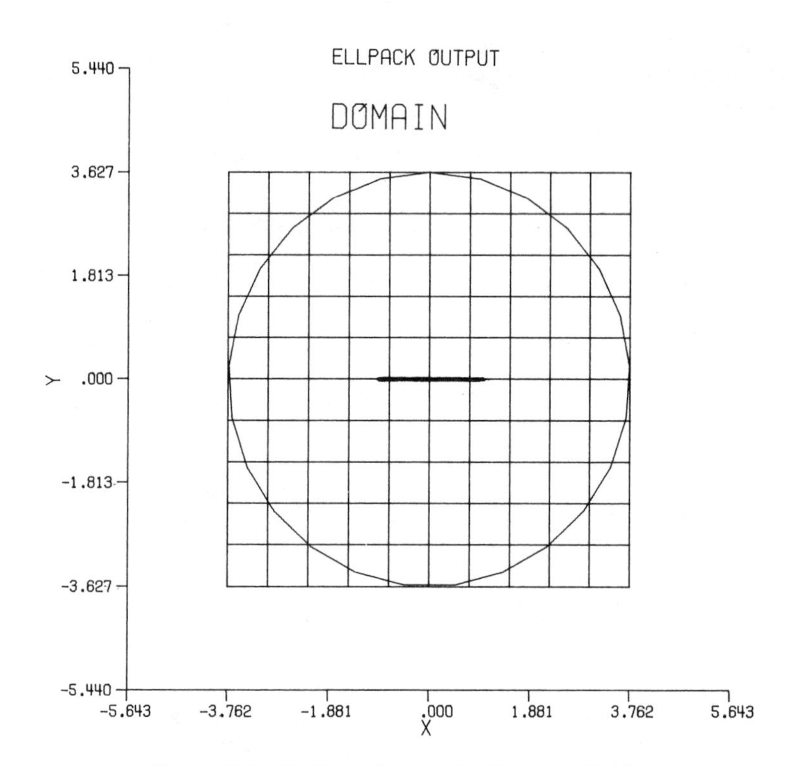

Figure 5.1. Problem domain for Program 5.A1.

```
*       ******************************************************
*       *                                                    *
*       *  EXAMPLE ELLPACK PROGRAM 5.A1                      *
*       *                                                    *
*       *      THIS PROGRAM USES THE ARC FEATURE IN ELLPACK. *
*       *      THE REGION IS BOUNDED BY CONFOCAL ELLIPSES.   *
*       *      THE SLIT IS A DEGENERATE ELLIPSE.             *
*       *                                                    *
*       ******************************************************
```

```
OPT.      TIME    $ CLOCKWISE = .TRUE.
EQ.       UXX  +  UYY  =  0.0
BOUND.    U = 0.  ON  X = COSH(2.0)*SIN(T),Y = SINH(2.0)*COS(T) &
                                          FOR  T = 0.0 TO 2*PI
ARC.      U = 1.  ON  LINE -1.0, 0.0  TO  1.0, 0.0
GRID.     10 X POINTS   -COSH(2.0) TO COSH(2.0)
          10 Y POINTS   -SINH(2.0) TO SINH(2.0)
DIS.      5-POINT STAR
SOL.      BAND GE
OUT.      SUMMARY(U)    $ MAX(ERROR)
          TABLE DOMAIN  $ TABLE BOUNDARY
          PLOT DOMAIN   $ PLOT(U)
SUBPROGRAMS.
       FUNCTION TRUE(X,Y)
          R1   = SQRT((X-1.0)**2+Y**2)
          R2   = SQRT((X+1.0)**2+Y**2)
          U    = ACOSH(0.5*(R1+R2))
          TRUE = (2.0-U)/2.0
          RETURN
       END
       FUNCTION ACOSH(X)
          ACOSH = ALOG(X+SQRT(X**2-1.0))
          RETURN
       END
END.
```

Output of ELLPACK Run (The contour plot is shown in Figure 5.2.)

```
---------------------
 PREPROCESSOR OUTPUT
---------------------
    3.48 SECONDS TO INITIALIZE PREPREPROCESSOR
    1.83 SECONDS TO PARSE ELLPACK PROGRAM
    5.70 SECONDS TO CREATE ELLPACK CONTROL PROGRAM
   11.02 SECONDS TOTAL
          ELLPACK VERSION --- APRIL 5, 1984

------------------
 DOMAIN PROCESSOR
------------------
    BOUNDARY POINTS FOUND            33
    BOUNDARY PIECES FOUND             1
    GRID SIZE                 10 BY  10
    EXECUTION SUCESSFUL

-------------------
 DOMAIN  PROCESSOR
-------------------
     H O L E   P R O C E S S O R
    BOUNDARY POINTS FOUND             3
    BOUNDARY PIECES FOUND             1
    GRID SIZE                 10 BY  10
    EXECUTION SUCESSFUL

-----------------------
 DISCRETIZATION MODULE
-----------------------
    5 - P O I N T   S T A R
    DOMAIN                        NONRECTANGULAR
    UNIFORM GRID                         10 X  10
    NUMBER OF EQUATIONS                        60
```

```
    MAX NO. OF UNKNOWNS PER EQ.                5
    EXECUTION SUCCESSFUL

--------------------
 SOLUTION MODULE
--------------------

   B A N D   G E

   NUMBER OF EQUATIONS          60
   LOWER BANDWIDTH               8
   UPPER BANDWIDTH               8
   REQUIRED WORKSPACE         1560
   EXECUTION SUCCESSFUL

------------------
 ELLPACK OUTPUT
------------------

            ++++++++++++++++++++++++++++++++++++++++++++++++
            +                                              +
            +     TABLE OF U       ON   10 X   10   GRID   +
            +                                              +
            ++++++++++++++++++++++++++++++++++++++++++++++++

    X-ABSCISSAE ARE
    ---------------

  -3.762196E+00   -2.926152E+00   -2.090109E+00   -1.254065E+00
  -4.180217E-01    4.180217E-01    1.254066E+00    2.090109E+00
   2.926152E+00    3.762196E+00

       Y = 3.626861E+00
       ----------------

   0.000000E+00    0.000000E+00    0.000000E+00    0.000000E+00
   0.000000E+00    0.000000E+00    0.000000E+00    0.000000E+00
   0.000000E+00    0.000000E+00

       Y = 2.820892E+00
       ----------------

   0.000000E+00    0.000000E+00    1.775396E-02    6.791662E-02
   1.006834E-01    1.006835E-01    6.791663E-02    1.775396E-02
   0.000000E+00    0.000000E+00

       Y = 2.014923E+00
       ----------------

   0.000000E+00    1.741062E-02    9.460503E-02    1.734560E-01
   2.342803E-01    2.342803E-01    1.734560E-01    9.460503E-02
   1.741064E-02    0.000000E+00

       Y = 1.208954E+00
       ----------------

   0.000000E+00    6.530494E-02    1.699166E-01    2.957487E-01
   4.244042E-01    4.244044E-01    2.957487E-01    1.699166E-01
   6.530496E-02    0.000000E+00

       Y = 4.029849E-01
       ----------------

   0.000000E+00    9.541320E-02    2.255071E-01    4.154175E-01
   7.340941E-01    7.340941E-01    4.154174E-01    2.255071E-01
   9.541324E-02    0.000000E+00

       Y =-4.029844E-01
       ----------------

   0.000000E+00    9.541321E-02    2.255071E-01    4.154176E-01
   7.340943E-01    7.340942E-01    4.154174E-01    2.255071E-01
   9.541325E-02    0.000000E+00

       Y =-1.208953E+00
       ----------------

   0.000000E+00    6.530496E-02    1.699167E-01    2.957488E-01
   4.244044E-01    4.244044E-01    2.957488E-01    1.699167E-01
   6.530499E-02    0.000000E+00
```

```
     Y =-2.014922E+00
     ----------------
   0.000000E+00    1.741065E-02    9.460511E-02    1.734562E-01
   2.342804E-01    2.342805E-01    1.734561E-01    9.460511E-02
   1.741068E-02    0.000000E+00
     Y =-2.820891E+00
     ----------------
   0.000000E+00    0.000000E+00    1.775399E-02    6.791668E-02
   1.006835E-01    1.006835E-01    6.791668E-02    1.775399E-02
   0.000000E+00    0.000000E+00
     Y =-3.626860E+00
     ----------------
   0.000000E+00    0.000000E+00    0.000000E+00    0.000000E+00
   0.000000E+00    0.000000E+00    0.000000E+00    0.000000E+00
   0.000000E+00    0.000000E+00
   +++++++++++++++++++++++++++++++++++++++++++++++++++++++++++++++++
   +                                                               +
   +  MAX( ABS(U      ) ) ON  10 X  10 GRID = 7.3409432E-01        +
   +  L1 NORM( U      )   ON  10 X  10 GRID = 2.0879418E-01        +
   +  L2 NORM( U      )   ON  10 X  10 GRID = 2.8064668E-01        +
   +                                                               +
   +++++++++++++++++++++++++++++++++++++++++++++++++++++++++++++++++

---------------
ELLPACK OUTPUT
---------------
   +++++++++++++++++++++++++++++++++++++++++++++++++++++++++++++++++
   +                                                               +
   +  MAX( ABS(ERROR ) ) ON  10 X  10 GRID = 1.5980986E-01         +
   +  L1 NORM( ERROR )   ON  10 X  10 GRID = 4.3808527E-02         +
   +  L2 NORM( ERROR )   ON  10 X  10 GRID = 5.8138087E-02         +
   +                                                               +
   +++++++++++++++++++++++++++++++++++++++++++++++++++++++++++++++++

---------------
ELLPACK OUTPUT
---------------
   +++++++++++++++++++++++++++++++++++++++++++++++++++++++++++++++++
   +                                                               +
   +    TABLE  OF  THE  POINT  TYPES  ON  10 X  10  GRID           +
   +                                                               +
   +++++++++++++++++++++++++++++++++++++++++++++++++++++++++++++++++
        THE POINT R1GRDX(1), R1GRDY(1) IS AT THE LOWER LEFT.
        --------------------------------------------------

10      0      0  -4031  -4032  -6001 -12001  -4003  -4004      0      0
 9      0  -6029   9030   1032   1033   1002   1003   3004 -12005      0
 8  -2028   9028    999    999    999    999    999    999   3006  -8007
 7  -2027   8027    999    999    999    999    999    999   2008  -8008
 6  -2026   8026    999    999   4036   4037    999    999   2009  -8009
 5  -2025   8025    999    999   1036   1037    999    999   2010  -8010
 4  -2024   8024    999    999    999    999    999    999   2011  -8011
 3  -2023  12022    999    999    999    999    999    999   6012  -8012
 2      0  -3021  12020   4019   4018   4017   4016   6014  -9013      0
 1      0      0  -1020  -1019  -1018  -1017  -1016  -1015      0      0
  ***   1      2      3      4      5      6      7      8      9     10

---------------
ELLPACK OUTPUT
---------------
   +++++++++++++++++++++++++++++++++++++++++++++++++++++++++++++++++
   +                                                               +
   +  TABLE OF THE BOUNDARY POINT TYPES ON  10 X  10  GRID         +
   +                                                               +
   +++++++++++++++++++++++++++++++++++++++++++++++++++++++++++++++++
```

NUMBER	R1XBND	R1YBND	R1BPAR	PECE	BPTY	BGRD	BNGH
1	0. E+00	3.626E+00	0. E+00	1	HORZ	10005	0
2	4.180E-01	3.604E+00	1.113E-01	1	VERT	9006	4
3	1.254E+00	3.419E+00	3.398E-01	1	VERT	9007	4
4	2.090E+00	3.015E+00	5.890E-01	1	VERT	9008	4
5	2.364E+00	2.820E+00	6.796E-01	1	HORZ	9008	8
6	2.926E+00	2.279E+00	8.911E-01	1	VERT	8009	4
7	3.128E+00	2.014E+00	9.817E-01	1	HORZ	8009	8
8	3.547E+00	1.209E+00	1.231E+00	1	HORZ	7009	8
9	3.738E+00	4.029E-01	1.459E+00	1	HORZ	6009	8
10	3.738E+00	-4.029E-01	1.682E+00	1	HORZ	5009	8
11	3.547E+00	-1.209E+00	1.910E+00	1	HORZ	4009	8
12	3.128E+00	-2.014E+00	2.159E+00	1	HORZ	3009	8
13	2.926E+00	-2.279E+00	2.250E+00	1	VERT	2009	1
14	2.364E+00	-2.820E+00	2.461E+00	1	HORZ	2008	8
15	2.090E+00	-3.015E+00	2.552E+00	1	VERT	1008	1
16	1.254E+00	-3.419E+00	2.801E+00	1	VERT	1007	1
17	4.180E-01	-3.604E+00	3.030E+00	1	VERT	1006	1
18	-4.180E-01	-3.604E+00	3.252E+00	1	VERT	1005	1
19	-1.254E+00	-3.419E+00	3.481E+00	1	VERT	1004	1
20	-2.090E+00	-3.015E+00	3.730E+00	1	VERT	1003	1
21	-2.364E+00	-2.820E+00	3.821E+00	1	HORZ	2002	2
22	-2.926E+00	-2.279E+00	4.032E+00	1	VERT	2002	1
23	-3.128E+00	-2.014E+00	4.123E+00	1	HORZ	3001	2
24	-3.547E+00	-1.209E+00	4.372E+00	1	HORZ	4001	2
25	-3.738E+00	-4.029E-01	4.601E+00	1	HORZ	5001	2
26	-3.738E+00	4.029E-01	4.823E+00	1	HORZ	6001	2
27	-3.547E+00	1.209E+00	5.052E+00	1	HORZ	7001	2
28	-3.128E+00	2.014E+00	5.301E+00	1	HORZ	8001	2
29	-2.926E+00	2.279E+00	5.392E+00	1	VERT	8002	4
30	-2.364E+00	2.820E+00	5.603E+00	1	HORZ	9002	2
31	-2.090E+00	3.015E+00	5.694E+00	1	VERT	9003	4
32	-1.254E+00	3.419E+00	5.943E+00	1	VERT	9004	4
33	-4.180E-01	3.604E+00	6.171E+00	1	VERT	9005	4
34	0. E+00	3.626E+00	6.283E+00	1	JUMP	10005	0
35	-1.000E+00	0. E+00	0. E+00	2	INTE	5004	0
36	-4.180E-01	0. E+00	2.909E-01	2	VERT	5005	5
37	4.180E-01	0. E+00	7.090E-01	2	VERT	5006	5

```
----------------
 ELLPACK OUTPUT
----------------
     P L O T   D O M A I N
     EXECUTION SUCCESSFUL

----------------
 ELLPACK OUTPUT
----------------
     CONTOUR PLOT OF      U
     GRID                20   BY    20
     EXECUTION SUCCESSFUL

----------------
 ELLPACK OUTPUT
----------------
          +++++++++++++++++++++++++
          +                       +
          +    EXECUTION  TIMES   +
          +                       +
          +++++++++++++++++++++++++

             SECONDS      MODULE NAME
                0.82      DOMAIN
                0.07      ARC
                0.17      5-POINT STAR
                0.07      BAND GE SETUP
                0.20      BAND GE
```

```
0.57     SUMMARY
0.18     MAX
0.23     TABLE-DOMAIN
0.50     TABLE-BOUNDARY
0.22     PLOT-DOMAIN
4.48     PLOT
7.60     TOTAL TIME
```

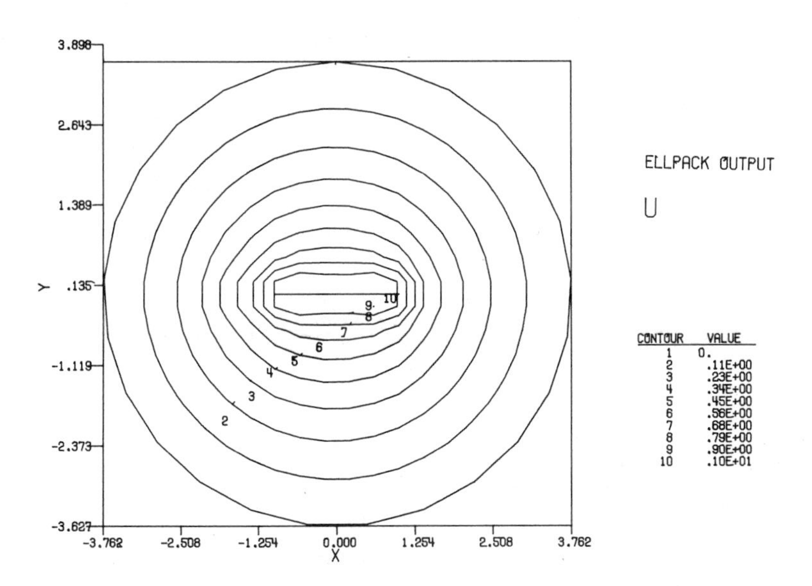

Figure 5.2. Contour plot of solution for Program 5.A1.

We modify this example to show how to solve problems with **double-valued boundary conditions**. One uses the HOLE segment to place a very thin hole in the domain and then specifies boundary conditions on each side of the hole. The domain used is seen in Figure 5.3; the hole is so narrow (length 2 and width 0.02) that it appears to be a slit. Care must be taken at the ends of the hole so that the domain processor can follow the boundary. One should make the ends of the hole pointed and be the ends of different boundary pieces.

```
* * * * * * * * * * * * * * * * * * * * * * * * * * * * * * * * * * * * * * * * * * * * * * * * * * * * *
*                                                                                             *
*  EXAMPLE ELLPACK PROGRAM 5.A2                                                               *
*                                                                                             *
*     THIS PROGRAM IS FOR A PROBLEM WITH AN INTERIOR DOUBLE-                                  *
*     VALUED BOUNDARY CONDITION ON A SLIT.  THE ARC FACILITY                                  *
*     OF ELLPACK DOES NOT APPLY SO A HOLE IN THE SHAPE OF A                                   *
*     LONG, VERY THIN DIAMOND IS USED INSTEAD.  CARE MUST BE                                  *
*     TAKEN IN DEFINING THE SLITS THIS WAY SO THE ELLPACK                                     *
*     DOMAIN PROCESSOR DOES NOT GET LOST.  DEFINING THIS SLIT                                 *
*     AS A LONG, VERY THIN RECTANGLE OR ELLIPSE WILL PROBABLY                                 *
*     FAIL.  THE ELLPACK PLOT ROUTINES ALSO ARE INACCURATE IN                                 *
*     THE NEIGHBORHOOD OF TWO-VALUED BOUNDARY CONDITIONS.                                     *
*                                                                                             *
* * * * * * * * * * * * * * * * * * * * * * * * * * * * * * * * * * * * * * * * * * * * * * * * * * * * *
```

```
OPT.      TIME   $  CLOCKWISE = .TRUE.
EQ.       UXX  +  UYY  =  0.0
BOUND.    U = 0.   ON  X = COSH(2.0)*SIN(T),Y = SINH(2.0)*COS(T) &
                                              FOR  T = 0.0 TO 2*PI
HOLE.     U = 1.       ON LINE -1.0, 0.0 TO 0.0, -0.01 TO  1.0, 0.0
                                                       TO  1.0, 0.0
        U = 2.-X**2  ON LINE  1.0, 0.0 TO 0.0,  0.01 TO -1.0, 0.0
                                                       TO -1.0, 0.0
GRID.     21 X POINTS   -COSH(2.0) TO COSH(2.0)
          21 Y POINTS   -SINH(2.0) TO SINH(2.0)
DIS.      5-POINT STAR
SOL.      BAND GE
OUT.      MAX(U) $ PLOT(U)
END.
```

Output of ELLPACK Run (The contour plot is shown in Figure 5.3.)

```
 ---------------------
 PREPROCESSOR OUTPUT
 ---------------------
     2.28 SECONDS TO INITIALIZE PREPREPROCESSOR
     1.33 SECONDS TO PARSE ELLPACK PROGRAM
     4.28 SECONDS TO CREATE ELLPACK CONTROL PROGRAM
     7.90 SECONDS TOTAL
          ELLPACK VERSION --- APRIL 5, 1984

 ------------------
 DOMAIN PROCESSOR
 ------------------
     BOUNDARY POINTS FOUND            68
     BOUNDARY PIECES FOUND             1
     GRID SIZE                21 BY  21
     EXECUTION SUCESSFUL

 --------------------
 DOMAIN  PROCESSOR
 --------------------
      H O L E   P R O C E S S O R
     BOUNDARY POINTS FOUND            12
     BOUNDARY PIECES FOUND             4
     GRID SIZE                21 BY  21
     EXECUTION SUCESSFUL

 ------------------------
 DISCRETIZATION MODULE
 ------------------------
     5 - P O I N T   S T A R
     DOMAIN                        NONRECTANGULAR
     UNIFORM GRID                          21 X  21
     NUMBER OF EQUATIONS                        300
     MAX NO. OF UNKNOWNS PER EQ.                  5
     EXECUTION SUCCESSFUL

 -----------------
 SOLUTION MODULE
 -----------------
    B A N D   G E
    NUMBER OF EQUATIONS       300
    LOWER BANDWIDTH            19
    UPPER BANDWIDTH            19
```

```
    REQUIRED WORKSPACE        17700
    EXECUTION SUCCESSFUL

-----------------
 ELLPACK OUTPUT
-----------------

     +++++++++++++++++++++++++++++++++++++++++++++++++++++++++++++
     +                                                           +
     +  MAX( ABS(U       ) ) ON  21 X  21 GRID = 1.4883974E+00   +
     +  L1 NORM( U       )   ON  21 X  21 GRID = 2.9009587E-01   +
     +  L2 NORM( U       )   ON  21 X  21 GRID = 3.9842123E-01   +
     +                                                           +
     +++++++++++++++++++++++++++++++++++++++++++++++++++++++++++++

-----------------
 ELLPACK OUTPUT
-----------------

     CONTOUR PLOT OF      U
     GRID                20   BY   20
     EXECUTION SUCCESSFUL
-----------------
 ELLPACK OUTPUT
-----------------
         +++++++++++++++++++++++++++
         +                         +
         +     EXECUTION  TIMES    +
         +                         +
         +++++++++++++++++++++++++++

          SECONDS       MODULE NAME
             1.98       DOMAIN
             0.13       HOLE
             0.42       5-POINT STAR
             0.28       BAND GE SETUP
             3.45       BAND GE
             0.38       MAX
             5.30       PLOT
            11.98       TOTAL TIME
```

5.B TWO-PHASE DIFFUSION PROBLEM

We consider the problem of steady-state heat conduction in a system in which two layers of material are sandwiched together. Each layer is homogeneous and isotropic, but have different heat conduction properties. Suppose the region $0 < x < 1$, $-1/2 < y < 0$ is of one material (which we denote as material A), while the region $0 < x < 1$, $0 < y < 1$ is of another material (material B). The material is insulated at $x = 0$ and $x = 1$, and held at fixed, but different, temperatures at $y = -1/2$ and $y = 1$.

Let the functions u and v represent the temperatures of materials B and A respectively. We then have the following models for the steady conduction of heat in the two regions:

In material A:

$$\nabla^2 v = 0.0 \quad \text{for} \quad 0 < x < 1, \quad -1/2 < y < 0,$$

$$\partial v/\partial x = 0.0 \quad \text{for} \quad x = 0.1, \quad -1/2 < y < 0,$$

$$v = 0.0 \quad \text{for} \quad 0 < x < 1, \quad y = -1/2.$$

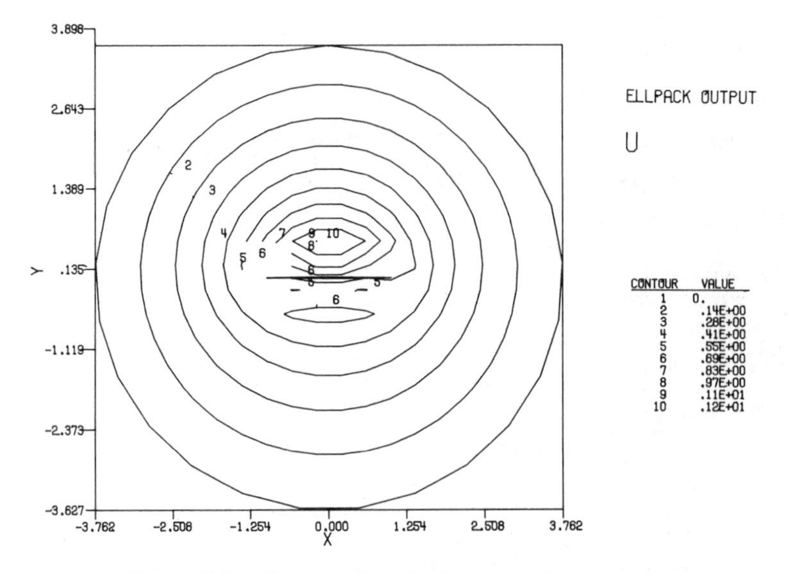

Figure 5.3. Contour plot of solution for Program 5.A2.

In material B:

$$\begin{aligned} \nabla^2 u &= f(x,y) && \text{for} \quad 0 < x < 1, \ 0 < y < 1, \\ \partial u/\partial x &= 0.0 && \text{for} \quad x = 0.1, \quad 0 < y < 1, \\ u &= 1.0 && \text{for} \quad 0 < x < 1, \ y = 1. \end{aligned}$$

The function f is a heat source term. For this example we take $f(x,y) = 4\, y(1-y)\sin(3x+1/2)$. The diffusion problems are coupled by two continuity conditions along the material interface (for $0 < x < 1, y = 0$):

$$u = v, \qquad \partial u/\partial y = k(\partial v/\partial y).$$

The constant k is the ratio of thermal conductivity of one material to the other. We take $k = 1/2$.

If we ignore the jump condition, then this problem is equivalent to a single phase steady state diffusion problem and is easily solved by the following ELLPACK program. Note that there is no need to distinguish between u and v.

```
EQUATION.     UXX + UYY = F(X,Y)

BOUNDARY.     UX = 0.0 ON X = 0.0
              UX = 0.0 ON X = 1.0
              U  = 0.0 ON Y = 1.0
              U  = 1.0 ON Y = -0.5

GRID.          9 X POINTS
              13 Y POINTS

DIS.           5-POINT STAR
SOL.           BAND GE
OUTPUT.        TABLE(U) $ PLOT(U)
```

```
SUBPROGRAMS.
        REAL FUNCTION F(X,Y)
        IF (Y .GE. 0.0) THEN
          F = 4.*Y*(1.-Y)*SIN(1.57080*(3.*X+0.5))
        ELSE
          F = 0.0
        ENDIF
        RETURN
        END
END.
```

This program produces the temperature distribution given in Figure 5.4.

One way to incorporate the jump condition is to modify the output of the 5 POINT STAR module. We wish to change the finite difference equations generated for points along the line $y = 0$. We write a subprogram ADJUMP to do this and insert the following code after the existing OUTPUT statement.

```
FORTRAN.
        CALL ADJUMP(R1COEF,I1IDCO,I1MNEQ,I1MNCO)
        L1SYMM = .FALSE.
SOL.        BAND GE
OUTPUT.     TABLE(U) $ PLOT(U)
```

This causes the equations in R1COEF to be modified to create a new discretization and the problem is solved again with the BAND GE. To write the subprogram ADJUMP one must be familiar with the difference equations produced by 5 POINT STAR as well as with the sparse matrix storage scheme used by ELLPACK (see Chapter 13). The changes made to the equations destroy the symmetry of the matrix and so the internal ELLPACK variable L1SYMM is set to .FALSE. (the discretization modules flag symmetry of their matrix by setting L1SYMM).

Let (x_i, y_j), $1 \leqslant i \leqslant 12$, denote the uniformly spaced grid point locations and let $U_{i,j}$, $V_{i,j}$, and $F_{i,j}$ denote functions evaluated at the point (x_i, y_j). The difference equation written by 5-POINT STAR for the point (x_i, y_j) in the interior of the domain is as follows: In material B,

$$U_{i+1,j} + U_{i,j+1} + U_{i-1,j} + U_{i,j-1} - 4U_{i,j} = h^2 F_{i,j} \tag{1}$$

In material A,

$$V_{i+1,j} + V_{i,j+1} + V_{i-1,j} + V_{i-1,j} + V_{i,j-1} - 4V_{i,j} = 0 \tag{2}$$

where $h = 1/8$ in this example. Along the left and right sides of the domain these equations must also incorporate the boundary conditions $\partial u/\partial x = 0$. For material B these equations become:

At the point (x_1, y_j),

$$U_{1,j+1} + 2U_{2,j} + U_{1,j-1} - 4U_{1,j} = h^2 F_{1,j} \tag{3}$$

At the point (x_9, y_j),

$$U_{9,j+1} + 2U_{8,j} + U_{9,j-1} - 4U_{9,j} = h^2 F_{9,j}. \tag{4}$$

with similar equations for material A.

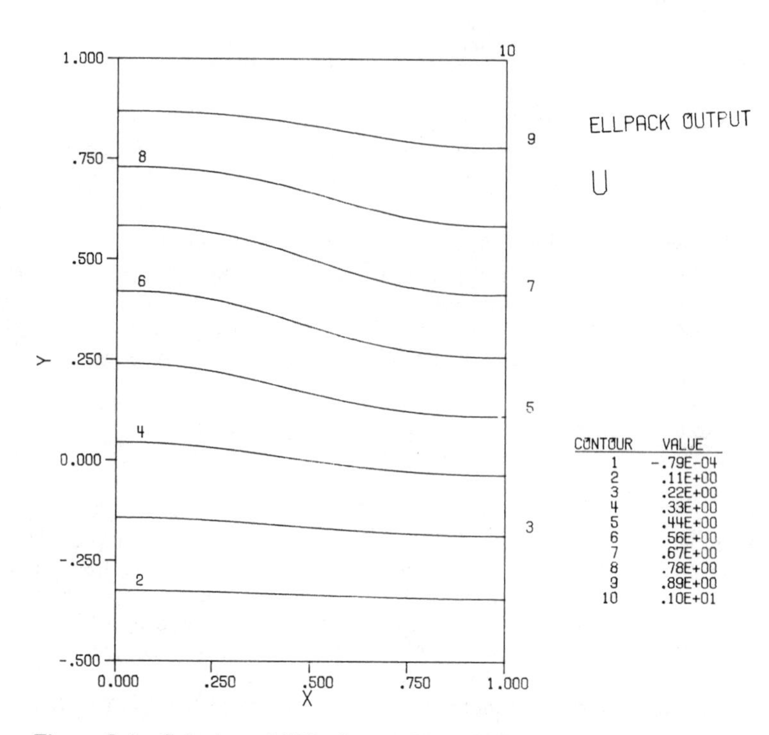

Figure 5.4. Solution of diffusion problem without jump condition.

Along the line $y = 0$ we also wish these finite difference equations to satisfy the jump condition $\partial u/\partial y = k\partial v/\partial y$. Replacing the derivatives by central differences at the point (x_i, y_5), we get the discrete analogue

$$U_{i,6} - U_{i,4} = k(V_{i,6} - V_{i,4}). \tag{5}$$

Note that we have introduced two fictitious quantities, $U_{i,4}$ and $V_{i,6}$, representing a material B temperature in material A and a material A temperature in material B respectively. We eliminate these using the relations (1) and (2) and use the continuity condition $U_{i,5} = V_{i,5}$ to get

$$(1+k)\,U_{i+1,5} + 2\,U_{i,6} + (1+k)\,U_{i-1,5} + 2kV_{i,4} - 4(1+k)\,U_{i,5} = 0.$$

For $i = 1$ and $i = 9$ we must use the boundary finite difference equations (3) and (4) and their material A analogues to eliminate the fictitious points from (5). Doing this, and using continuity, we obtain at the point (x_1, y_5),

$$2\,U_{1,6} + 2(1+k)\,U_{2,5} + 2kV_{1,4} - 4(1+k)\,U_{1,5} = 0$$

and at the point (x_9, y_5),

$$2\,U_{9,6} + 2(1+k)\,U_{8,5} + 2kV_{9,4} - 4(1+k)\,U_{9,5} = 0.$$

Note again that there is no need to distinguish between u and v in the ELLPACK program since exactly one value is defined at each grid point.

In ELLPACK each equation and unknown is given a single index number from one to the number of equations and unknowns. Thus we must also know how 5-POINT STAR maps the double subscripts used above into the single subscripts used in ELLPACK (equivalently, how grid points are numbered). The equation number corresponding to the (i,j)th grid point is given by I35PNU(i,j), where the array I35PNU(9,13) is available in the common block C35PNU after 5-POINT STAR is called. (See the description of 5 POINT STAR in Chapter 6).

The coefficients of the kth finite difference equation are stored into the kth row of the array R1COEF. The indices of the unknowns that these coefficients multiply are stored into the corresponding locations of the array I1IDCO. (See Chapter 13 for details.) Note that as a result of the way in which we defined the function F, we need not modify the right-hand sides of these equations. The following subprogram performs all these operations.

```
      SUBROUTINE ADJUMP (COEF,IDCO,MNEQ,MNCO)
C
C -----------------------------------------------------------
C CHANGE EQUATIONS ALONG Y=0 TO ACCOUNT FOR JUMP CONDITION
C -----------------------------------------------------------
C
C SOME IMPORTANT VARIABLES :
C
C    NX       = NUMBER OF GRID LINES IN X
C    H        = GRID SPACING
C    INTER    = GRID LINE IN Y CORRESPONDING TO INTERFACE
C    CO       = DIFFERENCE EQUATION COEFFICIENT (CENTER)
C    CN       = DIFFERENCE EQUATION COEFFICIENT (NORTHERN NEIGHBOR)
C    CS       = DIFFERENCE EQUATION COEFFICIENT (SOUTHERN NEIGHBOR)
C    CE       = DIFFERENCE EQUATION COEFFICIENT (EASTERN NEIGHBOR)
C    CW       = DIFFERENCE EQUATION COEFFICIENT (WESTERN NEIGHBOR)
C    COEF     = DIFFERENCE EQUATIONS IN ELLPACK FORMAT
C    IDCO       (NON-ZEROS IN ITH ROW OF MATRIX ARE STORED IN THE
C               ITH ROW OF COEF, AND THE CORRESPONDING ENTRIES IN
C               IDCO GIVE THEIR COLUMN INDICES)
C    I35PNU(J,K) = INDEX OF EQUATION CORRESPONDING TO (J,K)TH
C               GRID POINT
C
      REAL COEF(MNEQ,MNCO)
      INTEGER IDCO(MNEQ,MNCO)
      COMMON / C35PNU /  I35PNU(9,13)
C
      NX = 9
      H = 1.0/FLOAT(NX-1)
      H2 = H*H
      RK = 0.5
      RK1 = 1.0 + RK
      INTER = 5
C
C COMPUTE NEW DIFFERENCE EQUATION COEFFICIENTS
C
      CO = - 4.0*RK1
      CE = RK1
      CW = RK1
      CN = 2.0
      CS = 2.0*RK
```

```
C
C LOAD COEFFICIENTS FOR MODIFIED INTERIOR POINTS
C
      DO 100 J=1,NX
         I = I35PNU(J,INTER)
         IN = I35PNU(J,INTER+1)
         IS = I35PNU(J,INTER-1)
         IF (J .EQ. NX) THEN
           IE = 0
         ELSE
           IE = I35PNU(J+1,INTER)
         ENDIF
         IF (J .EQ. 1) THEN
           IW = 0
         ELSE
           IW = I35PNU(J-1,INTER)
         ENDIF
         COEF(I,1) = CO
         IDCO(I,1) = I
         COEF(I,2) = CN
         IDCO(I,2) = IN
         COEF(I,3) = CS
         IDCO(I,3) = IS
         COEF(I,4) = CE
         IDCO(I,4) = IE
         COEF(I,5) = CW
         IDCO(I,5) = IW
  100 CONTINUE
C
C FIX COEFFICIENTS FOR BOUNDARY POINTS ALONG INTERFACE
C
C  (THESE ARE EQUATIONS ILEFT AND IRIGHT)
      ILEFT = I35PNU(1,INTER)
      IRIGHT= I35PNU(NX,INTER)
      COEF(ILEFT,4)  = 2.0*CE
      COEF(ILEFT,5)  = 0.0
      IDCO(ILEFT,5)  = 0
      COEF(IRIGHT,5) = 2.0*CW
      COEF(IRIGHT,4) = 0.0
      IDCO(IRIGHT,4) = 0
C
C PRINT MODIFIED EQUATIONS
C
      DO 150 J=1,NX
        I = I35PNU(J,INTER)
        WRITE(6,3000) I
        DO 150 K=1,MNCO
          IF (IDCO(I,K) .NE. 0)
     +        WRITE(6,3001) K,IDCO(I,K), COEF(I,K)
  150 CONTINUE
      RETURN
C
 3000 FORMAT(/' EQUATION ',I3 ' *************')
 3001 FORMAT( '  K=',I2,' ID=',I3,' COEF='1PE15.8)
      END
```

The result of the solution of the modified finite difference equations is shown in Figure 5.5.

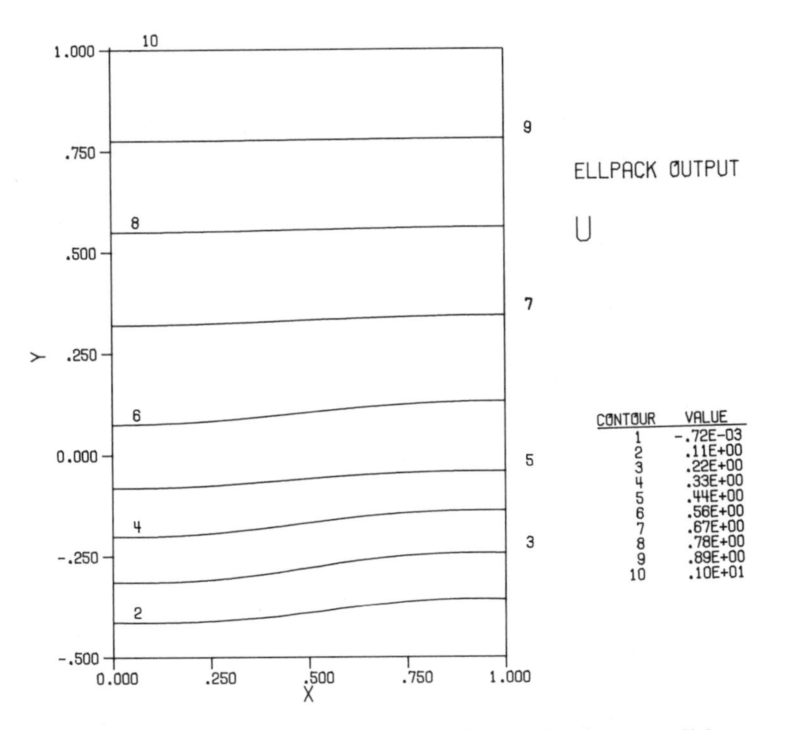

Figure 5.5. Solution of diffusion problem with jump condition.

5.C NEWTON ITERATION FOR NONLINEAR PROBLEMS

Program 4.D3 in the previous chapter illustrates how to solve nonlinear problems in ELLPACK using fixed point iteration (also known as Picard's method). This has a rate of convergence that is rarely fast. Newton's method usually converges very rapidly once one gets reasonably close to the solution and is very efficient when it works. [Ortega and Rheinboldt, 1970].

To derive Newton's method we consider the PDE as an operator equation,

$$F(u) = 0$$

where F is a nonlinear operator from one function space to another. Expanding F about some point (function) u_0 and discarding all but the first two terms we obtain

$$F(u) \approx F'(u_0) + F'(u_0)(u - u_0),$$

where $F'(u_0)$ is a linear operator known as the Frechet derivative of F. In our case $F'(u_0)$ is simply a linear elliptic partial differential operator which we denote as $L(u_0)$. If we set $F(u) = 0$ we can rewrite the equation above as

$$L(u_0)u = -(F(u_0) - L(u_0)u_0).$$

This suggests the following iteration to find u satisfying $F(u) = 0$.

Guess u_0
REPEAT
  Solve $L(u_0)u = -(F(u_0) - L(u_0)u_0)$ for u
  Set $u_0 \leftarrow u$
UNTIL converged

At each stage u_0 is known, and hence $L(u_0)u = -(F(u_0) - L(u_0)u_0)$ is simply a linear elliptic PDE. This is exactly what ELLPACK is prepared to solve.

Thus, to implement Newton's method in ELLPACK, one must first determine the Frechet derivative $L(u)$ for the problem. $L(u)$ is the linearized perturbation of $u+S$ in $F(u)$, and hence its derivation is straightforward, though often quite tedious. We later give an example of computing $L(u)$ using the MACSYMA system.

The technique is illustrated first for the simple example

$$F(u) = u_{xx} + u^2 u_{yy} - e^u - f = 0 \qquad 0 \leqslant x, y \leqslant 1$$

If we make a perturbation δ of u in this example and discard all powers of δ beyond the first, we obtain the following equation in u and δ:

$$\begin{aligned} F(u+\delta) &= (u+\delta)_{xx} + (u+\delta)^2(u+\delta)_{yy} - e^{u+\delta} - f \\ &= (u_{xx} + \delta_{xx}) + (u^2 u_{yy} + 2u\delta u_{yy} + u^2\delta_{yy}) - (e^u + \delta e^u) - f \\ &= (u_{xx} + u^2 u_{yy} - e^u - f) + [\delta_{xx} + u^2\delta_{yy} + (2uu_{yy} - e^u)\delta] \\ &= F(u) + L(u)\delta. \end{aligned}$$

For this example, then, the equation $L(u_0)u = -(F(u_0) - L(u_0)u_0)$ becomes, after simplification,

$$u_{xx} + (u_0)^2 u_{yy} + (2u_0 u_{0yy} - e^{u_0})u = 2(u_0)^2 u_{0yy} - e^{u_0}(u_0 - 1) + f.$$

An ELLPACK program for this example follows where $f(x,y)$ and the boundary conditions are chosen to make the true solution be $u(x,y) = \sin(x)\cos(y)$. It uses the initial guess $u_0(x,y) = 0$, solves the linearized problem by collocation with Hermite bi-cubics, and limits the method to 5 iterations. Various other features of the program are explained in the comments. The only output we give is the table produced by the subroutine SUMARY; it shows the convergence is quite fast.

Note that in the ELLPACK program, both $u_0(x,y)$ and $u(x,y)$ are denoted by u. The u's in the coefficients of L are the previous estimate u_0; the estimate u produced by solving the linearized problem becomes the u_0 for the next iteration.

```
*************************************************************
*                                                           *
*  EXAMPLE ELLPACK PROGRAM 5.C1                             *
*                                                           *
*      APPLY NEWTON'S METHOD TO THE NONLINEAR PROBLEM       *
*                                                           *
*      UXX + U**2 UYY = EXP(U) + F(X,Y)                     *
*                                                           *
*************************************************************
```

```
*
DECLARATIONS.
        REAL ERRMAX(100)
*    USE THE PDE FOR THE NEW U(X,Y) OBTAINED BY LINEARIZING THE
*    NONLINEAR PROBLEM
EQUATION. UXX + (U(X,Y)**2)*UYY                                    &
          + (2.*U(X,Y)*UYY(X,Y)-EXP(U(X,Y)))*U =                   &
      2*(U(X,Y)**2)*UYY(X,Y) - EXP(U(X,Y))*(U(X,Y)-1.) + F(X,Y)
BOUNDARY.  U = TRUE(X,Y) ON X = 0.
                         ON X = 1.
                         ON Y = 0.
                         ON Y = 1.
GRID.      5 X POINTS $ 5 Y POINTS
*               INITIALIZE THE NEWTON ITERATION BY U = 0
TRIPLE.    SET( U = ZERO )
*               USE FORTRAN TO CONTROL ITERATION AND OUTPUT
FORTRAN.
        I1LEVL = 1
        NITERS = 5
        DO 10 NITER = 1, NITERS
C                     SOLVE THE LINEARIZED PROBLEM
DISCRETIZATION.       HERMITE COLLOCATION
SOLUTION              BAND GE
FORTRAN.
C          COMPUTE INTERMEDIATE MAX ERROR, SAVE FOR TABLED OUTPUT
           CALL MAXERR(ERRMAX(NITER))
C          TURN OFF ELLPACK OUTPUT
           I1LEVL = 0
   10   CONTINUE
C       PROCESS FINAL RESULTS
        CALL SUMARY(ERRMAX,NITERS)
SUBPROGRAMS.
      FUNCTION F(X,Y)
C            F IS CHOSEN TO MAKE THE TRUE SOLUTION SIN(X)*COS(Y)
         TRUE   = SIN(X)*COS(Y)
         TRUEXX = -TRUE
         TRUEYY = -TRUE
         F = TRUEXX + TRUE**2*TRUEYY
     +         + (2.*TRUE*TRUEYY-EXP(TRUE))*TRUE
     +         - 2.*TRUE**2*TRUEYY + EXP(TRUE)*(TRUE-1.)
         RETURN
      END
      FUNCTION TRUE(X,Y)
         TRUE = SIN(X)*COS(Y)
         RETURN
      END
      SUBROUTINE MAXERR (ERRMAX)
C        COMPUTE THE MAXIMUM ERROR ON THE GRID, SAVE FOR LATER
         ERRMAX = 0.
         DO 20 I = 1, 5
            X = 0.25*(I-1)
            DO 10 J = 1, 5
               Y = 0.25*(J-1)
               ERRMAX = AMAX1(ERRMAX,ABS(TRUE(X,Y)-U(X,Y)))
   10       CONTINUE
   20    CONTINUE
         RETURN
      END
      SUBROUTINE SUMARY (ERRMAX, NITERS)
C                   PRINT SUMMARY OF RESULTS
```

```
          REAL ERRMAX(NITERS)
          PRINT 100
          DO 10 NITER = 1, NITERS
             PRINT 110, NITER, ERRMAX(NITER)
   10     CONTINUE
          RETURN
  100     FORMAT('1         EXAMPLE ELLPACK PROGRAM 5.C1'//
     A             T12,'ITER',T23,'MAX ERROR',/T11,6('-'),5X,11('-'))
  110     FORMAT(T11,I3,1,1P1E16.1)
        END
END.
```

The table produced by program 5.C1 is

```
EXAMPLE ELLPACK PROGRAM 5.C1
        ITER          MAX ERROR
        ----          ----------
          1            3.0E-02
          2            6.4E-04
          3            7.9E-06
          4            3.5E-06
          5            3.6E-06
```

To further illustrate Newton's method and to show that ELLPACK can solve difficult real world problems, we provide two more examples. See also [Birkhoff and Lynch 1984] for the solution of Plateau's problem using this approach. The second example is from nonlinear, laminar, non-Newtonion flow [Ames, 1965].

The first problem is

$$w(u)(u_{xx}+u_{yy})+w_x(u)u_x+w_y(u)u_y = g(x,y)$$

$$u_x = 0 \quad \text{on } x=0,1, \qquad u=b(x) \quad \text{on } y=0,1,$$

where the function $w(u)$ varies depending on the application. Set $a(u)=\sqrt{u_x^2+u_y^2}$, then physically meaningful cases of $w(u)$ are

$$w(u) = [a(u)]^\alpha \qquad\qquad w(u) = 1/(\alpha+\beta a(u))$$

$$w(u) = e^{[a(u)/(\alpha+\beta a(u))]}/a(u) \qquad w(u) = \alpha \tanh(\beta a(u))/a(u)$$

This nonlinear problem is the source of problems 19 and 23 in the PDE population given in Appendix A. We take one of the simplest possible cases here, $w(u)=a(u)$ (i.e. $\alpha=1$). We choose $f(x,y)$ and $b(x)$ so that the true solution of the problem is

$$u(x,y) = (1+e^{-y})\cos(\pi x)$$

Since linearizing the nonlinear operator can be tedious (and error prone) we give a **MACSYMA** program that produces the linear operator $L(u_0)$ automatically for this problem. This program can be easily adapted for other nonlinear problems. If one has nonlinear boundary conditions, then a linear operator $B(u_0)$ analogous to $L(u_0)$ can be obtained by a similar program. See [Rand, 1984] for more information about MACSYMA.

```
(C1) /*        EXAMPLE PROGRAM 5.C2                                  */
     /*   MACSYMA PROGRAM TO COMPUTE THE LINEARIZED EQUATION        */
     /*   FROM APPLYING NEWTON'S METHOD                              */
     /*   (WRITTEN BY WAYNE DYKSEN)                                  */
```

```
      /*       LIST THE NONLINEAR PDE COEFFICIENTS   */
      A(U)  := W(U)$
(C2)  B(U)  := 0$
(C3)  C(U)  := W(U)$
(C4)  D(U)  := DIFF(W(U),X)$
(C5)  E(U)  := DIFF(W(U),Y)$
(C6)  F(U)  := 0$
(C7)  G(U)  := 0$
(C8)  W(U)  := SQRT ( DIFF(U,X)**2 + DIFF(U,Y)**2 )$
(C9)  /*   THIS ENDS THE PROBLEM DEPENDENT INPUT FOR MOST PDES */
      /*   DEFINE DERIVATIVES OF U0, U0X, ETC. */
       GRADEF(U0,X,U0X)    $
(C10)  GRADEF(U0X,X,U0XX)  $
(C11)  GRADEF(U0X,Y,U0XY)  $
(C12)  GRADEF(U0Y,X,U0XY)  $
(C13)  GRADEF(U0,Y,U0Y)    $
(C14)  GRADEF(U0Y,Y,U0YY)  $
(C15)  GRADEF(U1,X,U1X)    $
(C16)  GRADEF(U1X,X,U1XX)  $
(C17)  GRADEF(U1X,Y,U1XY)  $
(C18)  GRADEF(U1Y,X,U1XY)  $
(C19)  GRADEF(U1,Y,U1Y)    $
(C20)  GRADEF(U1Y,Y,U1YY)  $
(C21)  /*       WRITE THE NONLINEAR PDE */
       PDE(U)  := A(U)*DIFF(U,X,2) + B(U)*DIFF(DIFF(U,X),Y)
                + C(U)*DIFF(U,Y,2)
                + D(U)*DIFF(U,X) + E(U)*DIFF(U,Y) + F(U)*U + G(U)$
(C22)  /*       DIFFERENTIATE AND COLLECT TERMS */
       DERIVATIVE: DIFF(PDE(U0+EPS*(U1-U0)),EPS) $
(C23)  TSERIES: PDE(U0) + EV(DERIVATIVE,EPS=0)   $
(C24)  RATSIMP(TSERIES)                $
(C25)  TSERIES: EXPAND(TSERIES)        $
(C26)  COEU1XX: COEFF(TSERIES,U1XX)  $
(C27)  COEU1XY: COEFF(TSERIES,U1XY)  $
(C28)  COEU1YY: COEFF(TSERIES,U1YY)  $
(C29)  COEU1X:  COEFF(TSERIES,U1X)   $
(C30)  COEU1Y:  COEFF(TSERIES,U1Y)   $
(C31)  COEU1 :  COEFF(TSERIES,U1)    $
(C32)  RS: (COEU1XX*U1XX + COEU1XY*U1XY + COEU1YY*U1YY
       + COEU1X*U1X + COEU1Y*U1Y + COEU1*U1)-TSERIES $
(C33)  RATSIMP(RS) $
(C34)  /* DISPLAY COEFFS OF LINEARIZED PDE FOR NEWTONS METHOD */
       COEU1XX: RATSIMP(COEU1XX * W(U0));
                                               2        2
(D34)                                       U0Y  + 2 U0X
(C35)  COEU1XY: RATSIMP(COEU1XY * W(U0));
(D35)                                          2 U0X U0Y
(C36)  COEU1YY: RATSIMP(COEU1YY * W(U0));
                                                 2      2
(D36)                                       2 U0Y  + U0X
(C37)  COEU1X  : RATSIMP(COEU1X  * W(U0));
           3                  3                      2          3
       U0X  U0YY + 2 U0XY U0Y  + 3 U0X U0XX U0Y  + 2 U0X  U0XX
(D37)  --------------------------------------------------------
                                   2      2
                                U0Y  + U0X
(C38)  COEU1Y  : RATSIMP(COEU1Y  * W(U0));
                  3          2                      3          3
          (2 U0Y  + 3 U0X  U0Y) U0YY + U0XX U0Y  + 2 U0X  U0XY
(D38)     ----------------------------------------------------
```

```
                                          2        2
                                       U0Y   + U0X
(C39) COEU1   : RATSIMP(COEU1      * W(U0));
(D39)                                        0
(C40) RS      : RATSIMP(RS         * W(U0));
                  2         2                  2
(D40)    (2 U0Y   + U0X  ) U0YY + U0XX U0Y   + 2 U0X U0XY U0Y
                   2
          + 2 U0X   U0XX
```

The algebra to derive the linearized problem is formidable even with the simplest possible choice of $w(u) = a(u)$. The results from the MACSYMA program are used in program 5.C3.

```
*      ***********************************************************
*      *                                                         *
*      *   EXAMPLE ELLPACK PROGRAM 5.C3                          *
*      *                                                         *
*      *       APPLY NEWTON'S METHOD TO THE NONLINEAR PROBLEM    *
*      *                                                         *
*      *       W(U)(UXX + UYY) + WX(U)UX + WY(U)UY = G           *
*      *                                                         *
*      *           W(U) = SQRT (UX**2 + UY**2)                   *
*      *                                                         *
*      ***********************************************************
DECLARATIONS.
        REAL ERRMAX(100)
*     USE THE PDE FOR THE NEW U(X,Y) OBTAINED BY LINEARIZING THE
*     NONLINEAR PROBLEM
EQUATION.        A(X,Y)*UXX + B(X,Y)*UXY + C(X,Y)*UYY   &
               + D(X,Y)*UX  + E(X,Y)*UY  + F(X,Y)*U    = RS(X,Y)
BOUNDARY.  UX = 0.                          ON X = 0.
                                            ON X = 1.
           U  =              2.*COS(PI*X)   ON Y = 0.
           U  = (1.+EXP(-1.))*COS(PI*X)     ON Y = 1.
GRID.      5 X POINTS $ 5 Y POINTS
*          INITIALIZE THE NEWTON ITERATION BY INTERPOLATING THE
*          BOUNDARY CONDITIONS WITH BLENDING FUNCTIONS
TRIPLE.  SET U BY BLENDING
FORTRAN.
C                 USE FORTRAN TO CONTROL ITERATION AND OUTPUT
        I1LEVL = 1
        NITERS = 5
        DO 10 NITER = 1, NITERS
C                 SOLVE THE LINEARIZED PROBLEM
DISCRETIZ.        HERMITE COLLOCATION
SOLUTION          BAND GE
FORTRAN.
C            COMPUTE INTERMEDIATE MAX ERROR, SAVE FOR FINAL OUTPUT
             CALL MAXERR(ERRMAX(NITER))
             I1LEVL = 0
   10   CONTINUE
C                 PROCESS FINAL RESULTS
        CALL SUMARY(ERRMAX,NITERS)
SUBPROGRAMS.
      FUNCTION W(X,Y)
         W = SQRT(UX(X,Y)**2 + UY(X,Y)**2)
         RETURN
      END
```

```
      FUNCTION WX(X,Y)
         WX = 2.*UX(X,Y)*UXX(X,Y)/SQRT(UX(X,Y)**2 + UY(X,Y)**2)
         RETURN
      END
      FUNCTION WY(X,Y)
         WY = 2.*UY(X,Y)*UYY(X,Y)/SQRT(UX(X,Y)**2 + UY(X,Y)**2)
         RETURN
      END
C
      FUNCTION RS(X,Y)
C              RS IS CHOSEN TO MAKE THE TRUE SOLUTION
C                  U(X,Y) = (1.+EXP(-Y))*COS(PI*X)
C
      RS = A(X,Y)*TRUEXX(X,Y) + B(X,Y)*TRUEXY(X,Y) +
     A     C(X,Y)*TRUEYY(X,Y) + D(X,Y)*TRUEX(X,Y)  +
     B     E(X,Y)*TRUEY(X,Y)  + F(X,Y)*TRUE(X,Y)
      RETURN
      END
      FUNCTION TRUE(X,Y)
C               ACCESS PI = 3.14159... FROM ELLPACK COMMON
         COMMON  / C1RVGL /  R1EPSG, R1EPSM, PI
         TRUE = (1.+EXP(-Y))*COS(PI*X)
         RETURN
      END
      SUBROUTINE   MAXERR(ERRMAX)
C
C         COMPUTE THE MAXIMUM ERROR ON THE GRID, SAVE FOR LATER
C         ACCESS INTERNAL ELLPACK VARIABLES
      COMMON  / C1IVGR /  I1NGRX, I1NGRY, I1NGRZ, I1NBPT, I1MBPT
      COMMON  / C1GRDX /  R1GRDX(1)
      COMMON  / C1GRDY /  R1GRDY(1)
         ERRMAX = 0.
         DO 20 I = 1, I1NGRX
            X = R1GRDX(I)
            DO 10 J = 1, I1NGRY
               Y = R1GRDY(J)
               ERRMAX = AMAX1(ERRMAX,ABS(TRUE(X,Y)-U(X,Y)))
   10       CONTINUE
   20    CONTINUE
         RETURN
      END
      SUBROUTINE SUMARY (ERRMAX, NITERS)
C               PRINT SUMMARY OF RESULTS
      REAL ERRMAX(NITERS)
         PRINT 100
         DO 10 NITER = 1, NITERS
            PRINT 110, NITER, ERRMAX(NITER)
   10    CONTINUE
         RETURN
  100    FORMAT('1        EXAMPLE ELLPACK PROGRAM 5.C2'//
     A              T8,'ITER',T16,'MAX ERROR',/T7,6('-'),2X,10('-'))
  110    FORMAT(T8,I4,1X,1P1E12.4)
      END
C************** NOTE NOTE NOTE NOTE ************************
C    ALSO NEED FORTRAN PROGRAMS FROM THE MACSYMA OUTPUT FOR
C         A(X,Y)  B(X,Y)  C(X,Y)  D(X,Y)  E(X,Y)  F(X,Y)
C    ALSO PROVIDE FORTRAN CODE FOR
C         TRUEX(X,Y)    TRUEY(X,Y)
          TRUEXX(X,Y)   TRUEXY(X,Y)   TRUEYY(X,Y)
C**********************************************************
END.
```

The last nonlinear problem discussed here comes from gas lubrication. This is the effect that keeps high speed tapes and disks from making physical contact with read/write heads. Two views of the physical situation are shown in Figure 5.6. The high speed of the disk pulls the air into the gap; it is compressed as it goes through and this builds up a pressure to keep the two parts separated. The separation between the disk and head is only a few thousandths of an inch.

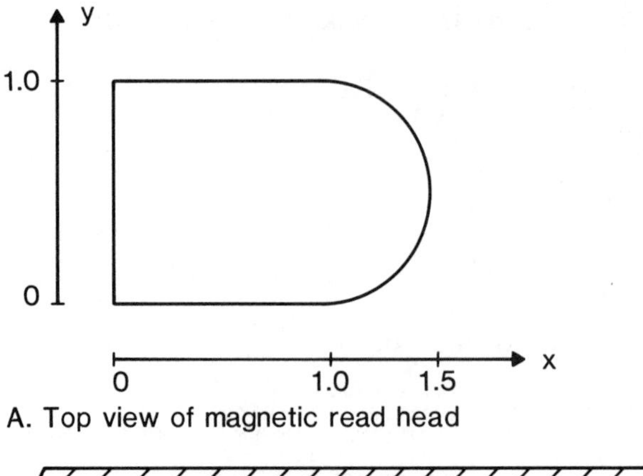

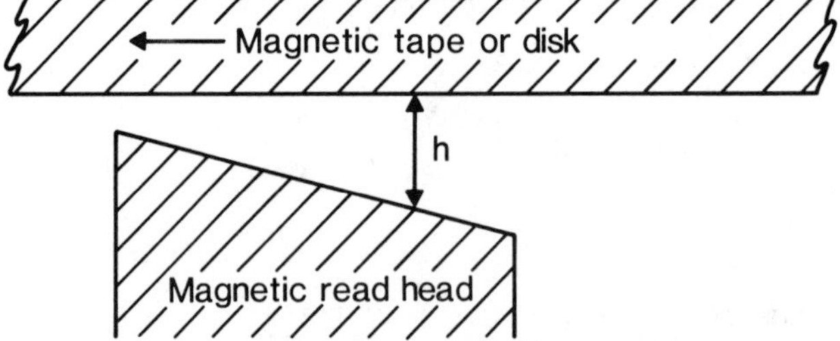

Figure 5.6. Top view A of a magnetic read head and side view B of the space between the head and disk.

The nonlinear elliptic problem to be solved for the pressure u on the domain shown in Figure 5.6a is

$$\begin{aligned} &(uh^3 u_x)_x + (uh^3 u_y)_y + c(uh)_x = 0 \qquad &&\text{in } R, \\ &u(x,y) = 1 &&\text{on } \partial R, \end{aligned}$$

where $h(x,y) = 1+2x$ and c is a physical constant. We first rewrite the elliptic equation as

$$uu_{xx} + uu_{yy} + (u_x + \frac{3h_x u}{h} + \frac{c}{h^2})\, u_x + (u_y + \frac{3h_y u}{h})\, u_y + \frac{ch_x}{h^3}\, u = 0$$

The linearized equation to be solved is then

$$u_0 u_{xx} + u_0 u_{yy} + [2u_{0x} + 3h_x u_0/h + c/h^2]u_x$$
$$+ [2u_{0y} + 3h_y u_0/h]u_y + [u_{0xx} + u_{0yy} + 3(u_{0x} + u_{0y})/h + ch_x/h^3]u$$
$$= u_0(u_{0xx} + u_{0yy}) + u_{0x}^2 + u_{0y}^2 + 3h(u_{0x} + u_{0y})/u_0$$

Program 5.C4 uses this linearized equation to solve this problem with Newton's method. The result of physical interest in this problem is the integral of $u(x,y)$ over the domain which is the load that the lubricant supports.

```
*      ************************************************************
*      *                                                          *
*      *   EXAMPLE ELLPACK PROGRAM 5.C4                           *
*      *                                                          *
*      *        APPLY NEWTON'S METHOD TO THE NONLINEAR PROBLEM    *
*      *                                                          *
*      *           3            3                                 *
*      *        (UH U )  +  (UH U )   + C(UH)   = 0              *
*      *             X X          Y Y        X                    *
*      *                                                          *
*      *   REYNOLD'S EQUATION FOR COMPRESSIBLE FLUID LUBRICATION *
*      *                                                          *
*      ************************************************************
*
DECLARATIONS.
         REAL DIFNWT(20),DIFGRD(20),UGRID(20,20)
         COMMON  / PARAM / CC,UOLD(20,20)
         DATA UOLD / 400*0.0 /, UGRID / 400*0.0 /
*                   USE THE PDE FOR THE NEW U(X,Y) OBTAINED BY
*                   LINEARIZING THE NONLINEAR PROBLEM
EQUATION.           A(X,Y)*UXX + C(X,Y)*UYY                              &
                 + D(X,Y)*UX  + E(X,Y)*UY  + F(X,Y)*U    = RS(X,Y)
BOUND.   U = 1. ON LINE 1,0 TO 0,0 TO 0,1 TO 1,1
         ON X = 1.+.5*SIN(T), Y = .5+.5*COS(T) FOR T = 0. TO PI
*                   SET MAXIMUM GRID SIZES
OPT.     MAX X POINTS = 15 $ MAX Y POINTS = 10 $ CLOCKWISE=.TRUE.
FORTRAN.
C                   SET PHYSICAL PARAMETER CC = 4.0
      CC = 4.0
C                USE FORTRAN TO LOOP OVER GRID SIZES
      DO 20 NGRD = 1,4
C                    SET NUMBER OF X AND Y GRID POINTS
         NYPT = 2 + 2*NGRD
         NXPT = (3*NYPT)/2
GRID.     NXPT X POINTS 0.0 TO 1.5 $ NYPT Y POINTS 0.0 TO 1.0
*                  INITIALIZE THE NEWTON ITERATION BY GUESSU = 1.0
TRIPLE.   SET( U = GUESSU )
FORTRAN.
C                    USE FORTRAN TO CONTROL NEWTON ITERATION
         DO 10 NITER = 1, 5
C                     SOLVE THE LINEARIZED PROBLEM
DISCRETIZATION.       COLLOCATION
SOLUTION              BAND GE
FORTRAN.
C                    COMPUTE AND SAVE MAX DIFFS IN NEWTON ITERATION
                     CALL MAXDIF(DIFNWT(NITER))
C                    SUPPRESS OUTPUT AFTER 1ST ITERATION
                     I1LEVL = 0
```

```
  10        CONTINUE
            PRINT *,'  TABLE OF NEWTON DIFFERENCES FOR NGRD =',NGRD
            CALL SUMARY(DIFNWT,5)
C                  PUT UGRID VALUES INTO UOLD FOR MAXDIF ON GRID
                DO 14 I = 1,20
                   DO 14 J = 1,20
  14                  UOLD(I,J) = UGRID(I,J)
C            COMPUTE AND SAVE MAX DIFF FOR CHANGING GRIDS
            CALL MAXDIF(DIFGRD(NGRD))
C                  PUT UOLD VALUES INTO UGRID FOR NEXT MAXDIF ON GRID
                DO 18 I = 1,20
                   DO 18 J = 1,20
  18                  UGRID(I,J) = UOLD(I,J)
  20  CONTINUE
      PRINT *,' * DIFFERENCES ON 18X12 GRID AS X,Y GRIDS CHANGE'
      CALL SUMARY(DIFGRD,4)
SUBPROGRAMS.
C       THE FUNCTIONS REQUIRED ARE NOT LISTED HERE, THEY ARE
C       SIMILAR TO THOSE OF PROGRAM 5.C3. THESE FUNCTIONS ARE
C         GUESSU  MAXDIF  SUMARY  H       HX      HY
C         A       C       D       E       F       G
END.
```

Figure 5.7 shows the contour plot of the pressure using a 12 by 9 grid. The value $c = 12$ used corresponds to a moderate speed in a real application. As c is increased the problem becomes more difficult and near $c = 25$ or 30 the initial guess $u = 1.0$ becomes too far away for Newton's method to converge. The changes in $u(x,y)$ measured as the grid is refined suggest that this solution is accurate within 0.001.

To judge the accuracy of the computed solution for the case $c = 4$ we solve the problem on a sequence of four increasingly finer grids. One expects the solution to be more accurate as the grid is made finer, and hence one can use the change in u from a coarse grid to a fine grid as a conservative estimate of the accuracy on the fine grid. On each grid we iterate Newton's method until the maximum absolute change in the Newton iterates is less than 2×10^{-5}. The table below shows the maximum absolute change in the solution obtained in going from one grid to the next (this change is measured on a fixed 18×12 grid). The results indicate that the solution on the finest grid probably has error no greater than 2×10^{-3}. This also indicates that fewer Newton iterations could have been taken on each grid.

Grid	*Maximum absolute change from previous grid*
6× 4	—
9× 6	0.01824
12× 8	0.00479
15×10	0.00137

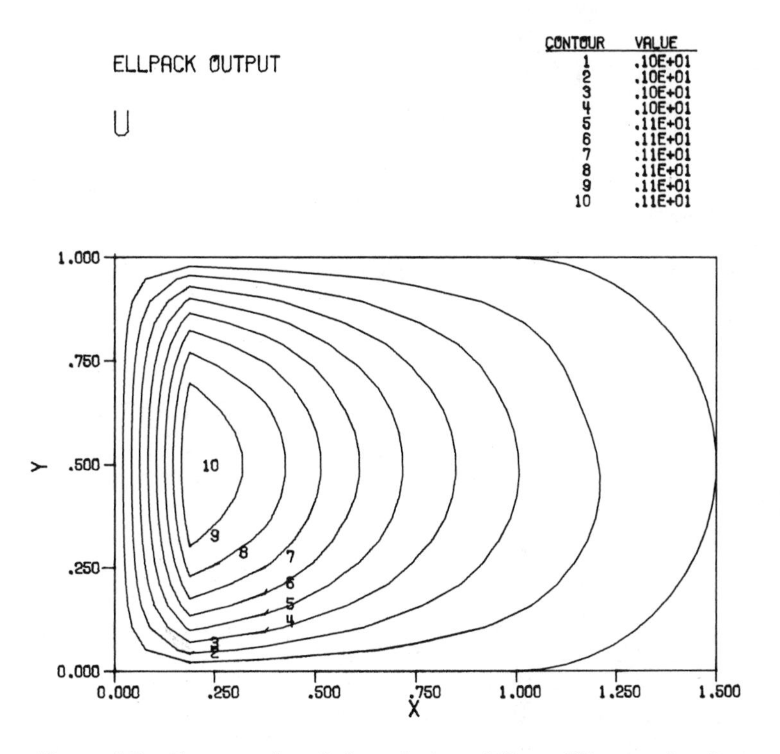

Figure 5.7. Contour plot of the solution of Reynold's equation for % c ^ = ^ 12 %.

5.D TIME DEPENDENT PROBLEM

ELLPACK can be used fairly directly for the following time dependent problem on a two-dimensional domain R:

$$
\begin{array}{lll}
u_t = \mathrm{L}u + f(x,y,t) & \text{on } \mathrm{R}, & t > 0 \\
u = g(x,y,t) & \text{on } \partial\mathrm{R}, & t > 0 \\
u = u_0(x,y) & \text{on } \mathrm{R} \cup \partial\mathrm{R}, & t = 0
\end{array}
$$

where L is a linear elliptic operator; for example,

$$
\mathrm{L}u = 4y^2 u_{xx} + u_{yy} + [2 + \tan(x + y + t)]\,u_y + u
$$

Note that the coefficients in L can depend on x, y and t. ELLPACK does not automatically discretize the u_t term, so this must be done in the program explicitly. The simplest discretization is

$$u_t(t-\Delta t) \approx \frac{u(t)-u(t-\Delta t)}{\Delta t}$$

which leads to the discrete equation

$$u(x,y,t) = u(x,y,t-\Delta t) + \Delta t[\mathrm{L}\, u(x,y,t-\Delta t) + f(x,y,t-\Delta t)]$$

ELLPACK can be used to discretize the L $u(x,y,t-\Delta t\,)$ term, but this is not an attractive use of the ELLPACK facilities. It is better in most cases to use the more accurate and stable Crank-Nicolson time discretization [Richtmyer and Morton, 1967].

The **Crank-Nicolson discretization** uses the same approximation to u_t, but it is viewed as estimating u_t at $t-\Delta t/2$ instead of at $t-\Delta t$. The partial differential equation is then discretized to be

$$u(x,y,t) = u(x,y,t-\Delta t)$$
$$+\Delta t[\mathrm{L}u(x,y,t) + f(x,y,t) + \mathrm{L}u(x,y,t-\Delta t) + f(x,y,t-\Delta t)]/2$$

This discretization in time is unconditionally stable so that time steps Δt are not restricted by stability considerations. For each time step one must solve the elliptic equation

$$\mathrm{L}u(x,y,t) - (2/\Delta t)u(x,y,t)$$
$$= -(2/\Delta t)u(x,y,t-\Delta t) - \mathrm{L}(x,y,t-\Delta t) - f(x,y,t) - f(x,y,t-\Delta t)$$

The terms on the right are known and on the left we have a linear elliptic operator; thus, this is an elliptic equation which ELLPACK can solve to find $u(x,y,t)$.

Note that any ELLPACK method can be used to solve this problem, but there should be an interaction between the method chosen and the choice of Δt. In particlar, one should choose Δt so that the spatial and temporal discretization errors are about the same. To discretize space we choose an x, y grid, and, for simplicity, we assume that x and y spacings are the same, h. We are essentially applying the method of lines with one line (in time) for each grid node. However, we do not need to examine these lines individually or label the corresponding line solutions. The time discretization error from Crank-Nicolson is proportional to $(\Delta t)^2$. If 5-POINT STAR is used the discretization error is proportional to h^2 and so one should have h and Δt of about the same size. At least, if they are decreased, they should be decreased proportionally. If HERMITE COLLOCATION or SPLINE GALERKIN (DEGREE=3,NDERV=2) is used, then since their discretization errors are proportional to h^4 one should have h^2 and Δt about the same size. With fourth order discretizations one can take many fewer time steps for a given accuracy.

We give an ELLPACK program to solve the problem at the beginning of this section. The functions $u_0(x,y)$, $g(x,y,t)$ and $f(x,y,t)$ are chosen so that the true solution is

$$u(x,y,t) = \sin\left(\frac{x+y+t}{4}\right) e^{-y^2-t} \quad \text{for } 1 \leqslant t \leqslant 2,\ 0 \leqslant x,y \leqslant 1.$$

The spatial problem is solved with HERMITE COLLOCATION which uses bicubic Hermite polynomials and has error of order h^4. We set $\Delta t = h^2/2$ and put the elliptic problem in a simple loop for the time steps.

We do not comment on the programming details here because a **template** for solving such problems is given below and the comments there explain most of these details. The bulk of the code is to evaluate $Lu(x,y,t-\Delta t)$ for both the previous time value and the initial conditions (which is a similar, but separate case). A small routine TMXEMX to measure the maximum error is included and the results are listed in Table 5.1 for grids of 3×3 ($\Delta t=1/8$), and 5×5 ($\Delta t=1/32$). The dominant discretization error is in the space variables. The fact that the 5×5 grid computation becomes unstable at $t=2.1$ shows that this cannot be applied blindly.

Table 5.1. Behavior of the error in solving a time dependent problem with Crank-Nicolson and HERMITE COLLOCATION.

	Maximum Relative Error $\times 10^4$ (X,Y)-*Grid*	
Time	3×3	5×5
1 + 1/8	1.9	.21
1 + 1/4	3.0	.13
1 + 3/8	2.9	.22
1 + 1/2	2.7	.20
1 + 5/8	2.4	.17
1 + 3/4	2.2	.23
1 + 7/8	2.0	.31
2	1.9	.88
2 + 1/4	1.3	6580
2 + 1/2	1.1	1.7×10^5

```
*     ***************************************************************
*     *                                                             *
*     *   EXAMPLE ELLPACK PROGRAM 5.D1                              *
*     *                                                             *
*     *       TIME DEPENDENT PROBLEM                                *
*     *       SEE THE ELLPACK PROGRAM TEMPLATE FOR GENERAL          *
*     *       COMMENTS. COMMENTS ARE GIVEN FOR STATEMENTS           *
*     *       SPECIAL TO THIS PROGRAM.                              *
*     *                                                             *
*     ***************************************************************
OPTIONS.            PAGE = 0   $  TIME
*       DECLARE FORTRAN ARRAYS FOR USE IN SUMARY AT END.
DECLARATIONS.
        REAL TRUMAX(100), ERRMAX(100)
GLOBAL.
        COMMON / GCOMON / T, DELTAT, NSTEP, HX, HY
EQUATION.       (4.*Y**2)UXX + UYY + (2.+TAN((X+Y+T)/4.))UY &
                 + (1.-2./DELTAT)U = PDERS(X,Y)
BOUNDARY.       U = G(X,Y)   ON X = 0.
                             ON X = 1.
                             ON Y = 0.
                             ON Y = 1.
GRID.           3 X POINTS $ 3 Y POINTS
FORTRAN.
        TSTART = 1.
        TSTOP  = 2.
        HY     = 0.5
        HX     = 0.5
```

```
C                 CHOOSE DELTA T = (DELTA X)**2 OVER 2
         DELTAT = HX**2/2
         NSTEPS = INT((TSTOP-TSTART)/DELTAT + .5)
         DELTAT = (TSTOP-TSTART)/NSTEPS
C                 LOOP FOR TIME STEPPING
         DO 10 NSTEP = 1, NSTEPS
               T = TSTART + NSTEP*DELTAT
DISCRETI.      INTERIOR COLLOCATION
SOLUTION.      BAND GE
FORTRAN.
C              COMPUTE MAX ERROR, SAVE FOR SUMMARY OUTPUT AT END
               CALL TMXEMX(TRUMAX(NSTEP),ERRMAX(NSTEP))
C              SUPPRESS OUTPUT AFTER FIRST ITERATION
               I1LEVL = 0
   10    CONTINUE
C           PRINT SUMMARY OF RESULTS FOR THIS EXAMPLE
         CALL SUMARY(TRUMAX,ERRMAX,TSTART,NSTEPS)
SUBPROGRAMS.
      FUNCTION PDERS(X,Y)
         COMMON / GCOMON / T, DELTAT, NSTEP, HX, HY
         T = T - DELTAT
         IF (NSTEP .EQ. 1) THEN
               UOFT = U0(X,Y)
            ELSE
               UOFT = U(X,Y)
         ENDIF
         PDERS = - (2./DELTAT)*UOFT - (RLUXYT(X,Y) + F(X,Y,T))
     A            - F(X,Y,T+DELTAT)
         T = T + DELTAT
         RETURN
      END
      FUNCTION RLUXYT(X,Y)
         REAL COEFOF(6)
         COMMON / GCOMON / T, DELTAT, NSTEP, HX, HY
         INTEGER CUXX, CUXY, CUYY, CUX, CUY, CU
         DATA    CUXX, CUXY, CUYY, CUX, CUY, CU
     A          /   1,    2,    3,   4,   5,  6/
C
C        OBTAIN PDE COEFFICIENTS FROM ELLPACK
         CALL Q1PCOE(X,Y,COEFOF)
C
         IF (NSTEP .EQ. 1) THEN
               RLUXYT = COEFOF(CUXX) * U0XX(X,Y)
     A                + COEFOF(CUYY) * U0YY(X,Y)
     B                + COEFOF(CUY)  * U0Y(X,Y)
     C                + (COEFOF(CU) + 2./DELTAT) * U0(X,Y)
            ELSE
               RLUXYT = COEFOF(CUXX) * UXX(X,Y)
     A                + COEFOF(CUYY) * UYY(X,Y)
     B                + COEFOF(CUY)  * UY(X,Y)
     C                + (COEFOF(CU) + 2./DELTAT) * U(X,Y)
         ENDIF
         RETURN
      END
      FUNCTION F(X,Y,T)
         T1 = .25*(X+Y+T)
         T2 = EXP(-Y**2-T)
         F = - (.25*COS(T1) - 2.*Y*SIN(T1)) * T2 * (TAN(T1)+2.)
     A       + (.0625 - 3.75*Y**2) * T2 * SIN(T1)
     B       + (.25 + Y) * T2 * COS(T1)
         RETURN
      END
      FUNCTION U0XX(X,Y)
         COMMON / GCOMON / T, DELTAT, NSTEP, HX, HY
```

```
            U0XX = -( EXP(-Y**2-T) * SIN((X+Y+T)/4.))/16.
            RETURN
         END
         FUNCTION U0YY(X,Y)
            COMMON / GCOMON / T, DELTAT, NSTEP, HX, HY
            U0YY = EXP(-Y**2-T) * ((4.*Y**2-2.0625)*SIN((X+Y+T)/4.)
     A             - Y*COS((X+Y+T)/4.) )
            RETURN
         END
         FUNCTION U0Y(X,Y)
            COMMON / GCOMON / T, DELTAT, NSTEP, HX, HY
            U0Y =   EXP(-Y**2-T)/4. * (COS((X+Y+T)/4.)
     A            - 8.*Y*SIN((X+Y+T)/4.))
            RETURN
         END
         FUNCTION U0(X,Y)
            COMMON / GCOMON / T, DELTAT, NSTEP, HX, HY
            U0 = SIN((X+Y+T)/4.) * EXP(-Y**2-T)
            RETURN
         END
         FUNCTION G(X,Y)
            COMMON / GCOMON / T, DELTAT, NSTEP, HX, HY
            UBOUND = TRUE(X,Y)
           RETURN
         END
         FUNCTION TRUE(X,Y)
            COMMON / GCOMON / T, DELTAT, NSTEP, HX, HY
            TRUE = SIN((X+Y+T)/4.) * EXP(-Y**2-T)
            RETURN
         END
         SUBROUTINE TMXEMX (TRUMAX, ERRMAX)
C
C THIS ROUTINE FINDS THE MAX ABSOLUTE VALUE OF TRUE AND ERROR
C ON A 5 BY 5 GRID AT THE CURRENT TIME STEP.
C
         TRUMAX = 0.
         ERRMAX = 0.
         DO 20 I = 1, 5
            X = I/6.
            DO 10 J = 1, 5
               Y = J/6.
               TRUXYT = TRUE(X,Y)
               TRUMAX = AMAX1(TRUMAX,TRUXYT)
               ABSERR = ABS(TRUXYT-U(X,Y))
               ERRMAX = AMAX1(ERRMAX,ABSERR)
   10       CONTINUE
   20    CONTINUE
            RETURN
            END
         SUBROUTINE SUMARY (TRUMAX, ERRMAX, TSTART, NSTEPS)
C
C THIS ROUTINE PRINTS A TABLE OF SOLUTION AND RELATIVE ERROR
C AT EACH TIME STEP.  THESE VALUES HAVE BEEN SAVED IN THE
C ARRAYS TRUMAX AND ERRMAX.
C
         REAL TRUMAX(1), ERRMAX(1)
C
C ACCESS  INFORMATION FROM ELLPACK PROGRAM
         COMMON / GCOMON / T, DELTAT, NSTEP, HX, HY
C
C PRINT PROBLEM/METHOD INFORMATION
C
         TSTOP = TSTART + NSTEPS*DELTAT
         PRINT 100,HX, HY, TSTART, TSTOP, DELTAT
C
```

```
C PRINT HEADING
C
      PRINT 110
      DO 10 NSTEP = 1, NSTEPS
         T = TSTART + NSTEP*DELTAT
         IF (TRUMAX(NSTEP) .NE. 0.) THEN
               PRINT 120, NSTEP, T, TRUMAX(NSTEP),
     A                    ERRMAX(NSTEP)/TRUMAX(NSTEP)
            ELSE
               PRINT 120, NSTEP, T, 0.
         ENDIF
   10 CONTINUE
      RETURN
  100 FORMAT('1      TIME DEPENDENT PROBLEM'//
     A        T7,'HX      =',1P1E12.4/
     B        T7,'HY      =',1P1E12.4/
     C        T7,'TSTART  =',1P1E12.4/
     D        T7,'TSTOP   =',1P1E12.4/
     E        T7,'DELTA T =',1P1E12.4//)
  110 FORMAT(T8,'STEP',T18,'TIME',T28,'MAX TRUE',T39,
     A        'MAX RELERR'/T7,6('-'),3(2X,10('-')))
  120 FORMAT(T8,I4,1X,1P3E12.4)
      END
END.
```

Output of ELLPACK Run:

```
---------------------
 PREPROCESSOR OUTPUT
---------------------
    2.17 SECONDS TO INITIALIZE PREPREPROCESSOR
    1.53 SECONDS TO PARSE ELLPACK PROGRAM
    4.60 SECONDS TO CREATE ELLPACK CONTROL PROGRAM
    8.30 SECONDS TOTAL

         ELLPACK VERSION --- APRIL 5, 1984

-----------------------
 DISCRETIZATION MODULE
-----------------------

    I N T E R I O R   C O L L O C A T I O N
    DOMAIN                                  RECTANGLE
    X INTERVAL               0.    E+00, 0.100E+01
    Y INTERVAL               0.    E+00, 0.100E+01
    GRID                                      3 X  3
    HX                                      0.500E+00
    HY                                      0.500E+00
    OUTPUT LEVEL                                    1
    NUMBER OF EQUATIONS                            16
    MAX NO. OF UNKNOWNS PER EQ.                    16
    EQUATIONS ARE NOT SYMETRIC
    EXECUTION SUCCESSFUL

-----------------
 SOLUTION MODULE
-----------------

   B A N D   G E

   NUMBER OF EQUATIONS         16
   LOWER BANDWIDTH              9
   UPPER BANDWIDTH              9
   REQUIRED WORKSPACE         464
   EXECUTION SUCCESSFUL

     TIME DEPENDENT PROBLEM
```

```
HX       =  5.0000E-01
HY       =  5.0000E-01
TSTART   =  1.0000E+00
TSTOP    =  2.0000E+00
DELTA T  =  1.2500E-01
  STEP        TIME        MAX TRUE     MAX RELERR
 ------    ----------    ----------    ----------
      1    1.1250E+00    1.5997E-01    2.7435E+00
      2    1.2500E+00    1.4861E-01    4.2489E+00
      3    1.3750E+00    1.3758E-01    4.4835E+00
      4    1.5000E+00    1.2698E-01    5.1813E+00
      5    1.6250E+00    1.1685E-01    5.7713E+00
      6    1.7500E+00    1.0726E-01    6.5438E+00
      7    1.8750E+00    9.8209E-02    7.4035E+00
      8    2.0000E+00    8.9723E-02    8.3283E+00

       +++++++++++++++++++++++++++
       +                         +
       +    EXECUTION   TIMES    +
       +                         +
       +++++++++++++++++++++++++++

          SECONDS      MODULE NAME
             0.20      INTERIOR COLLOCATION
             0.05      BAND GE SETUP
             0.05      BAND GE
             0.17      INTERIOR COLLOCATION
             0.03      BAND GE SETUP
             0.05      BAND GE
             0.18      INTERIOR COLLOCATION
             0.02      BAND GE SETUP
             0.05      BAND GE
             0.18      INTERIOR COLLOCATION
             0.02      BAND GE SETUP
             0.07      BAND GE
             0.18      INTERIOR COLLOCATION
             0.02      BAND GE SETUP
             0.07      BAND GE
             0.17      INTERIOR COLLOCATION
             0.02      BAND GE SETUP
             0.05      BAND GE
             0.23      INTERIOR COLLOCATION
             0.02      BAND GE SETUP
             0.05      BAND GE
             0.17      INTERIOR COLLOCATION
             0.02      BAND GE SETUP
             0.05      BAND GE
             2.62      TOTAL TIME
```

General Template

We end this section with a general **template for solving time dependent problems** in ELLPACK. The template is heavily commented to explain its use.

```
*    ELLPACK TIME DEPENDENT PROBLEM TEMPLATE
*    U  = LU + F(X,Y,T)
*     T
*    U = U0(X,Y) FOR T = TSTART,  0 < X,Y < 1
*    U = UBOUND(X,Y,T) FOR TSTART < T < TSTOP,  (X,Y) ON BOUNDARY
*    WHERE
*       L IS A LINEAR ELLIPTIC OPERATOR
*       U0 SPECIFIES THE INITIAL VALUES
*       UBOUND SPECIFIES THE BOUNDARY VALUES
```

```
* GLOBAL.  COMMON BLOCK GIVES FUNCTIONS ACCESS TO CURRENT TIME T,
*          TIME SPACING DELTAT, AND CURRENT STEP NUMBER NSTEP.
GLOBAL.
        COMMON / GCOMON / T, DELTAT, NSTEP
* EQUATION.  DEFINE EQUATION FOR EACH TIME T.
*            L IS THE LINEAR OPERATOR.
*            DEFINE RIGHT SIDE PDERS(X,Y) BELOW.
EQUATION.     LU - (2./DELTAT)U = PDERS(X,Y)
* BOUNDARY.  SPECIFY BOUNDARY VALUES.  DEFINE UBOUND(X,Y) BELOW.
BOUNDARY.     U = UBOUND(X,Y) ON X = 0.
                              ON X = 1.
                              ON Y = 0.
                              ON Y = 1.
* GRID.  CHOOSE GRID LINES FOR PROBLEM.
GRID.         5 X POINTS $ 5 Y POINTS
FORTRAN.
C     FORCE ELLPACK TO EVALUATE COEFFICIENTS OF L FOR EACH TIME T
C     IF SOME COEFFICIENTS DEPEND ON T BUT NOT X OR Y.
         L1CSTC = .FALSE.
C
C     SET TSTART, TSTOP.  SET DELTAT, DEPENDING ON DISCRETIZATION
C     METHOD.   IN THIS EXAMPLE, DELTAT IS SET TO HX (THE ELLPACK
C     VARIABLE R1HXGR).  COMPUTE NSTEPS (NUMBER OF STEPS), THEN
C     RECOMPUTE DELTAT TO MAKE THE STEPS COME OUT EVEN.
C
        I1LEVL = 1
        TSTART = 0.
        TSTOP  = 1.
        DELTAT = R1HXGR
        NSTEPS = INT((TSTOP-TSTART)/DELTAT + .5)
        DELTAT = (TSTOP-TSTART)/NSTEPS
C
C MAIN LOOP OVER TIME.  T IS THE TIME FOR THE CURRENT STEP.
C
        DO 10 NSTEP = 1, NSTEPS
                 T = TSTART + NSTEP*DELTAT
C
C CHOOSE MODULES.  ONE OF MANY POSSIBLE COMBINATIONS IS SHOWN.
C
DISCRETIZATION.    5 POINT STAR
SOLUTION.          LINPACK BAND
FORTRAN.
C
C SET LEVEL=0 TO AVOID REPEATED OUTPUT FROM EVERY TRIP THRU LOOP
C
                 I1LEVL = 0
   10   CONTINUE
C
C SUBPROGRAMS. DEFINE PDERS, RLUXYT, INITIAL AND BOUNDARY VALUES
C
SUBPROGRAMS.
      FUNCTION PDERS(X,Y)
C
C THIS FUNCTION EVALUATES THE PDE'S RIGHT SIDE AT CURRENT TIME T
C NOTE T IS PASSED IN GCOMON.
C PDERS =    (-2/DELTAT)*U(X,Y,T -DELTAT) - LU(X,Y,T -DELTAT)
C           - F(X,H,T-DELTAT) - F(X,Y,T)
C
C VARIABLES:
C    X,Y     -   SPACE VARIABLES AT WHICH TO EVALUATE RIGHT SIDE
C    T       -   TIME AT WHICH TO EVALUATE RIGHT SIDE
C    DELTAT  -   TIME SPACING
C    UOFT    -   TEMPORARY VARIABLE; HOLDS U(X,Y) AT LAST TIME T.
C    PDERS   -   RETURNED VALUE OF RIGHT SIDE
```

```
C
      COMMON / GCOMON / T, DELTAT, NSTEP
C
C NEED U, LU, AND F AT (X,Y,T -DELTAT).  MOVE TIME T BACK 1 STEP
C SO ALL FUNCTIONS ARE EVALUATED AT THE PREVIOUS TIME STEP.
C
      T = T - DELTAT
C
C FIND U(X,Y,T -DELTAT); IT'S EITHER U0(X,Y) FOR THE INITIAL
C STEP, OR U(X,Y) WHERE U IS THE ELLPACK FUNCTION WHICH GIVES
C THE RESULT AT THE PREVIOUS TIME STEP.
C
      IF (NSTEP .EQ. 1) THEN
         UOFT = U0(X,Y)
          ELSE
         UOFT = U(X,Y)
      ENDIF
C
C EVALUATE RIGHT SIDE USING RLUXYT FOR LU AT PREVIOUS TIME STEP
C
      PDERS = - (2./DELTAT)*UOFT
     A        - (RLUXYT(X,Y) + F(X,Y,T))
     B        - F(X,Y,T +DELTAT)
C
C RESTORE T TO CURRENT VALUE.
C
      T = T + DELTAT
      RETURN
      END
      FUNCTION RLUXYT(X,Y)
C
C THIS FUNCTION EVALUATES LU(X,Y,T).  NOTE THAT T IS PASSED IN
C GCOMON, AND THAT (2/DELTAT) MUST BE ADDED TO THE COEFFICIENT
C OF U BECAUSE ELLPACK THINKS THE (-2/DELTAT)U IS PART OF LU.
C
C VARIABLES:
C    X,Y                -  SPACE VARIABLES AT WHICH TO EVALUATE LU
C    COEFOF(6)          -  COEFFICIENTS OF L, EVALUATED AT TIME T
C    T                  -  TIME AT WHICH TO EVALUATE LU
C    NSTEP              -  CURRENT STEP
C    CUXX,CUXY,...      -  INDICES INTO COEFOF
C    RLUXYT             -  RETURNED VALUE OF LU(X,Y,T)
C
      REAL COEFOF(6)
      COMMON / GCOMON / T, DELTAT, NSTEP
      INTEGER CUXX, CUXY, CUYY, CUX, CUY, CU
      DATA    CUXX, CUXY, CUYY, CUX, CUY, CU
     A        /  1,    2,    3,   4,   5,  6/
C
C CALL ELLPACK ROUTINE Q1PCOE TO EVALUATE THE COEFFICIENTS
C OF THE PDE AT TIME T AND FILL COEFOF.
C
      CALL Q1PCOE(X,Y,COEFOF)
C
C IF ON 1ST STEP, NEED INITIAL VALUES (U0 AND ITS DERIVATIVES).
C OMIT TERMS WITH ZERO COEFFICIENTS IN AN ACTUAL CASE.
C
         IF (NSTEP .EQ. 1) THEN
              RLUXYT = COEFOF(CUXX) * U0XX(X,Y)
     A                + COEFOF(CUXY) * U0XY(X,Y)
     B                + COEFOF(CUYY) * U0YY(X,Y)
     C                + COEFOF(CUX)  * U0X(X,Y)
     D                + COEFOF(CUY)  * U0Y(X,Y)
     E                + (COEFOF(CU) + 2./DELTAT) * U0(X,Y)
C
```

```
C ELSE, NEED RESULTS OF PREVIOUS TIME STEP (U,UX,UY,...).
C OMIT TERMS WITH ZERO COEFFICIENTS IN AN ACTUAL CASE.
C
          ELSE
                RLUXYT = COEFOF(CUXX)  * UXX(X,Y)
     A                 + COEFOF(CUXY)  * UXY(X,Y)
     B                 + COEFOF(CUYY)  * UYY(X,Y)
     C                 + COEFOF(CUX)   * UX(X,Y)
     D                 + COEFOF(CUY)   * UY(X,Y)
     E                 + (COEFOF(CU) + 2./DELTAT) * U(X,Y)
      ENDIF
      RETURN
      END
C
C DEFINE THE FUNCTION F.
C
      FUNCTION F(X,Y,T)
      F = . . .
      RETURN
      END
C
C DEFINE INITIAL VALUE U0 AND NECESSARY DERIVATIVES.
C OMIT DERIVATIVE FUNCTIONS NOT APPEARING IN PDE.
C
      FUNCTION U0XX(X,Y)
      COMMON / GCOMON / T, DELTAT, NSTEP
      U0XX = . . .
      RETURN
      END
      FUNCTION U0XY(X,Y)
      COMMON / GCOMON / T, DELTAT, NSTEP
      U0XY = . . .
      RETURN
      END
      FUNCTION U0YY(X,Y)
      COMMON / GCOMON / T, DELTAT, NSTEP
      U0YY = . . .
      RETURN
      END
      FUNCTION U0X(X,Y)
      COMMON / GCOMON / T, DELTAT, NSTEP
      U0X = . . .
      RETURN
      END
      FUNCTION U0Y(X,Y)
      COMMON / GCOMON / T, DELTAT, NSTEP
      U0Y = . . .
      RETURN
      END
      FUNCTION U0(X,Y)
      COMMON / GCOMON / T, DELTAT, NSTEP
      U0 = . . .
      RETURN
      END
C
C         DEFINE THE BOUNDARY VALUES UBOUND(X,Y,T).
C         NOTE THAT T IS PASSED IN GCOMON.
C
      FUNCTION UBOUND(X,Y)
      COMMON / GCOMON / T, DELTAT, NSTEP
      UBOUND = . . .
      RETURN
      END
END.
```

5.E SYSTEMS OF ELLIPTIC PROBLEMS

We consider the solution of k simultaneous elliptic problems in the unknown functions $u_j(x,y)$, $j=1,\dots,k$ in which the i-th problem can be written as a quasilinear elliptic boundary value problem in the unknown function u_i. We attempt to solve such problems using the following algorithm:

```
Initialize u_i, i = 1,2,...,k
REPEAT
    FOR i = 1 TO k DO
        Solve equation i as a linear problem for u_i using current values of u_j,
          j = 1,...,k
        Save computed u_i
    END FOR
UNTIL converged
```

This iterative procedure is similar to Gauss-Seidel iteration for a set of linear algebraic equations. As in the linear algebra analogue, the iteration might or might not converge, and the convergence is directly affected by how the equations and unknowns are ordered. The advantage of this procedure is that a single elliptic problem is solved at each step, which is exactly what ELLPACK does automatically.

We note that an alternative approach has been used. This alternative may be summarized as follows:

```
Initialize u_i, i = 1,...,k
REPEAT
    (a) FOR i = 1,...,k DO
            Discretize the ith equation using an ELLPACK module using u_j,
            j = 1,...,k from the last iteration) and save the resulting
            coefficient matrix.
    (b) Combine all coefficient matrices (using Fortran code)
    (c) Solve the resulting (large) linear algebraic system using an ELLPACK
        SOLUTION module
UNTIL converged
```

This approach requires considerably more knowledge of the internal operation of the ELLPACK system than the approach we describe here.

In order to cycle through the solution of k separate elliptic problems we introduce an equation index, KEQN, as a Fortran variable in our ELLPACK program. For example, the system of two equations

$$v_{xx} + (1+w)v_{yy} + y(1-w)v = g_1$$
$$(1+v)w_{xx} + w_{yy} + y(1+w)w = g_2$$

may be written as

$$a(x,y)u_{xx} + c(x,y)u_{yy} + f(x,y)u = g(x,y)$$

where

$$a(x,y)=\begin{cases}1 & \text{if KEQN}=1\\ 1+v(x,y) & \text{if KEQN}=2\end{cases}$$

$$c(x,y)=\begin{cases}1+w(x,y) & \text{if KEQN}=1\\ 1 & \text{if KEQN}=2\end{cases}$$

$$f(x,y)=\begin{cases}y(1-w(x,y) & \text{if KEQN}=1\\ y(1+v(x,y) & \text{if KEQN}=2\end{cases}$$

$$g(x,y)=\begin{cases}g_1(x,y) & \text{if KEQN}=1\\ g_2(x,y) & \text{if KEQN}=2\end{cases}$$

$$u(x,y)=\begin{cases}v(x,y) & \text{if KEQN}=1\\ w(x,y) & \text{if KEQN}=2\end{cases}$$

In ELLPACK these equations can be written as

```
EQUATION.   A(X,Y)*UXX + C(X,Y)*UYY + F(X,Y)*U = G(X,Y)
```

where Fortran functions for A, C, F, and G are placed in the SUBPROGRAMS segment. These functions have the general form

```
FUNCTION A(X,Y)
COMMON / SWITCH / KEQN
IF (KEQN .EQ. 1) THEN
     A = UXX COEFFICIENT OF FIRST EQUATION
   ELSE
     A = UXX COEFFICIENT OF SECOND EQUATION
ENDIF
RETURN
END
```

The Fortran variable KEQN is passed to each of these functions via a Fortran COMMON block (named SWITCH here). The ELLPACK program must also have this block declared as follows.

```
DECLARATIONS.
      COMMON / SWITCH / KEQN
```

All user-defined functions in the EQUATION and BOUNDARY segments can be switched from problem to problem in this way. The technique applies straightforwardly to any number of equations.

In order to complete our definition of the function A(X,Y) we need a function which returns the current estimate of $w(x,y)$, the second solution component. In general we will need functions which return values of each of the solution components and possibly their partial derivatives. The built-in ELLPACK functions U(X,Y), UX(X,Y), etc. refer to the solution of the most recently solved elliptic problem, and we do not have direct access to them in order to insert the switch KEQN. Even if these programs were available, one would find it difficult to modify them because of their inherent complexity. Instead, we will create our own copies of the computed solution and manipulate them independently of U(X,Y), etc., In particular, we will create functions U1(X,Y), U2(X,Y), ..., Uk(X,Y) which return the most recently computed values of $u_1(x,y)$, $u_2(x,y)$, ..., $u_k(x,y)$, respectively. If the i-th PDE contains derivative terms of other than the i-th unknown function, or if the derivatives of the i-th unknown function appear nonlinearly, then we will also need to create functions such as U1X, U1Y, U2X, etc., which return the derivatives of solution components. We illustrate the procedure for the case of two equations and two unknown functions in two dimensions. The extension to k functions or three dimensions is straightforward.

The internal representation of the solution in ELLPACK is a one-dimensional array called R1UNKN. The meaning of the individual elements of R1UNKN varies with each discretization or triple module. For finite element modules R1UNKN contains the computed coefficients of an expansion of the solution in some finite set of basis functions, while for finite difference modules R1UNKN contains computed estimates of u at some subset of the grid points. Each discretization and triple module provides a Fortran function which evaluates the solution and its derivatives for any (x,y) in the domain given the array R1UNKN. Finite difference modules may use ELLPACK interpolation utilities to do this. The input vector for use by these routines must be in a standard format. Since many modules do not use this format for R1UNKN, they provide an initialization subprogram which loads a table, called R1TABL, in this standard format. The array R1TABL is then passed to the interpolation utilities.

In either case, to save several distinct computed solutions, we must save copies of the arrays R1UNKN or R1TABL. We allocate arrays UNKN1 and UNKN2 in Fortran COMMON blocks SAVEU1 and SAVEU2 to hold these solutions and provide the following utility for saving the KEQN-th copy when it is computed. The following subroutine saves these arrays.

```
      SUBROUTINE SAVE (UNKN,NUNKN)
C
C     SAVE UNKN AS THE KEQN-TH SOLUTION COMPONENT
C
      DIMENSION UNKN(NUNKN)
C
      COMMON / SWITCH / KEQN
      COMMON / SAVEU1 / UNKN1(1)
      COMMON / SAVEU2 / UNKN2(1)
C
      IF (KEQN .EQ. 1) THEN
            DO 10 I = 1,NUNKN
            UNKN1(I) = UNKN(I)
   10       CONTINUE
      ELSE
            DO 20 I = 1,NUNKN
            UNKN2(I) = UNKN(I)
   20       CONTINUE
      ENDIF
      RETURN
      END
```

The actual parameters passed to this routine, as well as the sizes of the COMMON blocks SAVEU1 and SAVEU2 depend upon whether the discretization or triple module uses R1UNKN or R1TABL. Table 5.2 lists each of these modules and the array that it uses.

When R1UNKN is used, then the following DECLARATION statements are included in the ELLPACK program.

Table 5.2. Internal names of ELLPACK subprograms for evaluating $u(x,y)$

Discretization	*Must save*	*Evaluation of $u(x,y)$*
5 POINT STAR (rectangular domain)	R1TABL	Initialization CALL Q35PVL values = R1QD2I(X,Y,R1TABL,IDERIV)
5 POINT STAR (general domain)	R1TABL	Initialization CALL Q35GVL Initialization CALL Q2XTMN(R1TABL) values = R1QD2I(X,Y,R1TABL,IDERIV)
7 POINT 3D	R1TABL	Initialization CALL Q37PVL values = R1QD3I(X,Y,Z,R1TABL,IDERIV)
CMM1	R1TABL	values = R1QD2I(X,Y,R1TABL,IDERIV)
CMM2	R1TABL	values = R1QD2I(X,Y,R1TABL,IDERIV)
CMM3	R1TABL	values = R1QD2I(X,Y,R1TABL,IDERIV)
COLLOCATION	R1UNKN	values = R3CGEV(X,Y,R1UNKN,IDERIV)
DYAKANOV CG	R1TABL	values = R1QD2I(X,Y,R1UNKN,IDERIV)
DYAKANOV CG 4	R1TABL	values = R1QD2I(X,Y,R1UNKN,IDERIV)
FFT 9 POINT	R1TABL	values = R1QD2I(X,Y,R1TABL,IDERIV)
FISHPAK HELMHOLTZ	R1TABL	values = R1QD2I(X,Y,R1TABL,IDERIV)
HERMITE COLLOCATION	R1UNKN	values = R3H1EV(X,Y,R1UNKN,IDERIV)
HODIE	R1TABL	Initialization CALL Q3HDVL values = R1QD2(X,Y,R1UNKN,IDERIV)
HODIE 27 POINT 3D	R1TABL	values = R1QD2I(X,Y,R1UNKN,IDERIV)
HODIE HELMHOLTZ	R1TABL	Initialization CALL Q3HHVL values = R1QD2I(X,Y,R1TABL,IDERIV)
HODIE FFT	R1TABL	Initialization CALL Q7H2VL values = R1QD2(X,Y,R1TABL,IDERIV)
HODIE FFT 3D	R1TABL	Initialization CALL Q7H3VL values = R1QD2I(X,Y,Z,R1TABL,IDERIV)
INTERIOR COLLOCATION	R1UNKN	values = R3H0EV(X,Y,R1UNKN,IDERIV)
MARCHING ALGORITHM	R1TABL	values = R1QD2I(X,Y,R1UNKN,IDERIV)
MULTIGRID MG00	R1TABL	values = R1QD2I(X,Y,R1UNKN,IDERIV)
P2C0 TRIANGLES	R1TABL	values = R1QD2I(X,Y,R1UNKN,IDERIV)
SET U	R1TABL	values = R1QD2I(X,Y,R1TABL,IDERIV)
SET U BY BICUBICS	—	values = R6HBVL(X,Y,IDERIV) (does not generate R1UNKN or R1TABL)
SET U BY BLENDING	—	values = R7IBVL(X,Y,IDERIV) (does not generate R1UNKN or R1TABL)
SPLINE GALERKIN	R1UNKN	values = R3SGPR(X,Y,R1UNKN,IDERIV)

```
DECLARATIONS.
        COMMON / SWITCH  / KEQN
        COMMON / SAVEU1  / UNKN1($I1MUNK)
        COMMON / SAVEU2  / UNKN2($I1MUNK)
```

The notation $I1MUNK tells the ELLPACK preprocessor to allocate the same space for these arrays as for the array R1UNKN (see Section 4.C). We then include the following Fortran statement after the SOLUTION or TRIPLE segment.

```
FORTRAN.
        CALL  SAVE(R1UNKN, I1NEQN)
```

The ELLPACK variable I1NEQN gives the actual number of elements in R1UNKN.

When R1TABL is used the following DECLARATIONS are included in the ELLPACK program.

```
DECLARATIONS.
        COMMON / SWITCH / KEQN
        COMMON / SAVEU1 / UNKN1($I1NGRX,$I1NGRY)
        COMMON / SAVEU2 / UNKN2($I1NGRX,$I1NGRY)
```

The notation $I1NGRX,$I1NGRY tells the ELLPACK preprocessor to allocate the same space for these arrays as for the array R1TABL (see Section 4.C). We then include the following Fortran statements after the SOLUTION or TRIPLE segment.

```
FORTRAN.
        CALL ROUTINE TO LOAD R1TABL,  IF NECESSARY
        CALL SAVE(R1TABL, I1NGRX*I1NGRY)
```

The actual number of elements in R1TABL is given by the expression I1NGRX*I1NGRY which involves two internal ELLPACK variables. The name of the subprogram which loads R1TABL can be found in Table 5.2. Note that some TRIPLE modules provide this array directly so no explicit loading is required.

The functions U1 and U2 are now easy to write. Suppose, for example, that the HERMITE COLLOCATION discretization were used. Then we would write

```
      FUNCTION U1 (X,Y)
      COMMON / SAVEU1 / UNKN1(1)
      U1 = R1H1EV(X,Y,UNKN1,6)
      RETURN
      END
```

```
      FUNCTION U2 (X,Y)
      COMMON / SAVEU2 / UNKN2(1)
      U2 = R1H1EV(X,Y,UNKN2,6)
      RETURN
      END
```

The function R1H1EV was found in Table 5.2. The fourth argument of this function is used to select the desired function value or partial derivative to be returned. There is a standard numbering for this in ELLPACK as follows (IDERIV is the number used in R1H1EV to indicate the derivative).

IDERIV	*Value returned*	IDERIV	*Value returned*
1	UXX	6	U
2	UXY	7	UZZ
3	UYY	8	UXZ
4	UX	9	UYZ
5	UY	10	UZ

Thus, if a function returning the first partial with respect to x of the second solution component were required, we would write a function U2X which would return R1H1EV(X,Y,UNKN2,4).

Example 5.E1

We consider the following boundary value problem for the unknown functions $v(x,y)$ and $w(x,y)$.

$$v_{xx} + (1+w)v_{yy} + y(1-w)v = g_1$$
$$(1+v)w_{xx} + w_{yy} + y(1+v)w = g_2$$

For boundary conditions

$$v_x = h_1, \quad w_x = h_3 \quad \text{on } x = 0,1$$
$$v = h_2, \quad w = h_4 \quad \text{on } y = 0,1$$

where g_1, g_2, h_1, h_2, h_3, and h_4 are chosen to make the solution exactly:

$$v(x,y) = (x+y)^4 \qquad w(x,y) = \exp(x-y)$$

We give two complete ELLPACK programs for solving this problem. Example 5.E1 solves the problem using the HERMITE COLLOCATION discretization and Example 5.E2 solves the problem using the 5 POINT STAR discretization. In each case the iterative procedure is started with the initial guesses $v(x,y) = w(x,y) = 0$, and the functions U1 and U2 are modified slightly so that the initial guesses will be used on the first iteration.

```
* * * * * * * * * * * * * * * * * * * * * * * * * * * * * * * * * * * * * * * * * * * * * * * * * * * *
*                                                                                                      *
*          EXAMPLE ELLPACK PROGRAM 5.E1                                                                *
*                                                                                                      *
*          TWO EQUATIONS   - SIMPLE INTERACTION BETWEEN THEM                                           *
*                                                                                                      *
*          L1(V,W)  =          VXX + (1+W)VYY + Y(1-W)V = G1(X,Y)                                      *
*          L2(V,W)  =  (1+V)WXX +          WYY + Y(1+V)W = G2(X,Y)                                     *
*                                                                                                      *
*                     V,W GIVEN ON Y=0,1                                                               *
*                     VX, WX GIVEN ON X=0,1                                                            *
*                                                                                                      *
*          KNOWN SOLUTION IS   V = (X+Y)**4     W = EXP(X-Y)                                           *
*                                                                                                      *
* * * * * * * * * * * * * * * * * * * * * * * * * * * * * * * * * * * * * * * * * * * * * * * * * * * *
*                                                                                                      *
*                          IMPLEMENTATION DETAILS                                                      *
*                          ----------------------                                                      *
*                                                                                                      *
*        ALGORITHM ...                                                                                 *
*                                                                                                      *
*             GUESS W                                                                                  *
*             REPEAT                                                                                   *
*                 SOLVE PROBLEM 1 FOR V                                                                *
*                 SOLVE PROBLEM 2 FOR W                                                                *
*             UNTIL CONVERGED                                                                          *
*                                                                                                      *
*        FUNCTIONS ...                                                                                 *
*                                                                                                      *
*             U1    = EVALUATES COMPUTED V FROM UNKN1                                                  *
*             U2    = EVALUATED COMPUTED W FROM UNKN2                                                  *
```

```
*          U01   = INITIAL GUESS FOR V                                *
*          U02   = INITIAL GUESS FOR W                                *
*                                                                     *
*      VARIABLES ...                                                  *
*                                                                     *
*          KEQN    = EQUATION SWITCH                                  *
*          UNKN1   = VECTOR FOR SAVING COMPUTED V                     *
*          UNKN2   = VECTOR FOR SAVING COMPUTED W                     *
*          INIT    = INITIALIZATION SWITCH                            *
*                    (IF INIT(I)=0, USE GUESS FOR ITH SOLUTION)       *
*                                                                     *
*      CONSTANTS ...                                                  *
*                                                                     *
*          NEQNS  = NUMBER OF EQUATIONS                               *
*          NITERS = NUMBER OF ITERATIONS                              *
*                                                                     *
***********************************************************************
* SOLUTION WITH HERMITE COLLOCATION / BAND GAUSS ELIMINATION          *
***********************************************************************
OPTION.          TIME     $    PAGE = 0     $ LEVEL = 0
DECLARATIONS.
      PARAMETER (NEQNS=2,NITERS=7)
      COMMON / SWITCH / KEQN
      COMMON / INITU  / INIT(NEQNS)
      COMMON / SAVEU1 / UNKN1($I1MNEQ)
      COMMON / SAVEU2 / UNKN2($I1MNEQ)
EQUATION.
            A(X,Y)*UXX + C(X,Y)*UYY + F(X,Y)*U = G(X,Y)
BOUNDARY.
            UX = TRUEX(X,Y)    ON X=0.0
            UX = TRUEX(X,Y)    ON X=1.0
            U  = TRUE (X,Y)    ON Y=0.0
            U  = TRUE (X,Y)    ON Y=1.0
GRID.     5 X POINTS
          5 Y POINTS
FOR.
C         ---------------
C         INITIALIZATIONS
C         ---------------
C
        DO 10 I=1,NEQNS
           INIT(I) = 0
  10    CONTINUE
C
C         -------------------
C         MAIN ITERATION LOOP
C         -------------------
C
        DO 200 NITER = 1, NITERS
           DO 100 KEQN = 1,NEQNS
              PRINT 20, NITER,KEQN
  20          FORMAT(//' - - - ITERATION ',I2,3X,
     +                 ' - - - EQUATION ',I1)
C
C              -------------------------
C              SOLVE KEQN-TH PDE PROBLEM
C              -------------------------
DIS.           HERMITE COLLOCATION
SOL.           BAND GE
FOR.
```

```
C                 ------------------------------------
C                 UPDATE  SAVED  SOLUTION  AND  RESULTS
C                 ------------------------------------
C
                CALL  SAVE(R1UNKN, I1NEQN)
                INIT(KEQN)  =  1
OUT.             MAX(ERROR)
FOR.
 100          CONTINUE
 200       CONTINUE
C
C       -------------
C       PLOT  SOLUTION
C       -------------
C
OUT.    PLOT(U1)
        PLOT(U2)
SUBPROG.
C                                                       -----------------
C                                                       PDE  COEFFICIENTS
C                                                       -----------------
      FUNCTION A(X,Y)
      COMMON  /  SWITCH  /  KEQN
      IF(  KEQN  .EQ.  1  )  THEN
                A  =  1.
         ELSE
                A  =  1.0  +  U1(X,Y)
      ENDIF
      RETURN
      END
      FUNCTION C(X,Y)
      COMMON  /  SWITCH  /  KEQN
      IF(  KEQN  .EQ.  1  )  THEN
                C  =  1.  +  U2(X,Y)
         ELSE
                C  =  1.0
      ENDIF
      RETURN
      END
      FUNCTION F(X,Y)
      COMMON  /  SWITCH  /  KEQN
      IF  (KEQN  .EQ.  1)  THEN
                F  =  (1.0  -  U2(X,Y))*Y
      ELSE
                F  =  (1.0  +  U1(X,Y))*Y
      ENDIF
      RETURN
      END
      FUNCTION G(X,Y)
      COMMON  /  SWITCH  /  KEQN
      V  =  (X+Y)**4
      W  =  EXP(X-Y)
      VXX  =  12.0*(X+Y)**2
      IF  (KEQN  .EQ.  1)  THEN
                G  =  (2.0+W)*VXX  +  (1.0-W)*Y*V
      ELSE
                G  =  (1.0  +  (1.0+Y)*(1.0+V))*W
      ENDIF
      RETURN
      END
C                                                       ---------------
C                                                       INITIAL  GUESSES
C                                                       ---------------
```

```
      FUNCTION U01(X,Y)
               U01 = 0.0
      RETURN
      END
      FUNCTION U02(X,Y)
               U02 = 0.0
      RETURN
      END
C                                                 ---------------
C                                                 KNOWN SOLUTION
C                                                 ---------------
      FUNCTION TRUE(X,Y)
      COMMON / SWITCH / KEQN
      IF (KEQN .EQ. 1) THEN
               TRUE = (X+Y)**4
      ELSE
               TRUE = EXP(X-Y)
      ENDIF
      RETURN
      END
      FUNCTION TRUEX(X,Y)
      COMMON / SWITCH / KEQN
      IF (KEQN .EQ. 1) THEN
               TRUEX = 4.0*(X+Y)**3
      ELSE
               TRUEX = EXP(X-Y)
      ENDIF
      RETURN
      END
C
C----------------------------------------------------------------------
C
C  UTILITIES FOR SOLVING SYSTEMS OF EQUATIONS WITH ELLPACK
C
C          CASE :  2 EQUATIONS
C                  HERMITE COLLOCATION DISCRETIZATION
C
C----------------------------------------------------------------------
C
      SUBROUTINE SAVE(UNKN,NUNKN)
C
C      SAVE UNKN AS KEQN-TH SOLUTION COMPONENT
C
      DIMENSION UNKN(NUNKN)
C
      COMMON / SWITCH / KEQN
      COMMON / SAVEU1 / UNKN1(1)
      COMMON / SAVEU2 / UNKN2(1)
C
      IF( KEQN .EQ. 1 )  THEN
         DO 10 I = 1,NUNKN
            UNKN1(I) = UNKN(I)
   10    CONTINUE
      ELSE
         DO 20 I = 1,NUNKN
            UNKN2(I) = UNKN(I)
   20    CONTINUE
      ENDIF
      RETURN
      END
      FUNCTION  U1(X,Y)
C
C      RETURNS VALUE OF FIRST SOLUTION COMPONENT
C
```

```
      COMMON / INITU  / INIT(2)
      COMMON / SAVEU1 / UNKN1(1)
      IF (INIT(1) .EQ. 0) THEN
         U1 = U01(X,Y)
      ELSE
         U1 = R3H1EV(X,Y,UNKN1,6)
      ENDIF
      RETURN
      END
      FUNCTION  U2(X,Y)
C
C       RETURNS VALUE OF SECOND SOLUTION COMPONENT
C
      COMMON / INITU  / INIT(2)
      COMMON / SAVEU2 / UNKN2(1)
      IF (INIT(2) .EQ. 0) THEN
         U2 = U02(X,Y)
      ELSE
         U2 = R3H1EV(X,Y,UNKN2,6)
      ENDIF
      RETURN
      END
END.
```

Example 5.E2

The following program solves the same elliptic problem as Example 5.E2. The HERMITE COLLOCATION discretization is replaced by 5 POINT STAR.

```
******************************************************************
*                                                                *
*       EXAMPLE ELLPACK PROGRAM 5.E2                             *
*                                                                *
*          ... SEE COMMENTS IN ELLPACK PROGRAM 5.E1              *
*                                                                *
******************************************************************
*  SOLUTION WITH  5 POINT STAR   /   BAND GAUSS ELIMINATION      *
******************************************************************

OPTION.           TIME     $    PAGE = 0     $ LEVEL = 0

DECLARATIONS.
      PARAMETER (NEQNS=2,NITERS=7)

      COMMON / SWITCH / KEQN
      COMMON / INITU  / INIT(NEQNS)
      COMMON / SAVEU1 / UNKN1($I1NGRX,$I1NGRY)
      COMMON / SAVEU2 / UNKN2($I1NGRX,$I1NGRY)

EQUATION.
              A(X,Y)*UXX + C(X,Y)*UYY + F(X,Y)*U = G(X,Y)
BOUNDARY.
              UX = TRUEX(X,Y)    ON X=0.0
              UX = TRUEX(X,Y)    ON X=1.0
              U  = TRUE (X,Y)    ON Y=0.0
              U  = TRUE (X,Y)    ON Y=1.0

GRID.     17 X POINTS
          17 Y POINTS
FOR.
C
C         ---------------
C         INITIALIZATIONS
C         ---------------
C
```

```
        DO 10 I=1,NEQNS
           INIT(I) = 0
  10    CONTINUE
C
C        -------------------
C        MAIN ITERATION LOOP
C        -------------------
C
        DO 200 NITER = 1, NITERS
           DO 100 KEQN = 1,NEQNS
              PRINT 20, NITER,KEQN
  20          FORMAT(//' - - - ITERATION ',I2,3X,
    +                 ' - - - EQUATION ',I1)
C
C              -------------------------
C              SOLVE KEQN-TH PDE PROBLEM
C              -------------------------
DIS.           5 POINT STAR
SOL.           BAND GE
FOR.
C              ----------------------------------
C              UPDATE SAVED SOLUTION AND RESULTS
C              ----------------------------------
C
C              ... FIRST LOAD R1TABL
              CALL Q35PVL
C
              CALL SAVE(R1TABL,I1NGRX*I1NGRY)
              INIT(KEQN) = 1
OUT.          MAX(ERROR)
FOR.
 100       CONTINUE
 200    CONTINUE
C
C      -------------
C      PLOT SOLUTION
C      -------------
C
OUT.   PLOT(U1)
       PLOT(U2)
SUBPROG.
C                                                ----------------
C                                                PDE COEFFICIENTS
C                                                ----------------
        ... SAME AS EXAMPLE 5.E1 ...
C                                                ---------------
C                                                INITIAL GUESSES
C                                                ---------------
        ... SAME AS EXAMPLE 5.E1 ...
C                                                 --------------
C                                                 KNOWN SOLUTION
C                                                 --------------
        ... SAME AS EXAMPLE 5.E1 ...
C
C-----------------------------------------------------------------------
C
C  UTILITIES FOR SOLVING SYSTEMS OF EQUATIONS WITH ELLPACK
C
C           CASE :  2 EQUATIONS
C                   5 POINT STAR DISCRETIZATION
C
C-----------------------------------------------------------------------
C
```

```
      SUBROUTINE SAVE (UNKN,NUNKN)
C
C       ... SAME AS EXAMPLE 5.E1 ...
C
      END
      FUNCTION  U1(X,Y)
C
C       RETURNS VALUE OF FIRST SOLUTION COMPONENT
C
      COMMON / INITU   / INIT(2)
      COMMON / SAVEU1  / UNKN1(1)
      IF (INIT(1) .EQ. 0) THEN
         U1 = U01(X,Y)
      ELSE
         U1 = R1QD2I(X,Y,UNKN1,6)
      ENDIF
      RETURN
      END
      FUNCTION  U2(X,Y)
C
C       RETURNS VALUE OF SECOND SOLUTION COMPONENT
C
      COMMON / INITU   / INIT(2)
      COMMON / SAVEU2  / UNKN2(1)
      IF (INIT(2) .EQ. 0) THEN
         U2 = U02(X,Y)
      ELSE
         U2 = R1QD2I(X,Y,UNKN2,6)
      ENDIF
      RETURN
      END
END.
```

The following results are obtained when these ELLPACK programs are run on a Sperry 1100/82 computer (about eight decimal digits of precision) using the FTN compiler with options OZ.

	HERMITE COLLOCATION		5 POINT STAR	
Iteration	*v-error*	*w-error*	*v-error*	*w-error*
1	2.88	0.421	2.90	0.429
2	0.361	0.0456	0.372	0.0469
3	0.0356	0.00472	0.0414	0.00538
4	0.00359	0.000508	0.00840	0.00107
5	0.000301	0.0000711	0.00502	0.000626
6	0.000109	0.0000271	0.00467	0.000580
7	0.000109	0.0000227	0.00463	0.000575

In both cases the convergence appears linear for the first five iterations, the error being reduced by a factor of about 10 at each step. Subsequent iterations fail to reduce the error further. The error that remains is the error in discretizing the PDEs. This error can only be made smaller by selecting a finer grid or a more accurate discretization method.

To further illustrate the behavior of this Gauss-Seidel like iteration we list the results of applying it to a variety of other problems. All the examples use HERMITE COLLOCATION (with 4×4 grid) and BAND GE and are defined on $0 \leqslant x, y \leqslant 1$ unless otherwise noted. All initial guesses are $u = v = 0$ unless otherwise noted. All these examples were run in single precision on a VAX 11/780, which uses about 7 decimal digits in the arithmetic.

Example 5.E3

$$u_{xx}+(1+v)u_{yy}+(v+y)u_x+y(1-v)u+(1-y)v_x=g_1$$
$$(1+u)v_{xx}+v_{yy}+(u+x)v_x+y(1+u)v-xu_y=g_2$$
$$\text{solution:}\quad u(x,y)=(x+y)^4 \qquad v(x,y)=e^{x-y}$$

Dirichlet boundary conditions.
Grid is 4 X POINTS $ 4 Y POINTS

Iteration	*u-error*	*v-error*
0	—	2.1
1	.693	.125
2	.120	.0302
3	.033	.00889
4	.0096	.00269
5	.00313	.00084
6	.00115	.000289
7	.00057	.000121
8	.00043	.000072

This example has coupling in derivative terms; they are placed on the right side along with g_1 and g_2. The maximum error is measured on a 10 by 10 interior grid rather than on the grid of the discretization.

Example 5.E4.

$$(1+2v)u_{xx}+(4-v)u_{yy}+(v+4y)u_x+xvu=g_1$$
$$(1+3u)v_{xx}+(1-u/40)v_{yy}+(4u+x)v_x+x^2yv=g_2$$
$$\text{solution:}\quad u(x,y)=(x+y)^4 \qquad v(x,y)=e^{x-y}$$

Dirichlet boundary conditions
Grid is 4 X POINTS $ 4 Y POINTS

Iteration	*u-error*	*v-error*
0	—	1.39
1	.717	.127
2	.0265	.00464
3	.000616	.000243
4	.000317	.0000861
5	.000319	.0000809
6	.000319	.0000805

Example 5.E5.

Same problem as Example 5.E4 with the boundary conditions at $x=0$ changed to the nonlinear boundary conditions.

$$(1+v)u+u_x = h_1 \qquad (1+u)v+uv_x = h_3$$

Iteration	*u-error*	*v-error*
0	—	—
1	.113	.0837
2	.00537	.00507
3	.000386	.00223
4	.000121	.00224
5	.000125	.00224

Example 5.E6

$$u_{xx} + [1+(\sin 2v)/2]\,u_{yy} + (v+4y)\,u_x + e^{vxy}\,u = g_1$$
$$v_{xx} + [1+u\sin(2xu)]v_{yy} + (4u+x)\,v_x + (\sin^2 u)\,yv = g_2$$

solution: $u(x,y) = (x+y)^4 \qquad v(x,y) = e^{x-y}$

Dirichlet boundary conditions
Grid is 4 X POINTS $ 4 Y POINTS

Iteration	*u-error*	*v-error*
0	—	2.72
1	70.8	1.78
2	8.68	2.15
3	6.68	1.34
4	19.08	.353
5	14.23	1.79
6	15.77	.511

This example shows that the simple iteration proposed here does not always converge. One would conjecture that it is due more to the highly nonlinear nature of the equations than due to the fact that there are two of them.

Example 5.E7

$$(1+w)\,u_{xx} + u_{yy} + (v+w)\,u_x + yxu = g_1$$
$$v_{xx} + (1+w)\,v_{yy} + (u+x)\,v_x + x^2yv = g_2$$
$$(1+u)\,w_{xx} + (1+v)\,w_{yy} + (u+v)\,w_x + xy^3w = g_3$$

solution: $u = (x+y)^2 \qquad v = e^{x-y} \qquad w = e^{x/2-y/4}$

Boundary conditions are

$$
\begin{aligned}
u, v, w \quad & \text{given for } g = 0, 1 \\
u_x, v_x, w_x \quad & \text{given for } x = 1 \\
u_x + u, v_x + v, w_x + w \quad & \text{given for } x = 0
\end{aligned}
$$

Initial guess: $u = x + y$, $v = 1$, $w = 1 + (x - y)/6$
Grid is 3 X POINTS $ 3 Y POINTS

Iteration	*u-error*	*v-error*	*w-error*
1	.137	.0358	.00596
2	.00342	.000415	.000166
3	.0000844	.000260	.0000099
4	.00000091	.000271	.000013
5	.0000024	.000271	.000013

5.F REFERENCES

Ames, W. F. [1965], *Nonlinear Partial Differential Equations in Engineering*, Academic Press, New York.

Birkhoff, G. and R. E. Lynch [1984], *Numerical Solution of Elliptic Problems*, SIAM, Philadelphia.

Ortega, J. M. and W. C. Rheinboldt [1970], *Iterative Solution of Nonlinear Equations in Several Variables*, Academic Press, New York.

Rand, R. H. [1984], *Computer Algebra in Applied Mathematics — An Introduction to MACSYMA*, Pitman Publishing, London.

Richtmyer, R. D. and K. W. Morton [1967], *Difference Methods for Initial Value Problems*, Second edition, Interscience, New York.

PART 2

CHAPTERS 6–7

THE ELLPACK MODULES

This part provides reference information about the 50 plus elliptic problem solving modules in ELLPACK. There is a general overview of the ITPACK software for iteration methods.

Chapter 6

THE ELLPACK MODULES

6.A INTRODUCTION

The ELLPACK language and system described so far is only half of the story; the other half is the heart and muscle of ELLPACK, the ELLPACK **modules.** No problem solving system is better than its underlying programs. The design of ELLPACK allows the collection of modules to grow or shrink, so a particular ELLPACK system may have more or fewer modules than discussed in Part 2 of this book, Chapters 6 and 7.

In 1984 the complete ELLPACK system had over 50 modules — so many that some users will find it difficult to choose among them. Part 1 of this book is written with reference to a smaller set, about 18 modules that constitute the basic set. This set includes the most well-known methods as well as one example of each "variety" of problem solving module. Part 3 of this book (Chapters 8, 9, 10 and 11) illustrates the performance of many ELLPACK modules on a set of nine model problems. This performance data gives some guidance in choosing modules for a particular problem, but one must keep in mind that it is not possible to predict reliably the relative performance of the modules for any untested problem. This is particularly so if a problem has some unusual features, as most real problems do.

Within the ELLPACK system there are two important subcollections of solution modules, the ITPACK software and the YSMP (Yale Sparse Matrix Package) software. Chapter 7 presents an overall view of the design, capabilities and methods in the ITPACK package.

The information about ELLPACK modules given in this chapter is intended to be a summary of the modules' properties and restrictions. However, there is not enough space to completely describe the design and methods for each module, so references are given to more detailed descriptions. The information for each module was written by the authors of that module, except for a few standard programs or simple methods that have been incorporated into ELLPACK. The format for each module description is:

Module The name of the module.

Author The name of the author or the source of the module.

Purpose A brief statement of what the module does.

Method A brief summary of the method used.

Mathematical formulation	Further details about the problems solved and methods used.
Parameters	Definition of the parameters (arguments) of the module.
Restrictions	Summary of the restrictions on the applicability of the method or module.
Performance estimates	Indicators of the amount of computer resources one can expect the module to use.
Acknowledgements	
References	

The descriptions are put into six groups, three of which — **DISCRETIZATION, INDEXING and SOLUTION** — correspond to a modular viewpoint for solving elliptic problems as illustrated in Figure 6.1. The fourth group describes **TRIPLES,** which are modules that solve an elliptic problem entirely by themselves. Triples correspond to methods where it is either inefficient or unnatural to divide the problem solutions into three separate phases. The next group describes **PROCEDURES,** which do not correspond to a step in solving the elliptic problem, but rather to some supporting computations. Examples include computing matrix eigenvalues (perhaps to analyze the convergence of an iterative method), displaying the pattern of nonzeros in a matrix, and initializing the unknowns (perhaps to initiate some iteration method). The final group describes **SYSTEM** facilities, which are easily identifiable components of the ELLPACK system. These facilities include the domain processor, the interpolation procedures and the plotting routine.

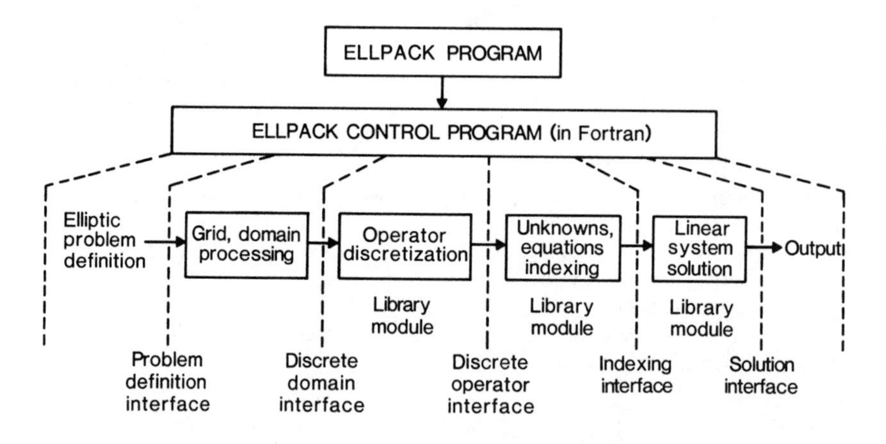

Figure 6.1. Modular viewpoint for solving elliptic problems with ELLPACK. The interfaces between the modules are precisely defined, which allows modules to be used in various combinations by moving back and forth to interfaces.

6.B DESCRIPTIONS OF THE MODULES

All of the modules available in the complete ELLPACK system are described briefly in this chapter. These modules are listed below. The numbers are not used in ELLPACK itself; they provide a simple designation of the modules.

DISCRETIZATION

1. COLLOCATION (general domain)
2. HERMITE COLLOCATION
3. HODIE
4. HODIE HELMHOLTZ
5. INTERIOR COLLOCATION
6. SPLINE GALERKIN
7. 5 POINT STAR
8. 5 POINT STAR (general domain)
9. 7 POINT STAR 3D

INDEXING

20. AS IS
21. HERMITE COLLORDER
22. INTERIOR COLLORDER
23. MINIMUM DEGREE
24. NESTED DISSECTION
25. RED BLACK
26. REVERSE CUTHILL MCKEE

SOLUTION

30. BAND GE
31. BAND GE NO PIVOTING
32. ENVELOPE LDLT
33. ENVELOPE LDU
34. LINPACK BAND
35. LINPACK SPD BAND
36. SOR
37. JACOBI SI
38. JACOBI CG
39. SYMMETRIC SOR SI
40. SYMMETRIC SOR CG
41. REDUCED SYSTEM SI
42. REDUCED SYSTEM CG
43. SPARSE GE NO PIVOTING
44. SPARSE LDLT
45. SPARSE LU COMPRESSED
46. SPARSE LU PIVOTING
47. SPARSE LU UNCOMPRESSED

TRIPLE

60. CMM EXPLICIT (capacitance matrix method)
61. CMM HIGHER ORDER (capacitance matrix method)
62. CMM IMPLICIT (capacitance matrix method)
63. DYAKANOV CG
64. DYAKANOV CG 4
65. FFT 9 POINT
66. FISHPAK HELMHOLTZ
67. HODIE FFT
68. HODIE FFT 3D
69. HODIE 27 POINT 3D
70. MARCHING ALGORITHM
71. MULTIGRID MG00
72. P2C0 TRIANGLES (general domain)
73. SET
74. SET U BY BICUBICS
75. SET U BY BLENDING

PROCEDURE

80. DISPLAY MATRIX PATTERN
81. DOMAIN FILL
82. EIGENVALUES
83. LIST MODULES
84. NONUNIQUE
85. PLOT COLLOCATION POINTS
86. REMOVE
87. REMOVE BICUBIC BC
88. REMOVE BLENDED BC
89. SET UNKNOWNS FOR HODIE HELMHOLTZ
90. SET UNKNOWNS FOR 5 POINT STAR

SYSTEM

100. DOMAIN PROCESSOR
101. INTERPOLATION
102. PLOT

The modular viewpoint of solving elliptic problems used in ELLPACK is designed for the easy interchange of software modules for performing each of the steps required to solve a problem. Practically, however, each module places restrictions on the types of problems it solves. Thus, one cannot always use modules of a given type interchangably. The most reliable way of checking module compatibility is to read the detailed descriptions given in this chapter. We do present several tables which summarize this information in what follows.

PROBLEM-DISCRETIZATION COMPATIBILITY

Each discretization or triple module in ELLPACK will successfully discretize the Dirichlet problem for Poisson's equation on a rectangle, but if the problem has a complicated domain or boundary condition one may find only a very small number of applicable modules. Table 6.1 indicates some of the common restrictions that apply to the use of discretization and triple modules.

MODULE SEQUENCING COMPATIBILITY

Table 6.2 lists the more common legal combinations of discretization, indexing, and solution modules in ELLPACK. Module numbers in parentheses indicate modules which have restrictions which are not always satisfied by the discretization; for example, a solution module may require a symmetric matrix, but the discretization only produces a symmetric matrix when the PDE is in self-adjoint form. One should read the module description in detail before using these. Note, however, that when they are applicable these modules can often be considerably more efficient than those that do not take advantage of the special property of the problem. The fourth column lists other solution modules which can be used, but which are not necessarily efficient for use with the discretization; for example, if the discretization module always produces a symmetric matrix, then one should use a symmetric system solver rather than one that takes no advantage of symmetry.

The most common module incompatibility occurs when an indexing or solution module requires a symmetric matrix, but the discretization module does not produce one. Table 6.3 lists discretization modules which produce symmetric discretizations and Table 6.4 indicates indexing and solution modules which require a symmetric matrix. Note that the ITPACK solution modules do not formally require symmetry, but convergence is not guaranteed unless the system is symmetric positive definite.

Table 6.1 Problem restrictions for the discretization and triple modules. The codes are as follows.

On domain	A = two-dimensional problem
	B = three-dimensional problem
	C = rectangular domain
	D = general domain
On PDE	E = nonself-adjoint form
	F = self-adjoint form
	G = constant coefficient PDE
	H = Poisson equation
	I = Helmholtz equation
	J = no mixed derivatives, e.g., u_{xy}
	K = no low order derivatives, e.g., u_x
On boundary conditions	L = Dirichlet conditions
	M = Uncoupled conditions
	N = No tangential derivative components
	O = constant coefficient mixed
	P = no periodic conditions
Other	Q = uniform grid in x, y (z) separately
	R = uniform grid with $h_x = h_y$ ($= h_z$)
	S = problem must have unique solution

					Comments
COLLOCATION	AD	E	P	S	no holes, arcs
HERMITE COLLOCATION	AC	E	P	S	
HODIE	ABC	E	L	Q	
HODIE HELMHOLTZ	AC	I	NO	Q	
INTERIOR COLLOCATION	AC	E	P		
SPLINE GALERKIN	AC	F	LP		
5 POINT STAR	AC	J	N		
5 POINT STAR	AD	EJ	P		
7 POINT 3D	BC	F	NP		
CMM HIGHER ORDER	AD	H	NP	R	$1{+}2^k$ X grid
CMM EXPLICIT	AD	I	P	R	$1{+}2^k$ X grid
CMM IMPLICIT	AD	I	NP	R	$1{+}2^k$ X grid
DYAKANOV CG	AC	F	P	Q	
DYAKANOV CG 4	AC	F	P	Q	odd X grid
FFT 9 POINT	AC	GJK	LP	Q	$1{+}2^k$ X and Y grids
FISHPAK HELMHOLTZ	AC	I	MN	Q	
HODIE FFT	AC	I	MN	QR	
HODIE FFT 3D	BC	I	MN	QR	
HODIE 27 POINT 3D	BC	H	LP	R	
MARCHING ALGORITHM	AC	F	P	Q	separable problem
MULTIGRID MG00	AC	EJK	P	Q	
P2C0 TRIANGLES	AD	F	P		
SET					
SET U BY BYCUBICS	AC		MP		
SET U BY BLENDING	AC		NOP		

Table 6.2 Common module combinations.

Dis.	*Indexing*	*Solution*	*Less common solution*
1	20	30 34 43	
	23	43	
2	20	30 34 43	
	21	30 31 33 34 43 45-47	
3	20 (26)	30 31 (32) 33 34 (35–40) 43 (44) 45–47	
	23 24	43 (44) 45–47	
4	20 (26)	31 (32) 33 (35) (36–40) (44) 45–47	30 34 43
	23 24	(44) 45–47	43
5	20	30 34 43	
	22	30 31 33 34 43 45–47	
6	20 26	31 32 35–40 44	30 34 43
	23 24	44	43 45–47
7	20 (26)	31 (32) 33 (35–40) (44) 45–47	30 34 43
	23 24	(44) 45–47	43
	25	(36–42)	
8	20	30 (31) (33) 34 43 (45–47)	(36–40)
	23 (24)	43 (45–47)	
9	20 26	31 32 35 36–40 44	30 33 34 43 45–47
	23 24	44	45–47
	25	36–42	

Table 6.3 Discretization modules producing symmetric matrices.

Discretization	*Conditions*
HODIE HELMHOLTZ	Constant coefficients, uncoupled boundary conditions
SPLINE GALERKIN	
5 POINT STAR	Rectangular domains, PDE in self-adjoint form
7 POINT STAR	PDE is in self-adjoint form

Table 6.4 Indexing and solution modules requiring symmetry.

Indexing	REVERSE CUTHILL MCKEE
Solution	ENVELOPE LDLT
	LINPACK SPD BAND
	JACOBI CG
	JACOBI SI
	SOR
	SYMMETRIC SOR CG
	SYMMETRIC SOR SI
	REDUCED SYSTEM CG
	REDUCED SYSTEM SI
	SPARSE LDLT

MODULE: COLLOCATION

AUTHOR: William Mitchell

PURPOSE: Discretizes a general elliptic operator with general linear boundary conditions on a general two-dimensional domain.

METHOD: This module implements a finite element method to obtain a Hermite bicubic piecewise polynomial approximation to $u(x,y)$. The coefficients of the approximation are determined to satisfy the elliptic problem exactly at a set of collocation points. Similarly, the boundary conditions are satisfied exactly at a set of collocation points on the boundary.

MATHEMATICAL FORMULATION: The mathematical formulation for COLLOCATION is the same as that for HERMITE COLLOCATION except for the determination of collocation points. In the case of a rectangular domain (given in parametric form as a nonrectangular domain), COLLOCATION reduces to HERMITE COLLOCATION.

1. **Grid definition.** An element is defined to be the region between two adjacent x grid lines and two adjacent y grid lines. Elements which are completely outside the domain are not used. Elements E through which the boundary of the domain R passes are used if and only if the ratio (area of $E \cap R$)/(area of E) is greater than the parameter DSCARE.

2. **Interior collocation points.** For an element completely inside the domain, the four Gauss points are used. For elements through which the boundary passes, a mapping is defined from the element E to $E \cap R$, and the image under this map of the four Gauss points [Rice, 1983] is used. The map is defined by partitioning the boundary of $E \cap R$ into four parts, defining maps from the four sides of the element to these four parts and using a linear blending of the four maps to create a mapping from E to $E \cap R$.

3. **Boundary collocation points.** Boundary collocation points are defined in two passes. In the first pass, intermediate boundary collocation points are placed on the boundary sides of the grid elements. Four points are placed around each node on the boundary of the grid, one in each surrounding element. Those points which are in elements outside the grid are projected onto the boundary sides of the grid and become the intermediate boundary collocation points. In the second pass, a mapping is defined from the boundary sides of the element to the portion of the boundary of the domain associated with the element. The boundary collocation points are the images, under this map, of the intermediate boundary collocation points. An example of the placement of interior and boundary collocation points is illustrated in Figure 6.2.

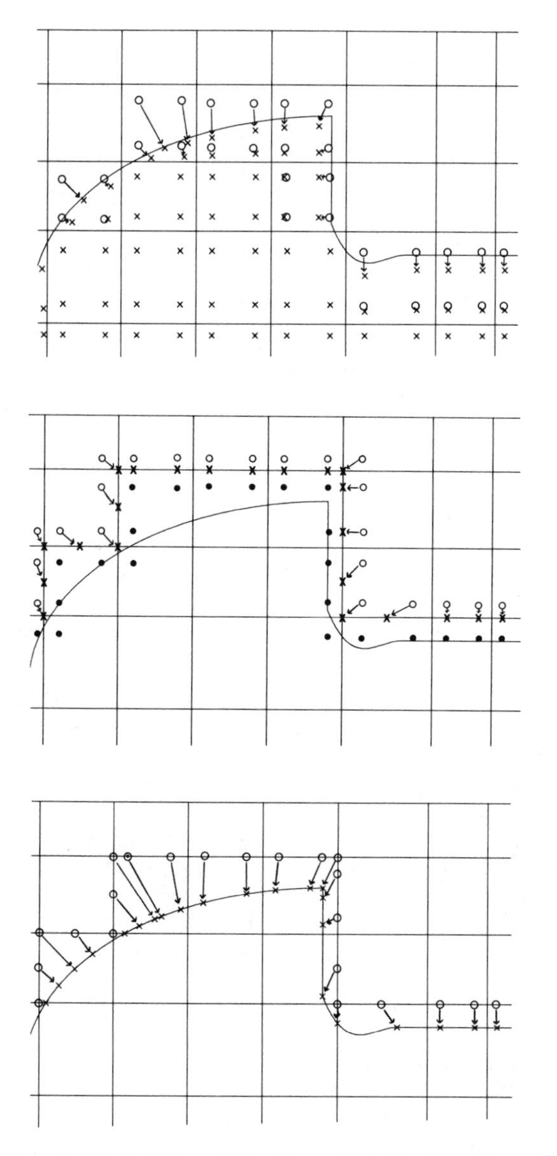

Figure 6.2. Determination of the collocation points for boundary elements.
Top: The mapping of the standard Gauss points (0's) to the interior of the domain (X's).
Middle: The mapping of exterior element Gauss points (0's) to the boundary of the rectangular elements (X's).
Bottom: The mapping to the collocation points (X's) on the boundary of the domain.

PARAMETERS:

BCP1, BCP2 — control placement of boundary collocation points. Boundary collocation points are placed on a boundary side as BCP1 and BCP2 are placed in the interval [0,1]. ($0 \leqslant \text{BCP}1 < \text{BCP}2 \leqslant 1$) and ($\text{BCP}1 \neq 0$ or $\text{BCP}2 \neq 1$). Defaults to Gauss points:

$$\text{BCP}1 = 0.5/(1-\sqrt{3}) \quad \text{and} \quad \text{BCP}2 = 1 - \text{BCP}1$$

DSCARE — the amount of an element that must be interior to the domain in order for the element to be used, $0 < \text{DSCARE} < 1$. Default = .05

GIVOPT — indicates what to do with the portion of the boundary of the domain which is inside an element that is not used.

GIVOPT = 1 (default): divide the portion up and associate subportions with neighboring elements.

GIVOPT = 2: do not use the boundary portion.

USECRN — allows the user to force a boundary collocation point at every corner of the boundary (joining of two pieces)

USECRN = .TRUE.: force collocation at boundary corners

USECRN = .FALSE. (default): no special action.

RESTRICTIONS: The domain must be specified CLOCKWISE and in the nonrectangular syntax with no arcs or holes. The equation cannot be given in self-adjoint form. The grid must be at least 3×3 and the problem must have a unique solution. No element can have all of its sides be boundary sides. At most two boundary pieces can be associated with one element, and an element cannot have exactly two boundary sides unless they are adjacent. If an element is not used in the grid and yet has part of the boundary in it, then no more than two of the neighboring four elements used in the grid can be without any boundary segment associated with them. The boundary can not enter an element more than once, except when it leaves the element and reenters without crossing any other grid line. The procedure NONUNIQUE option cannot be used in conjunction with this module.

PERFORMANCE ESTIMATES: For a domain with an NX by NY grid there are 16 coefficients per equation and at most $4\text{NX} \times \text{NY}$ equations. The matrix bandwidth is at most 4NY. Workspace of size $(\text{NX}-1) \times (\text{NY}-1)$ is used.

With LEVEL $\geqslant 2$, the actual number of equations and boundary points are printed so that the user can save memory space by setting array dimension parameters on future runs with the same domain and grid.

ACKNOWLEDGEMENTS: The development of this module was supported in part by the National Science Foundation under grants MCS77-01408 and MCS79-01437 and by the Department of Energy under contract DE-AC02-81ER-10997.

REFERENCES:

Houstis E. N., W. F. Mitchell and J. R. Rice [1983], "GENCOL: Collocation in general domains by Hermite bicubic polynomials", CSD-TR 444, Department of Computer Sciences, Purdue University.

Houstis E. N., W. F. Mitchell and J. R. Rice [1983a], "Collocation software for second order elliptic partial differential equations", CSD-TR 446, Department of Computer Science, Purdue University.

Rice J. R. [1983], *Numerical Methods, Software and Analysis*, McGraw-Hill, Chapter 10.

MODULE: HERMITE COLLOCATION

AUTHOR: Elias N. Houstis

PURPOSE: Discretizes a general elliptic operator with general linear boundary conditions on a two-dimensional rectangular domain.

METHOD: Let $\mathrm{L}\,u = f$ be the elliptic problem with solution $u(x,y)$, and let $U(x,y)$ denote the approximate solution obtained. This module implements a finite element method to obtain a Hermite bicubic piecewise polynomial approximation to $u(x,y)$. The coefficients are determined by making U satisfy the elliptic problem and the boundary conditions exactly at a set of collocation points. If the problem is well-behaved then the error of this discretization is of the order $\mathrm{HX}^4+\mathrm{HY}^4$ where HX and HY are average grid spacings in the x and y directions.

MATHEMATICAL FORMULATION: The approximate solution $U(x,y)$ is a linear combination of Hermite bicubic polynomials (see below). With $h_k(x)$ and $h_k^1(x)$ denoting the Hermite cubics associated with the k-th x-point of the grid and with $g_j(y)$, $g_j^1(y)$ defined similarly, we have

$$U(x,y) = \sum_{k=1}^{\mathrm{NGRX}} \sum_{j=1}^{\mathrm{NGRY}} a_{k,j} h_k(x) g_j(y) + b_{k,j} h_k(x) g_j^1(y) + c_{k,j} h_k^1(x) g_j(y) + d_{k,j} h_k^1(x) g_j^1(y)$$

where NGRX and NGRY are the number of x and y grid lines, respectively. The coefficients $a_{k,j}$, $b_{k,j}$, $c_{k,j}$, and $d_{k,j}$ are determined so that $U(x,y)$ satisfies $\mathrm{L}\,U = f$ exactly at the interior collocation points and U satisfies the boundary conditions exactly at the boundary collocation points. The tensor product nature of the approximation and the collocation points is exploited to reduce the computations. The boundary condition equations are scaled by an L_1 norm estimate of $\mathrm{L}\,U$; the scale factor is of the order $S = \dfrac{1}{\mathrm{HX}^2} + \dfrac{1}{\mathrm{HY}^2}$ for Dirichlet problems and of the order $\mathrm{HX} \times S$ and $\mathrm{HY} \times S$ for problems with u_x and u_y boundary conditions, respectively.

The Hermite cubic basis functions are defined as follows, where x_k is the k-th x grid point.

$$h_k(x) = \begin{cases} 0 & x \leqslant x_{k-1},\ x_{k+1} \leqslant x \\ -2\left(\dfrac{x - x_{k-1}}{x_k - x_{k-1}}\right)^3 + 3\left(\dfrac{x - x_{k-1}}{x_k - x_{k-1}}\right)^2 & x_{k-1} \leqslant x \leqslant x_k \\ -2\left(\dfrac{x_{k+1} - x}{x_{k+1} - x_k}\right)^3 + 3\left(\dfrac{x_{k+1} - x}{x_{k+1} - x_k}\right)^2 & x_k \leqslant x \leqslant x_{k+1} \end{cases}$$

$$h_k^1(x) = \begin{cases} 0 & x \leqslant x_{k-1},\ x_{k+1} \leqslant x \\ \left(\dfrac{x - x_{k-1}}{x_k - x_{k-1}}\right)^2 (x - x_k) & x_{k-1} \leqslant x \leqslant x_k \\ \left(\dfrac{x_{k+1} - x}{x_{k+1} - x_k}\right)^2 (x - x_k) & x_k \leqslant x \leqslant x_{k+1} \end{cases}$$

The positions and ordering of the collocation points are illustrated in Figure 6.3. The large dots are the boundary condition collocation points and the interior collocation points are indicated only by numbers. In the interior these are the Gauss points [Rice, 1983] for the grid rectangles.

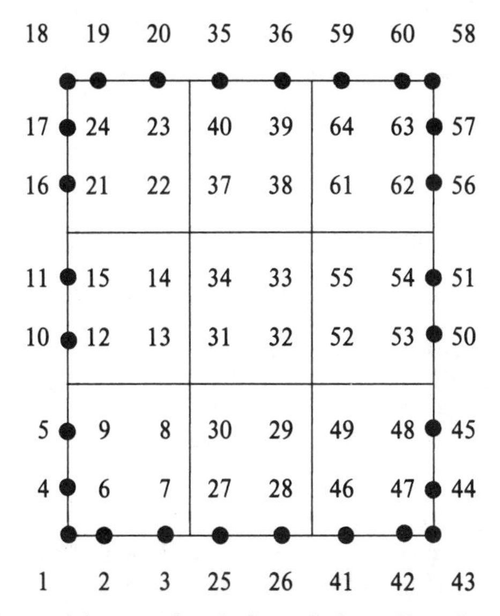

Figure 6.3. The positions and ordering of the collocation points used by HERMITE COLLOCATION for a 4 by 4 grid.

The numbering of the collocation points (equations) is done elementwise roughly from bottom to top and then from left to right. For an NGRX by NGRY grid we have 4NGRX × NGRY collocation points and 4NGRX × NGRY equations. The equations are generated by collocating the operator $Lu = f$ or the boundary conditions in the order in which the collocation points are numbered. In the interior, the $(m+l)$-th equation is in the i, j grid rectangle where $1 \leqslant l \leqslant 4$ and

$$m = 4(i-1)(\text{NGRY}-1) + 4j - 4 + l + \text{COR} + 2(\text{EDGEY} + \text{EDGEX})$$

where COR, EDGEY and EDGEX are the numbers of corners, y grid edges and x grid edges passed, respectively, in the numbering. We have

COR	= 1	for	$i = j = 1$
	2	for	$(i = 1, j = \text{NGRX} - 1)$ or $(2 \leqslant i \leqslant \text{NGRX} - 1)$
	3	for	$i = \text{NGRX} - 1,\ 1 \leqslant j \leqslant \text{NGRY} - 1$
	4	for	$i = \text{NGRX} - 1,\ j = \text{NGRY} - 1$
EDGEY	= j	for	$i = 1$
	NGRY	for	$2 \leqslant i \leqslant \text{NGRX} - 1$
	NGRY+j	for	$i = \text{NGRX} - 1$
EDGEX	= i	for	$j < \text{NGRY} - 1$
	$i+1$	for	$j = \text{NGRY} - 1$

The special case of homogeneous boundary conditions discussed below has

$$m = 4(i-1)(\text{NGRY}-1) + 4j - 4 + l$$

The values of $U(x,y)$ at the grid point (x_k, y_j) are (with $\text{N4} = 4(k-1)\text{NGRY} + j - 1$):

$$U(x_k, y_j) = a_{k,j} = \text{R1UNKN}(\text{N4}+1)$$
$$U_y(x_k, y_j) = b_{k,j} = \text{R1UNKN}(\text{N4}+2)$$
$$U_x(x_k, y_j) = c_{k,j} = \text{R1UNKN}(\text{N4}+3)$$
$$U_{xy}(x_k, y_j) = d_{k,j} = \text{R1UNKN}(\text{N4}+4)$$

These relations can be used to initialize the unknowns for this discretization.

PARAMETERS: BCP1, BCP2 control the placement of boundary collocation points. We have $0 \leqslant \text{BCP1} < \text{BCP2} \leqslant 1$ with $\text{BCP1}(1-\text{BCP2}) \neq 0$. The boundary collocation points are placed on a grid interval of the boundary as BCP1 and BCP2 are placed in [0,1]. If BCP1=0 or BCP2=1 then a special set of points are used at the corners which correspond to BCP1=1/3 and BCP2=2/3. The defaults are the Gauss points, $\text{BCP1}=0.5/(1-\sqrt{3})$ and BCP2=1−BCP1.

SPECIAL CASES: If there are homogeneous boundary conditions (Dirichlet or Neumann, not mixed), then a special version of this module is used which eliminates the basis functions and corresponding equations for the boundary conditions. This makes the approximation satisfy the boundary conditions everywhere and reduces the number of unknowns and equations from 4NGRX×NGRY to 4(NGRX-1)(NGRY-1).

RESTRICTIONS: The problem must have a rectangular domain in two dimensions with at least a 3×3 grid. The equation cannot be given in self-adjoint form. Periodic boundary conditions are not allowed. The procedure NONUNIQUE cannot be used with this module.

PERFORMANCE ESTIMATES: There are 16 coefficients per equation and 4NGRX×NGRY equations (4(NGRX−1)(NGRY−1) for homogeneous boundary conditions). The band width of the equations in the natural ordering is 4NGRY+7 (2NGRY+3 for homogeneous boundary conditions). The workspace required is 52(NGRY+NGRX−2) (28(NGRY−1) for the homogeneous case).

ACKNOWLEDGEMENTS: The development of this module was supported in part by the National Science Foundation under grants MCS77-01408, MCS79-01437 and MCS78-04878, and the US Army Research Office under contract DAAG29-33-K-0026.

REFERENCES:

Houstis E. N., W. F. Mitchell and J. R. Rice [1983], "Algorithms INTCOL and HERMCOL: Collocation on rectangular domains with bicubic Hermite polynomials", CSD-TR 445, Department of Computer Sciences, Purdue University.

Houstis E. N., W. F. Mitchell and J. R. Rice [1983], "Collocation Software for second order elliptic partial differential equations", CSD-TR 446, Department of Computer Sciences, Purdue University.

Rice J. R. [1983], *Numerical Methods, Software and Analysis*, McGraw-Hill, Chapter 10.

MODULE: HODIE

AUTHOR: Robert E. Lynch

PURPOSE: HODIE discretizes the Dirichlet problem for a two- or three-dimensional equation of the form

$$
\begin{aligned}
\mathrm{L}u = \quad & A200\, u_{xx} + A020\, u_{yy} + A002\, u_{zz} \\
+ \; & A110\, u_{xy} + A101\, u_{xz} + A011\, u_{yz} \\
+ \; & A100\, u_x + A010\, u_y + A001\, u_y + A000\, u = G,
\end{aligned}
$$

where the coefficients and G depend on x, y, and z. For the two-dimensional case, u_{zz}, u_{xz}, u_{yz}, u_z should not appear in the EQUATION segment and the coefficients should depend only on x and y. Depending on the user's EQUATION, the preprocessor selects the two- or three-dimensional HODIE program.

METHOD: In the two-dimensional case, the discrete equation is a HODIE equation of the form

$$\sum_{i=0}^{8} \alpha_i\, u(x_i, y_i) = h^2 \sum_{j=1}^{J} \beta_j\, G(x_j, y_j).$$

In the three-dimensional case, either 19 or 27 α's, 7 or 19 β's are used, and the summations are over three variables. For the two dimensional case, if $A110 \neq 0$, $A200 \neq 1$, $A020 \neq 1$, then $O(h^2)$ approximation is obtained. If $A110 = 0$, $A200 \neq 1$, $A020 \neq 1$, then $O(h^4)$ approximation is obtained. If $A110 = 0$, $A200 = 1$, $A020 = 1$, then at least $O(h^4)$ accuracy is obtained; $0(h^6)$ is obtained in certain cases, depending on $A100$ and $A010$, specifically, when it appears to the module that $A100_y$ is equal to $A010_x$. In the three-dimensional case similar accuracies are obtained, depending on whether the cross-derivative terms are nonzero or zero, if $A200$, $A020$, and $A002$ are unity, and so on.

PARAMETERS: IORDER = 2, 4, or 6, for the order of accuracy; default 4.

RESTRICTIONS:

Rectangular [box for three-dimensions] domain
Uniform grid of size at least 3 × 3 [3 × 3 × 3]
Dirichlet boundary conditions
$A200$ and $A020$ [and $A002$] must be nonzero.

ACKNOWLEDGEMENTS: This work was supported in part by National Science Foundation grants MCS77-01408 and MCS79-01437 and in part by Department of Energy contract DE-AC02-81ER-10997.

REFERENCES:

Lynch, R.E. and J. R. Rice [1978], "High accuracy approximations to solutions of elliptic partial differential equations", Proc. Nat. Acad. of Sci. 75, pp. 2541–2544.

DISCRETIZATION

MODULE: HODIE HELMHOLTZ

AUTHORS: Ronald F. Boisvert and John J. Nestor, III

PURPOSE: Discretizes the Helmholtz equation

$$L(u) = u_{xx} + u_{yy} + fu = g(x,y)$$

defined on a rectangular domain R, subject to the boundary conditions

$$\begin{aligned} Mu &= pu_x + su = \phi(x,y) \quad \text{on } R_1 \\ Nu &= qu_y + tu = \psi(x,y) \quad \text{on } R_2 \\ &\text{periodic} \quad \text{on } R_3 \end{aligned}$$

where $R_1 \cup R_2 \cup R_3$ is the boundary of R and f, g, ϕ, and ψ are functions of x and y and f, p, q, s, and t are constants. R_1 may at most consist of the left and right edges of R and R_2 may at most consist of the top and bottom edges of R. If R_3 contains one side of the domain then it must also contain the opposite side.

METHOD: Second, fourth, and sixth order accurate compact finite differences (HODIE). The default is fourth order which may be changed with the parameter IORDER. Note that the higher the order requested, the more restricted the set of problems allowed (see the table below for details).

<table>
<tr><th colspan="2"></th><th colspan="2">Dirichlet/periodic</th><th colspan="2">Mixed</th></tr>
<tr><th>IORDER</th><th>HX=HY</th><th>f constant</th><th>f variable</th><th>f constant</th><th>f variable</th></tr>
<tr><td rowspan="2">2</td><td>TRUE</td><td colspan="4" rowspan="2">(A)</td></tr>
<tr><td>FALSE</td></tr>
<tr><td rowspan="2">4</td><td>TRUE</td><td>(B)</td><td rowspan="2">(C)</td><td>(B)</td><td></td></tr>
<tr><td>FALSE</td><td>(C)</td><td></td><td></td></tr>
<tr><td rowspan="2">6</td><td>TRUE</td><td>(D)</td><td></td><td>(*)</td><td></td></tr>
<tr><td>FALSE</td><td></td><td></td><td></td><td></td></tr>
</table>

(A) Standard second order nine-point differences (see Collatz, 1960]).

(B) The fourth order 9-point compact finite difference technique (HODIE method) described in [Boisvert 1981a]. This is the same discretization as that used by the module HODIE FFT.

(C) The fourth order 9-point compact finite difference technique (HODIE method) called P(h,1,θ,f) in [Boisvert, 1981].

(D) The sixth order 9-point compact finite difference technique (HODIE method) presented in Theorem 8.1 of [Boisvert, 1981].

(*) Denotes problems not addressed by this software.

MATHEMATICAL FORMULATION: This module generates a system of equations in ELLPACK format whose solution yields values of the unknown function u at grid points on R and its boundary. The linear system is obtained by using high order compact difference formulas to approximate the PDE operator and the boundary conditions.

The type of finite difference approximation used depends upon the type of grid, the boundary conditions, and whether f is a constant. The user selects the order of accuracy desired (IORDER = 2, 4, 6), although the higher the order, the more restricted the class of problems. This is summarized in the table on the preceding page. (h_x = grid spacing in x, h_y = grid spacing in y, only uniform grids are allowed). For a discretization of order k, the maximum error at the grid points will be $O(h_x{}^k + h_y{}^k)$, provided the solution is smooth enough (i.e., has all partial derivatives of order k or less continuous). A high order method generally requires many fewer grid points to achieve a given accuracy than a lower order one.

A finite difference equation is written to approximate the differential equation and/or boundary conditions at each active grid point. We define active grid points as those grid points where the value of the solution u is unknown. All grid points are active *except* the following:

(a) points in R_1 or R_2 where p or q are zero,

(b) the point nearest that specified by the ELLPACK procedure NONUNIQUE

(c) the rightmost edge when periodic boundary conditions in x are specified, and

(d) the topmost edge when periodic boundary conditions in y are specified.

In cases (a) and (b) the solution is given explicitly and this value is used to eliminate the unknown associated with this point in the finite difference equations. In cases (c) and (d) the unknown is replaced by an active one using periodicity.

The finite difference approximation of the PDE operator takes the following general form. If p is an **interior point**, we have

$$L_p(u) = I_p(g)$$

where

$$L_p(u) = \sum_{i=1}^{9} \alpha_i u(x_i, y_i), \qquad I_p(g) = h^2 \sum_{j=1}^{13} \beta_j g(x_j, y_j)$$

and the associated point sets are shown at the top of the next page. The point X is the central stencil point. The same set of discretization points is used independent of IORDER. When IORDER = 2, $g(x,y)$ is only evaluated at the central stencil point; when IORDER = 4, all but the four corner points are used. All points are used when IORDER = 6.

If p is on the **right edge (not corner)**, we have

$$L_p(u) = I_p(g) + J_p(\phi)$$

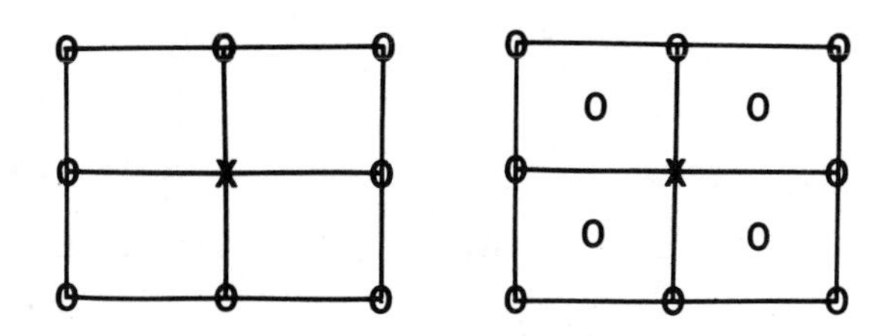

Discretization points (x_i, y_i) Evaluation points (x_j, y_j)

where

$$L_p(u) = \sum_{i=1}^{6} \alpha_i u(x_i, y_i), \qquad I_p(g) = h^2 \sum_{j=1}^{6} \beta_j g(x_j, y_j),$$

$$J_p(\phi) = h \sum_{k=1}^{3} \gamma_k \phi(x_k, y_k),$$

and the associated point sets are:

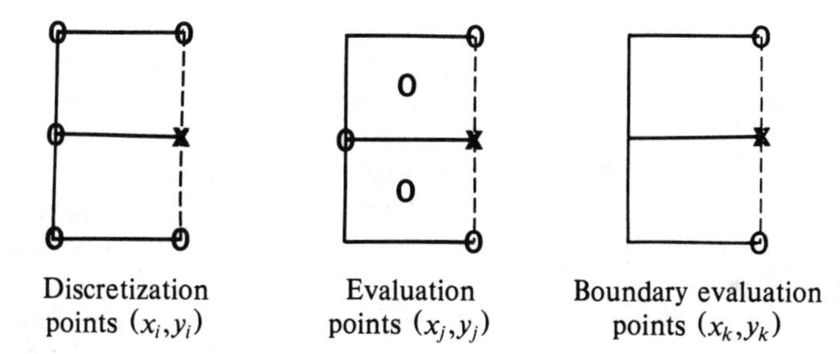

Discretization points (x_i, y_i) Evaluation points (x_j, y_j) Boundary evaluation points (x_k, y_k)

The point X is the central stencil point and the dashed line denotes the boundary. When IORDER = 2 only the point marked X is used as an evaluation and boundary evaluation point.

It p is at the **top right corner**, we have

$$L_p(u) = I_p(g) + J_p(\phi) + K_p(\psi)$$

where

$$L_p(u) = \sum_{i=1}^{4} \alpha_i u(x_i, y_i), \qquad I_p(g) = h^2 \sum_{j=1}^{4} \beta_j g(x_j, y_j)$$

$$J_p(\phi) = h \sum_{k=1}^{4} \gamma_k \phi(x_k, y_k), \qquad K_p(\psi) = h \sum_{l=1}^{4} \delta_l \psi(x_l, y_l)$$

and the associated point sets are:

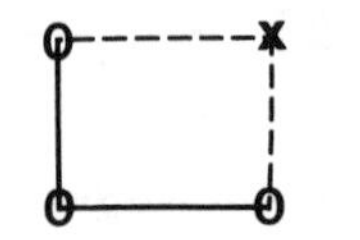

Discretization points (x_k, y_k)

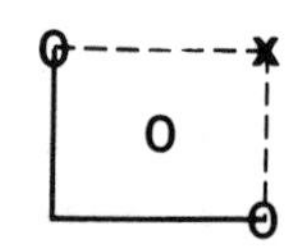

Evaluation points (x_l, y_l)

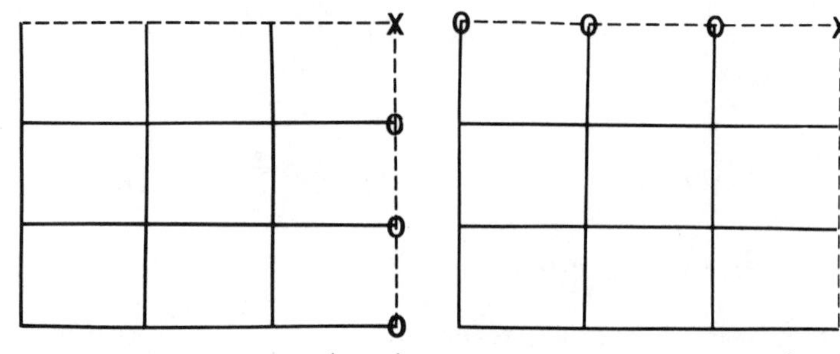

Boundary evaluation points (x_k, y_k) Boundary evaluation points (x_l, y_l)

The point X is the central stencil point and the dashed line denotes the boundary. When IORDER = 2 only the point marked X is used as an evaluation point, and only the point marked X and its two nearest neighbors on the boundary are used as boundary evaluation points.

The α, β, γ, and δ coefficients have been determined so that the discretization is exact on all polynomials whose degree is less than or equal to IORDER + 1 (IORDER for boundary points).

The equations and unknowns are numbered using the natural ordering (row-wise from bottom to top). After HODIE HELMHOLTZ has executed, the numbering of the equation (and unknown) associated with the (i,j)th grid point may be found in I3HHNU(i,j). A value of zero indicates that the point is not active. The array I3HHNU occupies the common block C3HHNU which is declared by the ELLPACK control program with dimension the same as the grid, i.e., I3HHNU(I1NGRX,I1NGRY).

In the case of f non-positive, $s/p = \text{sign}(d/dn)$ when p is not zero, $t/q = \text{sign}(d/dn)$ when q is not zero, and h_x, h_y sufficiently small, the discretization matrix is (a) block tridiagonal, (b) an L-matrix, (c) weakly diagonally dominant, (d) irreducible, and (e) of monotone type. If, in addition, f is constant and $ps = qt = 0$, then the matrix is also symmetric positive definite. In this case iterative solution techniques will perform well. The matrix bandwidth is I1NGRX+1.

PARAMETERS:

IORDER = the user requested order of accuracy of the discretization (integer). The default value is 4. The only acceptable values are 2, 4, or 6.

METHOD = the user specified discretization. If this is nonzero then the parameter IORDER is ignored. The possible values are

1 → discretization (A) above — second order.

2 → discretization (B) above — fourth order.
Requires a square grid and constant coefficients, but allows derivative boundary conditions. Evaluates g at non-grid points.

3 → discretization (C) above — fourth order.
Requires non-derivative boundary conditions, but allows variable f and HX $\neq$HY. Only evaluates g at grid points.

4 → discretization (D) above — sixth order.
Requires non-derivative boundary conditions, constant coefficients, and a square grid. Evaluates g at non-grid points.

RESTRICTIONS: This module is restricted to: Helmholtz equation. two dimensions, rectangular domain, no tangential derivative components in boundary conditions, boundary condition coefficients must be constants, grid size at least 4×4, and uniform grid spacing.

If IORDER = 4 then derivative boundary conditions may be specified only provided f is constant and the grid spacing in x equals the grid spacing in y ($h_x = h_y$).

If IORDER = 6 then f must be constant, only Dirichlet or periodic boundary conditions are allowed, and the grid spacing in x must equal the grid spacing in $y (h_x = h_y)$.

SPECIAL CASES: When IORDER = 2, then p, q, s, and t may be variable. The boundary conditions may switch from Dirichlet to non-Dirichlet at non-corner points in this case.

When IORDER = 4 boundary conditions may also switch from Dirichlet to non-Dirichlet at (a small number of isolated) non-corner, non-grid points without affecting the order of convergence.

Finally, if IORDER > 2 the solution computed by ELLPACK at non-grid points will be less accurate than those at grid points unless higher order accurate interpolation is performed. ELLPACK users may select accurate interpolation by writing INTERPOLATION = SPLINES in the OPTIONS segment.

PERFORMANCE ESTIMATES: Execution time for this module is roughly proportional to I1NGRX×I1NGRY (I1NGRX is the number of grid lines in the x direction and I1NGRY is the number of grid lines in the y direction). Running time is nearly independent of IORDER, except when the cost of computing the function g is very high, in which case IORDER = 2 will be faster. A workspace of size 9 + 5I1NGRX is required as well as a private common block (C3HHNU) of length I1NGRX×I1NGRY. The number of equations produced depends upon the boundary conditions, although this number is bounded by I1NGRX × I1NGRY. There are at most nine unknowns per equation and the resulting matrix has bandwidth no larger than I1NGRX+1. For smooth problems the error produced by this discretization is $O(h_x{}^k + h_y{}^k)$, where h_x and h_y are the grid spacings in the x and y directions respectively, and k = IORDER. For well-behaved problems a high order discretization requires much fewer grid points to achieve a given accuracy than a lower order method. The procedure module SET UNKNOWNS FOR HODIE HELMHOLTZ may be used to initialize the vector of unknowns in the linear system of equations produced by HODIE HELMHOLTZ. This is useful when the equations are to be solved by an iterative method. If used, this procedure must be invoked after HODIE HELMHOLTZ and before the solution module.

ACKNOWLEDGEMENTS: The development of this module was supported by the National Science Foundation grant MCS77-01408 and by the National Bureau of Standards.

REFERENCES:

Boisvert, F., [1981], "Families of high order accurate discretizations of some elliptic problems", SIAM J. Sci. Stat. Comp. 2, pp. 268–284.

Boisvert, F., [1981a], "High order compact difference formulas for elliptic problems with mixed boundary conditions", in *Advances in Computer Methods for Partial Differential Equations* IV, IMACS, New Brunswick, N. J., pp. 193–199.

Collatz, L., [1960], *The numerical treatment of differential equations*, Springer-Verlag, Berlin, 1960.

MODULE: INTERIOR COLLOCATION

AUTHOR: Elias N. Houstis

PURPOSE: Discretizes a general elliptic operator with uncoupled boundary conditions ($\alpha u + \beta u_n = \gamma$ where α or $\beta \equiv 0$ and $\alpha^2 + \beta^2 > 0$).

METHOD: Let $Lu = f$ be the elliptic problem with solution $u(x,y)$ and $U^0(x,y)$ denote a Hermite bicubic piecewise polynomial approximation to $u(x,y)$ which satisfies the boundary conditions $\alpha U^0 + \beta U_n^0 = \gamma'$ where γ' is a locally determined cubic interpolant of γ. The coefficients of U^0 are determined by forcing U^0 to satisfy the elliptic problem at a set of interior collocation points. If the problem is well-behaved, then the error of this discretization is of the order $h_x{}^4 + h_y{}^4$ where h_x and h_y are the average grid spacings in x and y.

MATHEMATICAL FORMULATION: The approximate solution $U(x,y)$ is defined as in the module HERMITE COLLOCATION. Some of the coefficients of U that correspond to boundary nodes of the mesh are predetermined locally so that U or U_n is an interpolant of the boundary conditions. The rest of the coefficients are determined by forcing U to satisfy $Lu = f$ at the interior collocation points defined as in the HERMITE COLLOCATION module.

PARAMETERS: None

RESTRICTIONS: The problem must have a rectangular domain in two dimensions with at least 3×3 grid. The equation cannot be given in self-adjoint form. The boundary conditions must be uncoupled and only change type at the boundary nodes. Periodic boundary conditions are not allowed. The procedure NONUNIQUE cannot be used with this module.

PERFORMANCE ESTIMATES: The number of coefficients to be determined is $4(\text{NGRX}-1)(\text{NGRY}-1)$. The bandwidth of the equations is the natural ordering is $2\text{NGRY}+5$. The workspaces required is $28(\text{NGRY}-1)$.

ACKNOWLEDGEMENTS: The development of this module was supported in part by the National Science Foundation under grants MCS77-01408, MCS79-01437 and MCS78-04878 and the US Army Research Office under contract DAA29-33-K-0026.

REFERENCES:

Houstis E. N., W. F. Mitchell and J. R. Rice [1983], "Collocation Software for second order elliptic partial differential equations", CSD-TR 446, Computer Sciences, Purdue University.

Houstis E. N., W. F. Mitchell and J. R. Rice [1983], "Algorithms INTCOL and HERMCOL: Collocation on rectangular domains with bicubic Hermite polynomials", CSD-TR 445, Computer Sciences, Purdue University.

MODULE: SPLINE GALERKIN

AUTHOR: A. Weiser (adapted for use in ELLPACK by John R. Rice)

PURPOSE: Discretizes a self-adjoint elliptic operator with Dirichlet boundary conditions.

METHOD: Uses Galerkin's method with basis functions which are tensor product B-splines of degree DEGREE (order DEGREE + 1) and have NDERV continuous derivatives. The default is smooth bicubic splines (DEGREE = 3, NDERV = 2). The boundary conditions are imposed via a least squares penalty method. Note that no advantage is taken of homogeneous boundary conditions or separable coefficient functions, and for nonsmooth splines (NDERV > DEGREE − 1), the natural ordering of the equations is inappropriate for this module.

PARAMETERS: DEGREE is the degree of the B-splines (default = 3) and NDERV is the global continuity (default = 2).

RESTRICTIONS:

Rectangular domain.
Two dimensions.
Self-adjoint.
Dirichlet boundary conditions.

REFERENCES:

Weiser, A., S. C. Eisenstat and M. H. Schutz [1980], "On solving elliptic equations to moderate accuracy", SIAM J. Numer. Anal. 17, pp. 908–919.

MODULE: 5-POINT STAR (rectangular domain version)

AUTHORS: Ronald F. Boisvert and John J. Nestor III

PURPOSE: Discretizes linear elliptic partial differential equations of the forms

$$Lu = au_{xx} + cu_{yy} + du_x + eu_y + fu = g \tag{1}$$

$$Lu = (au_x)_x + (cu_y)_y + fu = g \tag{2}$$

defined on a rectangular domain R, subject to the boundary conditions

$$Mu = s(\partial u/\partial n) + ru = pu_x + qu_y + ru = t \quad \text{on } R_1 \tag{3}$$

$$\text{periodic} \quad \text{on } R_2 \tag{4}$$

where $R_1 \cup R_2$ is the boundary of R and a, c, d, e, f, g, p, q, r, s, and t are functions of x and y. Note: the 5-POINT STAR module is implemented as two separate programs, one for rectangular domains and one for general domains. If the problem domain is rectangular, then the short form of the BOUNDARY segment for rectangular domains should be used, in which case this version of the module will be invoked.

METHOD: Classical second order central finite differences.

MATHEMATICAL FORMULATION: This module generates a linear system of equations whose solution yields approximate values of the unknown function u at grid points on R and its boundary. The linear system is obtained by using standard finite differences to approximate the derivatives in (1), (2), and (3).

We define active grid points as those grid points where the value of the solution u is unknown. All grid points are active *except* the following:

(a) points in R_1 where s is zero,

(b) the grid point where the user has specified the value of u in problems with non-unique solutions using the procedure NONUNIQUE (the nearest grid point is used if the point specified is not a grid point),

(c) points on the rightmost edge when periodic boundary conditions in x are specified,

(d) points on the topmost edge when periodic boundary conditions in y are specified.

A finite difference approximation to (1) or (2) is written for each active grid point based upon the following divided central difference approximations.

Case 1:

$$u_x = \frac{U_{i+1,j} - U_{i-1,j}}{h_x(1+\theta_x)}$$

$$u_{xx} = \frac{\theta_x U_{i+1,j} - (1+\theta_x) U_{i,j} + U_{i-1,j}}{h_x^2 \theta_x (1+\theta_x)/2}$$

where $h_x = x_{i+i} - x_i$ and $\theta_x = (x_i - x_{i-1})/h_x$.

Case 2:

$$(au_x)_x = \frac{z_1 U_{i+1,j} - (z_1+z_2) U_{i,j} + z_2 U_{i-1,j}}{h_x^2 \theta_x (1+\theta_x)/2}$$

where h_x and θ_x are given above and

$$z_1 = a((x_{i+i}+x_i)/2, y_j), \qquad z_2 = a((x_i+x_{i-1})/2, y_j).$$

Similar approximations are used for u_{yy} and u_y (in Case 1) and $(cu_y)_y$ (in Case 2). When inactive grid points are referenced in these approximations we either use the known solution values at that point [cases (a) and (b) above] or replace the referenced value by an unknown at a symmetrically placed point [cases (c) and (d) above].

When these approximations are used at active grid points in R_1 or R_2, values of U at points outside the domain (fictitious points) are introduced. In the R_1 case, a multiple of the finite difference approximation to (3) is added to the basic finite difference approximation to (1) or (2) in order to eliminate the term involving the fictitious grid point. In the R_2 case, the fictitious point is replaced by an active one using periodicity.

The equations and unknowns are numbered using the natural ordering (row-wise from bottom to top). After 5-POINT STAR has executed, the number of the equation (and unknown) associated with the (i,j)th grid point may be found in I35PNU(i,j). A value of zero indicates that the point is not active. The array I35PNU occupies the common block C35PNU which is declared by the ELLPACK Control Program with dimension the same as the grid, i.e., I35PNU(I1NGRX,I1NGRY).

Finally, the equations are scaled so that the generated linear system of equations is symmetric whenever the equation is in self-adjoint form (2). This is also true when the equation is trivially self-adjoint, i.e. (1) with constant coefficients and $d = e = 0$.

PARAMETERS: None

SPECIAL CASES: A symmetric difference approximation is used for problems expressed in the form (2) above, as well as for equations written in the form (1) with $d = e = 0$ and a, b, and f constant. In these cases solution modules requiring a symmetric matrix may be used. Processing time is reduced whenever the operator has constant coefficients or the forcing function g is identically zero.

RESTRICTIONS: The domain must be rectangular. The coefficient of u_{xy} must be zero. The grid must be at least 3 by 3. The boundary conditions may not involve derivatives in a direction tangent to the boundary (i.e., the coefficient of u_y must be zero along the left and right edges, and the coefficient of u_x must be zero along the top and bottom edges).

PERFORMANCE ESTIMATES: Execution time for this module is roughly proportional to I1NGRX×I1NGRY, (I1NGRX is the number of grid lines in the x direction and I1NGRY is the number of grid lines in the y direction). A fixed-size workspace of size 20 is required as well as a private common block (C35PNU) of length I1NGRX×I1NGRY. The number of equations produced depends upon the boundary conditions, although this number is bounded by I1NGRX×I1NGRY. There are at most five unknowns per equation and the resulting matrix has bandwidth no larger than I1NGRX. For smooth problems on uniformly spaced grids the error produced by this discretization is proportional to $h_x^2 + h_y^2$, where h_x and h_y are the grid spacings in the x and y directions respectively. The accuracy degrades to $O(h_x + h_y)$ when nonuniformly spaced grids are used (here h_x and h_y refer to average grid spacings).

RELATED MODULES: The procedure module SET UNKNOWNS FOR 5 POINT STAR may be used to initialize the vector of unknowns in the linear system of equations produced by 5-POINT STAR. This is useful when the equations are to be solved by an iterative method. If used, this procedure must be invoked after 5-POINT STAR and before the solution module.

ACKNOWLEDGEMENTS: The development of this module was supported by the National Science Foundation under grant MCS77-01408 and by the National Bureau of Standards.

REFERENCES:

Lapidus, L. and G. Pinder [1982], *Numerical Solution of Partial Differential Equations in Engineering and Science*, John Wiley, New York.

Varga, R. S., [1962], *Matrix Iterative Analysis*, Prentice-Hall, Englewood Cliffs, NJ, (Chapter 6).

MODULE: 5-POINT STAR (general domain version)

AUTHORS: Ronald F. Boisvert and John J. Nestor III

PURPOSE: Discretizes the linear elliptic partial differential equation

$$\mathrm{L}u = au_{xx} + cu_{yy} + du_x + eu_y + fu = g$$

on R subject to the boundary conditions

$$\mathrm{M}u = s(\partial u/\partial n) + ru = pu_x + qu_y + ru = t$$

where R is a general two-dimensional domain with boundary ∂R, and a, c, d, e, f, g, p, q, r, s and t are functions of x and y. Note: the 5-POINT STAR module is implemented as two separate programs, one for rectangular domains and one for general domains. If the problem domain is rectangular, then the short form of the BOUNDARY segment for rectangular domains should be used, in which case the rectangular version of this module will be invoked.

METHOD: Classical finite differences.

MATHEMATICAL FORMULATION: This module generates a system of linear equations whose solution yields values of the unknown function u at grid points in R and its boundary. The linear system is obtained using standard finite differences to approximate derivatives in Lu and Mu.

We say a grid point in R is active if the value of the solution there is unknown. All grid points in R are active *except* the following:

(a) grid points on ∂R where s is zero,

(b) the grid point where the user has specified the value of u in problems with non-unique solutions using the procedure NONUNIQUE (the nearest grid point is used if the point specified is not a grid point).

The points where ∂R intersects the grid and s is nonzero are called active boundary points.

A finite difference approximation to L$u = g$ is written for each active grid point, based upon the following divided central difference approximations.

$$u_x(x_j, y_j) = \frac{U_{i+1,j} - U_{i-1,j}}{h_x(1+\theta_x)}$$

$$u_{xx}(x_j, y_j) = \frac{\theta_x U_{i+1,j} - (1+\theta_x)\, U_{i,j} + U_{i-1,j}}{h_x^2 \theta_x (1+\theta_x)/2}$$

where $h_x = x_{i+1} - x_i$ and $\theta_x = (x_i - x_{i-1})/h_x$

Similar approximations are used for u_{yy} and u_y. These lead to finite difference equations involving (i,j) and its four neighbors $(i+1,j)$, $(i-1,j)$, $(i,j+1)$, and $(i,j-1)$ as shown in Figure 6.4a. When (i,j) is adjacent to ∂R one or more of its neighbors may be outside R. In these cases the grid point outside R is replaced by the nearest point in ∂R on the same grid line (see Figure 6.4b). The resulting difference formulae are similar to the ones given above with h_x and θ_x appropriately redefined. If inactive points are introduced in these approximations, the known value of the solution at those points is substituted.

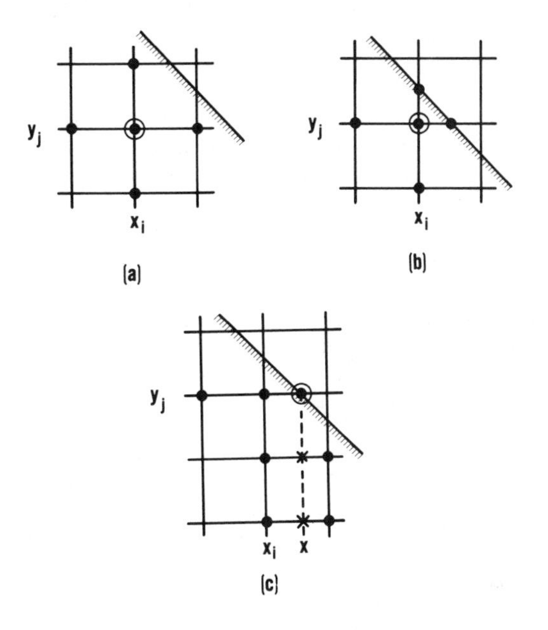

Figure 6.4. Finite difference stencils used by 5-POINT STAR for a general domain. It used (a) an interior point, (b) a point adjacent to the boundary, and (c) a boundary point. The circled point is the central stencil point in each case.

It is also necessary to generate finite difference approximations to the boundary conditions $Mu = t$ at each active boundary point. To do this we write one-sided difference approximations for each of u_x and u_y. For example, at the point $p_k = (x, y_j)$ of Figure 6.4c we use the approximations

$$u_x(p_k) = \alpha_0 U(p_k) + \alpha_1 U_{i-1,j} + \alpha_2 U_{i-2,j},$$

$$u_y(p_k) = \beta_0 U(p_k) + \beta_1 [z_1 U_{i,j-1} + z_2 U_{i+1,j-1}] + \beta_2 [z_1 U_{i,j-2} + z_2 U_{i+1,j-2}],$$

where

$$\alpha_0 = \frac{2x - x_i - x_{i-1}}{(x - x_i)(x - x_{i-1})}, \qquad \beta_0 = \frac{2y_j - y_{j-1} - y_{i-2}}{(y_j - y_{j-1})(y_j - y_{j-2})},$$

$$\alpha_1 = \frac{x - x_{i-1}}{(x_i - x_{i-1})(x_i - x)}, \qquad \beta_1 = \frac{y_j - y_{j-2}}{(y_{j-1} - y_{j-2})(y_{j-1} - y_j)},$$

$$\alpha_2 = \frac{x - x_i}{(x_{i-1} - x_i)(x_{i-1} - x)}, \qquad \beta_2 = \frac{y_j - y_{j-1}}{(y_{j-2} - y_{j-1})(y_{j-2} - y_j)},$$

$$z_1 = \frac{x_{i+1} - x}{x_{i+1} - x_i}, \qquad z_2 = \frac{x - x_i}{x_{i+1} - x_i}.$$

The formula for u_y is obtained by first writing one similar to that for u_x involving the point p_k and the two points marked X in Figure 6.4c. We then replace each of the X points by simple linear interpolation between grid points at the same y level.

Active grid points are numbered using the natural ordering (row-wise from bottom to top). Any active boundary points with $x_i \leqslant x < x_{i+1}$ or $y_j \leqslant y < y_{j+1}$ are numbered immediately after the point (x_i, y_j). After 5-POINT STAR has executed, the number of the equation (and unknown) associated with the (i,j)th grid point may be found in I35PNU(i,j). A value of zero indicates that the point is not active. The integer array I35PNU occupies the common block C35PNU declared in the ELLPACK Control Program with dimension (I1NGRX,I1NGRY). Similarly, the number of the equation associated with the kth boundary point may be found in I35GBN(k). An entry of zero indicates that the boundary point was not active. The integer array I35GBN occupies the common block C35GBN declared in the ELLPACK Control Program with dimension I1MBPT.

PARAMETERS: None

SPECIAL CASES: Processing time is reduced whenever the PDE has constant coefficients or the forcing function g is identically zero.

RESTRICTIONS: The domain must be two dimensional. The equation must be written in standard form. The coefficient of u_{xy} must be zero. The grid must be at least 3 by 3. The system of equations produced by this module is generally non-symmetric and hence solution modules requiring symmetric matrices cannot be used.

PERFORMANCE ESTIMATES: Execution time for this module is roughly proportional to I1NGRX×I1NGRY, where I1NGRX is the number of grid lines in the x direction and I1NGRY is the number of grid lines in the y direction. A fixed workspace of size 34 is required. Two private common blocks (C35PNU and C35GBN) of lengths I1NGRX×I1NGRY and I1NBPT are required where I1NBPT is the number of boundary points (this quantity is not known accurately until run time). The number of equations produced depends upon the boundary conditions, although this number is bounded by I1NGRX×I1NGRY + m, where m is the number of active boundary points. The resulting matrix has bandwidth no larger than I1NGRX for pure Dirichlet problems and 2*I1NGRX otherwise. For smooth problems on uniformly spaced grids the error produced by this discretization is $O(h_x^2 + h_y^2)$, where h_x and h_y are the grid spacings in the x and y directions respectively. The accuracy degrades to $O(h_x + h_y)$ when nonuniformly spaced grids are used (here h_x and h_y refer to average grid spacings) or when non-Dirichlet boundary conditions are present.

RELATED MODULES: The procedure module SET UNKNOWNS FOR 5 POINT STAR may be used to initialize the vector of unknowns in the linear system of equations produced by 5-POINT STAR. This is useful when the equations are to be solved by an iterative method. If used, this procedure must be called after 5-POINT STAR and before the solution module.

ACKNOWLEDGEMENTS: The development of this module was supported by the National Science Foundation under grant MCS77-01408 and by the National Bureau of Standards.

REFERENCES:

Forsythe, G., and W. Wasow [1960], *Finite Difference Methods for Partial Differential Equations*, John Wiley, New York.

Lapidus, L., and G. Pinder [1982], *Numerical Solution of Partial Differential Equations in Engineering and Science*, John Wiley, New York.

MODULE: 7 POINT 3D

AUTHOR: Roger Grimes and Mahesh Rathi

PURPOSE: Discretizes a general self-adjoint elliptic operator of the self-adjoint form

$$(au_x)_x + (bu_y)_y + (cu_z)_z + fu = g \tag{1}$$

or the general form

$$au_{xx} + bu_{yy} + cu_{zz} + du_x + eu_y + fu_z + gu = h, \tag{2}$$

with general linear boundary conditions on a rectangular box.

METHOD: Uses 7-point symmetric differences on a rectangular mesh for the self-adjoint case (1). A symmetric positive definite matrix is generated for all boundary conditions in this case. Uses ordinary second order finite differences for the general form (2). Nonuniform meshes are allowed.

RESTRICTIONS:

Rectangular domain
Three dimensions
Self-adjoint operator
Grid of size at least $3 \times 3 \times 3$
No u_{xy}, u_{xz}, or u_{yz} terms
No periodic boundary conditions
No tangential derivative components in boundary conditions

ACKNOWLEDGEMENTS: Work on this project was supported in part by the National Science Foundation grant MCS79-19829 with the University of Texas at Austin.

MODULE: AS IS

AUTHOR: Ronald F.Boisvert

PURPOSE: Indexes the equations as they were generated by the discretization module.

METHOD: Sets the indexing arrays in ELLPACK to the identity. This is the default indexing module and need not be specified explicitly.

RESTRICTIONS: None.

PERFORMANCE ESTIMATES: Requires no workspace storage.

MODULE: HERMITE COLLORDER

AUTHOR: Wayne R. Dyksen

PURPOSE: Reorders the linear system generated by HERMITE COLLOCATION so that the reordered system has a nonzero diagonal.

METHOD: The grid points are numbered in the natural way from south to north, west to east. The collocation points (equations, rows) are associated with the nearest grid point and are numbered in groups of four in the order of their corresponding grid point. The Hermite bicubic basis functions (unknowns, columns) are ordered in the natural way from south to north, west to east. Some of the collocation points are then reordered depending on the boundary conditions to insure a nonzero diagonal.

Besides having a nonzero diagonal, the reordered coefficient matrix exhibits other nice properties. It is a band matrix with band width 4NY+7 where NY is the number of horizontal grid lines. It is block symmetric in that it consists of 4×4 blocks such that if $B_{i,j} \neq 0$ then $B_{j,i} \neq 0$. There are at most 16 nonzero entries per row, occurring in 4 blocks. All the symmetric pairs of off-diagonal blocks can be stored in their natural order within one 4×4 block.

PARAMETERS: None.

RESTRICTIONS: Should be used only on rectangular domains in conjunction with the discretization module HERMITE COLLOCATION. To insure a nonzero diagonal, the coefficient functions in the PDE and the boundary conditions must not vanish. Also, the boundary conditions on any side may include only values of u and/or its normal derivative.

PERFORMANCE ESTIMATES: Requires no workspace storage.

REFERENCES: W.R. Dyksen and J.R. Rice [1984], "A new ordering scheme for the Hermite bicubic collocation equations", in *Elliptic Problem Solvers* II, (G. Birkhoff and A. Schoenstat, eds.), Academic Press.

ACKNOWLEDGEMENTS: Work on this module was supported in part by Department of Energy contract DE-AC02-81E10997 with Purdue University.

INDEXING

MODULE: INTERIOR COLLORDER

AUTHOR: Wayne R. Dyksen

PURPOSE: Reorders the linear system generated by INTERIOR COLLOCATION so that the reordered system has a nonzero diagonal.

METHOD: For general nonhomogeneous or mixed homogeneous boundary conditions, the grid points are numbered in the natural way from south to north, west to east. The collocation points (equations, rows) are associated with the nearest grid point and are numbered in groups of four in the order of their corresponding grid point. The Hermite bicubic basis functions (unknowns, columns) are ordered in the natural way from south to north, west to east. Some of the collocation points are then reordered depending on the boundary conditions to insure a nonzero diagonal.

Besides having a nonzero diagonal, the reordered coefficient matrix exhibits other nice properties. It is a band matrix with band width 4NY where NY is the number of horizontal grid lines. It is block symmetric in that it consists of 4×4 blocks such that if $B_{i,j} \neq 0$ then $B_{j,i} \neq 0$. There are at most 16 nonzero entries per row, occurring in 4 blocks. All the symmetric pairs of off-diagonal blocks can be stored in their natural order within one 4×4 block.

The grid points are numbered in the natural way from south to north, west to east. The collocation points (equations, rows) are associated with the nearest grid point and are numbered in groups of four or less in the order of their corresponding grid point. The Hermite bicubic basis functions (unknowns, columns) are ordered in the natural way from south to north, west to east.

PARAMETERS: None.

RESTRICTIONS: Should be used only on rectangular domains in conjunction with the discretization module INTERIOR COLLOCATION. To insure a nonzero diagonal, the coefficient functions in the PDE and the boundary conditions must not vanish. Also, the boundary conditions on any side may include only values of u and/or its normal derivative.

PERFORMANCE ESTIMATES: Requires no workspace storage.

REFERENCES: W.R. Dyksen and J. R. Rice [1984], "A new ordering scheme for the hermite bicubic collocation equations", in *Elliptic Problem Solvers* II, (G. Birkhoff and A. Schoenstat, eds.), Academic Press.

ACKNOWLEDGEMENTS: Work on this module was supported in part by Department of Energy contract DE-AC02-81E10997 with Purdue University.

MODULE: MINIMUM DEGREE

AUTHOR: Andrew H. Sherman

PURPOSE: Computes an indexing which attempts to minimize the number of nonzero elements introduced during Gauss elimination.

METHOD: Applies the minimum degree algorithm to $A + A^T$, where A is the coefficient matrix of the linear system.

PARAMETERS: None.

PERFORMANCE ESTIMATES: Requires a considerable workspace. Note: the default value may be a huge overestimate for small linear systems or too small for large ones. The actual length of workspace required for a specific problem is computed and printed by the module. After the first run, the workspace may be made larger or smaller as needed using the WORKSPACE options.

REFERENCES: Eisenstat, S.C., M. C. Gursky, M. H. Schultz and A. H. Sherman [1977], "Yale Sparse Matrix Package, I: The symmetric codes", Report 112. Computer Science, Yale University.

MODULE: NESTED DISSECTION

AUTHOR: Andrew H. Sherman

PURPOSE: Computes an indexing for grid-based equations and unknowns.

METHOD: The domain is assumed to be embedded in an $x-y$ or $x-y-z$ grid. A nested dissection ordering is performed on the smallest square in which the $x-y$ grid can be embedded and the actual points in the region are numbered in the same relative ordering as in this nested dissection ordering. The unknowns of the discretized equations are then numbered in the same relative order as the grid points with which they are associated, and the equations are given the same order as the unknowns. If several unknowns are associated with the same grid point, they are numbered consecutively in increasing order by the original unknown ordering.

PARAMETERS: NDTYPE specifies the type of nested dissection desired as follows:

NDTYPE	*Nested Dissection Type*
5	5 point (2D)
9	9 point (2D)
7	7 point (3D)
27	27 point (3D)

RESTRICTIONS: There must be some natural association between the equations/unknowns and the grid points.

PERFORMANCE ESTIMATES: For an NX by NY or NX by NY by NZ grid, NESTED DISSECTION requires a workspace as follows:

NDTYPE	*Length of workspace*
5	2×MAX(NX,NY)**2
9	2×NX×NY
7	2×NX×NY×NZ
27	2×NX×NY×NZ

REFERENCES: George, J. A., [1973], "Nested dissection of a regular finite element mesh", SIAM J. Numer. Anal. 10, pp. 345–363.

MODULE: RED BLACK

AUTHOR: Roger G. Grimes and John R. Respess

PURPOSE: Indexes the equations and unknowns using a red-black (checkerboard) ordering scheme.

METHOD: All of the grid points are labeled either red or black using a checkerboard pattern so that red points are adjacent only to black points and black points are adjacent only to red points. The equations and unknowns are ordered so that all red points are numbered first, followed by all black points. If a red-black ordering is not possible, an error message is printed.

PARAMETERS: LEVEL = Level of printed output. This parameter does not reset the global printing level as set in the OPTIONS statement and only affects this module. The default value is that value set in the OPTIONS statement.

LEVEL $\leqslant$ 3	Minimum output and error message
LEVEL $\geqslant$ 4	Print indexing arrays I1UNDX and I1ENDX. I1UNDX is the unknowns reordering permutation vector and I1ENDX is its inverse. Unknown I gets mapped to unknown I1UNDX(I) under the reordering.

MATHEMATICAL FORMULATION: See Chapter 7 for a detailed presentation of iterative methods in ELLPACK.

RESTRICTIONS: RED BLACK is designed to work on the equations obtained from discretization modules which use the standard 5-point or 7-point (3D) stencils for difference formulas.

ACKNOWLEDGEMENTS: Work on this project was supported in part by the National Science Foundation grant MCS79-19829 with the University of Texas at Austin.

MODULE: REVERSE CUTHILL MCKEE

AUTHOR: Andrew H. Sherman

PURPOSE: Computes an indexing to be used with envelope or band solvers.

METHOD: Applies the reverse Cuthill-McKee algorithm to the linear system.

RESTRICTIONS: The linear system must by symmetric.

PERFORMANCE ESTIMATES: Requires no workspace storage.

REFERENCES: Liu, W. H., and A. H. Sherman [1976], "A comparative analysis of the Cuthill-McKee and reverse Cuthill-McKee ordering algorithms for sparse matrices", SIAM J. Numer. Anal. 13, pp. 198–213.

MODULE: BAND GE
(Modified version of LINPACK BAND)

AUTHOR: Wayne R. Dyksen

PURPOSE: Solves a real banded system of linear equations.

METHOD: Solves the banded system by producing an *LU* factorization using Gauss elimination with scaled partial pivoting. An ELLPACK interface routine reformats the linear system into the LINPACK band storage mode in which diagonals are stored in rows. The solution phase of this module is a modified version of the general band solver of the LINPACK project supported by the National Science Foundation.

PARAMETERS: None.

RESTRICTIONS: None.

PERFORMANCE ESTIMATES: For a system of N equations with ML lower diagonals and MU upper diagonals, the generated band matrix has size (2ML + MU + 1) $\times$ N stored in workspace. An additional workspace of length N is also required.

REFERENCES:

Dongarra, J. J., et al., [1979], *LINPACK User's Guide*, SIAM, Philidelphia.

Rice, J. R., [1983], *Numerical Methods, Software and Analysis*, McGraw-Hill, New York, p. 139.

SOLUTION

MODULE: BAND GE NO PIVOTING
(Modified version of LINPACK BAND)

AUTHOR: William A. Ward

PURPOSE: Solves a real banded system of linear equations.

METHOD: Solves the banded system by producing an *LU* factorization using Gauss elimination without pivoting. An ELLPACK interface routine reformats the linear system into the LINPACK band storage mode in which diagonals are stored in rows. The solution phase of this module is a modified version of the general band solver of the LINPACK project supported by the National Science Foundation.

PARAMETERS: None.

RESTRICTIONS: The band system must be solvable by Gauss elimination **without** pivoting.

PERFORMANCE ESTIMATES: For a system of N equations with ML lower diagonals and MU upper diagonals, the generated band matrix has size (ML + MU + 1) × N stored in workspace.

REFERENCES:

Dongarra, J. J., et al., [1979], *LINPACK User's Guide*, SIAM, Philidelphia.

MODULE: ENVELOPE LDU
ENVELOPE LDLT

AUTHOR: Andrew H. Sherman

PURPOSE: Solve systems of sparse (envelope) linear equations. ENVELOPE LDLT is for symmetric systems, while ENVELOPE LDU is for nonsymmetric systems.

METHOD: Converts the coefficient matrix into envelope form and computes an LDU (LDL^T) decomposition of the resulting (symmetric) matrix. The triangular system is then solved to obtain the solution.

PARAMETERS: None.

RESTRICTIONS: The linear system must be solvable without pivoting. These modules also assume that any indexing done is symmetric; that is, the same permutations that are applied to the rows are also applied to the columns. In particular, ENVELOPE LDU cannot follow HERMITE COLLORDER.

PERFORMANCE ESTIMATES: Requires a workspace of length 2N + P for symmetric factorization and 3N + P + Q for nonsymmetric factorization where N is the number of unknowns, P is the size of the envelope of the strict upper triangle and Q is the size of the envelope of the strict lower triangle. (If MU is the upper half-bandwidth, then $P > N \times MU$. If ML is the lower half-bandwidth, then $Q < N \times ML$.) Note: The default value of the workspace may be a huge overestimate for small linear systems or too small for large ones. The actual length of workspace required for a specific problem is computed and printed by the module. After the first run, the workspace may be made larger or smaller as needed using the WORKSPACE option.

REFERENCES:

Eisenstat, S. C., and A. H. Sherman [1974], "Subroutines for envelope solution of sparse linear systems", Report 35, Computer Science, Yale University.

MODULE: LINPACK BAND

AUTHOR: The LINPACK group

PURPOSE: Solves a real banded system of linear equations.

METHOD: Solves the banded system by producing an *LU* factorization using Gauss elimination with partial pivoting. An ELLPACK interface routine reformats the linear system into the LINPACK band storage mode in which diagonals are stored in rows. The solution phase of this module is a product of the LINPACK project supported by the National Science Foundation.

PARAMETERS: None.

RESTRICTIONS: None.

PERFORMANCE ESTIMATES: For a system of N equations with ML lower diagonals and MU upper diagonals, the generated band matrix has size (2ML + MU + 1) $\times$N. A workspace of length N is also required.

REFERENCES:

Dongarra, J.J., et al., [1979], *LINPACK User's Guide*, SIAM, Philidelphia.

MODULE: LINPACK SPD BAND

AUTHOR: The LINPACK group

PURPOSE: Solves a real symmetric positive definite band system of linear equations.

METHOD: Solves the symmetric positive definite banded system by producing an LDL^T factorization using Cholesky decomposition. An ELLPACK interface routine reformats the linear system into the LINPACK symmetric band storage mode in which diagonals are stored in rows. The solution phase of this module is a product of the LINPACK project supported by the National Science Foundation.

PARAMETERS: None.

RESTRICTIONS: None.

PERFORMANCE ESTIMATES: For a system of N equations with MU upper diagonals, the generated band matrix has size $(ML+1) \times N$ stored in workspace.

REFERENCES: Dongarra, J. J., et al., [1979], *LINPACK User's Guide*, SIAM, Philidelphia.

SOLUTION

MODULE: JACOBI CG
JACOBI SI
SOR
SYMMETRIC SOR CG
SYMMETRIC SOR SI
REDUCED SYSTEM CG
REDUCED SYSTEM SI

AUTHORS: David R. Kincaid, Roger G. Grimes, David M. Young, Thomas C. Oppe, and John R. Respess. These modules were contributed by the ITPACK project of the Center for Numerical Analysis at the University of Texas at Austin.

PURPOSE: The ITPACK modules solve systems of symmetric positive definite or nearly symmetric linear equations.

METHOD: The ITPACK modules use iterative algorithms with automatic parameter determination and automatic stopping tests. The convergence of the basic iterative methods is accelerated by using either conjugate gradient (CG) or semi-iteration (SI).

MATHEMATICAL FORMULATION: See Chapter 7 for a detailed presentation of iterative methods in ELLPACK.

PARAMETERS: The following optional parameters allow a certain degree of control over the iterative algorithms.

Parameter	*Meaning or usage*	*Default value*
ITMAX	Maximum number of iterations allowed	100
LEVEL	Level of printed output. This parameter does not reset the global printing level as set in the OPTIONS statement and only affects this module. LEVEL < 0: No printing done = 0: Fatal error messages only = 1: Minimum output and warnings = 2: Reasonable summary = 3: Iteration parameter values = 4: Iteration solution vectors = 5: Original system	(a)
IADAPT	Switch indicating whether adaptive procedures should be used. IADAPT = 0: Fixed values for iteration parameters CME, SME, OMEGA, SPECR, and BETAB = 1: Adaptive procedures used for values of CME, SME, OMEGA, SPECR, and BETAB = 2: (SYMMETRIC SOR methods only) Fixed values for CME, BETAB, and OMEGA with adaptive values for SPECR	1

Parameter	*Meaning or usage*	*Default value*
	IADAPT = 3: (SYMMETRIC SOR methods only) Fixed value for BETAB and adaptive values for CME, OMEGA, SPECR	
ICASE	Switch indicating which case of the adaptive procedure to use for JACOBI SI and SYMMETRIC SOR methods. ICASE = 1: SME $\leqslant m(B)$ [SME fixed for JACOBI SI] = 2: $\lvert m(B)\rvert \leqslant M(B)$ [SME set to −CME]	1
IDGTS	Error analysis switch IDGTS = 0: Skip error analysis. = 1: Print DIGIT1: and DIGIT2, two measures of the approximate number of correct digits in the final solution. = 2: Print final solution vector = 3: Print final residual vector = 4: Print both final solution and residual vectors	0
ZETA	Tolerance level in stopping test	(b)
CME	Initial estimate of largest eigenvalue of the Jacobi matrix	0.0
SME	Initial estimate of smallest eigenvalue of the Jacobi matrix for Jacobi SI method	0.0
FF	Adaptive procedure damping factor. FF should be in the interval (0,1], with 1 causing the most frequent parameter changes when the fully adaptive switch IADAPT = 1 is used.	0.75
OMEGA	Overrelaxation parameter for SOR and SYMMETRIC SOR methods	1.0
SPECR	Initial estimate of spectral radius of the SYMMETRIC SOR matrix	0.0
BETAB	Initial estimate of the spectral radius of the LU matrix in SYMMETRIC SOR methods. Note that $A = I - L - U$ when A has been diagonally scaled to have a unit diagonal and L and U are respectively strictly lower and strictly upper triangular.	0.25

(a) As set in OPTIONS
(b) AMAX1(5.E−6, 5.E2*R1MACH(3))

The following parameters affect only certain modules as follows:

	Parameter				
Module affected	SME	OMEGA	SPECR	BETA	ICASE
JACOBI SI	X				X
SOR		X			
SYMMETRIC SOR CG		X	X	X	X
SYMMETRIC SOR SI		X	X	X	X

RESTRICTIONS: The REDUCED SYSTEM modules must be preceded by the RED-BLACK indexing module.

These iterative algorithms will converge for positive definite systems of linear equations. They can also be applied to nonsymmetric systems and are frequently effective for nearly symmetric systems. They are based on routines from ITPACK 2C which were modified especially for use as part of ELLPACK. In its original form, ITPACK 2C can be used as a stand alone package for solving large sparse systems of linear equations.

PERFORMANCE ESTIMATES: These modules require a workspace as follows:

Module	Length of workspace for N linear equations
JACOBI CG	5N + K×ITMAX
JACOBI SI	3N
SOR	2N
SYMMETRIC SOR CG	7N + K×ITMAX
SYMMETRIC SOR SI	6N
REDUCED SYSTEM CG	2N + 3NBLACK + K×ITMAX
REDUCED SYSTEM SI	2N + NBLACK

NBLACK = number of black grid points in red-black ordering
ITMAX = maximum number of iterations
K = 2 for symmetric systems
= 4 for nonsymmetric systems

REFERENCES:

Kincaid, D. R., J. R. Respess, D. M. Young and R. G. Grimes [1982], "ITPACK 2C: A Fortran package for solving large sparse linear systems by adaptive accelerated iterative methods", ACM Transactions on Math. Software 8, pp. 302–322.

Hageman, L. A. and D. M. Young [1981], *Applied Iterative Methods*, Academic Press, New York.

ACKNOWLEDGEMENTS: Work on this project was supported in part by the National Science Foundation grant MCS79-19829 with the University of Texas at Austin.

MODULE: SPARSE GE NO PIVOTING
SPARSE LU UNCOMPRESSED
SPARSE LU COMPRESSED
SPARSE LDLT

AUTHOR: Andrew H. Sherman

PURPOSE: These four modules solve sparse systems of linear equations.

METHOD: These routines compute an LU (LDL^T) decomposition of the sparse (symmetric) coefficient matrix of the linear system and solve the resulting triangular system to obtain the solution. These modules use storage conserving nonsymmetric, fast nonsymmetric, compressed nonsymmetric, and symmetric schemes, respectively.

PARAMETERS: The parameter NSP (= length of workspace storage) is computed by default if it is not specified in this module. This parameter is not present for SPARSE GE NO PIVOTING.

RESTRICTIONS: These routines assume that any indexing done is symmetric; that is, the same permutations that are applied to the rows are also applied to the columns. In particular, they cannot follow HERMITE COLLORDER.

PERFORMANCE ESTIMATES: These routines require a workspace of length NSP. Their run time is $O(N^{3/2})$ for an $N \times N$ linear system resulting from the standard 5-point difference formulas. Note: the default value of the parameter NSP may be a huge overestimate for small linear systems or too small for large ones. The actual length of workspace required for a specific problem is computed and printed by the module. After the first run, NSP may be made larger or smaller as needed.

REFERENCES:

Eisenstat, S. C., M. C. Gursky, M. H. Schultz and A. H. Sherman, [1977], "Yale Sparse Matrix Package, I: The symmetric codes", Report 112. Computer Science, Yale University, 1977.

Eisenstat, S. C., M. C. Gursky, M. H. Schultz, and A. H. Sherman, [1977], "Yale Sparse Matrix Package, II": The nonsymmetric codes, Report 114, Computer Science, Yale University.

Eisenstat, S. C., M. H. Schultz, and A. H. Sherman, [1981], "Algorithms and data structures for sparse symmetric Gaussian elimination", SIAM J. Sci. Stat. Comp. 2, pp. 225–237.

MODULE: SPARSE LU PIVOTING

AUTHOR: Andrew H. Sherman

PURPOSE: Solves systems of sparse linear equations.

METHOD: Converts the equations to sparse storage mode and solves them using sparse Gauss elimination with column pivoting.

PARAMETERS: MAXNZ is the maximum number of off-diagonal non-zero entries in the upper-triangular factor of the matrix. MAXNZ defaults to a value which is likely to be an excessive overestimate. is likely to be an excessive over estimate. After the first, the workspace can be adjusted using the WORKSPACE option.

PERFORMANCE ESTIMATES: The coefficient matrix is stored in a linear array of length I1MNEQ×I1MNCO. A workspace of length 3MAXNZ + 6I1MNEQ + 3 is required.

REFERENCES:

Sherman, A. H. [1978], "Algorithm 533: NSPIV, A Fortran subroutine for sparse Gaussian elimination with partial pivoting", ACM Trans. Math. Software 4, pp. 391–398.

MODULE: CMM EXPLICIT

AUTHOR: Wlodzimierz Proskurowski

PURPOSE: Uses the capacitance matrix method to solve the standard five-point difference approximation to the Helmholtz equation

$$-u_{xx} - u_{yy} + \lambda u = f(x,y)$$

in a bounded nonrectangular planar region subject to either the Dirichlet or Neumann boundary conditions.

METHOD: The **capacitance matrix method** solves the above problem as a lower order modification to a problem on an embedding rectangle, which in turn is solved using the standard five-point difference formula and a fast fourier transform.

This method extends the usefulness of fast rectangular solvers (FS) to nonrectangular regions. For CMM EXPLICIT, in addition to two calls of a FS, one needs to generate and solve a large and dense system of linear equations, called the **capacitance matrix system**. The capacitance matrix C comprises the modifications due to the irregular boundary. In EXPLICIT CMM the matrix C is generated and decomposed into triangular factors only once (in the preprocessing stage). Therefore for problems with multiple right hand sides, only the backsolving of the system with matrix C is needed. On the other hand, if the number of the mesh points adjacent to the boundary is large then the size of the matrix C increases rapidly and the fast memory is easily saturated.

The algorithm consists of the following steps:

1 Generate the capacitance matrix linear system,

2 Solve this system by a direct method,

3 Add this correction term to the right hand side of the difference equation, and use the fast solver to compute the final solution.

CMM EXPLICIT should be used whenever the capacitance matrix C can be stored in the fast memory. This CMM module can be used efficiently for problems with multiple right hand sides.

CMM EXPLICIT uses a special FS with periodic boundary conditions on an infinite strip. This choice of FS has its advantages and its limitations. Such an FS exploits the translation invariancy of the solution with these boundary conditions, which significantly decreases the cost for the generation of the matrix C. Moreover, it is designed so that either a sparse or fast Fourier transform is performed for the analysis and for the synthesis of data. This design makes this FS very flexible, which is advantageously exploited in the package. On the other hand, if the irregular region has one or more sides parallel to the coordinate axes, then it cannot match the prescribed boundary conditions on such sides, which leads to the unnecessary increase of the size of C.

RESTRICTIONS:

Two dimensions.
Nonrectangular bounded region.
Helmholtz equation.
Dirichlet or Neumann boundary conditions.
Uniform square grid, at least 17×17.
The number of X points is 1 plus a power of 2.

PARAMETERS: Let NX by NY be the grid size of the embedding rectangle, and IP the number of the grid points adjacent to the irregular boundary. CMM EXPLICIT has the following parameters:

Integer workspace	IIWORKI	$> 12\,IP$,
Real workspace	IIWORKR	$> (14+IP) \times IP$,
Number of right sides to the PDE	INUDATA	

It is advisable to specify ones own estimates of the workspace.

ACKNOWLEDGEMENTS: Work on this project was supported in part by National Science Foundation grant MCS-8003382 at the University of Southern California.

REFERENCES:

Proskurowski, W., [1983], Algorithm 593: "A Package for the Helmholtz equation in non-rectangular planar regions", ACM Trans. Math. Software 9, pp. 117–124 (and references therein).

TRIPLE

MODULE: CMM HIGHER ORDER

AUTHOR: Wlodzimierz Proskurowski

PURPOSE: Solves the standard five-point difference approximation to the Poisson equation

$$-u_{xx}-u_{yy}=f(x,y)$$

in a bounded nonrectangular planar region subject to the Dirichlet boundary conditions.

METHOD: The **capacitance matrix method** solves the above problem as a lower order modification to a problem on an embedding rectangle, which in turn is solved using the standard five-point difference formula and a Fast Fourier Transform.

This extends the usefulness of fast rectangular solvers (FS) to nonrectangular regions. CMM HIGHER ORDER is an explicit method; in addition to two calls of a FS, one needs to generate and solve a large and dense system of linear equations, called the **capacitance matrix system**. The capacitance matrix C comprises the modifications due to the irregular boundary. In CMM HIGHER ORDER the matrix C can be generated and decomposed into triangular factors only once (in the preprocessing stage); therefore for problems with multiple right hand sides, only the backsolving of the system with the matrix C is needed. This feature is efficiently exploited in the use of one or two deferred corrections which produces the solution with higher order accuracy.

The algorithm consists of the following steps:

1. Generate the capacitance matrix linear system.
2. Solve this system by a direct method.
3. Add this correction term to the right hand side of the difference equation, and use the fast rectangular solver to compute the second order solution.
4. Redo steps 2 and 3 to compute the deferred corrections to obtain the higher order accurate final solution.

This CMM module can be used efficiently for problems with multiple right hand sides.

CMM HIGHER ORDER uses a special FS with periodic boundary conditions on an infinite strip. This choice of FS has its advantages and its limitations. Such a FS exploits the translation invariance of the solution with these boundary conditions, which significantly decreases the cost for the generation of the matrix C. Moreover, it is designed so that either a sparse matrix method or fast Fourier transform is used to do the analysis and for the synthesis of data. This design makes this FS very flexible, which is advantageously exploited in the package. On the other hand, if the irregular region has one or more sides parallel to the coordinate axes, then it cannot match the prescribed boundary conditions on such sides, which leads to the unnecessary increase of the size of C.

RESTRICTIONS:

Two dimensions.
Non-rectangular bounded region.
Poisson equation.
Dirichlet boundary conditions.
Uniform square grid, at least 17×17.
The number of X points is 1 plus a power of 2.

PARAMETERS: Let NX by NY be the grid size of the embedding rectangle and IP the number of grid points adjacent to the irregular boundary. CMM HIGHER ORDER has the following parameters:

INTEGER WORKSPACE	IWORKI	$>$ 4 IP
REAL WORKSPACE	IWORKR	$>$ (IP+20) $\times$ P+2 NX $\times$ NY
NUMBER OF PDE RIGHT SIDES	NUDATA	

It is advisable to specify ones own estimates of the workspace.

ACKNOWLEDGEMENTS: Work on this project was supported in part by National Science Foundation Grant MCS-8003382 at the University of Southern California.

REFERENCES:

Proskurowski, W., [1983], "Algorithm 593: A package for the Helmholtz equation in non-rectangular planar regions", ACM Trans. Math. Software 9, pp. 117–124 (and references therin).

TRIPLE

MODULE: CMM IMPLICIT

AUTHOR: Wlodzimierz Proskurowski

PURPOSE: The Solves the standard five-point difference approximation to the Helmholtz equation.

$$-u_{xx} - u_{yy} + \lambda u = f(x,y)$$

in a bounded nonrectangular planar region subject to either Dirichlet or Neumann boundary conditions.

METHOD: The implicit capacitance matrix method solves the above problem as a lower order modification to a problem on an embedding rectangle, which in turn is solved using the standard five-point difference formula and a fast Fourier transform.

This method extends the usefulness of fast rectangular solvers (FS) to nonrectangular regions. For CMM IMPLICIT, in addition to two calls of a FS, one needs to solve a large and dense system of linear equations, called the capacitance matrix system, in an iterative way. The capacitance matrix C comprises the modifications due to the irregular boundary. In CMM IMPLICIT only a product of the matrix C and a given vector is required (without an explicit knowledge of C), at a cost of two calls of a FS per iteration. Moreover, the conjugate gradient iteration with a properly constructed matrix C converges almost superlinearly. Thus the final solution can be obtained at a cost proportional to several calls of a FS (on the order of 10) without generating, storing, and factoring the capacitance matrix.

The algorithm consists of the following steps:

1. Generate the right side to the capacitance matrix system.
2. Solve this system by the conjugate gradient method.
3. Add this correction term to the right hand side of the difference equations, and use the fast solver to compute the final solution.

CMM IMPLICIT uses a special FS with periodic boundary conditions on an infinite strip. This choice of FS has its advantages and its limitations. Such an FS exploits the translation invariance of the solution with these boundary conditions, which significantly decreases the cost for the generation of the matrix C. Moreover, it is designed so that either a sparse matrix method or a fast Fourier transform is used to do the analysis and the synthesis of data. This design makes this FS very flexible, which is advantageously exploited in the package. On the other hand, if the irregular region has one or more sides parallel to the coordinate axes, then it cannot match the prescribed boundary conditions on such sides, which leads to the unnecessary increase of the size C.

RESTRICTIONS:

Two dimensions.
Non-rectangular bounded region.
Helmholtz equation.
Dirichlet or Neumann boundary conditions.
Uniform square grid, at least 17×17.
The number of x points is 1 plus a power of 2.

PARAMETERS: Let NX by NY be the grid size of the embedding rectangle and IP the number of grid points adjacent to the irregular boundary. CMM IMPLICIT has the following parameters:

INTEGER WORKSPACE	IWORKI	> 21 IP
REAL WORKSPACE	IWORKR	> 13 IP

It is advisable to specify one's own estimates of the workspace.

ACKNOWLEDGEMENTS: Work on this project was supported in part by the National Science Foundation Grant MCS-8003382 at the University of Southern California.

REFERENCES:

Proskurowski, W., [1983], Algorithm 593: "A package for the Helmholtz equation in non-rectangular planar regions", ACM Trans. Math. Software 9, pp. 117–124 (and references therein).

MODULE: DYAKANOV CG

AUTHOR: R. E. Bank

PURPOSE: Solves nonseparable, self-adjoint elliptic operators with general boundary conditions.

METHOD: This module is based on a symmetric 5-point discretization. The resulting equations are solved using a preconditioned conjugate gradient iteration. A scaled, separable approximation of the nonseparable matrix is used as the preconditioning matrix. The generalized marching algorithm is used to solve the separable problems. The initial guess is zero.

PARAMETERS: ITMAX denotes the maximum number of CG iterations allowed; its default value is 100. DEMAND is a real parameter which instructs DYAKANOV CG to accept the solution if the estimated error in the linear equations problem is reduced by $10^{-\mathrm{DEMAND}}$; its default value is 3.

RESTRICTIONS:

Rectangular domain.

Two dimensions.

Self-adjoint, nonseparable equation.

Periodic boundary conditions are not allowed.

Uniform grid having at least 3, 4, or 5 x (y) grid lines if there are respectively 0, 1, or 2 Dirichlet boundary conditions in the x (y) direction.

PERFORMANCE ESTIMATES: The operation count for an NX by NY grid is of the order of $\mathrm{NX} \times \mathrm{NY} \times \log(\mathrm{NY}) \times \mathrm{ITNUM}$, where ITNUM is the number of iterations. The length of the storage required is of the order of $7 \times \mathrm{NX} \times \mathrm{NY} + \mathrm{NY} \times \log_2(\mathrm{NY})$.

ACKNOWLEDGEMENTS: This work was supported in part by the Office of Naval Research under contract N0014-75-C-0243. The original version of the code was written at Argonne National Laboratory.

REFERENCES:

Bank, R. E., [1978], "Algorithm 527: A Fortran implementation of the generalized marching algorithm", ACM Trans. Math. Software 4, pp. 165–176.

TRIPLE

MODULE: DYAKANOV CG 4

AUTHOR: R. E. Bank

PURPOSE: Solves nonseparable, self-adjoint elliptic operators with general boundary conditions.

METHOD: This module is based on a symmetric 5-point discretization. Fourth order accuracy is achieved by solving the problem on a coarse grid (NX+1)/2 by (NY+1)/2, and on a fine grid, NX by NY, and using Richardson extrapolation at grid points in the fine mesh. The resulting equations are solved using a preconditioned conjugate gradient iteration. A scaled, separable approximation of the nonseparable matrix is used as the preconditioning matrix. The generalized marching algorithm is used to solve the separable problems. The initial guess is zero.

PARAMETERS: ITMAX denotes the maximum number of CG iterations allowed; its default value is 100. DEMAND is a real parameter which instructs DYAKANOV CG 4 to accept the solution if the estimated error in the linear equations problem is reduced by $10^{-\mathrm{DEMAND}}$; its default value is 3.

RESTRICTIONS:
Rectangular domain.
Two dimensions.
Self-adjoint, nonseparable equation.
Periods boundary conditions are not allowed.
Uniform grid having at least 5, 7, or 9 x (y) grid lines if there are respectively 0, 1, or 2 Dirichlet boundary conditions in the x (y) direction.
The number of x or y grid lines must be odd integers.

PERFORMANCE ESTIMATES: The operation count is of the order of $\mathrm{NX} \times \mathrm{NY} \times \log(\mathrm{NY}) \times \mathrm{ITNUM}$, where ITNUM is the number of iterations. The length of the storage required is of the order of $8\mathrm{NX} \times \mathrm{NY} + \mathrm{NY} \times \log_2(\mathrm{NY})$.

ACKNOWLEDGEMENTS: This work was supported in part by the Office of Naval Research under contract N00014-75-C-0243 and the National Aeronautics and Space Administration under grant NSG-1632.

REFERENCES:

Bank, R. E., [1978], "Algorithm 527: A Fortran implementation of the generalized marching algorithm", ACM Trans. Math. Software 4, pp. 165–176.

TRIPLE

MODULE: FFT 9 POINT

AUTHORS: E. N. Houstis and T. S. Papatheodorou

PURPOSE: Discretizes an elliptic operator with constant coefficients of the form

$$a\,u_{xx} + b\,u_{yy} + c\,u = f$$

on a rectangle subject to Dirichlet boundary conditions, and solves the resulting linear system using the fast Fourier transform and cyclic reduction.

METHOD: This module discretizes the operator using a second, fourth, or sixth order 9-point difference formula. The resulting linear system is solved using the fast Fourier transform. The FFT 9 POINT algorithm consists of the following components:

1. Uniform grid
2. Discretization operator: second, fourth, or sixth order 9-point difference formula
3. Equation solver:
   (a) Odd/even reduction
   (b) Fourier analysis on even lines
   (c) Recursive cyclic reduction
   (d) Fourier synthesis on the even lines
   (e) Solution on the odd lines

PARAMETERS: IORDER = 2, 4, or 6 denotes the order of the method (default=4).

RESTRICTIONS: This module requires a rectangular domain two dimensions, and it must have the number of x (y) grid lines = $2^{i(j)+1}$ with $i, j \geqslant 3$.
Sixth order applies **only** if $a = b = 1$, $c = 0$, and $h_x = h_y$.
Further limitations on the domain are:

1. for IORDER = 2, one must have $h_x = h_y$, and
2. for IORDER = 4 or 6, the lower left corner of the domain must be (0,0).

PERFORMANCE ESTIMATES: For an NX by NY grid, FFT 9 POINT requires a workspace of length 7 max(NX,NY) + NX × NY + 3.

ACKNOWLEDGEMENTS: This work was supported in part by National Science Foundation grant MCS 76-10225.

REFERENCES:

Houstis E., and T. Papatheodorou [1979], "Algorithm 543. FFT9: Fast solution of Helmholtz type partial differential equations", ACM Trans. Math. Software 5, pp. 490–493.

Houstis E., R. Lynch and T. Papatheodorou [1980], "A sixth order fast direct Helmholtz equation solver", Math. and Comp. in Simulation, Trans. of IMACS 22, pp. 91–97.

MODULE: FISHPAK HELMHOLTZ

AUTHORS: John Adams, Paul Swarztrauber, and Roland Sweet
(Modified for use in ELLPACK by Wayne R. Dyksen)

PURPOSE: Solves the standard five-point finite difference approximation to the Helmholtz equation

$$u_{xx} + u_{yy} + \lambda u = f(x,y)$$

in a rectangle subject to Dirichlet or Neumann boundary conditions on any side of the rectangle. Periodic boundary conditions are also allowed.

METHOD: Invokes the Fishpak routine HWSCRT which uses the standard five-point difference formula and the fast Fourier transform to approximate the solution of a Helmholtz equation in a rectangle with Dirichlet or Neumann boundary conditions on any side. FISHPAK HELMHOLTZ is compatible with the June 1979 version of Fishpak (Version 3).

If Neumann boundary conditions are specified on all four sides of the rectangle for a Poisson equation ($\lambda = 0$), then a solution may not exist. A constant, PERTRB, is calculated and subtracted from the right side f, which insures that a solution exists. Fishpak then computes this solution which is a least squares solution to the original approximation. This solution plus any constant is also a solution. Hence, the solution is not unique. The value of PERTRB should be small compared to the right side f; otherwise a solution is obtained to an essentially different problem. This comparison should always be made to insure that a meaningful solution has been obtained. The value of PERTRB is printed if LEVEL $\geqslant 1$.

Fishpak will attempt to find a solution even if $\lambda > 0$, in which case a solution may not exist.

PARAMETERS: None.

RESTRICTIONS:

Rectangular domain.
Two dimensions.
Helmholtz equation.
Dirichlet, Neumann, or periodic boundary conditions only.
Uniform grid, at least 5×5.

PERFORMANCE ESTIMATES: For an NX by NY grid, FISHPAK HELMHOLTZ requires memory for one copy of the unknowns as well as a workspace of length $4NY + [13 + INT(\log_2(NX))] \times NX$. The execution time is roughly proportional to $NX \times NY \times \log_2(NY)$, but also depends on the boundary conditions.

REFERENCES:

Swarztrauber P. N., and R. A. Sweet [1979], "Algorithm 541: Efficient Fortran subprograms for the solution of separable elliptic partial differential equations", ACM Trans. Math. Software 5, pp. 352–364.

MODULE: HODIE FFT

AUTHOR: Ronald F. Boisvert
(Modified for ELLPACK use by John J. Nestor, III)

PURPOSE: To solve the Helmholtz equation

$$Lu = u_{xx} + u_{yy} + fu = g(x,y) \tag{1}$$

(f constant) on a rectangular domain R with any combination of the boundary conditions:

1. u prescribed (Dirichlet conditions)
2. u_n prescribed (Neumann condition)
3. periodicity in x, y, or both.

METHOD: Second and fourth order accurate 9-point compact finite differences. Linear algebraic equation solution via the Fourier method (FFT).

MATHEMATICAL FORMULATION: Let $\partial R = \partial R_1 \cup \partial R_2 \cup \partial R_3$ denote the boundary of R where ∂R_1, ∂R_2, and ∂R_3 are those portions of ∂R on which Dirichlet, Neumann, and periodic boundary conditions hold respectively. Each of ∂R_1, ∂R_2, and ∂R_3 are unions of sides of R. Place a uniform NX by NY grid with mesh width h over R (NX$-$1 subintervals of width h in x, NY$-$1 subintervals of width h in y), and first consider the most common finite difference stencil:

c	b	c
b	a	b
c	b	c

$u = dh^2 g$

where either $a = -4 + h^2 f$, $b = d = 1$, $c = 0$, or $a = -20 + 6h^2 f$, $b = 4$, $c = 1$, $d = 6$. Both are second order accurate. After incorporating the boundary conditions and numbering the unknowns with the natural ordering one obtains the linear system $Mu = g$ which we write in block form as

$$\begin{bmatrix} A/\mu & B & & & \theta B \\ B & A & B & & \\ & \cdot & \cdot & \cdot & \\ & & \cdot & \cdot & \cdot \\ & & & \cdot\ \cdot\ \cdot & \\ & & B & A & B \\ \theta B & & & B & A/\nu \end{bmatrix} \begin{bmatrix} u_1 \\ u_2 \\ \cdot \\ \cdot \\ \cdot \\ u_{m-1} \\ u_m \end{bmatrix} = \begin{bmatrix} g_1/\mu \\ g_2 \\ \cdot \\ \cdot \\ \cdot \\ g_{m-1} \\ g_m/\nu \end{bmatrix}$$

$$A = \begin{bmatrix} a/\eta & b & & & & \tau b \\ b & a & b & & & \\ & \cdot & \cdot & \cdot & & \\ & & \cdot & \cdot & \cdot & \\ & & & \cdot & \cdot & \cdot \\ & & & b & a & b \\ \tau b & & & & b & a/\sigma \end{bmatrix} \qquad B = \begin{bmatrix} b/\eta & c & & & & \tau c \\ c & b & c & & & \\ & \cdot & \cdot & \cdot & & \\ & & \cdot & \cdot & \cdot & \\ & & & \cdot & \cdot & \cdot \\ & & & c & b & c \\ \tau c & & & & c & b/\sigma \end{bmatrix} \qquad (2)$$

A and B are both n by n. The scalars μ, ν, η, σ depend on the boundary conditions on the top, bottom, left, and right sides of the domain respectively. They are 2 for Neumann conditions and 1 otherwise. The scalars ϕ and τ are 1 when u is periodic in y and x respectively and 0 otherwise.

A number of very efficient direct methods for solving $\mathrm{M}u = g$ have been developed (see [Buzbee, Golub, and Nielson, 1970] and [Swarztrauber, 1977]). This module is based on the fact that high order accurate discretizations may be obtained by a proper choice of a, b, and c, and a perturbation of the right hand side. The resulting matrix problem is then exactly the same as (2).

The general form of the perturbed difference equations is

$$\mathrm{L}_h u = h^2 \mathrm{I}_h g + h \mathrm{J}_h u_n$$

where

$$\mathrm{L}_h u = \sum_{i=0}^{8} a_i u(p_i), \qquad \mathrm{I}_h g = \sum_j \beta_j g(q_j), \qquad \mathrm{J}_h u_n = \sum_k \gamma_k u_n(r_k).$$

Here p_0 is the central stencil point. For $p_0 \in \mathrm{R}$ the discretization points $p_1, \ldots, p_8$ are the nearest neighbors of p_0 and J_h is dropped. The auxiliary evaluation points q_j are taken near p_0, but not necessarily at grid points. For $p_0 \in \partial\mathrm{R}_2$ we drop the p_i and q_j outside $\mathrm{R} \cup \partial\mathrm{R}$ and the points r_k are taken on $\partial\mathrm{R}_2$ near p_0. The number and location of the points q_j and r_k may be chosen to increase the order of accuracy of the formula. To obtain $O(h^N)$ accuracy one must select the coefficients a_i, β_j, and γ_k such that the truncation error $\mathrm{L}_h v - h^2 \mathrm{I}_h \mathrm{L} v - h \mathrm{J}_h v_n = 0$ for all $v \in P^{N+1}$ (or $v \in P^N$ when p_0 is a boundary point where u_n is given). Such discretizations are studied in [Lynch and Rice, 1978] and [Boisvert, 1981a] where they are called HODIE methods.

Our discretization of (1) results in the matrix equation (2) with $a = -(480 - 118F + F^2)/24$, $b = (192 + 8F + F^2)/48$, $c = (48 + 5F)/48$, $F = h^2 f$, and right hand side given in Table 1. Note that in addition to each grid point, the function g is evaluated at the midpoint of each subrectangle in the grid. $O(h^4)$ convergence is shown in [Boisvert, 1981] for the case of $f < 0$ and in the case $f = 0$ with $\partial\mathrm{R}_1$ not empty.

We use the Fourier method to solve equation (2). This technique relies on the fact that A and B have a common set of linearly independent eigenvalues (see Table 2). Let Q be the matrix of these eigenvectors. If we premultiply each block equation in (2) by Q^{-1} and scale each block of unknowns by Q^{-1}, the system decouples to the solution of n tridiagonal systems of size m by m. The algorithm may be summarized as follows (see [Buzbee, Golub, and Nielson, 1970] for details).

Step 1. Compute $g_j \leftarrow Q^{-1} g_j$, $j = 1, ..., m$

Step 2. Solve $T_i u_j = g_j$, $i = 1, ..., n$

Step 3. Compute $u_j \leftarrow Q u_j$, $j = 1, ..., m$

The vectors u_i and g_i are composed of the ith components of each of the u_j and g_j subvectors respectively. T_i has the same form as M, except that A and B are replaced by their ith eigenvalues. Note also that the vectors u and g may coincide in memory and that the matrix M need never explicitly be formed.

Table 1. Right Side of 2D Finite Difference Formulae*

Type	$\mathrm{I}_h g + \mathrm{J}_h u_n$
Interior points	$\beta_0 g(0,0) + \beta_1 [g(h,0) + g(0,h) + g(-n,0) + g(0,-h)]$ $+\beta_2 [g(k,k) + g(-k,k) + g(-k,-k) + g(k,-k)]$
Neumann sides (right)	$0.5\{\beta_0 g(0,0) + \beta_1 [g(0,h) + g(0,-h) + 2g(-h,0)]$ $+2\beta_2 [g(-k,k) + g(-k,-k)]$ $+\gamma_0 u_x(0,0) + \gamma_1 [u_x(0,h) + u_x(0,-h)]\}$
Neumann corner (top right)	$0.25\{\beta_0 g(0,0) + 2\beta_1 [g(-h,0) + g(0,-h)] + 4\beta_2 g(-k,-k)$ $+\delta_0 [u_x(0,0) + u_y(0,0)] + \delta_1 [u_x(0,-h) + u_y(-h,0)]$ $+\delta_2 [u_x(0,-2h) + u_y(-2h,0)] + \delta_3 [u_x(0,-3h) + u_y(-3h,0)]\}$

* $F = h^2 f$, $k = h/2$, $\beta_0 = (48 - 5F)/24$, $\beta_1 = F/48$, $\beta_2 = 1$, $\gamma_0 = -12 + 13F/12$, $\gamma_1 = F/12$, $\delta_2 = 2 - F/48$, $\delta_3 = [(6 - F)/3 - 4\delta_2]/18$, $\delta_1 = 8\delta_2 + 33\delta_3 - (10 - F)\beta_2$, $\delta_0 = -2(\beta_0 + 2\beta_1) - (\delta_2 + 4\delta_3) - (4 - F/2)\beta_2$

Table 2. Eigenvalues/eigenvectors of A and B*

	Case: Nonperiodic in x ($\theta = 0$)	
	$\sigma = 1$	$\sigma = 2$
$\eta = 1$	$\rho_j = \cos[j\pi/(n+1)]$ $Q_{i,j} = \sin[ij\pi/(n+1)]$ $n = \mathrm{NX}-2$	$\rho_j = \cos[(2j-1)\pi/(2n)]$ $Q_{i,j} = \sin[i(2j-1)\pi/(2n)]$ $n = \mathrm{NX}-1$
$\eta = 2$	$\rho_j = \cos[(2j-1)\pi/(2n)]$ $Q_{i,j} = \cos[(i-1)(2j-1)\pi/(2n)]$ $n = \mathrm{NX}-1$	$\rho_j = \cos[(j-1)\pi/(n-1)]$ $Q_{i,j} = \cos[(i-1)\pi/(n-1)]$ $n = \mathrm{NX}$
	Case: Periodic in x ($\eta = \sigma = \tau = 1$)	
	$\rho_j = \cos(2k\pi/n)$ where $k = \mathrm{floor}(j/2)$ $Q_{i,j} = \cos[j(i-1)\pi/n]$ even; $-\sin[(j-1)(i-1)\pi/n]$, odd $n = \mathrm{NX}-1$	

* Eigenvalues of A and B are $a + 2b\rho_j$ and $b + 2c\rho_j$, $j = 1, ..., n$ respectively. NX is the number of grid lines in x.

Multiplication by Q^{-1} is equivalent to performing a discrete Fourier transform, and thus may be computed using FFT techniques. Hence the algorithm has $0(mn \log_2 n)$ complexity on a scalar computer. We use the FFT software of [Swarztrauber, 1982] which contains all the special transforms required and places no restriction on vector lengths (although the transform will be fast only when NX−1 is highly composite).

PARAMETERS: IORDER specifies the order of accuracy of the discretization. If IORDER=2 the 9-point second order accurate discretization described above is used. If IORDER=4 (the default case), the fourth order accurate discretization described above is used.

RESTRICTIONS: The grid must be at least 4 by 4. The grid spacings in x and y must be uniform and equal, i.e., for the rectangle (AX,BX) × (AY,BY) with NX grid lines [(NX−1) subintervals] in x and NY grid lines [(NY−1) subintervals] in y, (BX−AX)/(NX−1) must equal (BY−AY)/(NY−1).

SPECIAL CASES: When $f = 0$ and only Neumann and/or periodic boundary conditions are specified an arbitrary constant may be added to the solution. In this case the minimum norm solution is returned. If the user specifies the value of the solution at a grid point using the procedure module NONUNIQUE, then the solution which interpolates this value is used instead.

PERFORMANCE ESTIMATES: The execution time of this module is highly dependent upon the prime factorization of NX−1, being most efficient when NX−1 is highly composite and least efficient when NX−1 is prime. The best case is when $\text{NX}-1 = 2^k$ for some k. In this case the execution time of this module is roughly proportional to NX × NY × k. When NX−1 is prime this degrades to $O(\text{NY} \times \text{NX}^2)$. The difference in running time between the cases IORDER=2 and 4 is small when evaluations of the function g are inexpensive. This module requires a workspace of size 2(NX+NY) + NX(IORDER × NY/2−1).

ACKNOWLEDGEMENTS: This work was supported by the National Bureau of Standards.

REFERENCES:

Boisvert, R. F., [1981], "High order compact difference formulas for elliptic problems with mixed boundary conditions", in *Advances in Computer Methods for Partial Differential Equations*, IV (R. Vichnevetsky and R.S. Stepleman, eds.), IMACS, Rutgers Univ., 193–199.

Boisvert, R. F., [1981a], "Families of high order accurate discretizations of some elliptic problems", SIAM J. Sci. Stat. Comp. 2, pp. 268–284.

Boisvert, R. F., [1983], "A fourth order fast direct mthod for the Helmholtz equation", in *Elliptic Problem Solvers* II (G. Birkhoff and A. Schoenstadt, eds.), Academic Press, New York.

Buzbee, B. L., G. H. Golub and C. W. Nielson [1970], "On direct methods for solving Poisson's equations", SIAM J. Numer. Anal. 7, pp. 627–655.

Lynch, R. E. and J. R. Rice [1978], "High accuracy finite difference approximations to solutions of elliptic partial differential equations", Proc. Nat. Acad. Sci. 75, pp. 2541–2544.

Swarztrauber, P. N., [1977], "The methods of cyclic reduction, Fourier analysis and the FACR algorithm for the discrete solution of Poisson's equation on a rectangle", SIAM Rev. 19, pp. 490–501.

Swarztrauber, P. N., [1982], in *Parallel Computation* (G. Rodrigue, ed.), Academic Press, New York.

MODULE: HODIE FFT 3D

AUTHOR: Ronald F. Boisvert

PURPOSE: To solve the Helmholtz equation

$$\mathrm{L}u = u_{xx} + u_{yy} + fu = g(x,y,z) \tag{1}$$

(f constant) on a rectangular parallelepiped R with any combination of the following boundary conditions.

1. u prescribed (Dirichlet condition)
2. u_n prescribed (Neumann condition)
3. periodicity in x, y, or z

METHOD: Second and fourth order accurate 19-point compact finite differences. Linear algebraic equation solution via the Fourier method (FFT). These are extensions to three dimensions of the techniques used by the module HODIE FFT.

MATHEMATICAL FORMULATION: Let R be the rectangular parallelepiped (AX,BX) × (AY,BY) × (AZ,BZ) and let $\partial R = \partial R_1 \cup \partial R_2 \cup \partial R_3$ denote the boundary of R, where ∂R_1, ∂R_2, and ∂R_3 are those portions of ∂R on which Dirichlet, Neumann, and periodic boundary conditions hold respectively. Each of ∂R_1, ∂R_2, and ∂R_3 are unions of sides of R. Place a uniform NX by NY by NZ grid with mesh width h over R (NX−1 subintervals of width h in x, NY−1 subintervals of width h in y, NZ−1 subintervals of width h in z), and first consider the most common finite difference stencil:

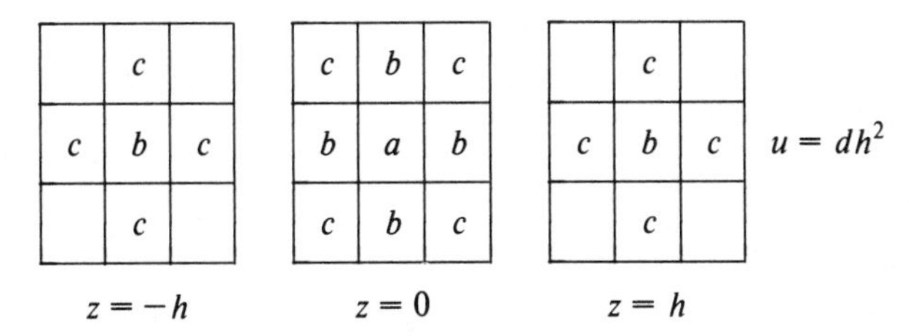

where either $a = -6 + h^2 f$, $b = d = 1$, $c = 0$, or $a = -36 + 6h^2 f$, $b = 4$, $c = 1$, $d = 6$. Both are second order accurate. After incorporating the boundary conditions and numbering the unknowns with the natural ordering one obtains the linear system $\mathrm{M}u = g$ which we write in block form as in (2) below, where $C = c\mathrm{I}$, E and F are block n by n, and A, B, and C are each l by l. The sizes n, m, and l are the number of unknowns in the x, y, and z directions respectively; these will depend upon the boundary conditions. The scalars m_1, n_1, μ_2, ν_2, μ_3, ν_3 depend upon the boundary consitions at $y = \mathrm{AY}$, $y = \mathrm{BY}$, $x = \mathrm{AX}$, $x = \mathrm{BX}$, $z = \mathrm{AZ}$, $z = \mathrm{BZ}$ respectively. They are 2 for Neumann conditions and 1 otherwise. The scalars τ_1, τ_2, and τ_3 are 1 when u is periodic in y, x, and z respectively and 0 otherwise.

$$\begin{bmatrix} D/\mu_1 & E & & & & \tau_1 E \\ E & D & E & & & \\ & \cdot & \cdot & \cdot & & \\ & & \cdot & \cdot & \cdot & \\ & & & \cdot & \cdot & \cdot \\ & & & E & D & E \\ \tau_1 E & & & & E & D/\nu_1 \end{bmatrix} \begin{bmatrix} u_1 \\ u_2 \\ \cdot \\ \cdot \\ \cdot \\ u_{m-1} \\ u_m \end{bmatrix} = \begin{bmatrix} g_1/\mu_1 \\ g_2 \\ \cdot \\ \cdot \\ \cdot \\ g_{m-1} \\ g_m/\nu_1 \end{bmatrix} \tag{2}$$

$$D = \begin{bmatrix} A/\mu_2 & B & & & & \tau_2 B \\ B & A & B & & & \\ & \cdot & \cdot & \cdot & & \\ & & \cdot & \cdot & \cdot & \\ & & & \cdot & \cdot & \cdot \\ & & & B & A & B \\ \tau_2 B & & & & B & A/\nu_2 \end{bmatrix} \qquad E = \begin{bmatrix} B/\mu_2 & C & & & & \tau_2 C \\ C & B & C & & & \\ & \cdot & \cdot & \cdot & & \\ & & \cdot & \cdot & \cdot & \\ & & & \cdot & \cdot & \cdot \\ & & & C & B & C \\ \tau_2 C & & & & C & B/\nu_2 \end{bmatrix}$$

$$A = \begin{bmatrix} a/\mu_3 & b & & & & \tau_3 b \\ b & a & b & & & \\ & \cdot & \cdot & \cdot & & \\ & & \cdot & \cdot & \cdot & \\ & & & \cdot & \cdot & \cdot \\ & & & b & a & b \\ \tau_3 b & & & & b & a/\nu_3 \end{bmatrix} \qquad B = \begin{bmatrix} b/\mu_3 & c & & & & \tau_3 c \\ c & b & c & & & \\ & \cdot & \cdot & \cdot & & \\ & & \cdot & \cdot & \cdot & \\ & & & \cdot & \cdot & \cdot \\ & & & c & b & c \\ \tau_3 c & & & & c & b/\nu_3 \end{bmatrix}$$

A number of very efficient direct methods for solving $\mathrm{M}u = g$ have been developed (see [Buzbee, Golub, and Nielson, 1970] and [Swarztrauber, 1977]). This module is based on the fact that high order accurate discretizations may be obtained by a proper choice of a, b, and c, and a perturbation of the right hand side. The resulting matrix problem is then exactly the same as (2).

The general form of the perturbed difference equations is

$$\mathrm{L}_h u = h^2 \mathrm{I}_h g + h \mathrm{J}_h u_n$$

$$\mathrm{L}_h u = \sum_{i=0}^{18} \alpha_i u(p_i), \qquad \mathrm{I}_h g = \sum_j \beta_j g(q_j), \qquad \mathrm{J}_h u_n = \sum_k \gamma_k u_n(r_k).$$

Here p_0 is the central stencil point. For $p_0 \in \mathrm{R}$ the discretization points $p_1, .., p_{18}$ are the nearest neighbors of p_0 and J_h is dropped. The auxiliary evaluation points q_j are taken near p_0, but not necessarily at grid points. For $p_0 \in \partial \mathrm{R}_2$ we drop the p_i and q_j outside $\mathrm{R} \cup \partial R$ and the points r_k are taken on $\partial \mathrm{R}_2$ near p_0. The number and location of the points q_j and r_k may be chosen to increase the order of accuracy of the formula. To obtain $O(h^N)$ accuracy we select coefficients α_i, β_j, and γ_k such that the truncation error $\mathrm{L}_h v - h^2 \mathrm{I}_h \mathrm{L} v - h \mathrm{J}_h v_n = 0$ for all $v \in P^{N+1}$ (or $v \in P^N$ when p_0 is a boundary

Table 1. Right Side of 3D Finite Difference Formulae[1]

Type	$\mathrm{I}_h g + \mathrm{J}_h u_n$
Interior points	$\beta_0 g(0,0,0) + \beta_1 [g(h,0,0) + g(0,h,0) + g(-h,0,0) + g(0,-h,0) + g(0,0,h) + g(0,0,-h)] + \beta_2 [g(k,k,k) + g(-k,k,k) + g(-k,-k,k) + g(k,-k,k) + g(k,k,-k) + g(-k,k,-k) + g(-k,-k,-k) + g(k,-k,-k)]$
Neumann faces[2] ($x = \mathrm{BX}$)	$\{\beta_0 g(0,0,0) + \beta_1 [g(0,h,0) + 2g(-h,0,0) + g(0,-h,0) + g(0,0,h) + g(0,0,-h)] + 2\beta_2 [g(-k,k,k) + g(-k,-k,k) + g(-k,k,-k) + g(-k,-k,-k)] + \gamma_0 u_x(0,0,0) + \gamma_1 [u_x(0,h,0) + u_x(0,-h,0) + u_x(0,0,h) + u_x(0,0,-h)]\}/2$
Neumann edges[2] ($x = \mathrm{BX}$) ($y = \mathrm{BY}$)	$\{\beta_0 g(0,0,0) + \beta_1 [2g(-h,0,0) + 2g(0,-h,0) + g(0,0,h) + g(0,0,-h)] + 4\beta_2 [g(-k,-k,-k) + g(-k,-k,-k)] + \delta_0 [u_x(0,0,0) + u_y(0,0,0)] + \delta_1 [u_x(0,-h,0) + u_y(-h,0,0)] + \delta_2 [u_x(0,-2h,0) + u_y(-2h,0,0)] + \delta_3 [u_x(0,-3h,0) + u_y(-3h,0,0)] + \delta_4 [u_x(0,0,h) + u_y(0,0,h) + u_x(0,0,-h) + u_y(0,0,-h)] + \delta_5 [u_x(0,-h,h) + u_y(-h,0,h) + u_x(0,-h,-h) + u_y(-h,0,-h)]\}/4$
Neumann corners[2] ($x = \mathrm{BX}$) ($y = \mathrm{BY}$) ($z = \mathrm{BZ}$)	$\{\beta_0 g(0,0,0) + 2\beta_1 [g(-h,0,0) + g(0,-h,0) + g(0,0,-h)] + 8\beta_2 g(-k,-k,-k) + \eta_0 [u_x(0,-h,0) + u_y(-h,0,0) + u_z(-h,0,0) + u_x(0,0,-h) + u_y(0,0,-h) + u_z(0,-h,0)] + \eta_1 [u_x(0,-2h,0) + u_y(-2h,0,0) + u_z(-2h,0,0) + u_x(0,0,-2h) + u_y(0,0,-2h) + u_z(0,-2h,0)] + \eta_2 [u_x(0,-3h,0) + u_y(-3h,0,0) + u_z(-3h,0,0) + u_x(0,0,-3h) + u_y(0,0,-3h) + u_z(0,-3h,0)] + \eta_3 [u_x(0,-h,-h) + u_y(-h,0,-h) + u_z(-h,-h,0)] + \eta_4 [u_x(0,-2h,-2h) + u_y(-2h,0,-2h) + u_z(-2h,-2h,0)] + \eta_5 [u_x(0,-h,-2h) + u_x(0,-2h,-h) + u_y(-h,0,-2h) + u_y(-2h,0,-h) + u_z(-2h,-h,0) + u_z(-h,-2h,0)]\}/8$

[1] $F = h^2 f$, $k = h/2$, $\beta_0 = 2 - F/4$, $\beta_1 = F/48$, $\beta_2 = 1/2$, $\gamma_0 = -12 + 11F/12$, $\gamma_1 = F/12$, $\delta_0 = (-23 + 125F/48)/3$, $\delta_1 = -6 + F/4$, $\delta_2 = 2 - F/48$, $\delta_3 = -(1 + F/24)/3$, $\delta_4 = (-1 + F/24)/2$, $\delta_5 = (1 + F/8)/2$, $\eta_0 = (-2328 + 245F)/144$, $\eta_1 = (141 - 14F)/18$, $\eta_2 = -(24 + F)/72$, $\eta_3 = (840 - 103F)/36$, $\eta_4 = (57 - 7F)/3$, $\eta_5 = (-1752 + 221F)/144$.

[2] Other points are obtained from symmetry.

point where u_n is given). Such discretizations are studied in [Lynch and Rice, 1978] and [Boisvert, 1981a] where they are called HODIE methods.

Our discretization of (1) results in the matrix equation (2) with $a = -24+5F-F^2/4$, $b = 2-F/24+F^2/48$, $c = 1+5F/48$, $d = 1$, $F = h^2 f$, and right hand side given in Table 1. Note that in addition to each grid point, the function g is evaluated at the center of each box in the grid.

We use the Fourier method to solve equation (2). This technique relies on the fact that A, B, and C have a common set of linearly independent eigenvectors (see Table 2 of the HODIE FFT description). Let Q be the matrix of these eigenvectors. If we premultiply each of the l by l blocks in (2) by Q^{-1} and scale each corresponding set of unknowns by Q^{-1}, then A, B, and C are replaced by diagonal matrices of their eigenvalues. If we then reorder the equations and unknowns, the system $\mathrm{M}u = g$ decouples to the solution of l two-dimensional block tridiagonal systems. At this point we may apply the Fourier algorithm given in the description of HODIE FFT l times.

Multiplication by Q^{-1} is equivalent to performing a discrete Fourier transform, and hence may be computed using FFT techniques giving this algorithm $O(l\,m\,n[\log^2 n + \log^2 l])$ complexity on a scalar computer. We use the FFT software of [Swarztrauber, 1982] which contains all the special transforms required and places no restriction on vector lengths (although the transforms will be fast only when both NX−1 and NZ−1 are highly composite).

PARAMETERS: IORDER specifies the order of accuracy of the discretization. If IORDER=2 the 19-point second order accurate discretization given above is used. If IORDER=4 (the default case), the fourth order accurate discretization given above is used.

RESTRICTIONS: The grid spacings in x, y, and z must be uniform and equal, i.e., (BX−AX)/(NX−1) = (BY−AY)/(NY−1) = (BZ−AZ)/(NZ−1). The grid must be at least 4 by 4 by 4.

SPECIAL CASES: When $f = 0$ and only Neumann and/or periodic boundary conditions are specified an arbitrary constant may be added to the solution. In this case the minimum norm solution is returned. If the user specifies the value of the solution at a grid point using the procedure module NONUNIQUE, then the solution which interpolates this value is used instead.

PERFORMANCE ESTIMATES: The execution time of this module is highly dependent upon the prime factorizations of NX−1 and NZ−1, being most efficient when they are highly composite and least efficient when they are primes. The best case is when NX−1 $= 2^{k_1}$ and NZ−1 $= 2^{k_2}$ for integers k_1 and k_2. In this case the execution time of this module is roughly proportional to NZ×NY×k_1 + NX×NY×k_2. When NX−1 is prime, for instance, this degrades to O(NZ×NY×NX^2). The difference in running time between the cases IORDER=2 and 4 is small when evaluations of the function g are inexpensive. This module requires a workspace of size

NX×NY×(IORDER×NZ/2−1) + 2(NX×NZ + NX×NY + NY×NZ).

REFERENCES: See the description of HODIE FFT.

TRIPLE

MODULE: HODIE 27 POINT 3D

AUTHOR: Robert E. Lynch
(Modified for use in ELLPACK by Ronald F. Boisvert)

PURPOSE: Discretizes the Dirichlet problem for the Poisson equation on a cube and solves the resulting equations.

METHOD: Uses a sixth-order 27-point compact finite difference discretization (HODIE). The resulting linear system is solved using tensor product methods.

RESTRICTIONS:
Cubical domain
Three dimensions
Poisson equation
Uniform grid of size at least $3 \times 3 \times 3$
$h_x = h_y = h_z$
Dirichlet boundary conditions

PERFORMANCE ESTIMATES: For an $N \times N \times N$ grid a workspace of length $2N^3 + 4N - 2$ is required and the number of unknowns is $(N-2)^3$.

REFERENCES:
R.E. Lynch [1977], "$O(h^6)$ accurate finite difference approximation to solutions of the Poisson equation in three variables", Report CSD-TR 221, Computer Sciences. Purdue University,

R.E. Lynch [1984], "$O(h^4)$ and $O(h^6)$ Finite difference approximations to the Helmholtz equation in n-dimensions", in *Advances in Computer Methods for Partial Differential Equations* V (R. VichneVetsky and R.S. Stepleman, eds.), IMACS, New Brunswick, N.J., pp. 199–202.

ACKNOWLEDGEMENTS: Work on this module was supported in part by National Science Foundation grants MCS77-01408 and MCS79-01437.

MODULE: MARCHING ALGORITHM

AUTHOR: R. E. Bank

PURPOSE: Solves separable, self-adjoint elliptic operators with general boundary conditions.

METHOD: This module is based on a symmetric 5-point finite difference discretization. The linear equations are solved using the generalized marching algorithm. A least-squares solution is computed for singular but semi-definite problems.

PARAMETERS: KGMA is the marching parameter; its default value is 2.

RESTRICTIONS: Rectangular domain
Two dimensions
Self-adjoint, separable equation
Periodic boundary conditions are not allowed
Uniform grid having at least 3, 4 or 5 x (y) grid lines if there are respectively 0, 1 or 2 Dirichlet boundary conditions in the x (y) direction.

PERFORMANCE ESTIMATES: The operation count for an NX by NY grid is of the order of $\mathrm{NX} \times \mathrm{NY} \times \log(\mathrm{NY}/\mathrm{KGMA})$.

ACKNOWLEDGEMENTS: This work was supported in part by the Office of Naval Research under contract N00014-75-C-0243. The original version of the code was written while the author was a summer visitor at Argonne National Laboratory.

REFERENCES:

Bank, R. E., [1978], "Algorithm 527: A Fortran implementation of the generalized marching algorithm", ACM Trans. Math. Software 4, pp. 165–176.

TRIPLE

MODULE: MULTIGRID MG00

AUTHORS: Hartmut Foerster and Kristian Witsch

PURPOSE: Solves the standard five-point finite-difference approximation to a second-order linear elliptic equation of the form

$$a(x,y)\,u_{xx} + b(x,y)\,u_{yy} + c(x,y)\,u = f$$

which is defined on a rectangle with general boundary conditions

$$\alpha(x,y)\,u + \beta(x,y)\,u_n = g$$

at each single side of the rectangular domain. (The subscript n denotes the derivative in the outward-pointing normal direction.)

METHOD: Version 1.0 of the MULTIGRID MG00 triple module implements different multigrid algorithms for different problems to allow a maximum of efficiency and robustness. In particular, isotropic $(a = b)$ and anisotropic differential operators are distinguished and different smoothing procedures and grid transfer operators are applied.

For the pure Neumann problem for Poisson's equation, MULTIGRID MG00 will calculate a solution if and only if the data satisfies the discrete compatibility condition. This condition may be checked and/or enforced by modifying the right-hand side f (see the description of the parameter INEUM). Other singular problems are not specially treated.

MULTIGRID MG00 attempts to find a solution even if ac or $\alpha\beta$ is not strictly positive; this might cause the coarse-grid correction process to magnify, instead of reduce, the error. (The parameter NMIN allows the coarsest grid to be finer, which is a remedy for at least some of these problems.)

MATHEMATICAL FORMULATION: The elliptic operator

$$Lu = au_{xx} + bu_{yy} + cu$$

is discretized by central difference approximations of order 2 on a uniformly spaced grid with grid lines matching the boundary. For non-Dirichlet boundary conditions the normal derivative is approximated by central differences. Function values beyond the boundary are eliminated.

In order to reduce storage requirement and to save operational work the resulting algebraic equations are transformed if and only if $b \equiv 1$, $\beta \equiv 1$, or if the grid has different grid spacings in x and y.

Corresponding discretizations are also used on coarser grids. Coarse subgrids are automatically generated by successively doubling the grid spacings as long as the grid lines match the boundary. It is this hierarchy of coarser and coarser grids, with a geometrically decreasing number of grid points, which is an essential element of multigrid algorithms. MULTIGRID MG00 provides two algorithmic approaches; one is by cyclic multigrid iterations, the other is called **full multigrid** and yields an approximate solution with an error which is comparable to the discretization error.

Multigrid iterations (MGI; relevant parameters are METHOD, UINIT, ITER, IGAMMA, and NMIN) start with some arbitrary (zero or user-supplied) approximation on the given grid and reduce the error in the linear system by cycling between that grid and coarser grids. V-type (IGAMMA = 1) and W-type cycles (IGAMMA = 2) are distinguished. For both schemes the total computational work to achieve a fixed accuracy is a small multiple of the number of unknowns. MGI reduces the error by about 1/20 for each iteration (V-cycles) independent of the grid size. ITER = 0 gives a single iteration which is cheaper by about 25 percent (30–40 percent for anisotropic operators, but less accurate with error reduction of about 1/6 to 1/10.) W-cycles require 50% more work than V-cycles but have proved to be especially robust. Experimentally measured error reductions, per iteration are 1/30 and 1/8 – 1/14, respectively. A small degradation may occur for problems which are not positive. Other sources that may possibly cause a degradation of the convergence rates are discontinuities in the coefficients and operator singularities.

The MULTIGRID MG00 iteration schemes consist of

* adaptive error smoothing by decoupling pattern Gauss-Seidel relaxation schemes,
* fine-to-coarse residual transfer by an accommodative weighting,
* coarse-to-fine transfer of corrections by linear interpolation,
* direct solution on coarsest grid by Gaussian elimination.

Cyclic multigrid iterations on each level of discretization are applied to **full multigrid.** The MG00 FMG algorithm solve a given problem in one cycle per level up to an error which is comparable to the discretization error, the error of the exact solution of the discretized problem. The size of the coarsest grid, the starting level, is implicitly defined by the number of grid lines in x and y, and the parameter NMIN. The FMG interpolation uses the discrete differential equations to obtain initial approximations from coarse-grid approximate solutions. This process incorporates an inherent partial smoothing. The FMG interpolation is of order four, diminished to three at boundary points in case of non-Dirichlet boundary conditions. If ITER > 1 is specified, then ITER–1 MGI steps on the finest level (= the given grid) follow the FMG algorithm to further reduce the algebraic error, i.e. the error to solving the difference equations. (The discretization error in the exact continuous solution usually cannot be diminished by additional iterations!)

PARAMETERS:

METHOD selects the multigrid algorithm:
= 0 for the Full Multigrid algorithm (FMG)
= 1 for cyclic Multigrid iterations (MGI)
The default is METHOD = 0

UINIT specifies the initial guess of U(X,Y) for MGI:
= 0 for zero as initial guess
= 1 for the current U(X,Y) as initial approximation.

Note that TRIPLE. SET(U=⟨name⟩) allows one to provide this as desired. Also consider the triples SET U BY BLENDING and SET U BY BICUBICS. The default is UINIT = 0.

NMIN is the minimum of the number of grid spacings the coarsest grid must have in both x and y. The default is NMIN = 2.

INEUM selects the action for pure Neumann boundary conditions.

- = 0 to ignore the discrete compatibility check and continue
- = 1 to orthogonalize the data so that the discrete compatibility condition becomes satisfied
- = 2 to verify that the discrete compatibility check is satisfied
- = 3 to test the discrete compatibility condition, and if it is not satisfied, orthogonalize the data so it becomes satisfied. The default is INEUM = 0.

ITER $\geqslant 0$ determines the number of multigrid iterations. With METHOD = 1, max {1, ITER} MGI cycles are performed, and with METHOD = 0, max {0, ITER − 1} MGI cycles follow the FMG algorithm. The default is ITER = 0.

IGAMMA determines the type of multigrid iterations: V-type (IGAMMA = 1) or W-type cycles (IGAMMA = 2). The default is IGAMMA = 1.

RESTRICTIONS: The partial differential equation is to be written in standard form with no cross derivative and first-order terms present. Periodic boundary conditions are not allowed. The domain must be rectangular in two dimensions with at least a 5 by 5 grid which is uniform in x and in y. There are at most 9 levels involved in the multigrid solution process (the minimum number is 2).

MULTIGRID MG00 is not well suited for rapidly changing coefficients. (Multigrid software with Galerkin-type approximations to coarse-grid differential operators, for example, performs better instead.) For problems with highly oscillatory solutions, the FMG method (METHOD = 0) should have ITER = 1 or 2 instead of the usual choice ITER = 0. To solve the pure Neumann problem for Poisson's equation on machines with short word length, round-off errors might force the orthogonalization of the data also on coarse grids. (One extra subroutine call can be inserted after the residual transfer to satisfy the discrete compatibility condition on a coarser level with sufficient accuracy.)

REFERENCES:

Foerster, H., and K. Witsch [1982], "Multigrid software for the solution of elliptic problems on rectangular domains: MG00 (Release 1)", in *Multigrid Methods* (W. Hackbusch and U. Trottenberg, eds.), Lecture Notes in Mathematics, Vol. 960, Springer-Verlag, Berlin.

TRIPLE

MODULE: P2C0 TRIANGLES [†]

AUTHOR: Granville Sewell

PURPOSE: Discretizes a general elliptic operator with general boundary conditions and solves the resulting systems of linear equations.

METHOD: Uses Galerkin's method with 6-node quadratic triangular elements, user-controlled grading of the triangular mesh, and the frontal method to organize out-of-core storage of the matrix if necessary.

The grid is used to define an initial triangularization which will have approximately 4 triangles per grid square which intersects the domain. The number of triangles desired in the final triangularization, NTRI (see parameters below), must be larger than this number. The closure of the intersection of any grid square with the domain must be convex or nearly so. Thus it is necessary in general that any domain corner with exterior angle less than 180° be cut by a grid line which divides the exterior angle into two parts. In the case of a 90° exterior angle with edges parallel to the axes, it is sufficient to put a grid point at that corner.

The Fortran function D3EST(X,Y) must be user-supplied. P2C0 TRIANGLES grades the initial triangularization so that the final triangularization is most dense where D3EST is largest. In particular, it attempts to distribute D3EST$(x_j,y_j) \times h_j^3$ uniformly, where (x_j,y_j) is the center of triangle j and with diameter h_j. Ideally, D3EST should be an estimate of the function

$$\max_{i+j=3} || D_x^i D_y^j u ||_\infty$$

in which case it is possible to obtain optimal order convergence to the solution of some singular problems. One may use D3EST = 1.0 to obtain a uniform grading of the triangulation.

PARAMETERS: NTRI is the number of triangles desired in the final triangularization. MEM is the length of the workspace storage and it set to the default value 46NTRI + 15NTRI**1.5 + 10. If external storage is used, MEM should be 71NTRI.

RESTRICTIONS: The problem must on on a two-dimensional domain specified by the general domain ELLPACK syntax. The equation must be given in self-adjoint form.

REFERENCES:

Sewell, G., [1979], "A finite element program with automatic user controlled mesh grading", in *Advances in Computer Methods for Partial Differential Equations* III, (R. Vichnevetsky and R.S. Stepleman, eds.), IMACS, Rutgers Univ., New Brunswick, N.J., pp. 8–10.

[†] This module was named 2DEPEP in earlier ELLPACK systems. It is not the software sold by IMSL under that name.

MODULE: SET

AUTHOR: John F. Brophy

PURPOSE: Sets the ELLPACK function U(X,Y) or U(X,Y,X) to a specified Fortran function. It is used to initialize iterations of various types.

METHOD: The evaluations of U, UX, UY, etc. are made from a table of values of the given function at the grid points. The system module for local quadratic interpolation is used to obtain values and derivatives. See the INTERPOLATION module description.

PARAMETERS: The parameter U = ⟨name⟩ specifies that the Fortran function ⟨name⟩ is used to create the table of values at the grid points. The default name is ZERO.

PERFORMANCE ESTIMATES: The table of values (of size NX × NY × NZ for a NX by NY by NZ grid) used by this module is used also by many other modules, so it is unlikely that this module will increase the memory requirements of an ELLPACK run. It requires NX × NY evaluations of the function ⟨name⟩ to fill the table and then a modest amount of arithmetic to compute a value of U, UX, etc. from that.

ACKNOWLEDGEMENTS: The development of this module was supported in part by the Department of Energy under contract DE-AC02-81ER-10997

MODULE: SET U BY BICUBICS

AUTHOR: Wayne R. Dyksen

PURPOSE: This module sets the function U(X,Y) to a Hermite bicubic piecewise polynomial which interpolates the boundary conditions and which is identically zero in the interior grid rectangles.

METHOD: The interpolation problem for Hermite bicubics is solved for the boundary values at the boundary collocation points (defined in the HERMITE COLLOCATION module). The support of the interpolating piecewise cubic is restricted to grid rectangles adjacent to the boundary. This method is the one used by the INTERIOR COLLOCATION module to homogenize the boundary conditions of the discrete elliptic problem (this technique was developed by E. N. Houstis and W. R. Mitchell)

PERFORMANCE ESTIMATES: The work to compute the interpolant is proportional to NX+NY for an NX by NY grid. The work to evaluate the resulting U(X,Y) is the same as for any Hermite bicubic. The workspace required is 26(NX+NY-2)+2 max(NX, NY).

RESTRICTIONS: The domain must be rectangular with a 3 by 3 grid and the boundary conditions must be uncoupled.

ACKNOWLEDGEMENTS: The development of this module was supported in part by the Department of Energy under contract DE-AC02-81ER-10997.

TRIPLE

MODULE: SET U BY BLENDING

AUTHOR: Linda C. Thiel

PURPOSE: This module determines U(X,Y) to be a smooth bivariate function which exactly matches given boundary conditions on a rectangular domain. On each of the four boundary segments, the boundary conditions may be any linear combination, with constant coefficients, of function value and normal derivative.

METHOD: The method is based on transfinite "blending function" interpolation. A sketch of the details is given below and a thorough account can be found in the reference given at the end. In brief, the boundary conditions lead to a set of two small (2×2) linear systems to be solved to determine appropriate blending functions. These blending functions are then used to construct a Boolean sum interpolant which exactly matches the given boundary conditions and is continuous over the domain of interest.

MATHEMATICAL FORMULATION: A general rectangular domain is mapped to the unit square. We only treat this case for simplicity and consider the boundary conditions as follows:

$$\begin{aligned}
\mathrm{L}_0[U] &= \alpha_0 U(0,y)+\beta_0 U_x(0,y) = g_0(y) && \text{along } x=0\\
\mathrm{L}_1[U] &= \alpha_1 U(1,y)+\beta_1 U_x(1,y) = g_1(y) && \text{along } x=1\\
\mathrm{M}_0[U] &= \zeta_0 U(x,0)+\eta_0 U_y(x,0) = h_0(x) && \text{along } y=0\\
\mathrm{M}_1[U] &= \zeta_1 U(x,1)+\eta_1 U_y(x,1) = h_1(x) && \text{along } y=1
\end{aligned}$$

where α_i, ζ_i, β_i and η_i for $i=0,1$ are constants.

We first consider the case in which the boundary conditions commute, i.e., the conditions are consistent at the corners:

$$\mathrm{L}_i\mathrm{M}_j[U] = \mathrm{M}_j\mathrm{L}_i[U] \qquad i,j=0,1.$$

We define two projectors as follows:

$$\begin{aligned}
\mathrm{P}_x[U] &= \phi_0(x)\mathrm{L}_0[U]+\phi_1(x)\mathrm{L}_1[U]\\
\mathrm{P}_y[U] &= \psi_0(y)\mathrm{M}_0[U]+\psi_1(y)\mathrm{M}_1[U]
\end{aligned}$$

where ϕ_i and ψ_j, called blending functions, satisfy the cardinality conditions:

$$\begin{aligned}
\mathrm{L}_i[\phi_k] &= \delta_{i,k} && \text{for } i,k=0,1 \\
\mathrm{M}_j[\psi_l] &= \delta_{j,l} && \text{for } j,l=0,1
\end{aligned} \qquad (\delta_{i,k}=\text{Kronecker } \delta).$$

Then the function U_0 obtained from the Boolean sum of P_x and P_y exactly satisfies all of the specified boundary conditions, i.e.

$$U_0 = (\mathrm{P}_x \oplus \mathrm{P}_y)[U] = \mathrm{P}_x[U]+\mathrm{P}_y[U]-\mathrm{P}_x\mathrm{P}_y[U].$$

The cardinality conditions above can be satisfied by polynomials of maximal degree three. The blending functions depend only on the constant coefficients α_i, ζ_i, β_j and η_j. They are determined by solving small linear systems and are of the form

$$\phi_i(x) = a_i + b_i x + c_i$$
$$\psi_j(x) = p_j + q_j x + r_j x^2 + s_j x^3.$$

For boundary operators L_i and M_j which do not commute:

$$L_i M_j[U] \neq M_j L_i[U] \qquad \text{for } i,j = 0,1,$$

the projection operators P_x and P_y are still determined as above. Now however,

$$P_x P_y[U] \neq P_y P_x[U]$$

and the interpolant $U_0 = (P_x \oplus P_y)[U]$ does not satisfy the boundary conditions. A special interpolant for corner inconsistencies is constructed as follows: At each corner (i,j) construct a function $w_{i,j}$ such that

$$L_k M_l[w_{i,j}] = \delta_{i,k}\delta_{j,l} L_i M_j[U]$$
$$M_l L_k[w_{i,j}] = \delta_{i,k}\delta_{j,l} M_j L_i[U].$$

The following functions $w_{i,j}$ satisfy the above conditions:

$$w_{00}(x,y) = \phi_0(x)\psi_0(y)\{T(\theta_{00})L_0M_0[U] + (1 - T(\theta_{00}))M_0L_0[U]\}$$
$$w_{01}(x,y) = \phi_0(x)\psi_1(y)\{T(\theta_{01})M_1L_0[U] + (1 - T(\theta_{01}))L_0M_1[U]\}$$
$$w_{10}(x,y) = \phi_1(x)\psi_0(y)\{T(\theta_{10})M_0L_1[U] + (1 - T(\theta_{10}))L_1M_0[U]\}$$
$$w_{11}(x,y) = \phi_1(x)\psi_1(y)\{T(\theta_{11})L_1M_1[U] + (1 - T(\theta_{11}))M_1L_1[U]\}$$

where

$$\theta_{00} = \arctan\left(\frac{y}{x}\right), \qquad \theta_{10} = \arctan\left(\frac{1-x}{y}\right),$$

$$\theta_{01} = \arctan\left(\frac{x}{1-y}\right), \qquad \theta_{01} = \arctan\left(\frac{1-y}{1-x}\right).$$

The functions

$$T(\theta_{i,j}) = 1 - \frac{2\theta_{i,j}}{\pi}, \qquad i,j = 0,1$$

are used for Dirichlet boundary conditions and

$$T(\theta_{i,j}) = \left(\frac{2\theta_{i,j}}{\pi} - 1\right)^2 \left(\frac{4\theta_{i,j}}{\pi} + 1\right), \qquad i,j = 0,1$$

are used for the more general operators L_i and M_j. Now, let

$$W(x,y) = w_{00}(x,y) + w_{01}(x,y) + w_{10}(x,y) + w_{11}(x,y)$$

and define

$$\begin{aligned} V(x,y) &= (P_x \oplus P_y)[U - W] \\ &= P_x[U - W] + P_y[U - W] \\ &= P_x[U] - P_x[W] + P_y[U] - P_y[W]. \end{aligned}$$

The interpolant

$$U_0(x,y) = W(x,y) + V(x,y)$$

satisfies the boundary conditions

$$L_i[U_0] = L_i[U] \qquad \text{and} \qquad M_j[U_0] = M_j[U].$$

The derivatives $\frac{\partial}{\partial x}$, $\frac{\partial}{\partial y}$, $\frac{\partial^2}{\partial x \partial y}$, $\frac{\partial^2}{\partial x^2}$, and $\frac{\partial^2}{\partial y^2}$ are determined numerically by taking a step size h that is a small fraction of the average grid size. For instance, to determine the value of $\frac{\partial}{\partial x}$ at the point (x,y), the value of the interpolant is calculated at $(x+h,y)$ and the derivative is evaluated using standard finite difference techniques.

Each of the four possible cases — Dirichlet consistent, Dirichlet inconsistent, mixed consistent and mixed inconsistent — is treated separately by the module.

PARAMETERS: None.

RESTRICTIONS: The module is restricted to a rectangular domain in two dimensions. The boundary conditions must be a linear combination of normal derivative and function value with constant coefficients. At least one side should be function value alone, so as to "pin down" the interpolant. An interpolant will be evaluated in any case; however, the user should be aware of possible nonuniqueness of the solution.

PERFORMANCE ESTIMATES: The workspace is fixed and independent of problem size. However, if the boundary conditions are inconsistent, the run time increases appreciably.

ACKNOWLEDGEMENTS: Both the research reported in the reference and the development of the code were supported by the Office of Naval Research and the Air Force Office of Scientific Research under contract N00014-80-C-0716 to Drexel University.

PROCEDURE

MODULE: DISPLAY MATRIX PATTERN

AUTHOR: Wayne R. Dyksen

PURPOSE: Prints the pattern of nonzeros in the coefficient matrix of the linear system.

METHOD: A matrix of characters is printed, one for each element in the coefficient matrix of the linear system. Different characters are printed for nonzero or zero elements, giving a picture of the nonzero/zero pattern of the coefficient matrix.

PARAMETERS: The following parameters can be used to format the output of DISPLAY MATRIX PATTERN. The default values are given in parenthesis.

MATZER = char	MATZER is printed to represent zero off-diagonal entries in the matrix. Default: 1H.
MATNZR = char	MATNZR is printed to represent nonzero off-diagonal entries in the matrix. Default: 1HX
MATDZR = char	MATDZR is printed to represent zero diagonal entries in the matrix. Default: 1H0
MATDNZ = char	MATDNZ is printed to represent nonzero diagonal entries in the matrix. Default: 1HD
MATBLK = integer	The matrix pattern will be printed in MATBLK by MATBLK blocks; that is, a blank row (column) will be printed every MATBLK rows (columns). Default: number of equations.
MATLNL = integer	The output line length is taken to be MATLNL; that is, the matrix pattern is printed MATLNL columns at a time. If the number of columns (including blank columns resulting from blocking) is greater than MATLNL, then the matrix is printed in strips of width MATLNL so that the resulting output can be pieced together. Default: 120
EPSMAT = real	Any matrix entry whose magnitude is less than EPSMAT will be considered zero. Default: 0.0
MATNBR = 0 or 1	MATNBR is 1, then the rows of the matrix will be numbered. Default: 0
MATNBC = 0 or 1	If MATNBC is 1, then the columns of the matrix will be numbered. Default: 0

MATOUT = integer The matrix pattern will be written on file MATOUT. This is useful if one wants to save the matrix pattern to be used as input to a plotting routine to produce reduced versions of the pattern. Default: standard output.

Since none of the display matrix parameters are used for calulating array dimensions, the integer and real parameters may be any legal Fortran expression containing ELLPACK predefined or user defined variables. For example, MATBLK may be given in terms of the number of X points as MATBLK = I1NGRX − 2.

PERFORMANCE ESTIMATES: For N equations, a workspace of length (N − 1)/MATBLK + N is required.

ACKNOWLEDGEMENTS: Work on this module was supported in part by Department of Energy contract DE-AC02-81E10997 with Purdue University.

PROCEDURE

MODULE: DOMAIN FILL

AUTHOR: John R. Rice

PURPOSE: The domain processor has a method to locate one interior point and then it tags grid points as being the interior using the **assumption that the interior is connected by the grid lines**. This assumption is not always satisfied, especially for domains with sharp corners. This procedure is used to specify additional points of the domain and the interior fill algorithm is restarted at each given initial point. It can also be used to fill the exterior (inside) of holes which might be incompletely filled for the same reason.

METHOD: The method is the same as in the domain processor, the subprograms used are almost identical.

A user can identify points that have not been marked properly by examining the two-dimensional type array printed by the domain processor with LEVEL=2. or by OUTPUT. TABLE DOMAIN. Interior grid points are indicated by positive integers in this table.

PARAMETERS: NFILL = number of additional initial points (default=1). The (x,y) coordinates of the additional points must be placed into the arrays

R7DFXX(I), I = 1, NFILL
R7DFYY(I), I = 1, NFILL

in a FORTRAN segment preceding this module. These two arrays are declared and dimensioned automatically. The values need not be given with great accuracy, as the procedure locates grid points close to the given coordinates.

EXTER switch for filling interior or exterior (inside holes).
.TRUE. for filling exterior
.FALSE. for filling interior
(default = .FALSE.)

RESTRICTIONS: The domain must be specified with the ELLPACK syntax for general domains. This procedure must follow the first GRID segment in the ELLPACK program.

PERFORMANCE ESTIMATES: This procedure uses storage that has already been used by the domain processor. The execution time should be very small.

ACKNOWLEDGEMENTS: The development of this module was supported in part by Department of Energy contract DE-AC01-81ER-10997.

PROCEDURE

MODULE: EIGENVALUES

AUTHORS: B. T. Smith, J. M. Boyle, B. S. Garbow, Y. Ikebe, V. C. Klema, and C. B. Moler
(Modified for use in ELLPACK by John F. Brophy)

PURPOSE: Computes and prints the eigenvalues of the coefficient matrix generated by the discretization module.

METHOD: The discretization matrix is stored as a full matrix; no advantage is taken of any symmetry or band structure which may be present. This full matrix is balanced and converted to upper Hessenberg form. The eigenvalues are then found using the QR algorithm. The only output from this module is a listing of the eigenvalues, multiplied by SCALE, in both rectangular and polar form.

PARAMETERS: The computed eigenvalues are multiplied by SCALE before being printed. The default value for SCALE is 1.0.

PERFORMANCE ESTIMATES: A workspace of NEQN × (NEQN + 4) is needed, where NEQN is the number of algebraic equations.

REFERENCES:

Smith, B.T., J.M. Boyle, B.S. Garin, Y. Ikebe, V.C. Klema and C.B. Moler [1976], *Matrix Eigensystems Routines – EISPACK Guide*, Lecture Notes in Computer Science 6, (2nd ed.), Springer-Verlag, New York.

PROCEDURE

MODULE: LIST MODULES

AUTHOR: Calvin J. Ribbens

PURPOSE: To list the modules in the present version of ELLPACK.

METHOD: This procedure has a list of all the modules in the master version of ELLPACK. As a copy is made for use elsewhere, this list is included. When invoked, this procedure lists the module names along with the names and default values of their arguments.

PARAMETERS: None.

MODULE: NON-UNIQUE

AUTHOR: Calvin J. Ribbens

PURPOSE: To supply a value of $u(x,y)$ or $u(x,y,z)$ at a single point in order to determine a unique solution for an elliptic problem.

METHOD: The value of the x, y (and z) coordinates are set along with a value for u at this point. Thus

PROCEDURE. NON-UNIQUE (X = 1. , Y = 2. , U = 3.)

indicates that $u(1,2) = 3$. The default values for the coordinates are the smallest x, y (and z) grid values, the default for u is zero.

There is no ELLPACK module for NON-UNIQUE, the Fortran code that implements this procedure is inserted in-line into the ELLPACK control program.

RESTRICTIONS: The discretization module used must be sensitive to the NON-UNIQUE procedure for this to have an effect. These include DYAKANOV CG, DYAKANOV CG4, HODIE FFT, HODIE HELMHOLTZ, MARCHING ALGORITHM, 5 POINT STAR (both rectangular and general domains).

PARAMETERS: X, Y, Z, and U, as indicated in the example above.

PROCEDURE

MODULE: PLOT COLLOCATION POINTS

AUTHOR: William F. Mitchell

PURPOSE: To produce a plot of the domain, grid, and collocation points as used by the discretization module COLLOCATION.

METHOD: PLOT COLLOCATION POINTS uses the same method as COLLOCATION to find a set of collocation points. An X is placed on the plot at each collocation point. If LEVEL $\geqslant 3$, it also plots (in each element through which the boundary passes) the image of a 10×10 grid under the mapping used to determine interior collocation points.

PARAMETERS: Same parameters as COLLOCATION, plus

PTSIZE Real variable, the size of the plot in inches. Default is 6.

IDPLOT Integer between 0 and 99999 to identify which plot goes with which printout. The date and IDPLOT will be on both the plot and the ELLPACK printout.

RESTRICTIONS: Same as for COLLOCATION.

PERFORMANCE ESTIMATES: For an NX by NY grid, workspace of $224 + (NX - 1)(NY - 1)$ is used.

ACKNOWLEDGEMENTS: The development of this module was supported in part by the Department of Energy under contract DE-AC02-81ER-10997.

PROCEDURE

MODULE: REMOVE
REMOVE BLENDED BC
REMOVE BICUBIC BC

AUTHORS: John F. Brophy, Mahesh Rathi, and John R. Rice

PURPOSE: These modules subtract a function $r(x,y)$ from the solution $u(x,y)$ to create a new problem with solution $v(x,y) = u(x,y) - r(x,y)$. This new problem is created internally and automatically; the function $r(x,y)$ is added back to the computed $V(x,y)$ to produce the computed solution function U(X,Y).

The REMOVE module operates with a user-supplied function $r(x,y)$.

The REMOVE BLENDED BC module uses $r(x,y)$ which interpolates the boundary conditions exactly. Thus the problem actually solved has homogeneous boundary conditions. This allows some discretization modules (e.g. INTERIOR COLLOCATION) to be applied that otherwise could not be. This $r(x,y)$ is sometimes such a good estimate of $u(x,y)$ that a coarser grid can be used on the transformed problem.

The REMOVE BICUBIC BC module uses $r(x,y)$ as a Hermite bicubic piecewise polynomial which interpolates the boundary conditions at many points and which is zero in all interior grid elements. This sometimes removes boundary layers in a way so that a coarser grid can be used on the transformed problem.

METHOD: The technique to create the new problem is straightforward provided that the derivatives of $r(x,y)$ are known. For REMOVE these are computed numerically with finite differences using a step size much smaller than the grid which discretizes the PDE. One sided difference formulae are used on the boundaries of the domain.

The derivatives of $r(x,y)$ for REMOVE BICUBIC BC are computed exactly. The derivatives of $r(x,y)$ for REMOVE BLENDED BC are computed partially numerically and partially by formulae. Finite differences are used to estimate the derivatives of the boundary condition functions $h(x,y)$ (see below) along the boundaries.

MATHEMATICAL FORMULATION: Let L denote the original PDE operator and M the boundary condition operator, so that the problem is

$$\begin{aligned} \mathrm{L}u &= g \quad \text{on domain} \\ \mathrm{M}u &= f \quad \text{on boundary} \end{aligned}$$

If we set $u(x,y) = v(x,y) + r(x,y)$ we see that $v(x,y)$ satisfies

$$\begin{aligned} \mathrm{L}v &= g - \mathrm{L}r \quad \text{on domain} \\ \mathrm{M}v &= f - \mathrm{M}r \quad \text{on boundary} \end{aligned}$$

where $\mathrm{L}r$ and $\mathrm{M}r$ are known (or computable) quantities. Thus $v(x,y)$ satisfies the same problem as $u(x,y)$ except that the right side functions $g(x,y)$ (R1PRHS in ELLPACK) and $f(x,y)$ (R1BRHS in ELLPACK) are modified.

The numerical computation of the derivatives uses the following technique. At a "normal" point (x,y) which is interior to the grid, we use central difference formulae with $h_0 =$ the cube root of the machine epsilon as an approximate optimum step value. The value h actually used is the maximum of h_0 and one hundreth the grid size. The basic formulas are

$$u_{xx}(0) \approx \frac{u(h) - 2u(0) + u(-h)}{h^2}$$

$$u_x(0) \approx \frac{u(h) - u(-h)}{2h}$$

$$u_{xy}(0) \approx \frac{u(h,h) - u(h,-h) - u(-h,h) + u(-h,-h)}{4h^2}.$$

Where one-sided differences are required, the formulae are

$$u_{xx}(0) \approx \frac{2u(0) - 5u(h) + 4u(2h) - u(3h)}{h^2}$$

$$u_x(0) \approx \frac{-3u(0) + 4u(h) - u(2h)}{2h}.$$

PARAMETERS: REMOVE has the name of the Fortran function which defines $r(x,y)$. The form is

REMOVE(R = ⟨name⟩)

with ⟨name⟩ = ZERO as the default name. ZERO is a standard ELLPACK function which returns 0.0 for all input.

There are parameters HXSTEP, HYSTEP and HZSTEP of REMOVE which supply values of h in the numerical differentiation. The default values for these three parameters is -1. which implies that the step h used in the differentiation is computed as described above.

RESTRICTIONS: REMOVE BLENDED BC requires that the triple SET U BY BLENDING be applicable (see the discussion of that module). REMOVE BICUBIC BC requires that the triple SET U BY BICUBICS be applicable (see the discussion of that module). In particular, REMOVE BLENDED BC and REMOVE BICUBIC BC are restricted to two-dimensional rectangular domains.

PERFORMANCE ESTIMATES: This procedure may make the evaluation of certain functions (the right sides and U, UX, UY, etc.) considerably more expensive.

ACKNOWLEDGEMENTS: The development of these modules was supported in part by the Department of Energy under contract DE-AC02-81ER-10997.

PROCEDURE

MODULE: SET UNKNOWNS FOR HODIE HELMHOLTZ

AUTHORS: John J. Nestor III and Ronald F. Boisvert

PURPOSE: To provide an estimate of the solution of the PDE problem for use as a starting point for linear algebraic solution modules employing iterative techniques. The estimate used is based upon the values of a user-defined function which approximates the solution.

METHOD: This module initializes the internal vector R1UNKN which gives the solution to the system of linear equations generated by the discretization module HODIE HELMHOLTZ. If the user's estimate of the solution is given by the function UEST(x,y), then UEST (x_i,x_j) is used to initialize the unknown associated with the grid point (x_i,y_i). The function UEST is evaluated at all grid points inside the domain as well as all grid points on the boundary where the solution is not explicitly prescribed by the boundary conditions. The correspondence between grid points and boundary points and the elements of R1UNKN is given by a numbering array in a HODIE HELMHOLTZ private common block (see the descriptions of the module HODIE HELMHOLTZ).

PARAMETERS: The parameter UEST gives the name of the real-valued user-defined function of two arguments (x,y) which is used to initialize the solution vector. The default is the ELLPACK standard function ZERO, which returns zero for all arguments. Note that the parameter assignment UEST=U may be used to set the solution vector to some previously computed solution.

RESTRICTIONS: This module must be preceded by the discretization module HODIE HELMHOLTZ.

PERFORMANCE ESTIMATES: Execution time is roughly proportional to NX×NY×T, where NX×NY is the grid size and T is the average time required to evaluate the user-defined function. No workspace is required.

ACKNOWLEDGEMENTS: This work was supported by the National Bureau of Standards.

PROCEDURE

MODULE: SET UNKNOWNS FOR 5-POINT STAR

AUTHORS: John J. Nestor III and Ronald F. Boisvert

PURPOSE: To provide an estimate of the solution of the PDE problem for use as a starting point for linear algebraic solution modules employing iterative techniques. The estimate used is based upon the values of a user-defined function which approximates the solution.

METHOD: This module initializes the internal vector R1UNKN which gives the solution to the system of linear equations generated by the discretization module 5-POINT STAR. If the user's estimate of the solution is given by the function UEST(x,y), then UEST(x_i,y_j) is used to initialize the unknown associated with the grid point (x_i,y_j). The function UEST is evaluated at all grid points inside the domain as well as at all boundary-grid intersections where the value of the solution is unknown. The correspondence between grid points and boundary points and the elements of R1UNKN is given by numbering arrays in 5-POINT STAR private common blocks (see the descriptions of the module 5-POINT STAR).

PARAMETERS: The parameter UEST gives the name of the real-valued user-defined function of two arguments (x,y) which is used to initialize the solution vector. The default is the ELLPACK standard function ZERO, which returns zero for all arguments. Note that the parameter assignment UEST=U may be used to set the solution vector to some previously computed solution.

RESTRICTIONS: This module must be preceded by the discretization module 5-POINT STAR.

PERFORMANCE ESTIMATES: Execution time is roughly proportional to NX$\times$NY$\times$T, where NX$\times$NY is the grid size and T is the average time required to evaluate the user-defined function. No workspace is required.

ACKNOWLEDGEMENTS: This work was supported by the National Bureau of Standards.

SYSTEM

MODULE: DOMAIN PROCESSOR

AUTHOR: John R. Rice

PURPOSE: The domain processor generates information to relate a two dimensional grid to a domain within it. This information is made available to discretization modules.

METHOD: The grid is specified by a pair of vectors XGRID(I vts), I=1 to NX, YGRID(J vts), J=1 to NY and the domain boundary is specified parametrically by

$$x_i(t), y_i(t) \qquad b_{1i} \leqslant t \leqslant b_{2i}, \quad i = 1,2, \ldots, NB$$

The domain processor produces seven vectors of information about the **boundary points** where the grid and domain boundary intersect. It also produces a two dimensional array that types all the grid points relative to the domain. Detailed definitions and examples are given in Chapter 13, Section B.

The structure of the domain processing is as follows:

```
LOCATE FIRST BOUNDARY POINT IN GRID

LOOP OVER PIECES OF BOUNDARY (I = 1 TO NB)
     WHILE NOT AT END OF I-TH BOUNDARY PIECE
          FIND NEXT BOUNDARY-GRID INTERSECTION
          DETERMINE TYPE OF INTERSECTION POINT
     CHECK CONTINUITY OF BOUNDARY

CHECK CLOSING OF BOUNDARY

LOOP OVER GRID TO TYPE POINTS IN RELATION TO BOUNDARY

IDENTIFY INTERIOR POINTS
     LOCATE FIRST INTERIOR POINT EXPAND INTERIOR ALONG
          GRID LINES

SET POINTERS FROM BOUNDARY TO INTERIOR NEIGHBORS
```

The domain processor is a substantial program with about 2000 executable statements. The numerical methods used are summarized in the references below. The output of the domain processor is described in detail in Section 13.B. Figure 6.5 shows the labeling of the boundary points for the domain of Figure 2.1. The pointer for point (7.4) is 6011 indicating that there are boundary points to the right and below (6 = 0110 in binary) and the lowest numbered one is 11.

PARAMETERS: None.

RESTRICTIONS: The methods used by the domain processor use the following **assumptions**:

1. The boundary does not enter a grid element several times.

2. The boundary is fairly smooth on the scale of the grid.
3. The boundary is parameterized by smooth, monotonic and well-behaved functions.
4. There are no corners on the boundary except at the ends of pieces.
5. The coordinates and parameters are all ordinary-sized numbers and of similar size.
6. The boundary definition starts where the domain is "fat" enough (it is wider than one grid square along the grid lines).

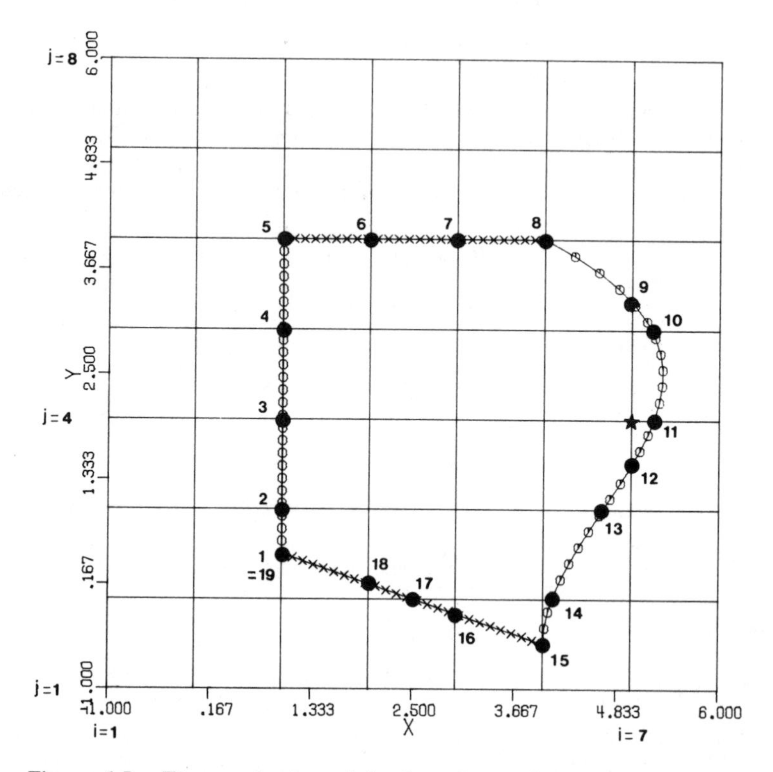

Figure 6.5. The numbering of the boundary points and grid points for the domain of Figure 2.1.

PERFORMANCE ESTIMATES: The work of most of the computation is proportional to the number of boundary points, that is, $O(\mathrm{NX} + \mathrm{NY})$ for an NX by NY grid and a reasonable domain. A second part of the work is proportional to the number of grid points or interior grid points, that is, $O(\mathrm{NX} \times \mathrm{NY})$. The second part usually does not dominate the time of the processing because the computations at each point are simple. The work space required is $510 + 28(\mathrm{NX} + \mathrm{NY}) + \mathrm{NX} \times \mathrm{NY}$.

ACKNOWLEDGEMENTS: This work was supported in part by National Science Foundation Grant MCS 79-26396 and Department of Energy contract DE-AC02-81ER-10997.

REFERENCES:

Rice, J. R., [1984], "Numerical computation with general two dimensional domains", ACM Trans. Math. Software 10, December.

Rice, J. R., [1984], "Algorithm: A two dimensional domain processor", ACM Trans. Math. Software 10, December.

MODULE: INTERPOLATION

AUTHORS: Ronald F. Boisvert and William A. Ward

PURPOSE: The interpolation software extends a numerical solution defined only on a grid to the whole domain. It also provides values for the first and second derivatives of the solution. Similar facilities are provided for the three-dimensional case. This software is only used in conjunction with finite difference discretizations.

METHOD: There are two distinct methods that can be selected using the OPTIONS segment. The default is **local quadratic interpolation**. Given a point (x,y) in the domain, a quadratic polynomial is found which interpolates the computed solution at neighboring grid points. This polynomial is then evaluated to give U(X,Y). This method is simple but not very accurate. The error due to interpolation is sometimes substantially larger than the error in the numerical solution. The interpolation error is zero at the grid points.

One may select **spline interpolation**, which computes a single spline approximation by interpolating the values of the computed solution given at the grid points. The polynomial degree of the spline interpolant is set by the discretization module. This degree is selected to make the interpolation error comparable to the solution error, but there is no guarantee that this is always so.

Both interpolation methods are less accurate near the boundaries of nonrectangular domains because some values for the computed solution at grid points outside the domain are computed by local extrapolation. Errors in these extrapolations might be much larger than the solution errors at the interior grid points.

MATHEMATICAL FORMULATION: The local quadratic polynomial

$$a + bx + cy + dx^2 + exy + fy^2 + gx^2y + hxy^2 + ix^2y^2$$

has nine coefficients. If the grid point (m,n) is the closest to (x,y) then the computed values of u at the grid points

$$(x_j, y_k) \qquad m-1 \leqslant j \leqslant m+1, \quad n-1 \leqslant k \leqslant n+1$$

are used. The computation uses a divided difference method. If the point (x_j, y_k) is on the edge of the grid, then the indices (j,k) are shifted appropriately to use computed values of u in the domain.

The spline interpolation uses tensor products of the spline interpolation software of Carl de Boor. The details of the mathematics behind this software are given in Chapter XVII of [de Boor, 1978]. This software has been incorporated into ELLPACK with minor interface modifications.

The extrapolation method used near the domain boundaries is illustrated for the situation in Figure 6.6. Points close to the (m,n) grid point need values at the exterior points labeled a, b, and c. Values at a and c are computed by one-dimensional quadratic extrapolation using values from the points (m,n), $(m,n-1)$, $(m,n-2)$ and (m,n), $(m-1,n)$, $(m-2,n)$. If the grid is so coarse that some of the grid points needed for the extrapolation are exterior to the domain, then the degree of the extrapolating polynomial is reduced to 1 (for linear extrapolation) or even 0 in rare instances.

Once a and c values are found, then a biquadratic $p(x,y)$ is found which interpolates at the eight points (m,n), $(m-1,n)$, $(m,n-1)$, $(m-1,n-1)$, $(m-1,n+1)$, $(m+1,n-1)$, a and c. The interpolated value at b is the value of $p(x,y)$ at this point.

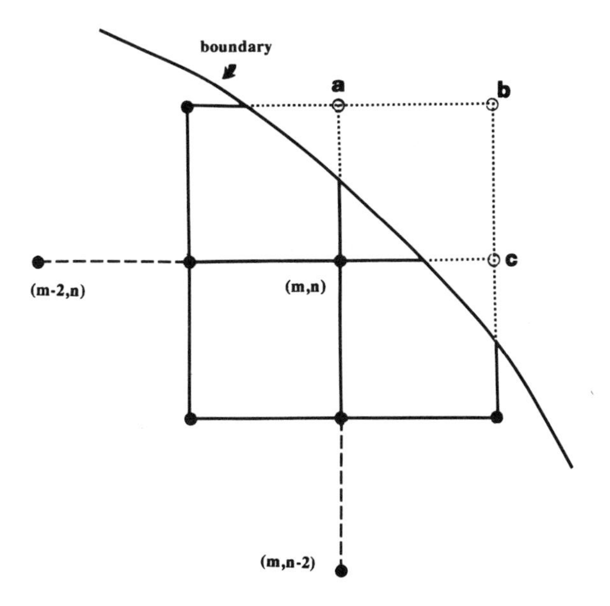

Figure 6.6. Illustration of interpolation near domain boundaries.

PARAMETERS: INTERPOLATION = SPLINES or QUADRATICS may be set in the OPTION segment.

SPECIAL CASES: The interpolation is much more reliable and accurate for rectangular domains.

PERFORMANCE ESTIMATES: Quadratic interpolation requires a modest amount of computation each time U(X,Y) is evaluated; only a small amount of extra storage is required. Spline interpolation requires a substantial computation the first time U(X,Y) is evaluated; the spline approximation is computed for the entire problem in this case. This computation may be as substantial as solving the PDE. The spline coefficients require $(K+1) \times NX \times NY$ storage, where K is set by the discretization module (K is usually the order of the discretization).

ACKNOWLEDGEMENTS: Work on this module was supported by the National Bureau of Standards, by the Department of Energy, and by the National Science Foundation.

REFERENCES:

de Boor, C., [1978], *A Practical Guide to Splines*, Applied Mathematical Sciences 27, Springer-Verlag, New York.

SYSTEM

MODULE: PLOT

AUTHORS: William J. Snyder
(Modified for use in ELLPACK by Ronald F. Boisvert, John F. Brophy, and William A. Ward)

PURPOSE: Produce a contour plot of a function of two variables over a general domain. Ten contours are given between the maximum and minimum of the function.

METHOD: The function $f(x,y)$ to be plotted is evaluated on a grid (not the ELLPACK grid) and then the algorithm of W.J. Snyder applied to produce the plot. If the domain is not rectangular, values of $f(x,y)$ are extrapolated to grid points near the domain, but outside it.

PARAMETERS: PLOT is invoked in the OUTPUT segment, where its syntax is PLOT(⟨f⟩, ⟨grid⟩), where ⟨f⟩ is the name of a function and ⟨grid⟩ is a pair of integers (e.g. PLOT(UX, 25, 25)). The function ⟨f⟩ may either be a standard ELLPACK function or a function defined in the SUBPROGRAMS segment. The default grid is 20 by 20. Usually, the finer the plotting grid, the smoother the contour lines appear.

RESTRICTIONS: There are no restriction on PLOT except for two dimensions, but two situations might give unexpected results. First, the contours might misbehave near the edges of curved domains for functions such as U(X,Y) which are not defined outside the domain. This is due to the simple extrapolation of values used. Second, the contours might not behave properly in the neighborhood of ARCs or narrow slits in the domain. This is due to the fact that the contours are plotted directly from the tabulated values and no attempt is made to incorporate other information about the geometry of the domain.

PERFORMANCE ESTIMATES: Contour plotting is computationally expensive and it is not unusual for a contour plot to take substantially longer than solving an elliptic problem. A typical plot might generate from 10 to 20 thousand plotter commands. The workspace requirement of PLOT is at most $17 + 21NX \times NY$ for a plotting grid of NX by NY. Machine dependencies can reduce this substantially.

ACKNOWLEDGEMENTS: The work of W.J. Snyder was sponsored by the National Aeronautics and Space Administration under contract NAS7-100 at the Jet Propulsion Laboratory. The modifications for ELLPACK were supported in part by the National Science Foundation under grant MCS76-10225.

REFERENCES:

Snyder, W. J., [1978], "Algorithm 531: Contour plotting, ACM Trans. Math. Software 4, pp. 290–294.

Chapter 7

ITPACK SOLUTION MODULES

David R. Kincaid

Thomas C. Oppe

John R. Respess

David M. Young

The University of Texas at Austin

The purpose of this chapter is to describe some of the solution modules which can be used within ELLPACK. We describe the ITPACK solution modules (JACOBI CG, JACOBI SI, SOR, SYMMETRIC SOR CG, SYMMETRIC SOR SI, REDUCED SYSTEM CG, REDUCED SYSTEM SI) which are based on the use of iterative algorithms. These modules are routines adapted from the ITPACK software package developed at The University of Texas at Austin. The objective of this discussion is to provide the ELLPACK user with sufficient information for intelligent use of these solution modules.

Section 7.A provides background on the iterative algorithms. Section 7.B gives an overview of the ITPACK modules and details their actual usage. Section 7.C contains a brief discussion of the ITPACK Project. Additional information on the iterative methods and algorithms can be found in the encyclopedia article [Kincaid and Young, 1979], in the book [Hageman and Young, 1981], and in the report [Grimes, Kincaid, and Young, 1979]. More information about the usage of the ITPACK software package can be found in the paper [Kincaid, Respess, Young, and Grimes, 1982].

7.A ITPACK ITERATIVE ALGORITHMS

In this section, we discuss iterative algorithms used in the ITPACK solution modules for solving linear systems of the form

$$Ax = b, \qquad (1)$$

where A is a given real $N \times N$ coefficient matrix, b is a given column vector of length N, and the N-component solution vector x is sought. To simplify the discussion in this section, we assume that the linear system has a symmetric positive definite (SPD) coefficient matrix A, even though the ITPACK modules are designed to handle both SPD and mildly nonsymmetric systems. While these solution modules can be called with any linear system containing positive diagonal elements, iterative methods are most often applied to linear systems in which A is sparse, i.e., a large number of the off-diagonal entries of A are zeros. The coefficient matrix A can be written as

$$A = D - C_L - C_U, \tag{2}$$

where D is a diagonal matrix, C_L is a strictly lower triangular matrix, and C_U is a strictly upper triangular matrix.

As the first step in each ITPACK module, the linear system (1) is changed to a system in which the diagonal elements are all 1's by scaling with the diagonal matrix $D^{1/2} = \text{Diag}\,[(a_{i,i})^{1/2}]$ as follows:

$$(D^{-1/2} A D^{-1/2})(D^{1/2} x) = D^{-1/2} b.$$

If we let

$$U = D^{-1/2} C_U D^{-1/2}, \qquad L = D^{-1/2} C_L D^{-1/2},$$
$$u = D^{1/2} x, \qquad c = D^{-1/2} b,$$

then we have the scaled "related system"

$$(I - B) u = c, \tag{3}$$

where

$$B = L + U.$$

This linear system is ideally suited for solution by an iterative method. Clearly, after u is determined, the solution to the original system (1) is $x = D^{-1/2} u$. The scaling and unscaling is done automatically when an ITPACK solution module is called.

In all but one case, each ITPACK algorithm consists of four components:

(a) a basic iterative method;

(b) an acceleration method which is either semi-Iteration (SI) (also called *Chebyshev acceleration*), or *conjugate gradient* (CG) *acceleration*;

(c) an adaptive procedure for computing acceleration parameters automatically where needed;

(d) a stopping procedure to determine whether the approximate solution is sufficiently close to the true solution.

Basic Iterative Methods

We now describe the basic linear stationary iterative methods used in the ITPACK modules. The basic iterative methods are of the form

$$u^{(n+1)} = G u^{(n)} + k, \tag{4}$$

where $u^{(0)}$ is an initial guess to the solution vector. The **Jacobi method** for the related system (3) is given by

$$u^{(n+1)} = Bu^{(n)} + c.$$

If at each iterative step, one uses the components of the new approximate solution vector immediately after they are available, i.e., if the components of $u^{(n+1)}$ are used with the lower triangular elements of B, and if an acceleration parameter ω is introduced, we have the **Successive OverRelaxation method (SOR)** given by

$$u^{(n+1)} = \omega(Lu^{(n+1)} + Uu^{(n)} + c) + (1-\omega)u^{(n)}.$$

Each step of the **Symmetric SOR method (SSOR)** consists of one forward SOR sweep followed by one backward SOR sweep. The symmetric SOR method can be written as

$$\begin{aligned} u^{(n+1/2)} &= \omega(Lu^{(n+1/2)} + Uu^{(n)} + c) + (1-\omega)u^{(n)}, \\ u^{(n+1)} &= \omega(Uu^{(n+1)} + Lu^{(n+1/2)} + c) + (1-\omega)u^{(n+1/2)}. \end{aligned}$$

The linear system (1) is said to be a **red-black system** if it can be permuted into the block form

$$\begin{bmatrix} D_R & H_R \\ H_B & D_B \end{bmatrix} \begin{bmatrix} x_R \\ x_B \end{bmatrix} = \begin{bmatrix} b_R \\ b_B \end{bmatrix}, \tag{5}$$

where D_R and D_B are square diagonal matrices of order N_R and N_B, respectively, $(N = N_R + N_B)$, x_R and b_R are N_R-component vectors, and x_B and b_B are N_B-component vectors. In this ordering, the components of the unknown vector x are considered as either "red" (x_R) or "black" (x_B) mesh points, respectively. A red-black ordering is any ordering such that every black unknown follows all of the red unknowns. A red-black ordering is possible when the coefficient matrix has *Property A* [Young, 1971], which is another way of saying that the map of the coefficient matrix is two-colorable. For example, this form is obtained from a 5-point discretization of a partial differential equation on a rectangular mesh by labeling the mesh points according to a checkerboard or red-black ordering. In ELLPACK, the indexing module RED-BLACK will form, if possible, an indexing array for reordering of the unknowns into a red-black system. With this ordering, the scaled related linear system (3) has the form

$$\begin{bmatrix} I_R & -F_R \\ -F_B & I_B \end{bmatrix} \begin{bmatrix} u_R \\ u_B \end{bmatrix} = \begin{bmatrix} c_R \\ c_B \end{bmatrix},$$

where

$$\begin{aligned} F_R &= -D_R^{-1/2} H_R D_B^{-1/2}, & F_B &= -D_B^{-1/2} H_B D_R^{-1/2}, \\ u_R &= D_R^{1/2} x_R, & u_B &= D_B^{1/2} x_B, \\ c_R &= D_R^{-1/2} b_R, & c_B &= D_B^{-1/2} b_B. \end{aligned}$$

The scaled related system can be written as

$$\begin{aligned} u_R &= F_R u_B + c_R, \\ u_B &= F_B u_R + c_B. \end{aligned} \tag{6}$$

Since the u_R can be eliminated in the system (6), we obtain the scaled reduced system

$$(I - F_B F_R) u_B = F_B c_R + c_B.$$

This gives us the **Reduced System method** (**RS**) defined by

$$u_B^{(n+1)} = F_B F_R u_B^{(n)} + F_B c_R + c_B.$$

It should be noted that the matrix product $F_B F_R$ does not need to be formed since it can be computed in two steps as given in (6). The reduced system method has the same general form (4) as the other basic methods with some modifications such as u_B replacing u, etc.

Each iterative method discussed corresponds to a "splitting" of the original unscaled matrix A. This is a representation of A in the form

$$A = Q - (Q - A),$$

where the nonsingular matrix Q is a **splitting matrix.** To derive the basic iterative method corresponding to Q, we rewrite (1) in the form

$$Qx = (Q - A)x + b.$$

We then have the iterative method

$$Qx^{(n+1)} = (Q - A)x^{(n)} + b,$$

which is equivalent to (4) with

$$G = I - Q^{-1}A,$$
$$k = Q^{-1}b.$$

For the basic methods discussed previously, the splitting matrices for the unscaled system are

Method	*Splitting matrix Q*
Jacobi	D
SOR	$\omega^{-1}(D - \omega C_L)$
Symmetric SOR	$[\omega(2-\omega)]^{-1}(D - \omega C_L)D^{-1}(D - \omega C_U)$
Reduced System	D_B

Here the matrices D, C_L, C_U are the diagonal, strictly lower triangular, and strictly upper triangular parts of the matrix A as given by (2). Moreover, the matrix D_B is the diagonal submatrix corresponding to the black points in the red-black system (5), and it is the splitting matrix for the unscaled reduced system. After scaling, the diagonal matrices D and D_B become identity matrices.

ACCELERATION METHODS

Many iterative methods can be speeded up by the use of a polynomial acceleration procedure such as Chebyshev acceleration or conjugate gradient acceleration. For this to be possible, it is sufficient for the iterative method to be "symmetrizable", i.e., that the matrix $W(I - G)W^{-1}$ is SPD for some nonsingular matrix W. If A is SPD, then the Jacobi, symmetric SOR (with $0 < \omega < 2$), and reduced system methods are symmetrizable. If an iterative method is symmetrizable, then the eigenvalues of the iteration matrix G are real and less than unity. In a typical case, W can be taken to be $Q^{1/2}$ when Q is SPD. If $W = Q^{1/2}$, then one does not actually need to compute $Q^{1/2}$, since all necessary formulae can be written in terms of $W^T W$ instead of W. For the basic methods, the symmetrization matrices are

Method	*Symmetrization matrix* W
Jacobi	$D^{1/2}$
Symmetric SOR	$\omega^{-1}D^{-1/2}(D-\omega C_U)$
Reduced system	${D_B}^{1/2}$

Both Chebyshev and conjugate gradient acceleration of the iterative method (4) can be written in the form

$$u^{(n+1)} = \rho_{n+1}[\gamma_{n+1}(Gu^{(n)}+k)+(1-\gamma_{n+1})u^{(n)}]+(1-\rho_{n+1})u^{(n-1)}$$

with $\rho_1 = 1$.

In the case of **Chebyshev acceleration**, we have

$$\rho_{n+1} = \begin{cases} 1, & n = s, \\ [1-2(\sigma/2)^2]^{-1}, & n = s+1, \\ [1-(\sigma/2)^2\rho_n]^{-1}, & n \geqslant s+2, \end{cases}$$

$$\gamma_{n+1} = 2/[2-M(G)-m(G)],$$

$$\sigma = [M(G)-m(G)]/[2-M(G)-m(G)].$$

Here $M(G)$ and $m(G)$ are the largest and smallest eigenvalues of G, respectively. If $M(G)$ and $m(G)$ are not known, estimates are computed automatically in the adaptive procedure. Moreover, the variable s is initially zero and is reset to the current iteration number whenever the acceleration process is restarted. At this point, new eigenvalue estimates are computed, the parameters γ_{n+1} and σ are recomputed, and the value of ρ_{n+1} begins again with 1.

In the **conjugate gradient acceleration** with both A and Q being SPD and with $W = Q^{1/2}$, we have

$$\rho_{n+1} = [1-\gamma_{n+1}(\gamma_n\rho_n)^{-1}(\delta^{(n)},Q\delta^{(n)})/(\delta^{(n-1)},Q\delta^{(n-1)})]^{-1},$$

$$\gamma_{n+1} = [1-(\delta^{(n)},QG\delta^{(n)})/(\delta^{(n)},Q\delta^{(n)})]^{-1},$$

where the **pseudo-residual vector** $\delta^{(n)}$ is given by

$$\delta^{(n)} = Gu^{(n)}+k-u^{(n)}.$$

ADAPTIVE PROCEDURES

We first describe the adaptive procedure for Chebyshev acceleration. The behavior of this acceleration process is very sensitive to the estimate M_E for $M(G)$. Thus, if $M(G) = 0.99990$ and if the (apparently accurate) estimate $M_E = 0.99989$ were used, the rate of convergence would be reduced by 30 percent! Except in special cases, it would not appear possible to estimate $M(G)$ to sufficient accuracy in advance. However, it is possible to use an adaptive procedure, as we now describe.

For simplicity, let us assume that a reasonably good lower bound m is known for the eigenvalues of G. We let $m_E = m$ and do not change m_E. We choose $M_E^{(1)}$ such that $m \leqslant M_E^{(1)} < M(G)$ and iterate on M_E. We measure the convergence rate based on the norms of the pseudo-residual vectors $\delta^{(n)}$. If the measured convergence rate is substantially less than it should be, we modify our estimate to obtain $M_E^{(2)}$. The new value is chosen to be the maximum of two estimates for M_E based on a "Chebyshev equation" and on a "Rayleigh quotient". The iteration parameter s is set to the iteration number at which the estimate M_E was last changed. The process is repeated, with new estimates $M_E^{(3)}, M_E^{(4)}, \ldots,$ being obtained, until convergence is achieved for $u^{(n)}$ (but not necessarily for the $M_E^{(i)}$). To take full advantage of Chebyshev acceleration, a new estimate for M_E is not computed on every iteration, even though a nonoptimum estimate may slow the convergence for a time. The test which is used to determine when the current value for M_E should be changed involves a "damping factor" FF. The value of FF $= 0.75$ is used, since it is a compromise between never changing (FF $= 0$) and changing on each iteration (FF $= 1.0$) and since it seems to work well in practice. This way a new estimate for M_E is not computed unless the current value is significantly less than it should be. The behavior of the adaptive procedure is rather insensitive to the exact value selected for FF.

In conjugate gradient acceleration, an estimate for M_E is not needed for determining the acceleration parameters but is needed in the stopping test. The value of M_E is computed as the largest eigenvalue of a tridiagonal matrix. New estimates for M_E are computed on each of the first several iterations until the values for M_E converge.

The use of conjugate gradient acceleration yields a convergence rate which is nearly always faster than Chebyshev acceleration and is often considerably faster. (It is always faster if the error is measured in a suitable norm.) Moreover, no estimates of $m(G)$ and $M(G)$ are required. On the other hand, the formulae for conjugate gradient acceleration are somewhat more complicated. However, by a modification of these formulae, the number of extra operations required can be minimized. For the case of the symmetric SOR method, even with conjugate gradient acceleration, the parameter ω must be estimated. An adaptive procedure for doing this is included in the ITPACK modules.

STOPPING PROCEDURES

The problem of determining when an approximate solution $u^{(n)}$ is close enough to the true solution $\bar{u}$ of (3) is one which is often treated too lightly. In many cases, the iterative process is stopped when the difference between $u^{(n+1)}$ and $u^{(n)}$, measured in some norm, is sufficiently small. This often results in either premature convergence or convergence long after the desired accuracy has been obtained. Alternatively, the process may be stopped when the **residual vector**

$$r^{(n)} = b - Au^{(n)}$$

or the **pseudo-residual vector**

$$\delta^{(n)} = Gu^{(n)} + k - u^{(n)}$$

is sufficiently small. Unfortunately, the **error vector**

$$\epsilon^{(n)} = u^{(n)} - \bar{u}$$

may be quite large even though the above quantities are small.

The stopping procedures used in the ITPACK modules for both Chebyshev and conjugate gradient acceleration are as follows. We continue to assume that the method is symmetrizable and that $W(I-G)W^{-1}$ is SPD for some matrix W. Ideally, if one knew the true solution $\bar{u}$, then the iterative solution $u^{(n)}$ could reasonably be accepted when

$$\frac{\|u^{(n)} - \bar{u}\|_W}{\|\bar{u}\|_W} < \zeta, \tag{7}$$

where ζ is a prescribed stopping tolerance typically in the range 10^{-4} to 10^{-8}. Since $\bar{u}$ is seldom known in advance, the following alternative test based on the pseudo-residual vector $\delta^{(n)}$ is used:

$$\frac{[1 - M(G)]^{-1} \|\delta^{(n)}\|_W}{\|u^{(n)}\|_W} < \zeta, \tag{8}$$

which can be shown to be related to the inequality (7). If we are using the adaptive Chebyshev procedure, we replace $M(G)$ by the latest estimate M_E. If we are using conjugate gradient acceleration, then we can estimate $M(G)$ as the largest eigenvalue of a certain tridiagonal matrix involving previously computed values for ρ_{n+1} and γ_{n+1}. Numerical studies have shown that for the adaptive methods, (8) is effective and usually stops the iterative algorithm soon after the ideal test (7) is satisfied. Furthermore, the pseudo-residual vector $\delta^{(n)}$ is determined in the acceleration procedure, and $\|\delta^{(n)}\|_W$ is computed in the test for changing parameters for Chebyshev acceleration. Thus, the additional computation involved in this stopping test is minimized. For the reduced system methods, the stopping test is similar to (8):

$$\frac{2^{1/2}[1 - M_E{}^2]^{-1} \|\delta_B{}^{(n)}\|_W}{\|u_B{}^{(n)}\|_W} < \zeta.$$

Here $M_E{}^2$ is an estimate for $M(F_B F_R)$. For the Jacobi and reduced system methods, the W-norm for the scaled system is actually the 2-norm.

The test (8) is very accurate for the adaptive Chebyshev procedure. For many numerical experiments, the number of iterations required for (8) to be satisfied was within one or two of the number of iterations required for (7) to be satisfied. For nonadaptive Chebyshev acceleration in which the optimum value of $M(G)$ was used from the outset, the agreement was not as close but was still within 10 to 15 percent. For conjugate gradient acceleration, similar agreement was obtained.

One should be careful not to specify too small a value for ζ, especially on short word-length computers. If ζ is too small the process may never converge because of roundoff errors. Thus, one cannot specify a ζ of less than approximately 500 times the basic machine rounding error.

Additional details on these adaptive and stopping procedures are given with flowcharts and formulae in the technical report [Grimes, Kincaid, and Young, 1979]. The theory and the derivation of these algorithms are developed in the book [Hageman and Young, 1981]. Questions concerning the matrix properties necessary for convergence of these and other iterative methods are also addressed in this book.

7.B ITPACK MODULES

The ITPACK modules refine an initial guess to the matrix solution with a series of iterative approximations until the convergence criterion is satisfied. The initial guess is the zero vector or, alternatively, whatever the user has specified with the ELLPACK statement

```
PROCEDURE.    SET UNKNOWNS FOR 5 POINT STAR
```

The algorithmic procedure can be controlled to some degree with the module parameter passing facility of the ELLPACK system, which is explained below.

The ITPACK modules are as follows:

ITPACK Module	*Method*
JACOBI CG	Jacobi conjugate gradient
JACOBI SI	Jacobi semi-iteration
SOR	Successive overrelaxation
SYMMETRIC SOR CG	Symmetric SOR conjugate gradient
SYMMETRIC SOR SI	Symmetric SOR semi-iteration
REDUCED SYSTEM CG	Reduced system conjugate gradient
REDUCED SYSTEM SI	Reduced system semi-iteration

and the ITPACK calling sequence is

```
SOLUTION.    ⟨ITPACK module⟩    (⟨module parameters⟩)
```

The ITPACK algorithms are not guaranteed to converge for all linear systems, but have been observed to work successfully for a large number of symmetric and nonsymmetric linear systems which arise from solving elliptic partial differential equations.

Overview of ITPACK Solution Modules

We now present an overview of the ITPACK solution modules in ELLPACK.

Basic iterative procedures such as the Jacobi method, the SOR method, the symmetric SOR method, and the reduced system method are combined, where possible, with acceleration procedures such as Chebyshev (semi-iteration, SI) and conjugate gradient (CG), for rapid convergence. Automatic selection of the acceleration parameters and the use of accurate stopping criteria are major features of these modules. While the ITPACK modules can be called with any linear system containing nonzero diagonal elements, they are most successful in solving systems with symmetric positive definite or mildly nonsymmetric coefficient matrices. The SOR method does not lend itself to acceleration, and therefore it is included with only an automatic procedure for selecting the relaxation factor ω. To be effective, the Chebyshev acceleration procedure requires accurate estimates for the bounds on the eigenvalues of the basic method, i.e., $M(G)$ and $m(G)$.

Adaptive procedures are used for computing and improving such estimates. Conjugate gradient acceleration does not require these estimates and leads to faster convergence than Chebyshev acceleration over a large class of linear systems. However, there is more work performed in the computation of the acceleration parameters on each iteration of the conjugate gradient acceleration and more storage is required than for Chebyshev acceleration.

Before an ITPACK module begins the iteration process, the elements of the linear system stored in the arrays COEF, RHS, and U are scaled, as described in Section 7.A, so that the system has a unit diagonal. Afterwards, these arrays are unscaled. Consequently, the values in the arrays COEF and RHS may change slightly due to roundoff errors in the computer arithmetic. Scaling the linear system reduces the number of arithmetic operations. Iterations based on these algorithms are carried out until convergence is reached, based on the relative accuracy requested via the stopping criterion for the scaled solution vector $u = D^{1/2}x$. The solution vector x is obtained by unscaling, and the linear system is returned to its original form subject to roundoff errors in the arithmetic.

The ITPACK iterative algorithms make no use of the structure of the resulting linear system. Because this code is not tailored to any particular class of partial differential equations or discretization procedure, but rather to general sparse linear systems, it is felt that these modules can be used to solve a wide class of problems.

Each of the ITPACK solution modules can be used with any ordering of a linear system except for the reduced system modules, which require a red-black system.

ACCELERATED JACOBI METHODS.

The **Jacobi CG method** does not require an estimate for bounds on the eigenvalues of the Jacobi iteration matrix **B** from the related system

$$u = Bu + c.$$

Such bounds are determined automatically, as are the iteration parameters, and these bounds are used in the stopping procedure.

The **Jacobi SI method** sets the lower bound on the eigenvalues of the Jacobi matrix B to a fixed value which is known for many systems derived from partial differential equations. The adaptive procedure for the parameter determination is based on the upper bound for the eigenvalues of B.

SOR METHOD.

The **SOR method** assumes a *consistently ordered system*, and a heuristic procedure developed in [Hageman and Young, 1981] is used so that the estimated optimum relaxation factor ω is less than the true optimum ω. In the SOR algorithm, the first iteration uses $\omega = 1$ and the stopping criterion is set to a large value so that at least one iteration with $\omega = 1$ is performed before an approximate value is computed for the optimum relaxation parameter. The SOR method has been observed to be more effective with the red-black ordering than with the natural ordering for some problems [Young, 1971].

SYMMETRIC SOR METHODS.

The **symmetric SOR methods** work best when the *SSOR condition* is satisfied or is slightly relaxed, i.e., the spectral radius of the matrix LU is less than or equal to 0.25, where $L + U = B$. The symmetric SOR procedures can be operated in a fully adaptive or a partially adaptive mode. The symmetric SOR methods generally require fewer iterations for convergence with the natural ordering than with the red-black ordering.

Both the SOR and SYMMETRIC SOR modules may move the off-diagonal elements of the COEF and JCOEF arrays around to enhance vectorization (see the discussion of sparse matrix storage given below). If room allows, these modules arrange COEF and JCOEF so that each column of COEF contains only diagonal elements, only elements in the upper triangle part, or only elements in the lower triangle part of the original scaled coefficient matrix. In other words, the elements in the columns of COEF are segregated to increase vectorization [Kincaid and Oppe, 1982].

REDUCED SYSTEM METHODS.

Before using a reduced system method, it is required that the linear system be reordered into a red-black system using

```
INDEX.    RED-BLACK.
```

When the scaled linear system is partitioned into a red-black form, the reduced system can be written as

$$u_B = F_B F_R u_B + F_B c_R + c_B .$$

The **reduced system CG method** requires no estimates for the bounds on the eigenvalues of $F_B F_R$. Automatic procedures are used to establish these bounds and the iteration parameters. This algorithm in a different form is equivalent to one developed in [Reid, 1972] which is called the **compressed Jacobi conjugate gradient method** in [Hageman and Young, 1981]. The reduced system SI method determines an upper bound on the eigenvalues of $F_B F_R$ adaptively. This method is equivalent to the **cyclic Chebyshev semi-iterative method** (CSSI) of [Golub and Varga, 1961].

The **reduced system SI method** converges twice as fast as the Jacobi SI method. This is a theoretical result and does not count the time involved in establishing the red-black indexing and the red-black partitioned system. Similarly, the reduced system CG method converges, theoretically, exactly twice as fast as the Jacobi CG method. Hence, the accelerated reduced system methods are preferable to the accelerated Jacobi methods when red-black indexing is possible.

The successful convergence of iterative methods may be dependent on conditions that are difficult to determine in advance unless some knowledge about the linear system is available. For many linear systems encountered during the solution of elliptic partial differential equations, one does know that the matrix A is SPD. However, determining whether a given coefficient matrix is positive definite can be as costly to check as solving the system. On the other hand, some conditions affecting convergence of an iterative method such as positive diagonal elements, diagonal dominance, and symmetry are relatively easy to verify. For some applications, the theory may not exist to guarantee the convergence of an iterative method. The ITPACK algorithms have been tested most extensively for linear systems arising from elliptic partial differential equations. These modules can be applied, formally, to any linear system which fits into high-speed memory. However, rapid convergence, and indeed convergence itself, cannot be guaranteed unless the matrix of the system is symmetric and positive definite. Success can be expected, though not guaranteed, for mildly nonsymmetric systems. In other words, iterative methods may not converge when applied to systems with coefficient matrices which are completely general with no special properties.

MODULE PARAMETERS

The parameters of the ITPACK modules and their defaults are listed in Chapter 6. For additional details see [Kincaid, et al., 1982] and [Hageman and Young, 1981].

ITPACK PRINTING

If LEVEL $\geqslant$ 3, two parameter arrays RPARM and IPARM will be printed both prior to commencement of the iteration and after convergence. These arrays will contain the adaptive variables CME, SME, OMEGA, SPECR, and BETAB as well as other parameters which the user may find valuable when rerunning the same problem with new parameters. A list of the relevant parameters follows:

Parameter	*Meaning or Usage*
IPARM(1)	ITMAX (Iteration counter). Input: Maximum number of iterations allowed. Output: Number of iterations performed.
IPARM(2)	LEVEL (Level of printed output).
IPARM(3)	Not available in ELLPACK.
IPARM(4)	Not available in ELLPACK.
IPARM(5)	ISYM (Symmetry indicator). = 0 if matrix is assumed to be symmetric, = 1 if matrix is not assumed to be symmetric.
IPARM(6)	IADAPT (Adaptive switch; see module parameters in Chapter 6).
IPARM(7)	ICASE (Adaptive strategy switch; see module parameters in Chapter 6).
IPARM(8)	NWKSP (Workspace requirements). Input: 0. Output: Amount of workspace required for this solution module. The ELLPACK control program usually overestimates the amount of workspace required, so this information can be used to reduce the amount of workspace allocated to the problem if it needs to be rerun.
IPARM(9)	NB (Red-black ordering indicator). = −2 if AS IS Ordering has been used,

Parameter	*Meaning or Usage*
	= NB if RED-BLACK ordering has been used. In this case, NB is the order of the black subsystem.
IPARM(10)	Not available in ELLPACK.
IPARM(11)	ITIME (Timing indicator). = 0 if TIME was specified in the OPTIONS statement, = 1 otherwise.
IPARM(12)	IDGTS (Error analysis switch; see module parameters in Chapter 6).
RPARM(1)	ZETA (Stopping criterion; see module parameters in Chapter 6). Input: ZETA. Output: If the method failed to converge in ITMAX iterations, RPARM(1) is reset to an estimate of the relative accuracy achieved. Otherwise, it is unchanged.
RPARM(2)	CME (Estimate of largest eigenvalue of Jacobi matrix). Input: Initial estimate of CME. Output: Final adaptive value of CME.
RPARM(3)	SME (Estimate of smallest eigenvalue of Jacobi matrix). Input: Initial estimate of SME. Output: Final adaptive value of SME.
RPARM(4)	FF (Adaptive procedure damping factor; see module parameters in Chapter 6).
RPARM(5)	OMEGA (Overrelaxation parameter; see module parameters in Chapter 6). Input: Initial estimate of OMEGA. Output: Final adaptive value of OMEGA.
RPARM(6)	SPECR (Spectral radius estimate for SSOR. See module parameters in Chapter 6). Input: Initial estimate of SPECR. Output: Final adaptive value of SPECR.
RPARM(7)	BETAB (Estimate of spectral radius of *LU* matrix; see module parameters in Chapter 6). Input: Initial estimate of BETAB. Output: Final adaptive value of BETAB.
RPARM(8)	Not available in ELLPACK.
RPARM(9)	TIME1 (Time of iterative process). Input: 0.0. Output: Total time in seconds from beginning of iterative algorithm until convergence, if TIME was specified in the OPTIONS statement.
RPARM(10)	TIME2 (Time for entire solution). Input: 0.0.

Output: Total time in seconds for the entire solution process including scaling and unscaling, if TIME was specified in the options statement.

RPARM(11) DIGIT1 (See discussion of IDGTS in module parameters in Chapter 6).
Input: 0.0.
Output: DIGIT1.

RPARM(12) DIGIT2 (See discussion of IDGTS in module parameters in Chapter 6).
Input: 0.0.
Output: DIGIT2.

ERROR CONDITIONS

The ITPACK modules assign to an integer variable IER certain values to indicate error conditions (or lack of them) upon exiting. The following values of M are used to distinguish the ITPACK module:

M = 10 Jacobi CG
M = 20 Jacobi SI
M = 30 SOR
M = 40 Symmetric SOR CG
M = 50 Symmetric SOR SI
M = 60 Reduced system CG
M = 70 Reduced system SI

The meanings of the values of IER are as given in the following table:

Value of IER	*Meaning*
0	No error detected.
1+M	Invalid order of linear system.
2+M	Insufficient workspace assigned. IPARM(8) is set to the amount required.
3+M	Failure to converge in ITMAX iterations. RPARM(1) has been reset to the last computed value of the stopping test.
4+M	Invalid order of black subsystem for red-black indexing.
401	Zero diagonal element.
402	Nonexistent diagonal element.
501, 502, 601	Difficulty encountered in eigenvalue estimation.
602	Matrix not positive definite.

SPECIAL CASES

Textbook methods such as the Jacobi (J) method, Gauss-Seidel (GS) method, successive overrelaxation (SOR) method with fixed relaxation factor ω, symmetric successive overrelaxation (SSOR) method with fixed relaxation factor ω, and the reduced system (RS) method can be obtained by setting appropriate module parameters.

Method	*Use module*	*Module parameter Values*
J	Jacobi SI	IADAPT = 0, ICASE = 2
GS	SOR	IADAPT = 0
SOR, fixed ω	SOR	IADAPT = 0, OMEGA = ω
SSOR, fixed ω	Symmetric SI	IADAPT = 0, OMEGA = ω
RS	Reduced system SI	IADAPT = 0

EXAMPLE

The iterative algorithms in ITPACK have been tested over a wide class of matrix problems arising from elliptic partial differential equations with Dirichlet, Neumann, and mixed boundary conditions on arbitrary two-dimensional regions (including cracks and holes) and on rectangular three-dimensional regions (see, for example, [Eisenstat, et al. 1979]). The small sample problem given below is given as an example of the use of ITPACK modules within ELLPACK. It was run on the CDC CYBER 170/750 at the University of Texas with the FTN5 compiler (OPT=3):

$$(xu_x)_x + (yu_y)_y = x + y, \qquad \text{on } R = [0,1]\times[0,1],$$
$$u = 1 + xy, \qquad \text{on boundary of } R,$$

We use the standard 5-point central difference approximation (5 POINT STAR) on a mesh with spacing 1/4, resulting in a symmetric positive definite system of nine linear equations. To illustrate the use of two ITPACK modules in ELLPACK for solving this system, we solve it first using Jacobi CG with the natural ordering and then using the reduced system CG with the red-black ordering.

```
*         EXAMPLE OF USING ITPACK MODULES
OPT.      TIME $ LEVEL=1
EQ.       (X*UX)X + (Y*UY)Y = X + Y
BOUND.    U = G(X,Y)   ON   X=0.0
                       ON   X=1.0
                       ON   Y=0.0
                       ON   Y=1.0
GRID.     5 X-POINTS
          5 Y-POINTS
DIS.      5 POINT STAR
*         JACOBI CG WITH NATURAL ORDERING
INDEX.    AS IS
SOL.      JACOBI CG (ZETA=1.E-8,LEVEL=3,IDGTS=1)
*         REDUCED SYSTEM CG WITH RED-BLACK ORDERING
INDEX.    RED-BLACK (LEVEL=5)
SOL.      REDUCED SYSTEM CG (ZETA=1.E-8,LEVEL=3,IDGTS=1)
SUB.
          REAL FUNCTION G(X,Y)
          G = 1.0 + X*Y
          RETURN
          END
END.
```

A portion of the output of this ELLPACK program follows:

```
-----------------
 ELLPACK OUTPUT
-----------------

        +++++++++++++++++++++++++
        +                       +
        +    EXECUTION  TIMES   +
        +                       +
        +++++++++++++++++++++++++

    MODULE NAME                        SECONDS
    ----------------------------------------

    5-POINT STAR                           .01
    AS IS                                  .01
    JACOBI CG SETUP                       0.00
    JACOBI CG                              .03
    RED-BLACK                              .01
    REDUCED SYSTEM CG SETUP               0.00
    REDUCED SYSTEM CG                      .03
    TOTAL TIME                             .11

-------------------------
 DISCRETIZATION  MODULE
-------------------------

    5 - P O I N T     S T A R

    DOMAIN                                  RECTANGLE
    X INTERVAL              0.         ,    .100E+01
    Y INTERVAL              0.         ,    .100E+01
    DISCRETIZATION                           UNIFORM
    GRID                                     5 X   5
    HX                                      .250E+00
    HY                                      .250E+00
    B.C.S ON PIECES 1,2,3,4                  1,1,1,1
    OUTPUT LEVEL                                   1
    NUMBER OF EQUATIONS                            9
    MAX NO. OF UNKNOWNS PER EQ.                    5
    EXECUTION SUCCESSFUL

-----------------
 INDEXING MODULE
-----------------

    A S   I S

    EQUATIONS  INDEXED          9
    UNKNOWNS  INDEXED           9

    EXECUTION SUCCESSFUL

-----------------
 SOLUTION MODULE
-----------------

I T P A C K     J A C O B I   C G

    INITIAL ITERATIVE PARAMETERS

         IPARM(1)   =              100      (ITMAX)
         IPARM(2)   =                3      (LEVEL)
         IPARM(3)   =                0      (IRESET)
         IPARM(4)   =                6      (NOUT)
         IPARM(5)   =                0      (ISYM)
         IPARM(6)   =                1      (IADAPT)
         IPARM(7)   =                1      (ICASE)
         IPARM(8)   =                0      (NWKSP)
         IPARM(9)   =               -2      (NB)
         IPARM(10)  =                0      (IREMOV)
         IPARM(11)  =                0      (ITIME)
         IPARM(12)  =                1      (IDGTS)
```

```
          RPARM(1)   =  .10000000E-07      (ZETA)
          RPARM(2)   = 0.                  (CME)
          RPARM(3)   = -.10000000E+01      (SME)
          RPARM(4)   =  .75000000E+00      (FF)
          RPARM(5)   =  .10000000E+01      (OMEGA)
          RPARM(6)   = 0.                  (SPECR)
          RPARM(7)   =  .25000000E+00      (BETAB)
          RPARM(8)   =  .35527137E-12      (TOL)
          RPARM(9)   = 0.                  (TIME1)
          RPARM(10)  = 0.                  (TIME2)
          RPARM(11)  = 0.                  (DIGIT1)
          RPARM(12)  = 0.                  (DIGIT2)
IN THE FOLLOWING, RHO AND GAMMA ARE ACCELERATION PARAMETERS, CME
IS THE ESTIMATE OF THE LARGEST EIGENVALUE OF THE JACOBI MATRIX

     INTERMEDIATE OUTPUT AFTER EACH ITERATION

NUMBER OF    CONVERGENCE      CME              RHO            GAMMA
ITERATIONS      TEST

     0       .94248E+01   0.               .10000E+01     .14418E+01
     1       .49364E+00   .30640E+00       .12943E+01     .13453E+01
     2       .44301E+00   .62484E+00       .12528E+01     .73430E+00
     3       .19661E+00   .70166E+00       .12581E+01     .12412E+01
     4       .17867E-01   .73578E+00       .10032E+01     .71654E+00
     5       .57674E-14   .73598E+00       .10032E+01     .71654E+00
JACOBI CG  HAS CONVERGED IN     5 ITERATIONS

          APPROX. NO. OF DIGITS IN STOPPING TEST = 14.2  (DIGIT1)
          APPROX. NO. OF DIGITS IN RATIO TEST    = 13.9  (DIGIT2)

     FINAL ITERATIVE PARAMETERS

          IPARM(1)   =               5     (ITMAX)
          IPARM(2)   =               3     (LEVEL)
          IPARM(3)   =               0     (IRESET)
          IPARM(4)   =               6     (NOUT)
          IPARM(5)   =               0     (ISYM)
          IPARM(6)   =               1     (IADAPT)
          IPARM(7)   =               1     (ICASE)
          IPARM(8)   =              55     (NWKSP)
          IPARM(9)   =              -2     (NB)
          IPARM(10)  =               0     (IREMOV)
          IPARM(11)  =               0     (ITIME)
          IPARM(12)  =               1     (IDGTS)
          RPARM(1)   =  .10000000E-07      (ZETA)
          RPARM(2)   =  .73598007E+00      (CME)
          RPARM(3)   = -.10000000E+01      (SME)
          RPARM(4)   =  .75000000E+00      (FF)
          RPARM(5)   =  .10000000E+01      (OMEGA)
          RPARM(6)   = 0.                  (SPECR)
          RPARM(7)   =  .25000000E+00      (BETAB)
          RPARM(8)   =  .35527137E-12      (TOL)
          RPARM(9)   =  .15000000E-01      (TIME1)
          RPARM(10)  =  .15000000E-01      (TIME2)
          RPARM(11)  =  .14239023E+02      (DIGIT1)
          RPARM(12)  =  .13904092E+02      (DIGIT2)
EXECUTION SUCCESSFUL

-----------------
 INDEXING MODULE
-----------------

     R E D - B L A C K

          ORDER OF BLACK SUBSYSTEM =      4

          I1UNDX(    1) =     1          I1ENDX(    1) =     1
          I1UNDX(    2) =     9          I1ENDX(    2) =     3
```

```
        I1UNDX(   3) =      2          I1ENDX(   3) =      5
        I1UNDX(   4) =      8          I1ENDX(   4) =      7
        I1UNDX(   5) =      3          I1ENDX(   5) =      9
        I1UNDX(   6) =      7          I1ENDX(   6) =      8
        I1UNDX(   7) =      4          I1ENDX(   7) =      6
        I1UNDX(   8) =      6          I1ENDX(   8) =      4
        I1UNDX(   9) =      5          I1ENDX(   9) =      2
EXECUTION SUCCESSFUL

-----------------
 SOLUTION MODULE
-----------------

I T P A C K    R E D U C E D   S Y S T E M   C G
    INITIAL ITERATIVE PARAMETERS

        IPARM(1)   =              100    (ITMAX)
        IPARM(2)   =                3    (LEVEL)
        IPARM(3)   =                0    (IRESET)
        IPARM(4)   =                6    (NOUT)
        IPARM(5)   =                0    (ISYM)
        IPARM(6)   =                1    (IADAPT)
        IPARM(7)   =                1    (ICASE)
        IPARM(8)   =                0    (NWKSP)
        IPARM(9)   =                4    (NB)
        IPARM(10)  =                0    (IREMOV)
        IPARM(11)  =                0    (ITIME)
        IPARM(12)  =                1    (IDGTS)
        RPARM(1)   =   .10000000E-07     (ZETA)
        RPARM(2)   =  0.                 (CME)
        RPARM(3)   =  -.10000000E+01     (SME)
        RPARM(4)   =   .75000000E+00     (FF)
        RPARM(5)   =   .10000000E+01     (OMEGA)
        RPARM(6)   =  0.                 (SPECR)
        RPARM(7)   =   .25000000E+00     (BETAB)
        RPARM(8)   =   .35527137E-12     (TOL)
        RPARM(9)   =  0.                 (TIME1)
        RPARM(10)  =  0.                 (TIME2)
        RPARM(11)  =  0.                 (DIGIT1)
        RPARM(12)  =  0.                 (DIGIT2)

        ORDER OF BLACK SUBSYSTEM =     4 (NB)

IN THE FOLLOWING, RHO AND GAMMA ARE ACCELERATION PARAMETERS, CME
IS THE ESTIMATE OF THE LARGEST EIGENVALUE OF THE JACOBI MATRIX

    INTERMEDIATE OUTPUT AFTER EACH ITERATION

NUMBER OF    CONVERGENCE      CME              RHO          GAMMA
ITERATIONS      TEST

      0        .10894E+02  0.              .10000E+01   .16826E+01
      1        .43641E+00   .63694E+00     .10872E+01   .14171E+01
      2        .79855E-14   .73598E+00     .10872E+01   .14171E+01

REDUCED SYSTEM CG  HAS CONVERGED IN      2 ITERATIONS

        APPROX. NO. OF DIGITS IN STOPPING TEST = 14.1   (DIGIT1)
        APPROX. NO. OF DIGITS IN RATIO TEST    = 14.1   (DIGIT2)

    FINAL ITERATIVE PARAMETERS

        IPARM(1)   =                2    (ITMAX)
        IPARM(2)   =                3    (LEVEL)
        IPARM(3)   =                0    (IRESET)
        IPARM(4)   =                6    (NOUT)
        IPARM(5)   =                0    (ISYM)
        IPARM(6)   =                1    (IADAPT)
        IPARM(7)   =                1    (ICASE)
        IPARM(8)   =               34    (NWKSP)
```

```
IPARM(9)   =              4     (NB)
IPARM(10)  =              0     (IREMOV)
IPARM(11)  =              0     (ITIME)
IPARM(12)  =              1     (IDGTS)
RPARM(1)   =  .10000000E-07     (ZETA)
RPARM(2)   =  .73598007E+00     (CME)
RPARM(3)   = -.10000000E+01     (SME)
RPARM(4)   =  .75000000E+00     (FF)
RPARM(5)   =  .10000000E+01     (OMEGA)
RPARM(6)   = 0.                 (SPECR)
RPARM(7)   =  .25000000E+00     (BETAB)
RPARM(8)   =  .35527137E-12     (TOL)
RPARM(9)   = 0.                 (TIME1)
RPARM(10)  =  .17000000E-01     (TIME2)
RPARM(11)  =  .14097698E+02     (DIGIT1)
RPARM(12)  =  .14118331E+02     (DIGIT2)
EXECUTION SUCCESSFUL
```

There are several remarks which should be made:

1. The use of LEVEL = 3 as a module parameter in the ITPACK calls does not reset the global printing level LEVEL = 1 set in the OPTIONS statement for ELLPACK. Hence it is possible to see a great deal of detail in the solution process without seeing much detail in the discretization or indexing process. For large problems, LEVEL should not be set higher than 3, since this will result in too much printing.

2. Natural indexing (AS IS) resulted in the following numbering of the unknowns:

7	8	9
4	5	6
1	2	3

Red-black indexing (RED-BLACK) resulted in the numbering:

4	6	5
8	3	7
1	9	2

Note that the grid was first color-coded in a checkerboard fashion with the "red" nodes being numbered before the "black" nodes.

R	B	R
B	R	B
R	B	R

Note also that the permutation vector I1UNDX gives the mapping between node numbers when going from natural to red-black ordering, while the permutation vector I1ENDX gives the inverse mapping. (See Chapter 13 for details.)

SPARSE MATRIX STORAGE

We now present a brief review of the sparse storage scheme for linear systems used in ELLPACK.

In ELLPACK, the A and b of the linear system (1) are generated by a discretization module, and then an ITPACK module solves the associated system without translation into another data structure. The equation coefficients and column identifiers are stored in the two-dimensional real and integer arrays COEF and JCOEF, respectively, of dimension N by MAXNZ. Here N is the number of linear equations and MAXNZ is the maximum number of nonzeros per row. The internal ELLPACK names for these variables are, respectively, R1COEF, I1IDCO, I1NEQN, and I1MNCO. The coefficients of a particular equation and its column identifiers are stored in corresponding rows of the arrays. For example, the coefficient matrix

$$A = \begin{bmatrix} 4. & -1. & -2. & 0. \\ -1. & 4. & 0. & -2. \\ -3. & 0. & 4. & -1. \\ 0. & 0. & -1. & 4. \end{bmatrix}$$

is represented by

$$\text{COEF} = \begin{bmatrix} 4. & -1. & -2. \\ 4. & -1. & -2. \\ 4. & -3. & -1. \\ 4. & -1. & 0. \end{bmatrix}, \qquad \text{JCOEF} = \begin{bmatrix} 1 & 2 & 3 \\ 2 & 1 & 4 \\ 3 & 1 & 4 \\ 4 & 3 & 0 \end{bmatrix}.$$

The nonzero coefficients in a particular row are not necessarily left justified and may be in any order. However, if the diagonal element is not in column 1, the ITPACK module will place it there without returning it to its original position upon exiting. Also, if the diagonal element is negative (the ELLPACK discretization module 5 POINT STAR usually constructs a system with negative diagonal elements), ITPACK will change the signs of the coefficient elements in that row of COEF and the corresponding right-hand-side element in RHS without changing them back upon exiting.

7.C ITPACK PROJECT

The seven ITPACK modules which are included in ELLPACK are routines for solving large sparse linear systems by adaptive accelerated iterative algorithms. They were adapted from the ITPACK 2C package described in [Kincaid, et al., 1981] by changing the data strucure used to store the coefficient matrix from three linear arrays (IA, JA, A) into two 2-dimensional arrays (COEF, JCOEF). This change resulted in increased vectorization of the package and improved efficiency on high performance computers in addition to compatibility with the ELLPACK storage scheme [Kincaid, Oppe, and Young, 1982].

Since ITPACK codes in ELLPACK are a modified subset of the entire ITPACK package, the user who does not need ELLPACK to generate the linear system is encouraged to obtain the complete ITPACK software package and to use it as a stand-alone package. For more information on the ITPACK project, see "ITPACK: Past, Present, and Future" [Kincaid and Young, 1983].

History

The ITPACK modules described here are the result of several years of research and development. The ITPACK project involves the development of research-oriented mathematical software, based on iterative algorithms, for solving large systems of linear algebraic equations with sparse coefficient matrices. The emphasis is on linear systems arising in the solution of partial differential equations by discretizations involving finite difference or finite element methods. The development of ITPACK software began in the early 1970's when Professor Garrett Birkhoff suggested that general purpose software for solving linear systems should be developed for iterative methods as well as for direct methods. Initially, prototype programs were written based on preliminary iterative algorithms involving adaptive selection of parameters and automatic stopping procedures. These routines were designed for solving self-adjoint elliptic partial differential equations. The ITPACK routines used iterative algorithms which were refined from the prototype programs. However, these routines were designed to solve large sparse linear systems of algebraic equations instead of partial differential equations. The ITPACK package was modified, improved, and enhanced through various version changes over a period of years. The 2C version of ITPACK was published in the ACM Transactions on Mathematical Software [Kincaid, et al., 1981] and is available from IMSL, Inc. The ITPACK routines in ELLPACK are a subset of this package with the storage scheme changed to be compatible with the ELLPACK data structure.

Vector versions of ITPACK 2C have been written — one using the vector syntax of the CDC CYBER 205 and one in standard Fortran for use on other supercomputers. Also, a double precision version of ITPACK 2C has been written for use on short-word length computers. Future plans include the expansion of ITPACK to handle more iterative methods and more general systems.

Additional information on the ITPACK project can be obtained by writing

Center for Numerical Analysis
RLM Bldg 13.150
University of Texas
Austin, Texas 78712-1067

7.D ACKNOWLEDGEMENTS

The authors wish to express their appreciation to the many people who contributed to the development and testing of the ITPACK software. In particular, Roger G. Grimes contributed to the development of earlier versions of this software package. Work on this project was supported in part by the National Science Foundation grant MCS-79-19829 and by the Department of Energy grant DE-AS05-81ER10954 to the University of Texas at Austin.

7.E REFERENCES

Eisenstat, S., A. George, R. Grimes, D. Kincaid, and A. Sherman [1979], "Some comparisons of software packages for large sparse linear systems", in *Advances in Computer Methods for Partial Differential Equations* III (R. Vichnevetsky and R. Stepleman, eds.), IMACS, Rutgers Univ., New Brunswick, N.J., pp. 98–106.

Golub, G. and R. Varga, [1961], "Chebyshev semi-iterative methods, successive overrelaxation iterative methods", and second-order richardson iterative methods, parts I & II, Numer. Math. 3, pp. 147–168.

Grimes, R., D. Kincaid, W. MacGregor, and D. Young [1978], "ITPACK report: adaptive iterative algorithms using symmetric sparse storage", CNA-139, Center for Numerical Analysis, Univ. of Texas, Austin, Texas.

Grimes, R., D. Kincaid and Young, D. [1979], "ITPACK 2.0 User's Guide", CNA-150, Center for Numerical Analysis, Univ. of Texas, Austin, Texas.

Grimes, R., D. Kincaid and D. Young [1980], "ITPACK 2A: A Fortran Implementation of Adaptive Accelerated Iterative Methods for Solving Large Sparse Linear Systems", CNA-164, Center for Numerical Analysis, Univ. of Texas, Austin, Texas.

Hageman, L. and D. Young [1981], *Applied Iterative Methods*, Academic Press, New York.

Hayes, L. and D. Young [1977], "The Accelerated SSOR Method for Solving Large Linear Systems: Preliminary Report", CNA-123, Center for Numerical Analysis, Univ. of Texas, Austin, Texas.

Kincaid, D. and R. Grimes [1977], "Numerical Studies of Several Adaptive Iterative Algorithms", CNA-126, Center for Numerical Analysis, Univ. of Texas, Austin, Texas.

Kincaid, D., R. Grimes, W. MacGregor and D. Young [1979], "ITPACK - adaptive iterative algorithms using symmetric sparse storage", in *Symposium on Reservoir Simulation*, Soc. Pet. Engr. of AIME, 6200 North Central Expressway, Dallas, Texas, pp. 151–160.

Kincaid, D., R. Grimes and D. Young [1979], "The use of iterative methods for solving large sparse PDE-related linear systems", *Mathematics and Computers in Simulation* XXI, North-Holland Publ. Co., New York, pp. 368–375.

Kincaid, D. and D. Young [1979], Survey of Iterative Methods, in *Encyclopedia of Computer Sciences and Technology* 13 (J. Belzer, A. Holzman, and A. Kent, eds.), Marcel Dekker, Inc., New York, pp. 354–391.

Kincaid, D. and D. Young [1981], "Adapting iterative algorithms developed for symmetric systems to nonsymmetric systems", in *Flliptic Problem Solvers* (M. Schultz, ed.), Academic Press, New York, pp. 353–359.

Kincaid, D. [1981], "Acceleration parameters for a summetric successive overrelaxation conjugate gradient method for nonsymmetric systems", in *Advances in Computer Methods for Partial Differential Equations* IV (R. Vichnevetsky and R. Stepleman, eds.), IMACS, Rutgers Univ., New Brunswick, N.J., pp. 294–299.

Kincaid, D., R. Grimes, J. Respess, and D. Young, D. [1981], "ITPACK 2B: A Fortran Implementation of Adaptive Accelerated Iterative Methods for

Solving Large Sparse Linear Systems", CNA-173, Center for Numerical Analysis, Univ. of Texas, Austin, Texas.

Kincaid, D., T. Oppe and D. Young [1982], "Adapting ITPACK Routines for Use on a Vector Computer", CNA-177, Center for Numerical Analysis, Univ. of Texas, Austin, Texas.

Kincaid, D. and T. Oppe [1982], ITPACK on Supercomputers, CNA-178, Center for Numerical Analysis, Univ. of Texas, Austin, Texas.

Kincaid, D., J. Respess, D. Young and R. Grimes [1982], "Algorithm 586 ITPACK 2C : A Fortran package for solving large sparse linear systems by adaptive accelerated iterative methods", ACM Trans. Math. Soft. 8, pp. 302–322.

Kincaid, D. and D. Young [1984], The ITPACK project: past, present, and future, in *Elliptic Problem Solvers* II, (G. Birkhoff and A. Schoenstadt, eds.), Academic Press, New York.

Lawson, C., R. Hanson, D. Kincaid and F. Krogh, "Basic linear algebra subprograms for fortran usage", ACM Trans. on Math. Software 5, pp. 308–323.

Reid, J. [1972], "The use of conjugate gradients for systems of linear equations possessing property A", SIAM J. Numer. Anal. 9, pp. 325–332.

Young, D. [1971], *Iterative Solution of Large Linear Systems*, Academic Press, New York.

Young, D. and D. Kincaid [1981], "The ITPACK package for large sparse linear systems", in *Elliptic Problem Solvers* (M. Schultz, ed.), Academic Press, New York, pp. 163–185.

Young, D., C. Jea and D. Kincaid [1984], "Accelerating nonsymmetrizable iterative methods", in *Elliptic Problem Solvers* II, (G. Birkhoff and A. Schoenstadt, eds.), Academic Press, New York.

PART 3

CHAPTERS 8–11

PERFORMANCE EVALUATION

Part 3 presents a short study of the performance of the software modules in ELLPACK. The approach and nine model problems are given in Chapters 8 and 9; then some performance studies of discretization modules and solution modules are presented in Chapter 10 and 11, respectively. There are very large differences in the efficiency between modules of the same type.

Chapter 8

PERFORMANCE AND ITS EVALUATION

8.A MEASURES AND PERFORMANCE

There are many factors that determine the quality of software, but we concentrate on only three here: accuracy, speed, and memory use. Other important qualities like ease of use, reliability, portability, documentation, etc. are ignored. (The latter are much more difficult to quantify, but we hope that the ELLPACK system and its modules perform well in all these respects).

Accuracy can be measured in several ways; we choose to measure it by

$$\max_{(x,y) \in G} |u(x,y) - U(x,y)|$$

where $u(x,y)$ is the true solution, $U(x,y)$ is the computed solution, and the grid G is the set of grid intersections in the domain defined by the GRID segment. This value is just an approximation to the maximum over the domain and can be inaccurate for a coarse grid; it has the advantage of simplicity. For example, finite difference methods only return values at the grid points and their error off the grid points is not easily defined in general.

We measure **speed** as the number of seconds that the ELLPACK control program executes the problem solving modules. The system provides detailed timing data and allows one to separate the time to make, for example, a contour plot from the time used by a problem solving module. Similarly, the ELLPACK Preprocessor time is usually not counted as part of the problem solving time. The total time referred to later is the sum of the discretization, indexing and solution times for solving a problem.

We measure **memory** as the number of words used by the ELLPACK control program. It is not so easy to assign the memory to any one module, because various arrays are used by several of them. Thus, we normally discuss the overall memory use by a particular combination of modules.

8.B ANALYTIC EVALUATION OF PERFORMANCE

The traditional and very useful approach to estimating the performance of methods is to combine an **operations count** (usually one counts multiplications or multiplications and additions) with an **asymptotic error analysis**. The simplest example is for 5-POINT STAR + BAND GE. Assume the grid is N by N, so the error for a smooth problem is asymptotically $O(1/N^2)$. The work to discretize the PDE is small; it takes $O(N^4)$ multiplications and additions to solve the linear system of equations using Gauss elimination for a band matrix. Thus we estimate work $t = O(N^4)$ is needed to obtain an error $\epsilon = O(1/N^2)$. One can express ϵ in terms of t by eliminating N to obtain

$$\epsilon = O(1/\sqrt{t}).$$

If one applies a similar analysis to HERMITE COLLOCATION + BAND GE, one obtains $\epsilon = O(1/N^4)$, since collocation with Hermite bicubics is fourth order. The work is still $t = O(N^4)$ since the linear system also has bandwidth proportional to N. The expression of accuracy in terms of work is

$$\epsilon = O(1/t).$$

which suggests that collocation is much more efficient than ordinary finite differences.

The conclusion reached from the above simple analysis is correct, but this example also illustrates the shortcomings of this approach. The collocation method has $4N^2$ unknowns on an N by N mesh, compared to N^2 unknowns for finite differences. Each collocation equation has 16 terms compared to 5 terms for finite differences. The bandwidth is $4N$ for collocation and N for finite differences. Thus, $t \sim 64N^4$ for collocation, compared to $t \sim N^4$ for finite differences. On the other hand, the constants in the ϵ formulas involve the maxima of second derivatives of $u(x,y)$ for finite differences and the maxima of fourth derivatives for collocation. These values are generally unavailable and cannot be compared. Thus, this analytic approach gives a general feel for the comparative efficiency of methods, but it rarely leads to very precise results.

There is another shortcomming of this approach; most methods are useful for a wider range of problems than we can analyze. The above analysis fails to provide any estimate if the solution $u(x,y)$ is not smooth, even though both methods can solve a variety of singular or nearly-singular problems. It spite of these shortcomings, the analytic approach has proved very useful in providing general guidance about the potential performance of methods. See Chapter 25 of [Forsythe and Wasow, 1960] for an early use of this approach and Chapter 10 of [Rice, 1983] for its application to a variety of methods.

8.C EXPERIMENTAL EVALUATION OF PERFORMANCE

It is now recognized that precise performance evaluation can be made only with actual measurements on working software. The variations in the performance of different implementations of the "same method" are just too large to allow one to rely only on analytic estimates of performance. The idea of the **experimental approach** is simple: one applies a program to a large set of representative problems, measures its performance, and analyzes the resulting data to estimate the expected value for performance in general. This is the standard approach in all of experimental science. There are many ways one can go wrong: the sample set might be too small, the sample might not be representative (what is a "typical" elliptic problem?), the measurements might be biased by the machine or computer system used, and so forth. Many of these issues are discussed in [Boisvert, Houstis and Rice, 1979], [Rice, 1979], and [Houstis and Rice, 1980].

A substantial collection of linear, second order elliptic problems on rectangular domains is given in Appendix A. This set is adequate for most experimental studies within this class of problems. Note that it usually requires 10 to 20 sample problems to achieve a high level of statistical significance when comparing two problem solving modules. The following chapters present some experimental results which are designed to provide some indication of the performance of the ELLPACK modules, but the experiments are based on too small a problem set to be any more than suggestive.

The influence of machines and computer systems on the performance of elliptic PDE software is analyzed in [Rice, 1982]. The principle conclusion is: relative differences of 50 percent in speed are not likely to be significant over a broad range of computer systems. This is for computers with the standard, single processor von Neumann architecture. Further, one can often construct problems which exploit the idiosyncrasies of a particular machine or system to run extraordinarily slow or fast. However, there are examples in the following chapters where one module solves a problem a thousand times faster than another; such a large difference is not likely to be changed significantly by changing from one machine to another.

Perhaps the most important fact to keep in mind is that **experimental studies only evaluate software modules; they do not evaluate numerical methods.** In fact, most numerical methods (such as described in text books) are not precisely defined. If a textbook description of a method is given to 10 different people to program, the resulting 10 programs are likely to have a very wide range of performances due to the details of the implementations. The software modules in ELLPACK are believed to be high quality implementations of the underlying numerical methods, better quality than even a knowledgeable person would make in a short time. Nevertheless, experience shows that from time to time new implementation techniques are discovered which substantially improve the performance of software based on even the best understood methods.

8.D REFERENCES

Boisvert, R. F., J. R. Rice and E. N. Houstis, [1979], "A system for performance analysis of partial differential equations software", IEEE Trans. Software Engng. 5, pp. 418–425.

Forsythe, G. E. and W. R. Wasow, [1960], *Finite Difference Methods for Partial Differential Equations*, John Wiley & Sons, New York.

Houstis, E. N. and J. R. Rice, [1980], "An experimental design for the computational evaluation of elliptic partial differential equation solvers", in *Production and Assessment of Numerical Software* (M. A. Hennell and L. M. Delves, eds.), Academic Press, London, pp. 57–66.

Rice, J. R., [1979], "Methodology for the algorithm selection problem", in *Performance Evaluation of Numerical Software* (L. Fosdick, ed.), North-Holland, Amsterdam, pp. 301–307.

Rice, J. R., [1982], "Machine and compiler effects on the performance of elliptic PDE software", in *Proceedings of the 10th IMACS World Congress on System Simulation and Scientific Computation*, vol. 1, IMACS, Rutgers Univ., New Brunswick, N. J., pp. 446–448.

Rice, J. R., [1983], *Numerical Methods, Software and Analysis*, McGraw-Hill, New York.

Chapter 9

THE MODEL PROBLEMS

9.A THE MODEL PROBLEMS

Nine elliptic problems are used to illustrate the performance of ELLPACK modules. Each is a model for an important class of problems. The first seven are from the PDE population of Appendix A with the identification as given. The generic functions $f(x,y)$ and $g(x,y)$ are used to denote complicated functions determined to make the given $u(x,y)$ functions the true solutions of the model problems.

MODEL PROBLEM A: Poisson problem on a square (Problem 4-1)

This problem represents the simplest, best behaved elliptic problems. Almost every method should be applicable to this problem and the best methods for it are those that take advantage of its special, simple nature:

$$u_{xx} + u_{yy} = 6xy\, e^{x+y}(xy + x + y - 3), \qquad 0 < x, y < 1,$$
$$u = 0 \quad \text{for} \quad x = 0, 1 \quad \text{and} \quad y = 0, 1,$$
$$u(x,y) = 3\, e^{x+y}(x - x^2)(y - y^2).$$

MODEL PROBLEM B: Helmholtz-type problem on a square (Problem 6-1)

This problem has some complexity, the coefficient of u is variable and the solution has several oscillations. Many different methods are applicable to it:

$$u_{xx} + u_{yy} - [100 + \cos(2\pi x) + \sin(3\pi y)]u = f(x,y), \qquad 0 < x, y < 1,$$
$$u = 0 \quad \text{for} \quad x = 0, 1 \quad \text{and} \quad y = 0, 1,$$
$$u(x,y) = -0.31[5.4 - C(x)]S(x)(y^2 - y)[5.4 - C(y)][(1 + v(x,y)^4)^{-1} - 0.5],$$
$$C(z) = \cos(4\pi z), \qquad S(x) = \sin(\pi x), \qquad v(x,y) = 4(x - 0.5)^2 + 4(y - 0.5)^2.$$

MODEL PROBLEM C: Helmholtz problem with a boundary layer (Problem 9-2)

This problem is fairly difficult to solve, all the variation in the solution occurs right next to two of the boundaries. This PDE is a simple type, so that many methods apply to it:

$$u_{xx}+u_{yy}-100\,u = 150\frac{\cosh 20y}{\cosh 20}, \qquad 0 < x,y < 1,$$
$$u = g(x,y) \quad \text{for} \quad x = 0,1 \quad \text{and} \quad y = 0,1,$$
$$u(x,y) = \frac{\cosh 10x}{2\cosh 10} + \frac{\cosh 20y}{2\cosh 20}.$$

MODEL PROBLEM D: General, but easy, problem on a square (Problem 30-2)

This problem is of Helmholtz-type with general variable coefficients. The solution and coefficients are very well behaved and the problem should be fairly easy to solve accurately for those methods that apply to it:

$$[2+(y-1)\,e^{-y^4}]u_{xx} + [1+\frac{1}{(1+4x^2)}]u_{yy}$$
$$+5[x(x-1)+(y-0.3)(y-0.7)]u = f(x,y), \qquad 0 < x,y < 1,$$
$$u = g(x,y) \quad \text{for} \quad x = 0,1 \quad \text{and} \quad y = 0,1,$$
$$u(x,y) = \frac{x+y^2}{1+2x} + (1+x)(y-1)e^{-y^4} + 5(x+y)\cos(xy).$$

MODEL PROBLEM E: General, difficult problem on a square (Problem 54-2)

This problem has variable coefficients for every term except the xy-derivative. The solution has a fairly sharp ridge and considerable complexity:

$$(1+x^2)\,u_{xx} + [1+a(y)^2]u_{yy} + 2x\,u_x + 16y\,a(y)\,u_y$$
$$-[1+(8y-x-4)^2]u = f(x,y), \qquad 0 < x,y < 1,$$
$$u = g(x,y) \quad \text{for} \quad x = 0,1 \quad \text{and} \quad y = 0,1,$$
$$a(y) = 4y^2 + 0.9,$$
$$b(x,y) = \max\,[0, (3 - x/a(y))^3],$$
$$c(x,y) = \max\,[0, x - a(y)],$$
$$d(x,y) = e^{-\min\,[50,\, b(x,y)/c(x,y)]},$$
$$u(x,y) = 2.25x[x-a(y)]^2\,[1-d(x,y)]\,a(y)^{-3} + [1+(8y-x-4)^2]^{-1}.$$

The elliptic operator can be written in self-adjoint form as

$$[(1+x^2)u_x]_x + ([1+a(y)^2]u_y)_y - [1+(8y-x-4)^2]u.$$

MODEL PROBLEM F: Mixed boundary conditions (Problem 31-1)

Most methods become more complex when the boundary conditions involve derivatives and many implementations do not handle such conditions. This problem is simple except for the boundary condition and has a reasonably well-behaved solution:

$$u_{xx}+u_{yy} = -1, \qquad -1 < x,y < 1,$$
$$u+u_x = g(x,y) \quad \text{on} \quad x=1, \qquad u-u_x = g(x,y) \quad \text{on} \quad x=-1,$$
$$u+u_y = g(x,y) \quad \text{on} \quad y=1, \qquad u-u_y = g(x,y) \quad \text{on} \quad y=-1.$$

The boundary condition is actually $u+u_n = g$, where the n-subscript indicates the derivative normal to the boundary. The solution is a 12-th degree polynomial determined to make $g(x,y)$ as close to zero as such a polynomial can be; $u(x,y)$ is given in terms of harmonic polynomials:

$$\begin{aligned}u(x,y) = &-0.25(x^2+y^2)+0.821564-0.01440(x^4-6x^2y^2+y^4)\\&+0.0000493(x^8-28x^6y^2+70x^4y^4-28x^2y^6+y^8)\\&-0.00000064(x^{12}-66x^{10}y^2+495x^8y^4-924x^6y^6\\&\qquad+495x^4y^8-66x^2y^{10}+y^{12}).\end{aligned}$$

Model Problem G: Singular Problem (Problem 47-2)

This problem has singularities in the derivatives along two edges of the domain; the second derivative is continuous and the third is infinite. The contour plot of the solution is smooth and regular:

$$\begin{aligned}&u_{xx}+u_{yy} = f(x,y), \qquad 0 < x,y < 1,\\&u = 0 \quad \text{on} \quad x = 0, y = 0,\\&u = x^{2.5} \quad \text{on} \quad y = 1,\\&u = y^{2.5} \quad \text{on} \quad x = 1,\\&u(x,y) = (xy)^{2.5}.\end{aligned}$$

Model Problem H: Helmholtz problem on a general domain

This is the same as model problem B except for the domain. Thus, the elliptic operator is of Helmholtz type with variable coefficients. The domain and boundary specification are shown in Figure 9.1.

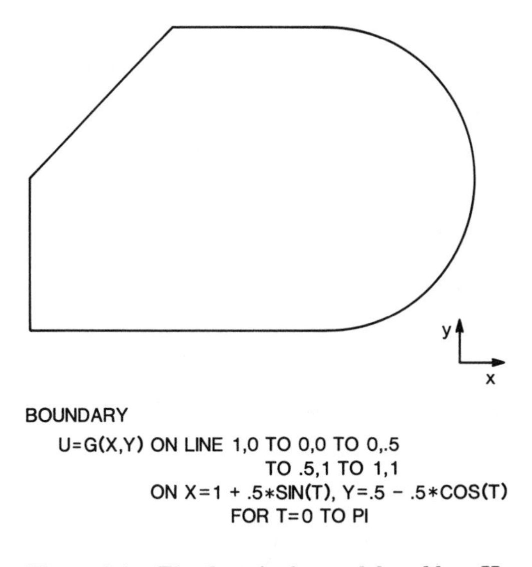


```
BOUNDARY
    U=G(X,Y) ON LINE 1,0 TO 0,0 TO 0,.5
                    TO .5,1 TO 1,1
            ON X=1 + .5*SIN(T), Y=.5 - .5*COS(T)
                FOR T=0 TO PI
```

Figure 9.1. The domain for model problem H. $G(x,y)$ is the true solution.

MODEL PROBLEM I: Poisson problem on a cube

This is a direct extension of model problem A to three dimensions; it is a very simple problem with a very well-behaved solution:

$$u_{xx} + u_{yy} + u_{zz} = f(x,y,z) \qquad 0 < x, y, z < 1$$

$$u = 0 \quad \text{on the surface of the cube}$$

$$u(x,y,z) = 14\,e^{x+y+z}(x-x^2)(y-y^2)(z-z^2).$$

Chapter 10

PERFORMANCE OF DISCRETIZATION MODULES

The goal of the performance evaluations in this chapter is to show how the discretization error decreases as the grid is refined and more computer time is used. Standard error analyses estimate the discretization error as a function of the grid size N (i.e., for an N by N grid), or, respectively, the grid step $h = 1/N$. Thus, a **second order method**, such as 5 POINT STAR uses, has

$$\text{error} = O(h^2) = O(1/N^2).$$

The order can be estimated from an experiment by measuring the average slope of error versus N on a log-log plot, see Figure 10.1. The **error** used in the graphs in Chapters 10 and 11 is the maximum error on the grid divided by the maximum of the solution on the grid; thus an error of 1 corresponds to a 100 percent error relative to the size of u. The data here show the modules have orders close to that predicted in theory. The measured slope is multiplied by two to compensate for the fact that the vertical scale has twice the size of the horizontal scale.

The relationship between error and total computing time depends on the solution method as well as the discretization method. The data in this chapter uses just two solution methods: Gauss elimination (BAND GE or LINPACK SPD BAND, as appropriate) or the iteration module JACOBI CG. Chapter 11 contains much broader data on the solution methods. We include the data on error versus total time here because the triple modules combine the discretization and solution.

The nine model problems from Chapter 9 are considered in order. The computations were all made on a VAX 11/780 using double precision arithmetic (about 13 decimal digits).

10.A POISSON PROBLEM

Model problem A is simple and all ELLPACK discretization modules are applicable. Figure 10.1 shows the **order of accuracy**, error versus N, where a uniform N by N grid is used. All of the modules based on second order finite differences,

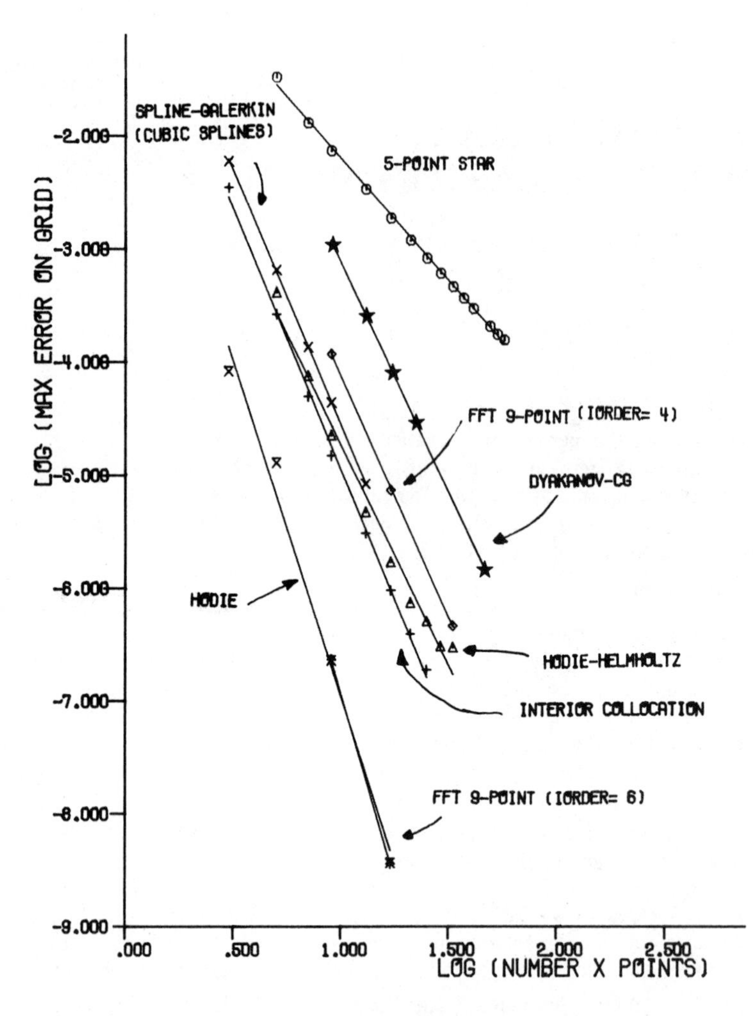

Figure 10.1. Order of convergence of discretization modules for model problem A. The maximum error on the grid is plotted with a log-log scale versus the number N of x points of a uniform grid.

DYAKANOV CG
FFT 9-POINT (IORDER=2)
HODIE
MARCHING ALGORITHM
MULTIGRID MG00
5 POINT STAR

have the same error; so only the 5 POINT STAR data are plotted in Figure 10.1. The slope is as close to -2 as one can measure.

The modules shown based on fourth order discretizations are

DYAKANOV CG 4
FFT 9-POINT (IORDER=4)
HODIE HELMHOLTZ
INTERIOR COLLOCATION
SPLINE GALERKIN (DEGREE=3, NDERV=2)

and the slopes of all but HODIE HELMHOLTZ are almost exactly the same (slightly more than −4). HODIE HELMHOLTZ starts out with some slope, but seems to flatten out; this might reflect some round-off problem starting to set in, although this would be a surprise with 13-digit computations. The differences between these methods are due to different constants in front of the dominant term in the error formula. The difference between the best, INTERIOR COLLOCATION, and the worst, DYAKANOV CG 4, is a factor of 81, which is quite significant.

Only two data points are shown for the sixth order module FFT 9-POINT(IORDER=6) because the convergence is so fast. The accuracy achieved by FFT 9-POINT(IORDER=6) with a 33 by 33 grid would require a grid of about 12500 by 12500 for one of the second order modules. The HODIE discretization is the same as used by FFT 9-POINT(IORDER=6) for this problem.

Figure 10.2 shows the **efficiency**, error versus total time, for the **second order** modules. There are three distinct groups:

1. 5 POINT STAR + BAND GE
   5 POINT STAR + JACOBI CG (with zero initial guess)
   DYAKANOV CG
2. FFT 9-POINT (IORDER=2)
   MARCHING ALGORITHM
   FISHPAK HELMHOLTZ
3. MULTIGRID MG00

The group 1 computations are much slower than the rest. The modules in group 2 are almost indistinguishable and slower than MULTIGRID MG00. Note that it appears that FISHPAK HELMHOLTZ and FFT 9-POINT (IORDER=2) have a steeper slope than MULTIGRID MG00. This suggests that for very large calculations (200–300 hours) there is a cross-over and MULTIGRID MG00 will no longer be the most efficient.

Figure 10.2 also shows a phenomenon that is confirmed by data in Chapter 11; Gauss elimination is more efficient than iteration for the 5 POINT STAR equations on coarse grids, but iteration becomes increasingly more efficient after $N = 12-15$ (i.e., more than about 200 equations).

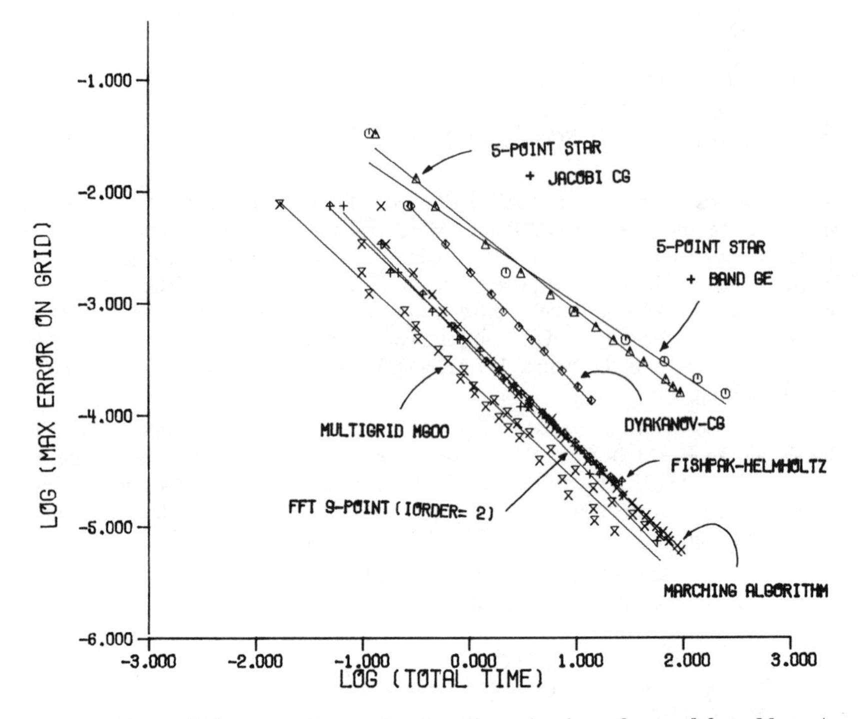

Figure 10.2. Efficiency of second order discretizations for model problem A. The maximum error is plotted on a log-log scale versus the total computing time to solve the problem.

Figure 10.3 shows the **efficiency**, error versus total time, for the **higher order** modules plus the best second order method. The high order FFT 9-POINT module obviously outperforms the others substantially. FFT 9-POINT (IORDER=6) achieves an accuracy in 3 seconds (on a 33 by 33 grid) that would require about 18 days with the second order accurate MULTIGRID MG00, 28 hours with INTERIOR COLLOCATION, 100 minutes with SPLINE GALERKIN (DEGREE=3, NDERV=4), or HODIE HELMHOLTZ, and 100 seconds with FFT 9-POINT (IORDER=4) or HODIE. The advantage of FFT 9-POINT (IORDER=6) over HODIE + LINPACK BAND is in the efficiency of solving the linear equations. The systems of equations generated by the higher order modules can differ substantially in nature and thus substantial variations are due to the solution modules' performance. The very fastest of the second order modules, MULTIGRID MG00, performs poorly in this problem compared to the high order modules.

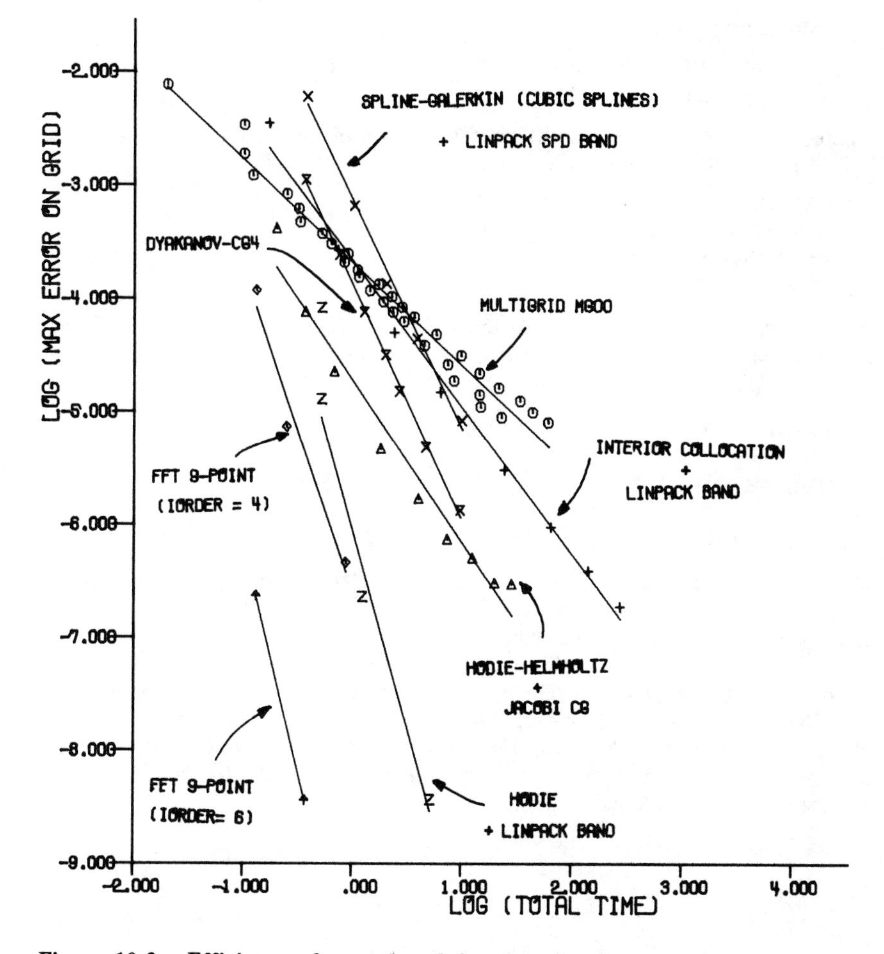

Figure 10.3. Efficiency of second and fourth order discretizations for model problem A. The maximum error is plotted on a log-log scale versus the total computing time to solve the problem.

10.B HELMHOLTZ PROBLEM

Model problem B is

$$u_{xx} + u_{yy} - [100 + \cos(2\pi x) + \sin(3\pi y)]u = f(x, y)$$
$$u = 0 \quad \text{for} \quad x = 0, 1 \quad \text{and} \quad y = 0, 1.$$

Figure 10.4 shows the **order** of the discretization error for the modules used. Four DISCRETIZATION modules based on second order finite differences,

DYAKANOV CG
FFT 9 POINT (IORDER=2)
MARCHING ALGORITHM
5 POINT STAR

apply and have the same error; only the 5 POINT STAR error is plotted in Figure 10.4. The slope is as close to -2 as one can measure.

The modules based on fourth order discretizations are

DYAKANOV CG4
HODIE
HODIE HELMHOLTZ
INTERIOR COLLOCATION
SPLINE GALERKIN (DEGREE=3,NDERV=2)

The slopes of all but SPLINE GALERKIN and HODIE are the same, just a bit less than -4. The five data points for SPLINE GALERKIN indicate an order of convergence order of about 3.2 instead of the expected 4.0. Additional data could not be obtained because SPLINE GALERKIN (DEGREE=3,NDERV=2) ran out of memory. The HODIE module has a more accurate, sixth order discretization than HODIE HELMHOLTZ has for this problem. The differences in the constants of the dominant error term of the fourth order discretizations lead to differences in the error of a factor of about 20 between the best, INTERIOR COLLOCATION, and the worst, DYAKANOV CG 4.

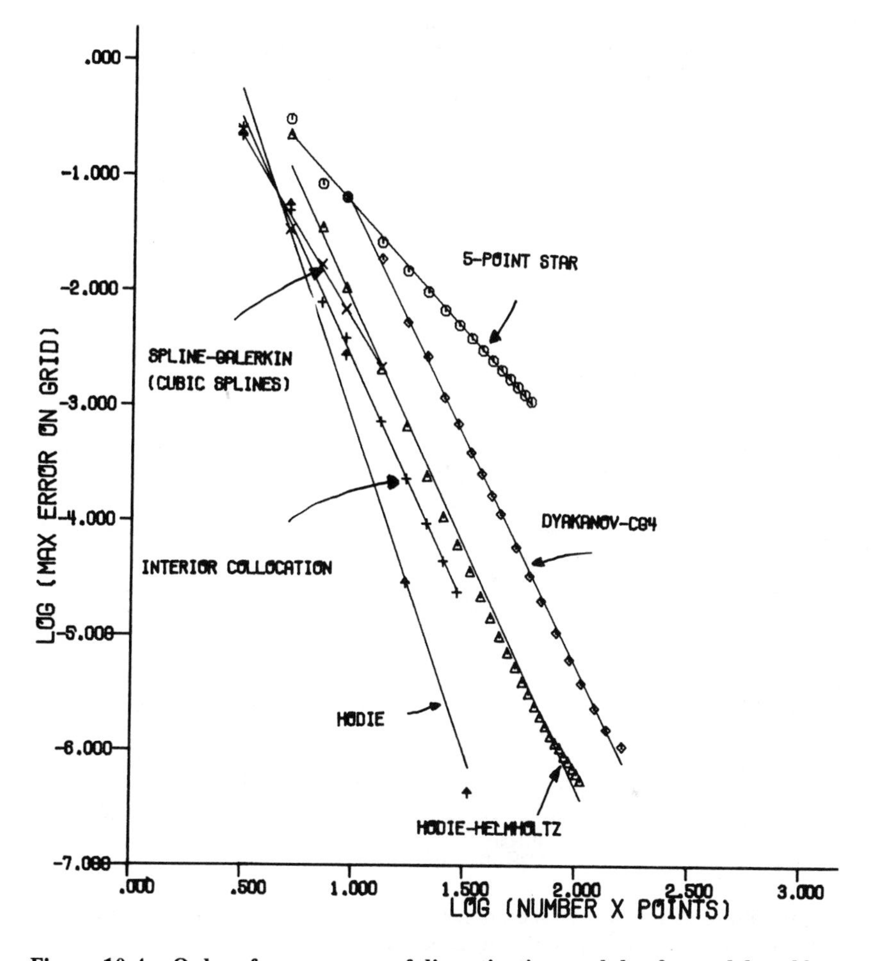

Figure 10.4. Order of convergence of discretization modules for model problem B. The maximum error on the grid is plotted with a log-log scale versus the number N of x points of a uniform grid.

Figure 10.5 shows the **efficiency**, error versus total time, of the **second order** modules. The relationship shown in Figure 10.2 for the Poisson problem is repeated here. MULTIGRID MG00 is clearly superior, while 5 POINT STAR with standard Gauss elimination or iteration solution is the worst. Again, one sees the cross-over between Gauss elimination and iteration for 5 POINT STAR at about 300 equations.

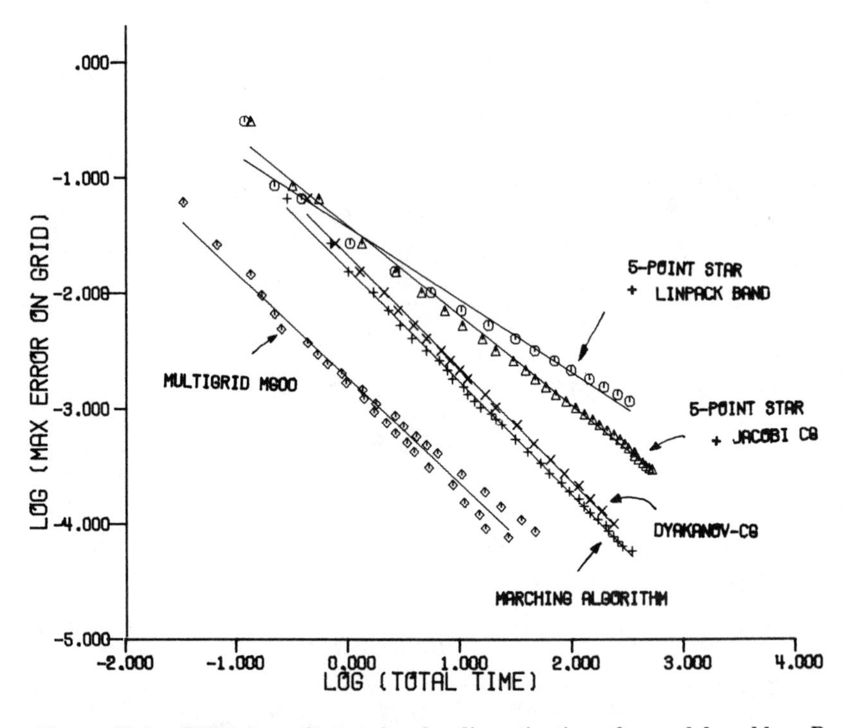

Figure 10.5. Efficiency of second order discretizations for model problem B. The maximum error is plotted on a log-log scale versus the total computing time to solve the problem.

Figure 10.6 shows the **efficiency** of the **higher order** modules plus the best second order method. Here the results have change significantly from model problem A; the second order module MULTIGRID MG00 is more efficient for an interesting range of errors, up to about 3.5 digits of accuracy (execution times up to about 200 seconds). The HODIE HELMHOLTZ and DYAKANOV CG 4 modules have considerably steeper slopes and become much more efficient than MULTIGRID MG00 if high levels of accuracy are needed. The HODIE module with LINPACK BAND is yet more efficient due to its sixth order error.

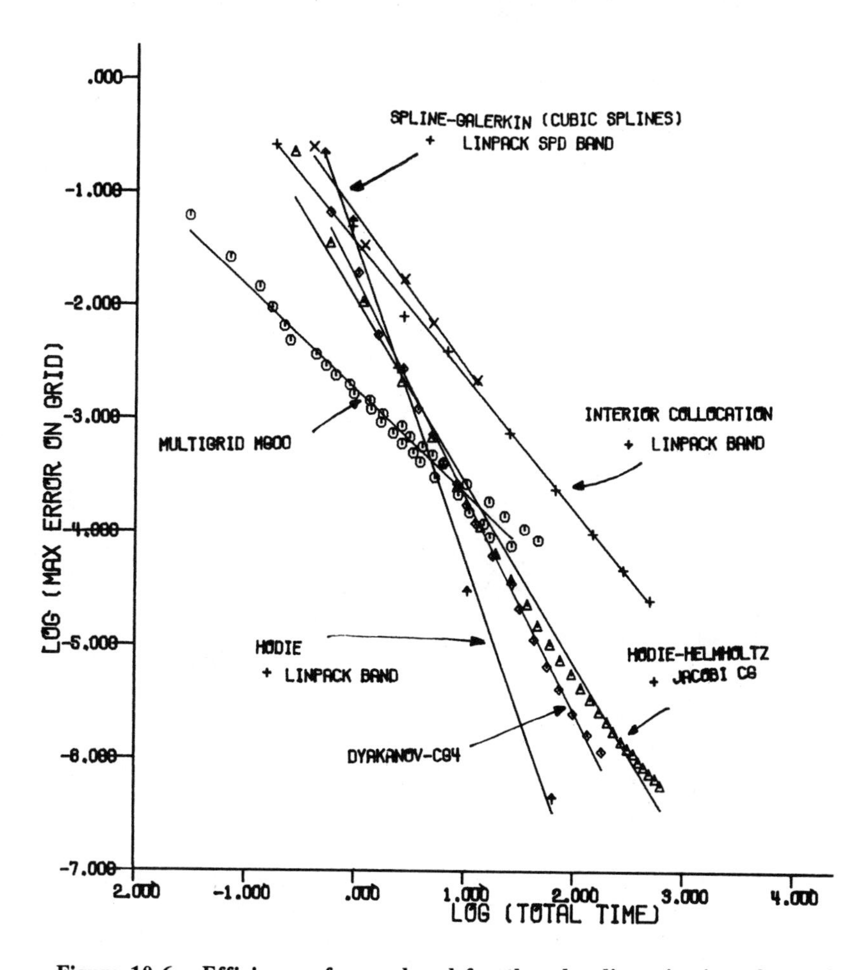

Figure 10.6. Efficiency of second and fourth order discretizations for model problem B. The maximum error is plotted on a log-log scale versus the total computing time to solve the problem.

10.C BOUNDARY LAYER PROBLEM

Model problem C is

$$u_{xx} + u_{yy} - 100\,u = 150\,\frac{\cosh 20y}{\cosh 20}$$

with boundary conditions chosen to make the solution

$$u(x,y) = \frac{\cosh 10x}{2\cosh 10} + \frac{\cosh 20y}{2\cosh 20}.$$

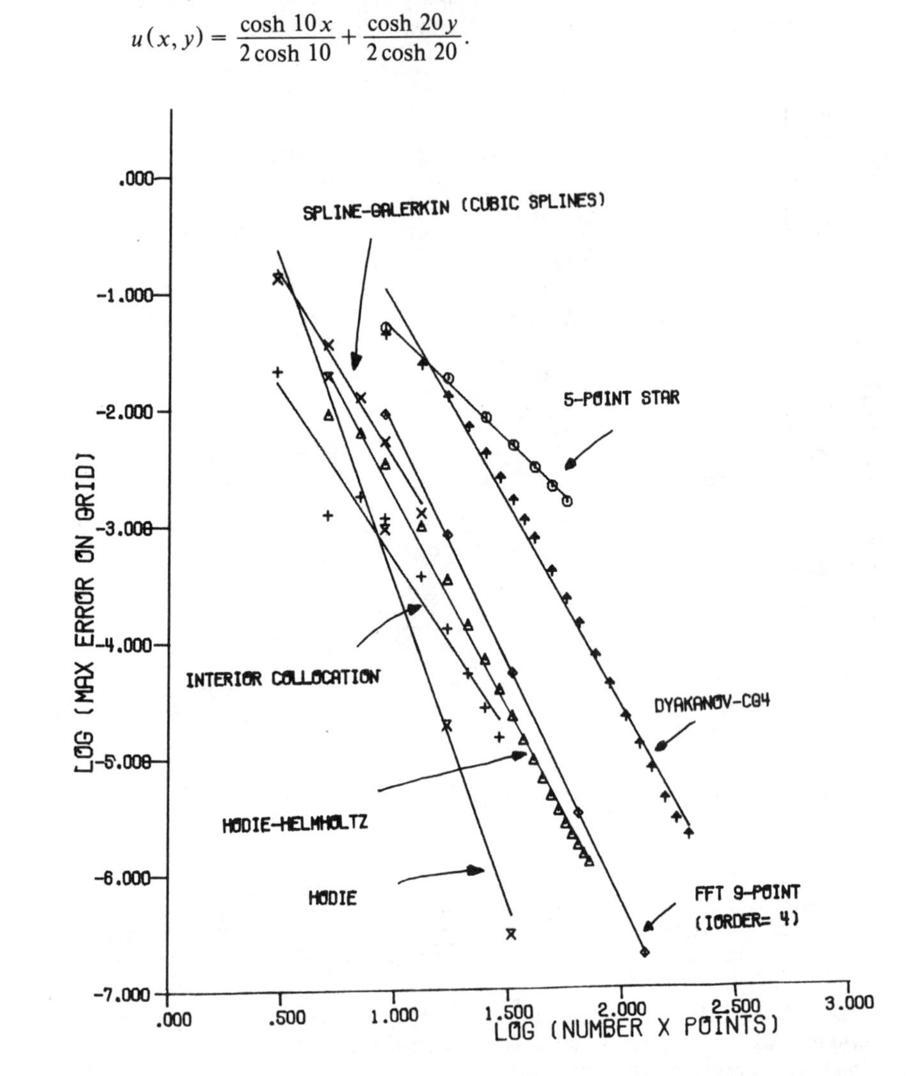

Figure 10.7. Order of convergence of discretization modules for model problem C. The maximum error on the grid is plotted with a log-log scale versus the number N of x points of a uniform grid.

The operator is similar to that of model problem B, but the solution has a pronounced boundary layer (see the contour plot of Problem 9-2 in Appendix A). The width of the boundary layer is about 8 percent of the domain. All but the sixth order FFT 9-POINT module are applicable to this problem.

Figure 10.7 shows the **order** of the discretization error for the modules used. Again all the modules based on second order finite differences have the same errors, and so only the 5 POINT STAR data are plotted in Figure 10.7. The slope is as close to -2 as one can measure, but only after an initial flat spot for grids coarser than 13 by 13. On the coarse grids the discretizations cannot resolve the boundary layer at all.

All the fourth order modules show some inability to resolve the boundary layer on coarse grids. Once the grid is comparable to the boundary layer width, then the expected fourth order rate of convergence appears (the measured slopes are almost exactly -4). The discretization used by the HODIE module is sixth order for this problem and the measured slope is about -5.5.

Figure 10.8 shows the **efficiency**, error versus total time, of the **second order** modules. The relationships seen in Figure 10.2 for the Poisson problem are almost exactly repeated here; there are a few initial perturbations for coarse grids.

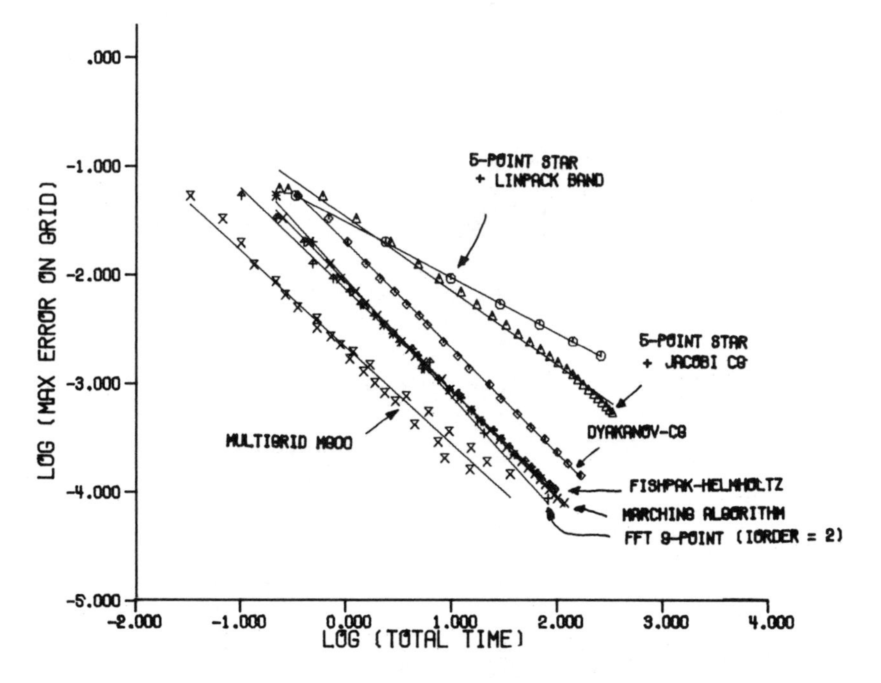

Figure 10.8. Efficiency of second order discretizations for model problem C. The maximum error is plotted on a log-log scale versus the total computing time to solve the problem.

MULTIGRID MG00 is clearly superior, while 5 POINT STAR with standard solution modules is the worst.

Figure 10.9 shows the **efficiency** of the **higher order** modules plus the best second order one. FFT 9-POINT (IORDER=4) is clearly superior, with HODIE, followed by HELMHOLTZ clearly next. The second order MULTIGRID MG00 is not competitive with the fourth order modules for this problem. For high enough accuracy, HODIE + LINPACK BAND is more efficient than FFT 9-POINT (IORDER=4) but few applications require such accuracy. A faster solution module (e.g., JACOBI CG) might make HODIE the most efficient for somewhat lower levels of accuracy.

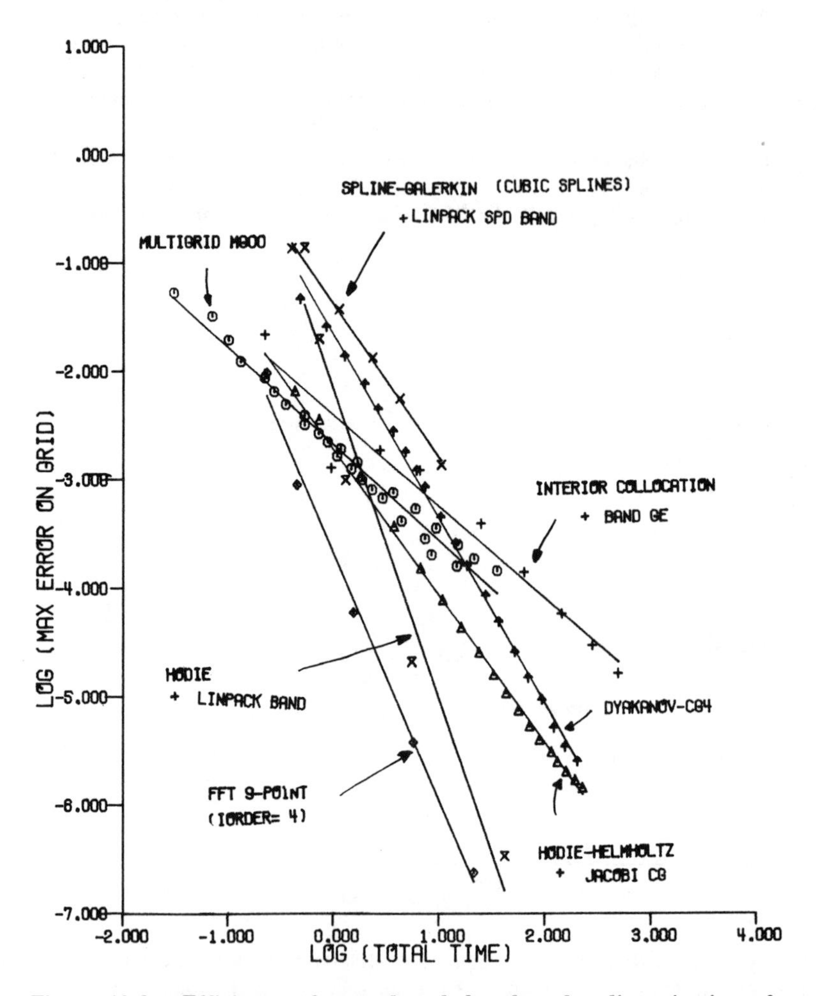

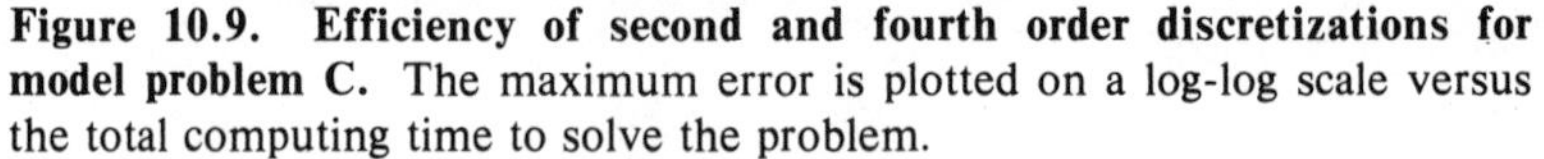
Figure 10.9. Efficiency of second and fourth order discretizations for model problem C. The maximum error is plotted on a log-log scale versus the total computing time to solve the problem.

10.D GENERAL ELLIPTIC OPERATOR

Model problem D is of the form

$$au_{xx} + cu_{yy} + fu = g$$

with a, c, f, and g general functions of x and y. The problem has Dirichlet boundary conditions chosen to make $u(x, y)$ a smooth function of general form. Only three DISCRETIZATION modules, 5 POINT STAR, HODIE, and INTERIOR COLLOCATION, are applicable. (HERMITE COLLOCATION is also applicable, but it gives the same approximation as INTERIOR COLLOCATION at slightly more computing expense.)

Figure 10.10 shows the **orders** of the discretization errors of the three modules. The orders of convergence, as estimated from the slopes, are 2 and 4, as one expects. Recall that the discretization used by the HODIE module is only fourth order for this more complicated problem. The coefficient in the error term for HODIE is about a factor of 5 larger than that for INTERIOR COLLOCATION. Figure 10.11 shows the **efficiency** of the modules: 5 POINT STAR is used both with LINPACK BAND and with JACOBI CG, but neither makes it competitive in efficiency with INTERIOR COLLOCATION or HODIE. One still sees slight evidence of the cross-over between Gauss elimination and iteration, but now it occurs for much larger systems of equations, at about 1500 equations. Note that the linear system of equations generated by 5 POINT STAR for this problem is nonsymmetric, and hence the theoretical properties of the iteration scheme used in JACOBI CG are less well understood. The larger error in HODIE discretization is more than compensated by the increased efficiency in solving the linear system of equations. Recall that for a given number N of x points, the HODIE discretization produces N^2 equations with bandwidth N while the collocation discretization produces $4N^2$ equations with bandwidth $2N$.

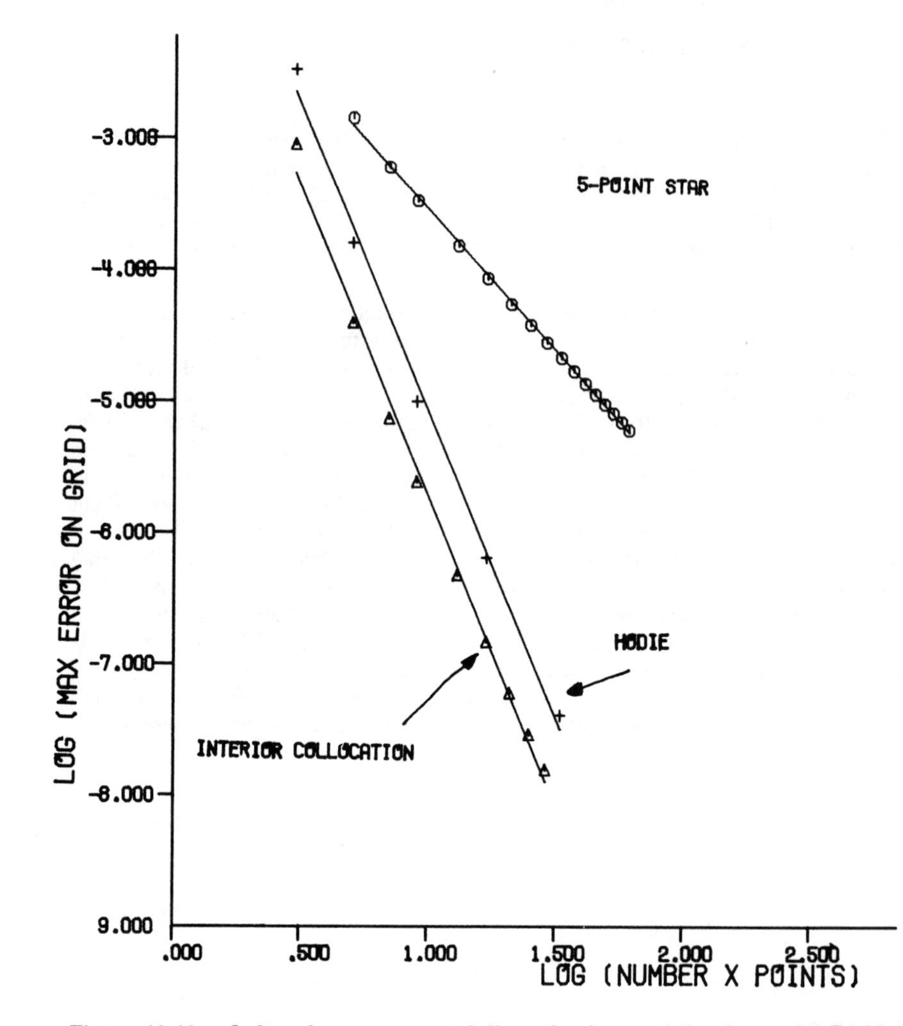

Figure 10.10. Order of convergence of discretization modules for model Problem D. The maximum error on the grid is plotted with a log-log scale versus the number N of x points of a uniform grid.

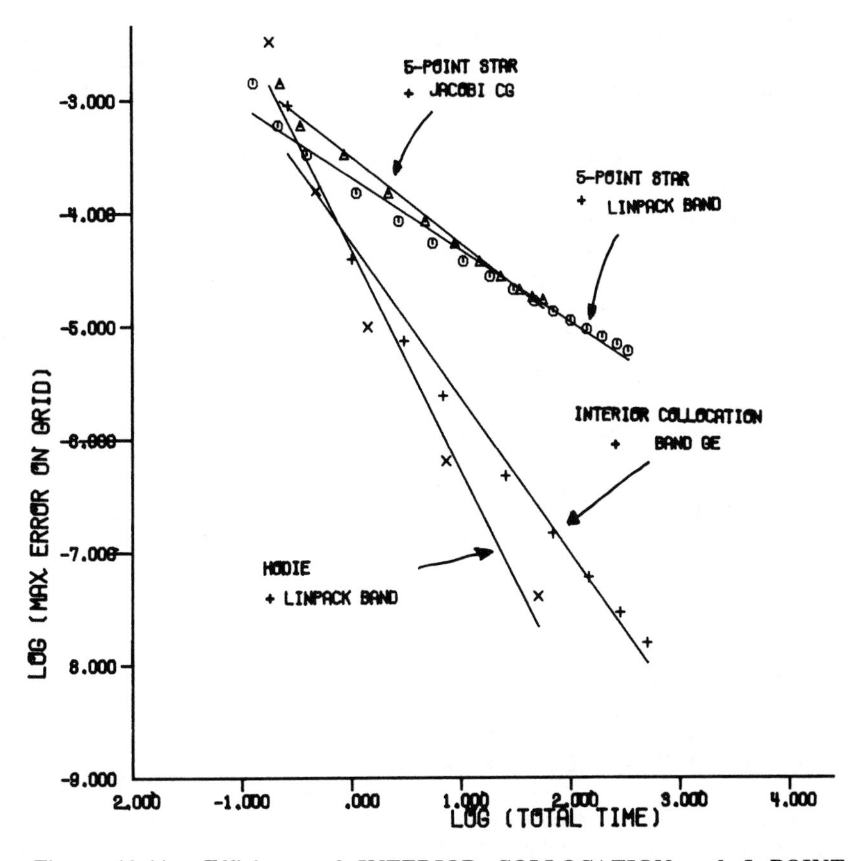

Figure 10.11. Efficiency of INTERIOR COLLOCATION and 5 POINT STAR for model problem D. The maximum error is plotted on a log-log scale versus the total computing time to solve the problem.

10.E GENERAL, DIFFICULT ELLIPTIC PROBLEM

Model problem E is a self-adjoint problem with general variable coefficients. Dirichlet boundary conditions are chosen to make the true solution have a sharp ridge with an attached gentle ridge (see the contour plot of Problem 54-2 in Appendix A).

The only second order DISCRETIZATION module applied to this problem is 5 POINT STAR, along with the following higher order modules:

HODIE (fourth order)
INTERIOR COLLOCATION (fourth order)
SPLINE GALERKIN (DEGREE=3, NDERV=1) (fourth order, C^1 cubics)
SPLINE GALERKIN (DEGREE=3, NDERV=2) (fourth order, C^2 cubics)
SPLINE GALERKIN (DEGREE=4, NDERV=2) (fifth order, C^2 quartics)

The modules DYAKANOV CG and DYAKANOV CG 4 are applicable but were not included in this study. Figure 10.12 shows the **order**, error versus N, for these modules; this problem is difficult enough that asymptotic rates of convergence are hard to estimate. The measured order is taken from these data with the initial, low accuracy points deleted. These orders are not taken from the slopes of the lines in Figure 10.17. The following orders of convergence are suggested by these data:

Module	*Measured Order*	*Expected Order*
5 POINT STAR	2.4	2.0
SPLINE GALERKIN (DEGREE=3, NDERV=2)	4.4	4.0
HODIE	5.3	4.0
INTERIOR COLLOCATION	5.4	4.0
SPLINE GALERKIN (DEGREE=3, NDERV=1)	5.5	4.0
SPLINE GALERKIN (DEGREE=4, NDERV=2)	5.6	5.0

All these are greater than the expected rate of convergence; one explanation is the performance data used here falls into two phases. First there is an initial, low accuracy, slow convergence phase where the grid is being refined so as to resolve the sharp ridge in $u(x,y)$. Then there is a normal, faster convergence phase where the ridge has been resolved and the error is being reduced by refining the grid. The second phase might appear to be faster just because the first phase is so slow. It is clear that the high order modules have much better accuracy, even initially, and that there are noticeable differences between them.

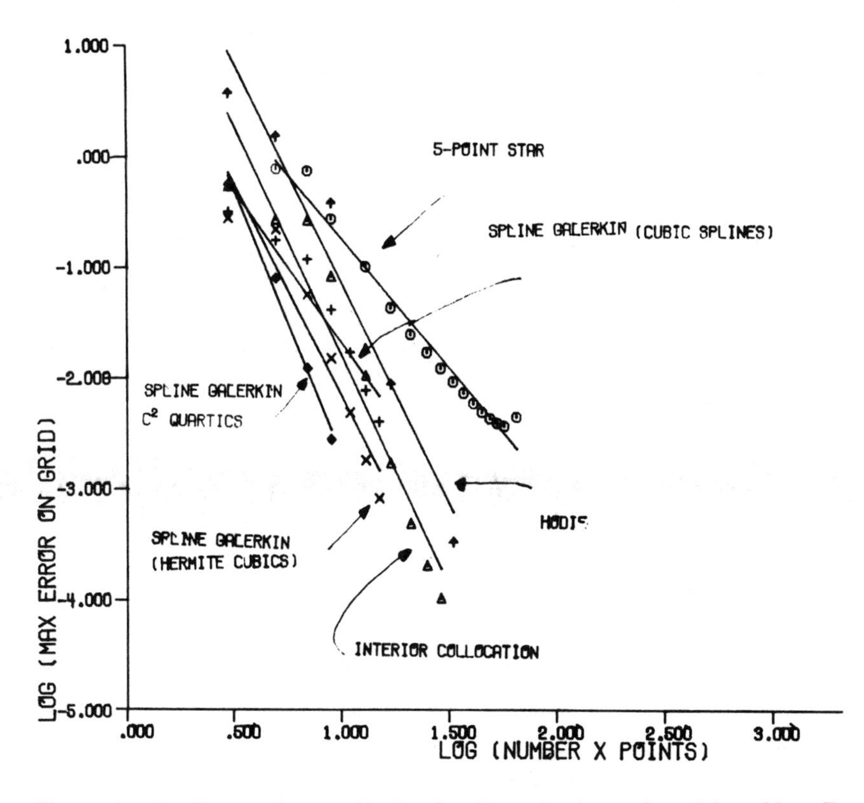

Figure 10.12. Error versus grid size for discretizations of model problem E. The maximum error on the grid is plotted with a log-log scale versus the number N of x points of a uniform grid.

Figure 10.13 shows the **efficiency** of these six modules. They are bunched somewhat closer than the rates of convergence. The data suggest the following:

1. The second order 5 POINT STAR is least efficient.
2. The fourth order HODIE module is the most efficient.
3. The three fourth order, finite element modules are comparable.
4. The fifth order, finite element module based on C^2 quartics starts out worst, but becomes more efficient than the finite element modules based on cubics.

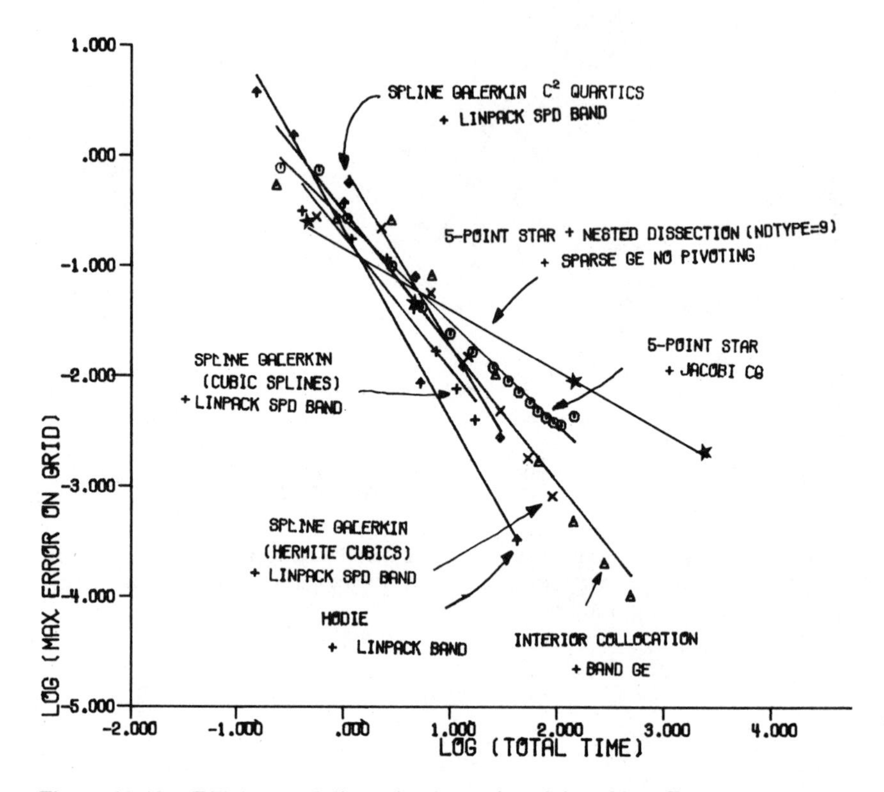

Figure 10.13. Efficiency of discretizations of model problem E. The maximum error is plotted on a log-log scale versus the total computing time to solve the problem.

10.F MIXED BOUNDARY CONDITIONS

Model problem F is a simple Poisson equation with mixed boundary conditions as follows:

$$u + u_x = g \quad \text{on} \quad x = 1, \qquad u - u_x = g \quad \text{on} \quad x = -1,$$
$$u + u_y = g \quad \text{on} \quad y = 1, \qquad u - u_y = g \quad \text{on} \quad y = -1.$$

These conditions are all $u + u_n = g$, where u_n is the normal derivative. The function g is chosen to make the solution a polynomial of degree 12; the problem

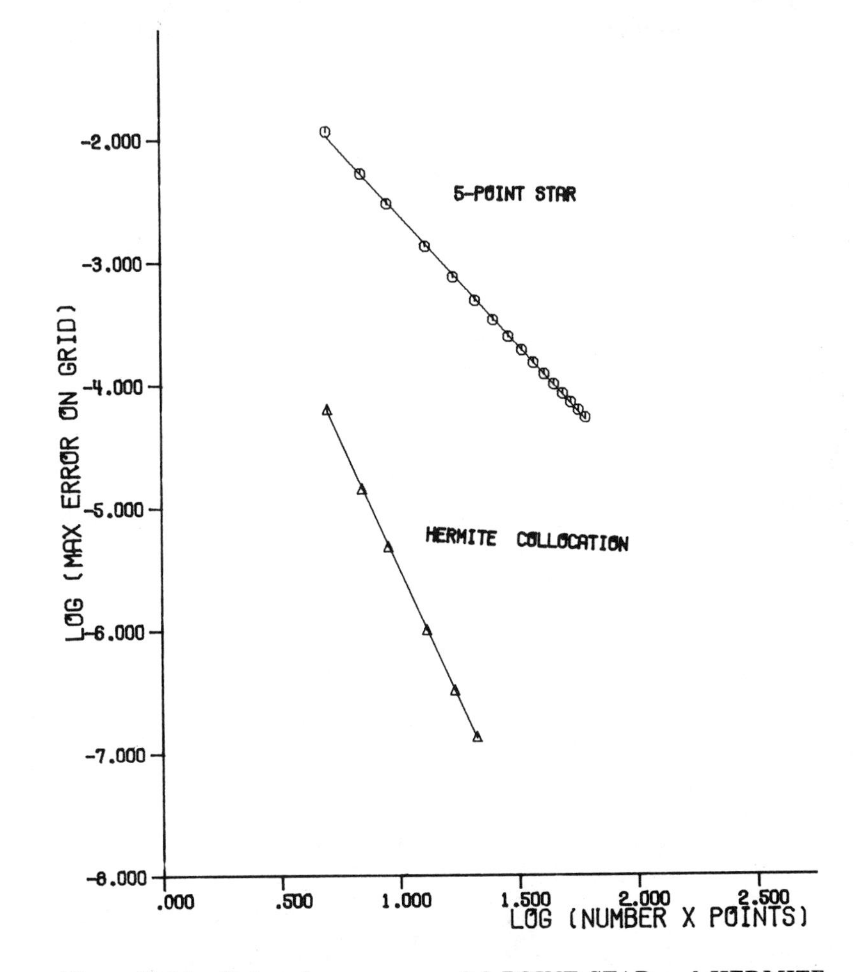

Figure 10.14. Order of convergence of 5 POINT STAR and HERMITE COLLOCATION for model problem F. The maximum error on the grid is plotted with a log-log scale versus the number N of x points of a uniform grid.

is not easy to solve accurately, see the contour plot for Problem 31-1 in Appendix A.

Figure 10.14 shows the **order** of convergence of the two modules, 5 POINT STAR and HERMITE COLLOCATION. The measured slopes agree very well with the theoretical slopes of -2 and -4. Compare this figure with Figure 10.11 which involves the same modules. The observed slopes are the same, but there is a much larger difference between them. It appears that HERMITE COLLOCATION approximates the derivative boundary conditions in a way that produces a much smaller constant factor in the error relative to that of 5 POINT STAR.

Figure 10.15 shows the **efficiency** of these modules; HERMITE COLLOCATION + BAND GE is much more efficient than 5 POINT STAR + LINPACK BAND.

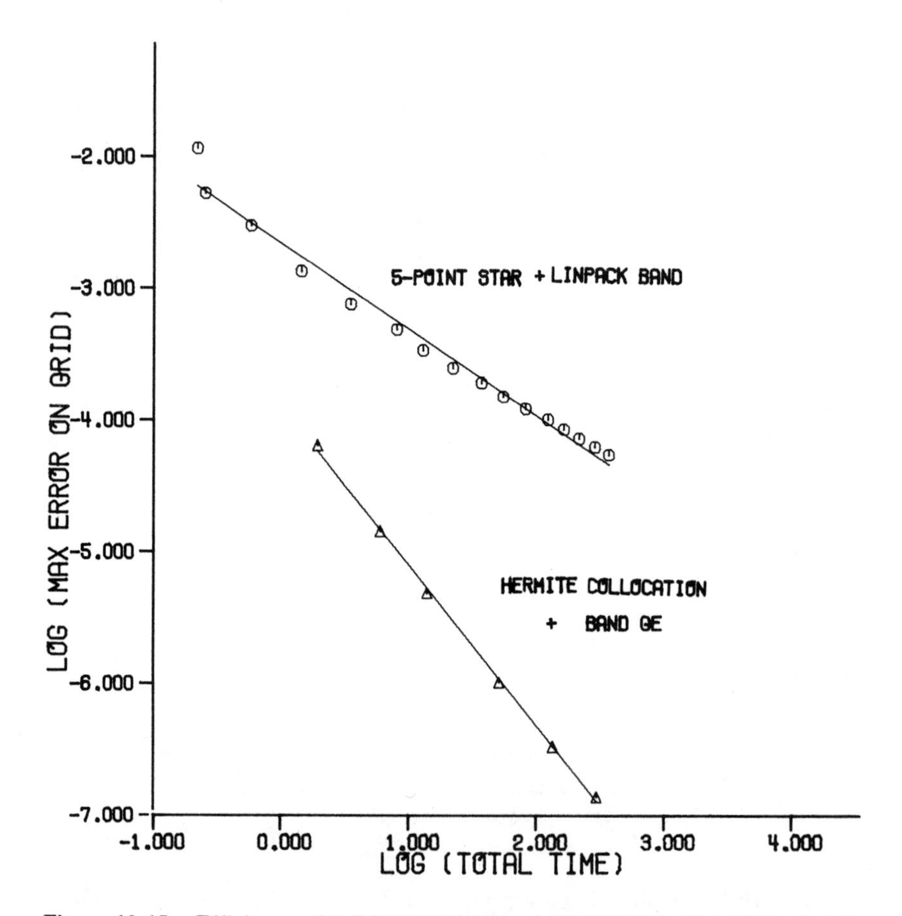

Figure 10.15. Efficiency of 5 POINT STAR and HERMITE COLLOCATION for model problem F. The maximum error is plotted on a log-log scale versus the total computing time to solve the problem.

10.G SINGULAR PROBLEM

This is a simple Poisson problem with Dirichlet boundary conditions on the unit square. The solution $u(x,y)$ is $(xy)^{2.5}$ which has square root singularities in its second derivative along two sides of the domain. A contour plot of u (see Problem 47-2 in Appendix A) shows it to appear very smooth and well behaved. However, the singularity in the solution slows down the order of convergence of all the high order methods; theoretically, the maximum possible order is 2.5.

Figure 10.16 shows the **order of convergence** for 5 POINT STAR and INTERIOR COLLOCATION. The order for 5 POINT STAR measured from the slope in Figure 10.16 is 1.86, slightly less than the expected value of 2.0 for a module based on a second order method. INTERIOR COLLOCATION is based on a fourth order method, so one could hope for an order of 2.5. The order for INTERIOR COLLOCATION measured from Figure 10.16 is 2.25, again slightly less than the theoretical maximum.

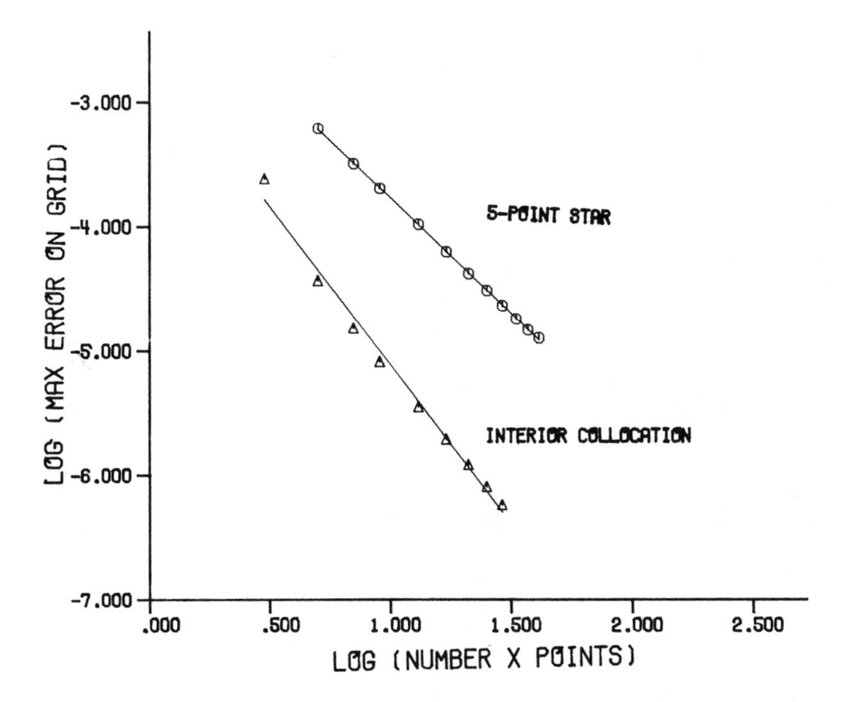

Figure 10.16. Order of convergence of 5 POINT STAR and INTERIOR COLLOCATION for model problem G. The maximum error on the grid is plotted with a log-log scale versus the number N of x points of a uniform grid.

Figure 10.17 shows the **efficiency** of these two modules coupled with standard SOLUTION modules (JACOBI CG and BAND GE). The efficiency slopes are about parallel, with INTERIOR COLLOCATION faster by a factor of about 5. However, a fast second order module such as MULTIGRID MGOO should be expected to outperform INTERIOR COLLOCATION here because the fourth order convergence of the collocation method cannot be achieved.

The paper [Houstis and Rice, 1982] gives extensive additional data on performance for singular problems.

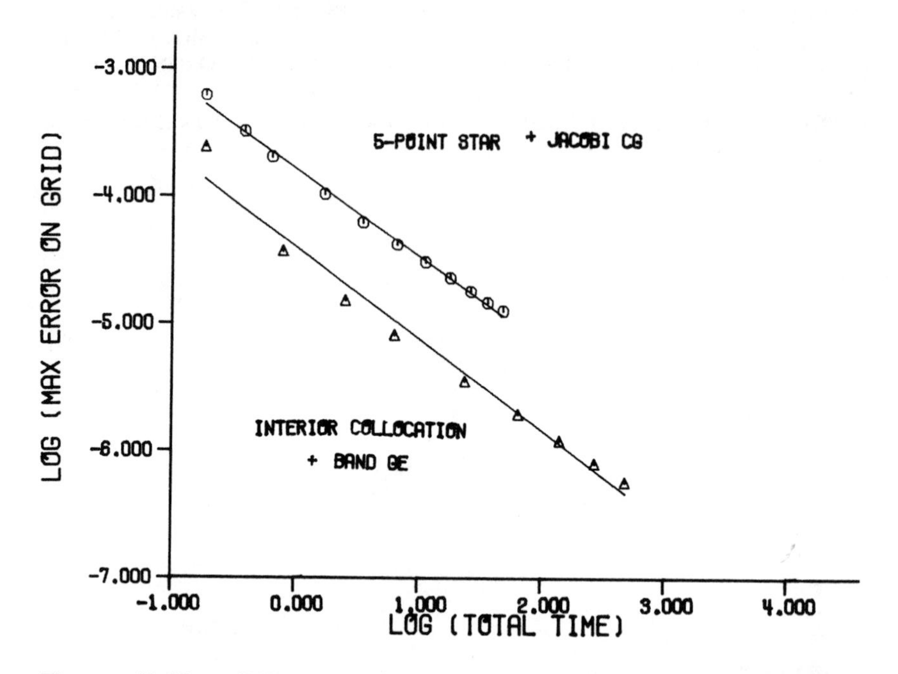

Figure 10.17. Efficiency of 5 POINT STAR and INTERIOR COLLOCATION for model problem F. The maximum error is plotted on a log-log scale versus the total computing time to solve the problem.

10.H NONRECTANGULAR DOMAIN

The domain for model problem H is shown in Figure 9.1; it is a square with the upper left corner clipped off and a semi-circular disk added on the right side. The operator is same as model problem B and the boundary conditions are chosen to give the same $u(x,y)$, see the contour plot of Problem 6-1 in Appendix A.

Figure 10.18 shows the **order of convergence**, error versus the number N of x grid points. There are substantial "local perturbations" depending upon just how the grid lines fit the domain, but the measured slopes of an average line are very close to 2 for 5 POINT STAR and 4 for COLLOCATION. These are the best orders of convergence that one could hope for by these modules. COLLOCATION is clearly superior to 5 POINT STAR in achieving high accuracy.

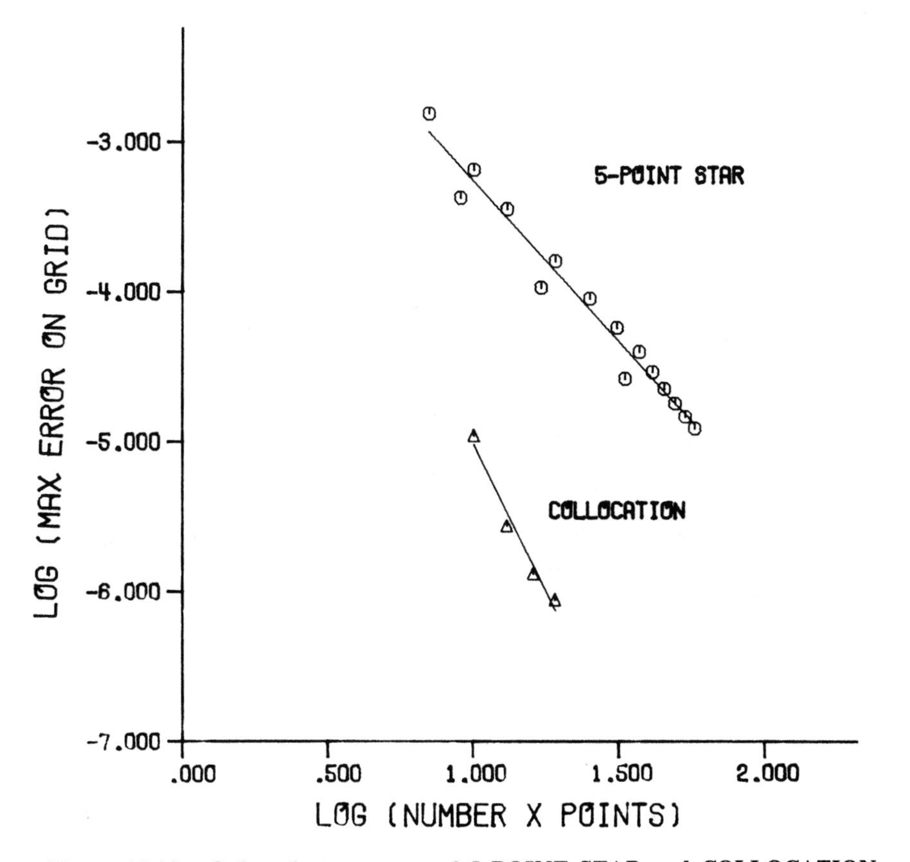

Figure 10.18. Order of convergence of 5 POINT STAR and COLLOCATION for model problem H. The maximum error on the grid inside the domain is plotted with a log-log scale versus the number N of x points of a uniform grid covering the domain.

Figure 10.19 shows the **efficiency** of these two modules when coupled with Gauss elimination and iteration. The efficiencies suggested by these data are quite good in comparison with Figure 10.11 or 10.15 which show efficiencies on rectangular domains.

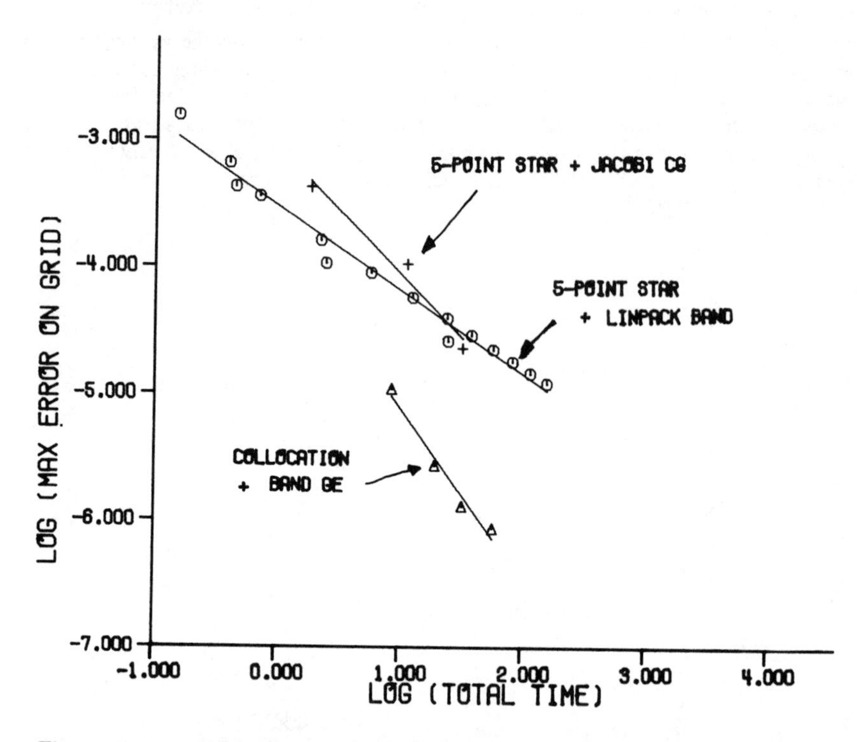

Figure 10.19. Efficiency of 5 POINT STAR and COLLOCATION for model problem H. The maximum error is plotted on a log-log scale versus the total time to solve the problem.

10.I THREE DIMENSIONAL POISSON PROBLEM

Model problem I is a direct extension to three dimensions of model problem A:

$$u_{xx} + u_{yy} + u_{zz} = f.$$

Dirichlet boundary conditions are chosen to give the solution

$$u(x, y, z) = 14\,e^{x+y+z}(x - x^2)(y - y^2)(z - z^2).$$

The second order finite differences module 7 POINT STAR and the sixth order module HODIE 27 POINT 3D are applicable to this problem. Figure 10.20A shows the **order** of convergence; the measured slopes agree well with the expected values of 2 and 6, respectively.

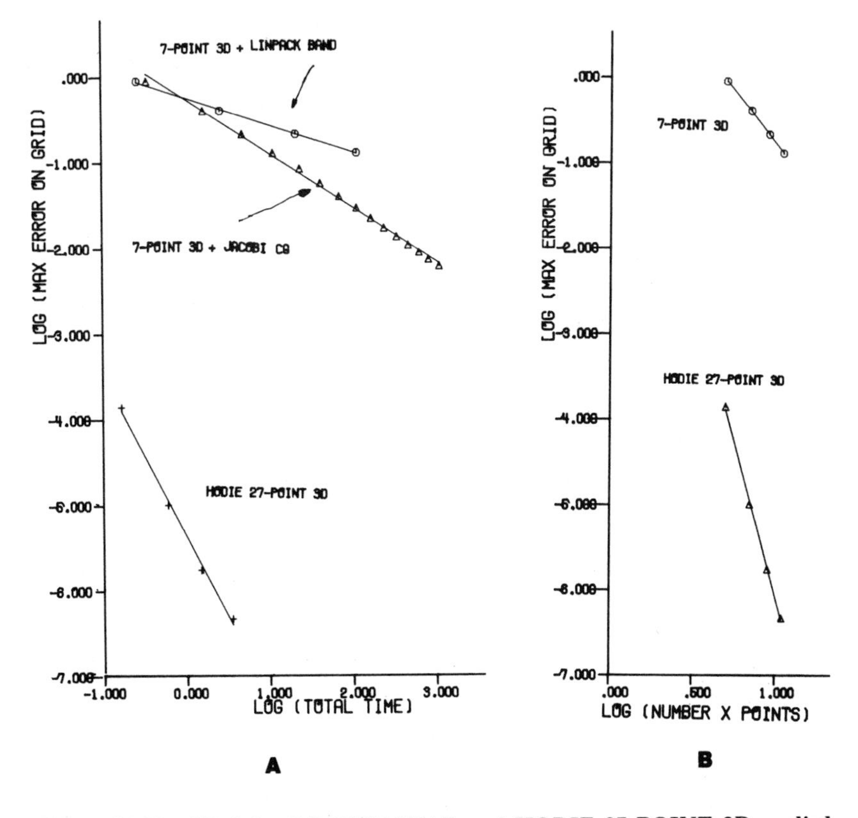

Figure 10.20. Modules 7 POINT STAR and HODIE 27 POINT 3D applied to problem I. The maximum error on the three dimensional grid is plotted with a log-log scale versus (A) the total computing time to solve the problem, and (B) the number of x points of a uniform grid. Plot (A) provides estimates of efficiency and plot (B) provides estimates of the order of convergence.

Figure 10.20B shows the **efficiency** of the two discretizations, with 7 POINT STAR coupled with both LINPACK BAND and JACOBI CG. The iteration method JACOBI CG is significantly more efficient for 7 POINT STAR, much more so than for 5 POINT STAR (compare with Figures 10.2, 10.5, 10.8 and 10.11). These data support the widely held belief that iteration methods should be used whenever possible in solving three dimensional elliptic problems. The triple module HODIE 27 POINT 3d uses a special tensor product technique to solve the linear system. This technique has the same work as JACOBI CG for a given grid, but the sharply smaller error of HODIE 27 POINT 3D produces a much more efficient module. One would expect that FFT-based modules would perform even better for three dimensional Poisson problems.

10.J REFERENCES

Houstis, E. N., and J. R. Rice [1982], "High order methods for elliptic partial differential equations with singularities", Inter. J. Numer. Meth. Engng. 18, pp. 737–754.

Chapter 11

PERFORMANCE OF SOLUTION MODULES

ELLPACK has three sets of general purpose modules for solving linear algebraic systems:

Band Elimination Modules

BAND GE	LINPACK BAND
BAND GE NO PIVOTING	LINPACK SPD BAND
ENVELOPE LDLT	ENVELOPE LDU

These all use variants of the classical Gauss elimination for banded matrices.

Iteration (ITPACK) Modules

SOR	
JACOBI SI	JACOBI CG
SYMMETRIC SOR SI	SYMMETRIC SOR CG
REDUCED SYSTEM SI	REDUCED SYSTEM CG

These modules are discussed in Chapter 7. Except for SOR, each uses some kind of acceleration of classical iterations; all required parameters are estimated adaptively. The RED-BLACK module must be used with the reduced system modules.

Sparse Matrix Elimination Modules

SPARSE LU PIVOTING	SPARSE GE NO PIVOTING
SPARSE LU UNCOMPRESSED	SPARSE LU COMPRESSED
SPARSE LDLT	

These all use variants of the classical Gauss elimination along with special data structures whose objectives are to reduce storage and arithmetic. These are normally best used with the indexing modules MINIMUM DEGREE or NESTED DISSECTION.

The goal in this chapter is to provide some data about the effectiveness of these methods for the principal types of algebraic systems that arise in elliptic PDE computation. We only examine the **time** to solve the linear systems as a function of the space discretization $h = 1/(N-1)$ where the unit square is partitioned with a rectangular grid of N by N grid lines. Questions of storage and accuracy are mostly ignored here, even though they may be critical in some cases. A similar study which includes storage data is described in [Eisenstat, et al., 1979].

Five types of discretizations are considered:

5-POINT STAR	(finite differences)
HODIE-HELMHOLTZ	(finite differences)
INTERIOR COLLOCATION	(finite element)
SPLINE GALERKIN	(finite element)
7-POINT 3D	(finite differences in 3D)

Some representative general purpose solution modules are applied to the linear systems created by the discretization modules for a particular PDE. This is done for a number of grids and the results plotted on a log-log scale. One expects the resulting curves to be straight lines (at least asymptotically as N increases) so that the solution time is

$$t = O(N^s) = O(h^{-s}),$$

where s is the slope of one of the lines. The value of s may be estimated from the plots given as $s = -1/(2r)$, where r is slope actually measured from the plot. The factor 2 is due to the fact that the time scale is twice as big as the h scale. The time t is in seconds on a VAX 11/780 with a floating point accelerator. One wants s to be small and one also wants the constant C in the relation

$$t = C N^s + o(N^s)$$

to be small. The values for s and C presented are to be considered suggestive only; the amount of data available is much to small to provide statistically significant results and the slopes are measured fairly crudely. However, when the measured s values agree well with theoretical asymptotic values, one has more confidence in both.

One knows that $s = 4$ for band elimination methods, and one expects that $s = 3$ for iteration methods (at least for 2D model finite difference discretizations). These expectations agree fairly well with the data presented here. The value of s for banded 3D methods should theoretically be 7; the observed value is about 8. The data suggest that the ITPACK modules are more efficient than one expects from theoretical results.

For sparse matrix modules the expected value for s is less certain. For some model problems and indexings (NESTED DISSECTION) it is known that the asymptotic value of s is also 3. This behavior is not seen here; the measured values of s vary widely but seem to average about 3.5 or so.

Note that the time for a band elimination module is generally independent of the PDE being solved so that there is little to be gained by obtaining data for them from other PDEs. This observation is true to a lesser extent for the sparse matrix modules, and the time for the ITPACK modules can change completely as the PDEs change. In fact, the performance of the ITPACK modules is quite consistent for finite difference discretization. The iterations diverge for the collocation discretizations and may produce inaccurate results for some SPLINE GALERKIN cases (probably due to poor scaling of the equations).

11.A SOLUTION OF THE 5-POINT STAR EQUATIONS

We compare the performance of various solution modules for the matrices generated by the 5 POINT STAR discretization module for problems B, E, and H. 5 POINT STAR is based on standard second order accurate central finite difference approximations which lead to matrices with a very simple structure. For Dirichlet problems on an N by N grid there are $(N-2)^2$ equations with at most five unknowns per equation. With the natural ordering of grid points (AS IS) there are five diagonal bands, two immediately adjacent to the main diagonal and two at a distance $N-1$ from the main diagonal. For each of the problems we consider the matrix is (irreducibly) diagonally dominant [Varga, 1962], and hence classical iterative methods are convergent. For problems B and E, the matrix is also symmetric positive definite.

Figure 11.1 shows the performance of four **band elimination** modules for model problem B, a simple problem on the unit square. There is no significant difference between the four modules and the slope s is as close to 4 as one can measure. Note that one expects BAND GE NO PIVOTING to execute twice as fast as BAND GE. This does not occur in practice because there is much more "fill-in" when one does not pivot. (BAND GE and BAND GE NO PIVOTING are essentially identical in this case because the problem happens to produce a positive definite linear system so that BAND GE does not pivot).

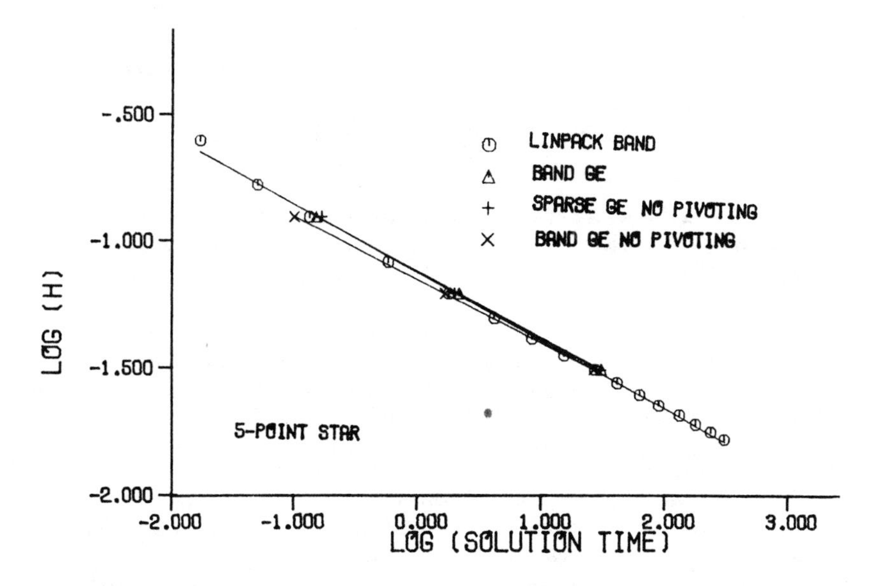

Figure 11.1. Performance of band elimination modules for model problem B. The four modules give $t = O(N^4)$ for the 5-POINT STAR equations.

Figure 11.2 shows the performance for model problem H, which has a nonrectangular domain. The slopes for the two **band elimination** modules is measured to be 3.9; the difference from 4.0 could well be measurement error. The **iteration** modules start off by taking more time, but they have much smaller slopes (2.9 for SOR and 2.2 for JACOBI CG), so they become more efficient at about $N = 25$ (about 530 equations or $t = 300$). This crossover between iteration and direct elimination is typical, see Figures 10.2 and 10.10 for previous examples. The crossover occurs later in this case than usual.

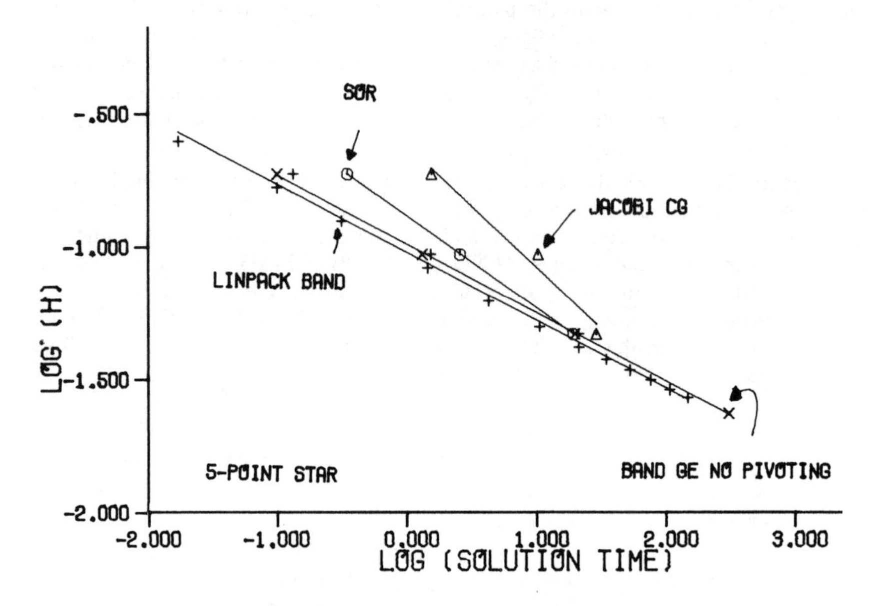

Figure 11.2. Performance of band elimination and iteration modules for model problem H. The crossover point for this nonrectangular domain problem is about N=25 (530 equations).

Figure 11.3 shows the performance of the seven **iteration** modules for model problem B. Their performances appear similar; the slopes varying from 2.3 (for RED-BLACK + REDUCED SYSTEM CG) to 2.8 (for SOR, JACOBI SI, JACOBI CG, and SYMMETRIC SOR SI). The largest system solved by all the modules has about 3800 equations and the solution times ranged from about 450 to 1000 seconds, a difference of more than 100 percent. Comparing Figures 11.1 and 11.3, one estimates that the crossover between these modules and the direct elimination modules occurs for N about 12 (about 100 equations).

Figure 11.4 shows the **band elimination** module LINPACK BAND along with three **iteration** modules for model problem E, a general, difficult elliptic problem on a square. The slope s for LINPACK BAND is 3.9 (about as one expects) and for the three iteration modules s is 2.6 (better than one expects). The crossover point is at about $N = 18$, which gives about 250 equations.

Figure 11.5 shows three **sparse matrix** modules with varying results. The best are MINIMUM DEGREE + SPARSE GE NO PIVOTING and MINIMUM DEGREE + SPARSE LU UNCOMPRESSED, which appear identical except for N very small. The slope s is estimated to be about 3.5, but the constant C is the smallest of all the modules. These two are always faster than the band elimination modules and faster than the iteration modules for N up to about 22 (about 400 equations).

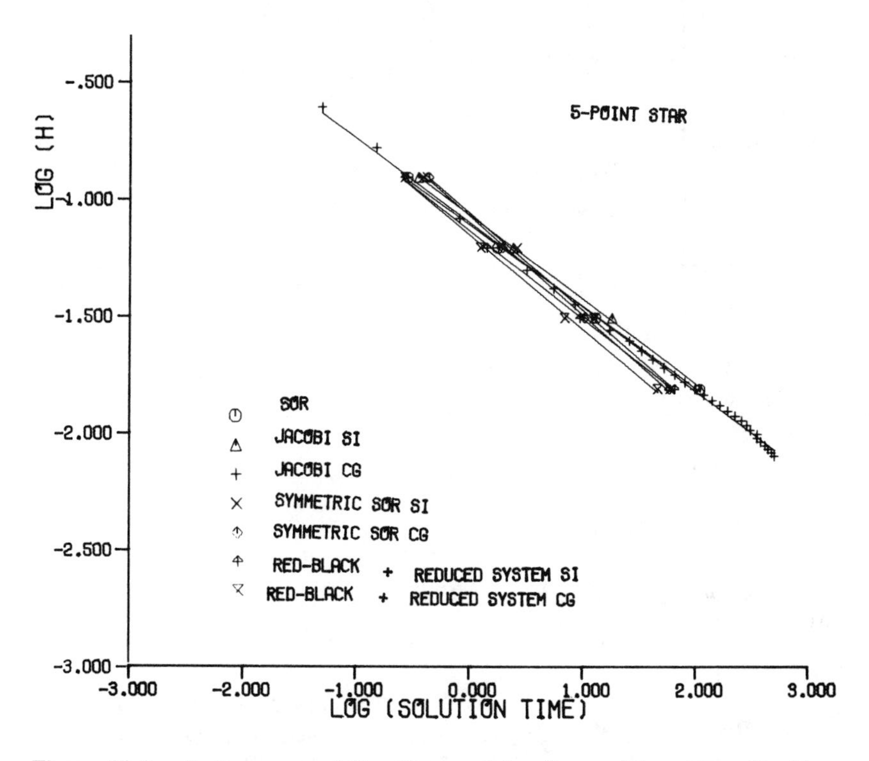

Figure 11.3. Performance of iteration modules for model problem B. The fastest is RED-BLACK + REDUCED SYSTEM CG, which has $t = O(N^{2.3})$, and the crossover point with band elimination modules is at $N = 12$.

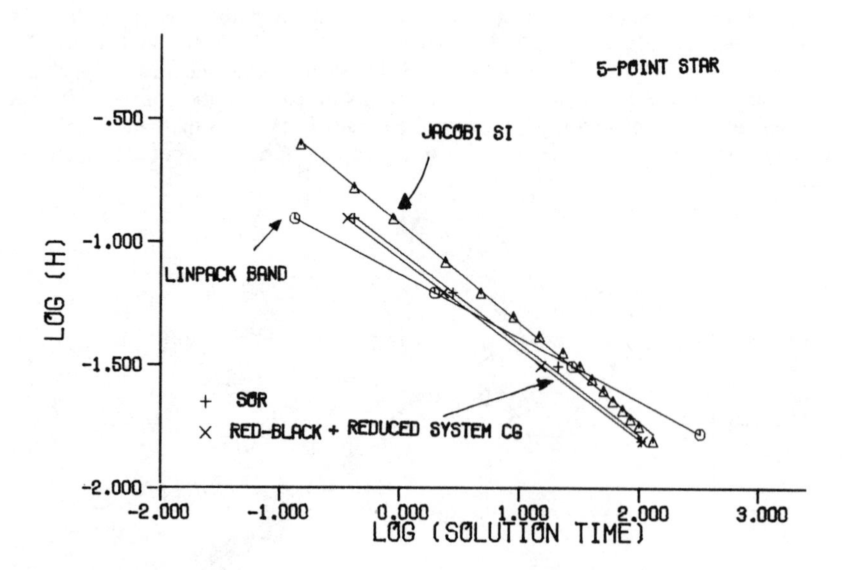

Figure 11.4. Performance of band elimination and iteration modules for model problem E. The crossover occurs at about $N = 18$ (250 equations)

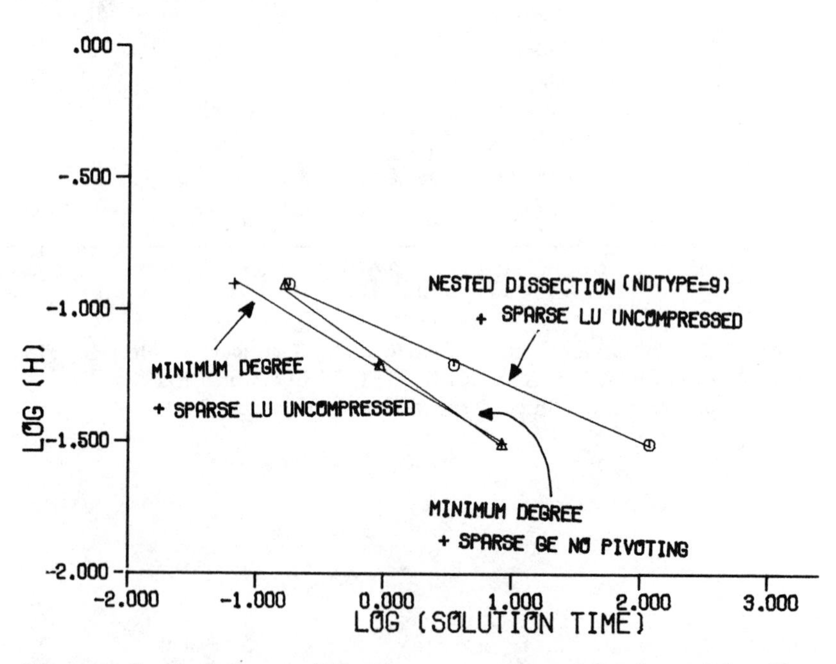

Figure 11.5. Performance of three sparse matrix modules for model problem B. The fastest two have $t = O(N^{3.5})$ for the 5-POINT STAR equations.

11.B SOLUTION OF THE HODIE-HELMHOLTZ EQUATIONS

We compare the performance of various solution modules for the matrix generated by the HODIE HELMHOLTZ discretization module for problem B. HODIE HELMHOLTZ produces a discretization which is an $O(h^2)$ perturbation of the following second order finite difference discretization of the Poisson equation.

$$-20u_{i,j} + 4(u_{i+1,j} + u_{i-1,j} + u_{i,j+1} + u_{i,j-1} + (u_{i+1,j+1} + u_{i-1,j+1} + u_{i-1,j-1} + u_{i+1,j-1}) = 6h^2 g_{i,j}$$

The perturbed difference equation is fourth order accurate. For Dirichlet problems on an N by N grid there are $(N-2)^2$ equations with at most nine unknowns per equation. With the natural ordering of grid points (AS IS), there are nine diagonal bands; two immediately adjacent to the main diagonal and two sets of three centered at distance $N-1$ from the main diagonal.

Since the structure of these matrices is quite similar to those generated by 5 POINT STAR, one expects the performance of band elimination and sparse matrix modules to be very similar to their performance for ordinary finite differences. The data in this study support this expectation. Figure 11.6 shows the performance of three **band elimination** modules and one **sparse matrix** module for the equations; in addition, data for 5 POINT STAR + LINPACK BAND are reproduced from Figure 11.1 for comparison. The performances appear quite similar, with the slope s varying from 3.7 to 3.9, just a little smaller than one expects.

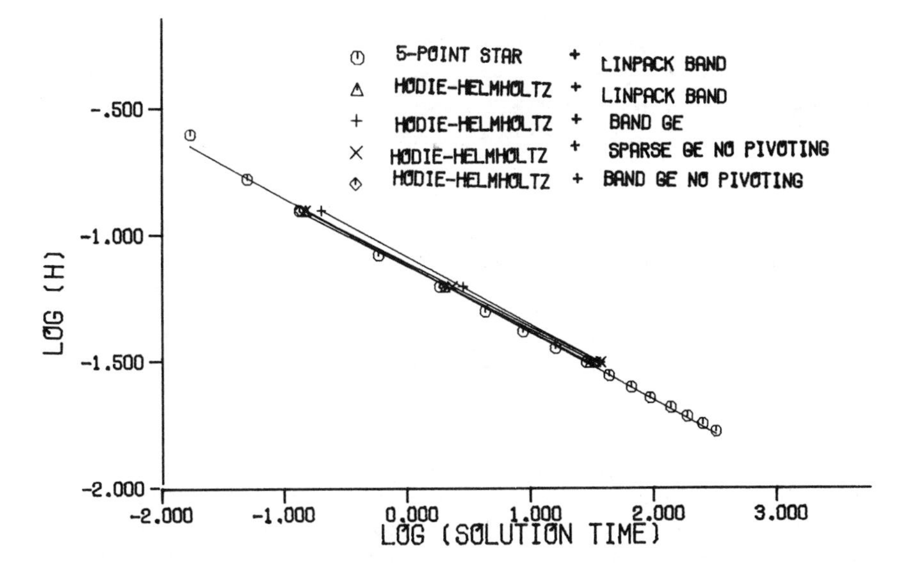

Figure 11.6. Performance of direct elimination modules for model problem B. The HODIE-HELMHOLTZ equations are solved with $t = O(N^{3.9})$.

Figure 11.7 shows the performance of three **sparse matrix** modules. Their relative performances are the same, as seen in Figure 11.5 for the 5 POINT STAR equations. The best of the three has $t = O(N^{3.6})$. MINIMUM DEGREE + SPARSE LU UNCOMPRESSED seems to be asymptotically faster than the band elimination modules, the crossover point being about $N = 33$ (about 960 equations) for this problem.

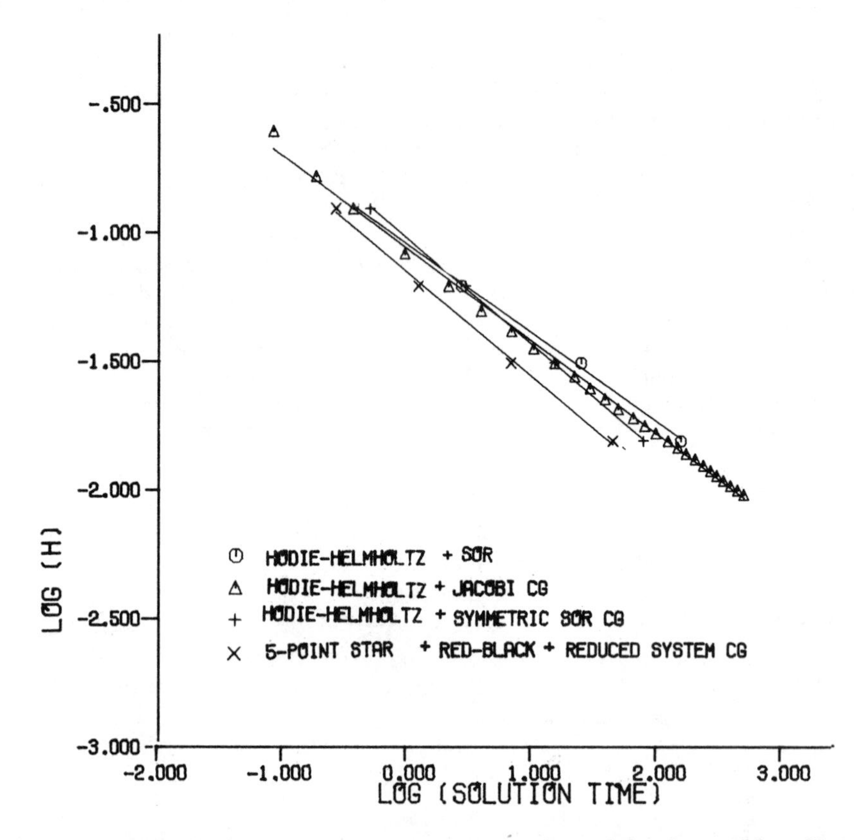

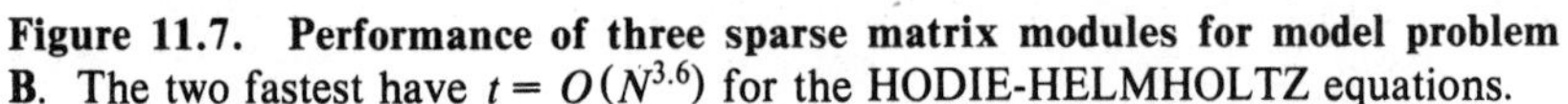

Figure 11.7. Performance of three sparse matrix modules for model problem B. The two fastest have $t = O(N^{3.6})$ for the HODIE-HELMHOLTZ equations.

The discretization matrix produced by HODIE HELMHOLTZ for model problem B is nonsymmetric, but it is (irreducibly) diagonally dominant [Varga, 1982], and hence classical iterative methods are convergent. However, since there are roughly twice as many nonzero coefficients per equation, one expects the **iteration modules** to require twice as much time as the 5 POINT STAR equations. Figure 11.8 shows the performance of three iteration modules; in addition, data for 5 POINT STAR + RED BLACK + REDUCED SYSTEM CG are reproduced from Figure 11.3 for comparison. The SYMMETRIC SOR CG module gives $t = O(N^{2.4})$, which is essentially the same as the best for the five point star equations. Notice that the difference in work per iteration is reflected in the constant difference between the corresponding lines. The slowest module, SOR, has $t = O(N^{2.9})$. The crossover between band elimination and iteration occurs at about $N = 11$ (80 equations).

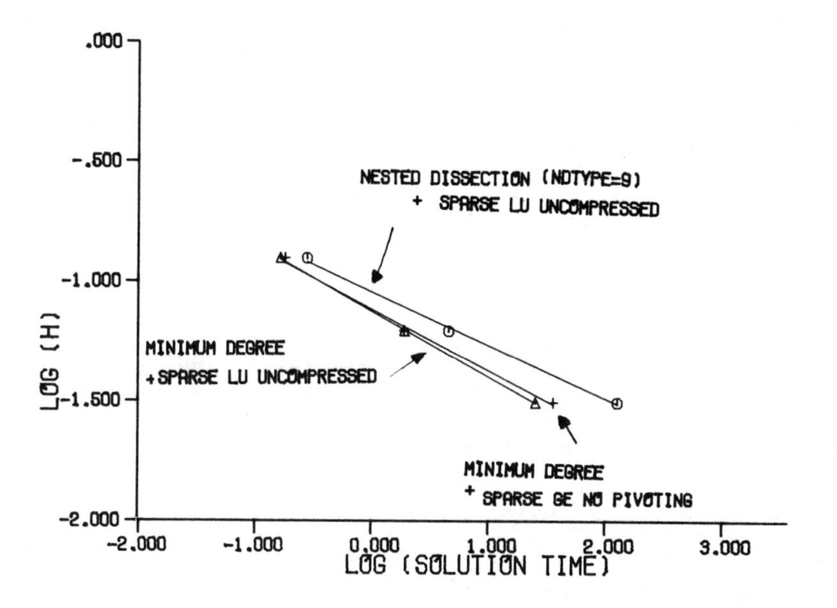

Figure 11.8. Performance of iteration modules for model problem B. The 9-point finite difference equations inherently take more work to solve than the 5-point equations.

11.C SOLUTION OF THE COLLOCATION EQUATIONS

The collocation equations generated in ELLPACK use Hermite bicubic polynomial basis functions which leads to a band matrix. The most common ordering is the finite element ordering which gives (for an N by N grid) a somewhat bi-diagonal matrix of order $4(N+1)^2$ and bandwidth about $4N$. There are 16 nonzero entries in each row and many diagonal elements are zero. The **collorder ordering** used by the HERMITE COLLORDER and INTERIOR COLLORDER module gives a somewhat tridiagonal matrix with nonzero diagonal entries. For detailed information on these equations, see [Dyksen and Rice, 1984] and [Dyksen, Houstis, Lynch and Rice, 1984].

Normal iteration methods diverge for either way of writing the collocation equations. Figure 11.9 shows the performance of **direct elimination** modules plus one **sparse matrix** module. The best three modules give $t = O(N^{3.3})$ while the other two give $t = O(N^{3.9})$. The collocation equations are difficult to scale properly so BAND GE is preferable to LINPACK BAND because it does scaled partial pivoting (the cost of which is negligible). Note that the sparse matrix module SPARSE LU UNCOMPRESSED outperforms the direct elimination modules for the equations in the collorder order. This module is not applicable with the finite element order (using AS IS with INTERIOR COLLOCATION) because it does not do pivoting.

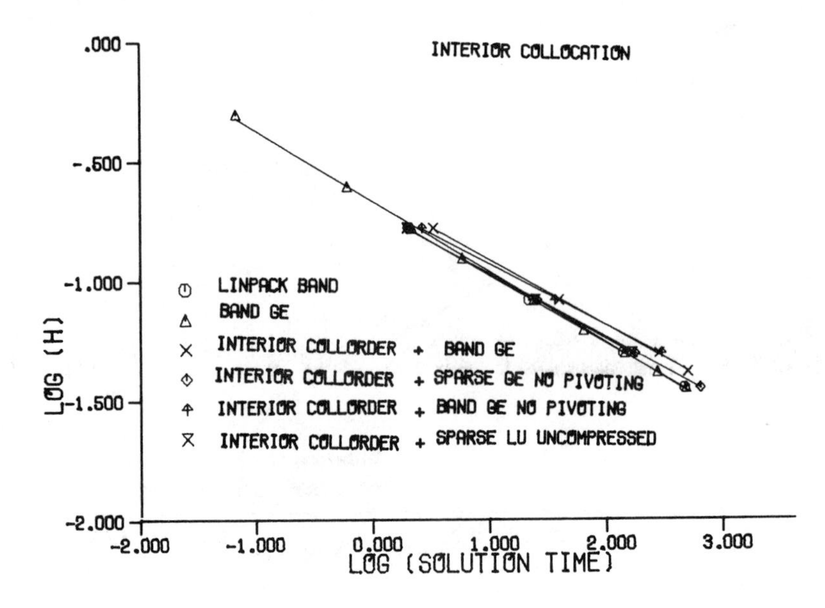

Figure 11.9. Performance of band elimination methods for the collocation equations and model problem B. The three best methods give $t = O(N^{3.3})$.

11.D SOLUTION OF THE GALERKIN EQUATIONS

The nature of the equations produced by the SPLINE GALERKIN module depends on the parameters DEGREE (polynomial degree) and NDERV (number of continuous derivatives), which specify the basis functions used. The coefficient matrix is in a **tensor product** ordering which is symmetric, positive definite and has several diagonal bands (depending on DEGREE and NDERV). The number of nonzero elements in a row is fairly high, e.g., 36 for Hermite bicubics (DEGREE=3, NDERV=1).

Figure 11.10 shows the performance of various cases for model problem B. There are three choices of basis functions (Hermite bicubics, bicubic splines and C^2 biquartics). One band elimination module is used along with several iteration modules. For the three cases using the bicubic spline basis functions, the LINPACK SPD BAND module is substantially faster. Its asymptotic efficiency is about $t = O(N^{2.6})$, which suggests that the iterative methods will be more efficient for larger N [JACOBI CG has $t = O(N^{1.8})$]. The crossover is for N about 30 (about 1000 equations). The JACOBI CG module gives $t = O(N^2)$ for the bicubic

Hermite basis functions and $t = O(N^{2.4})$ for the C^2 biquartics. Thus JACOBI CG (and presumably the other ITPACK modules) provides an efficient method to solve these equations for large N.

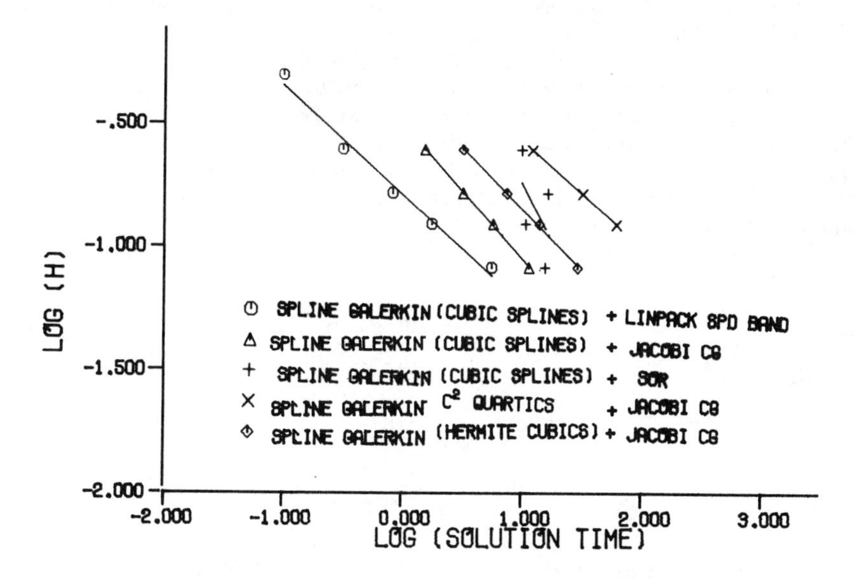

Figure 11.10. The performance of some solution modules for the SPLINE GALERKIN equations for model problem B.

The performance of SOR for these equations is an enigma, there is clearly something wrong. The trouble is probably due to the poor scaling or ill-conditioning of these equations. We were not able to include any results for model problem B using bicubic splines because the iteration always failed.

An extensive study of the performance of methods for the bicubic Hermite case is given in [Rice, 1983]. The principal conclusion reached is that the iteration methods are the most efficient for these equations and that LINPACK SPD BAND is next. All of the reasonable methods available in ELLPACK were included in this study. The crossover point between iteration and LINPACK SPD BAND is for N about 7 (observed range for 13 PDEs is 5 to 13).

11.E SOLUTION OF THE 7 POINT 3D EQUATIONS

It has been observed for many years that iteration methods are the best hope for solving the equations from 3D elliptic problems. For an N by N by N grid the work of band elimination methods is $O(N^7)$, which severely limits N even for very fast computers. For special problems, fast direct modules such as HODIE 27 POINT 3D (see Figure 10.20) or HODIE FFT 3D give reasonable execution times.

One conjectures that iteration methods converge in the same number of iterations in 3D as in 2D, there is both experimental and theoretical evidence to support this. The work per iteration is higher in 3D, but one would still obtain $t = O(N^4)$ if this conjecture is correct.

Figure 11.11 shows the performance of five **iteration** modules and one **band elimination** module (LINPACK BAND) for model problem I. The measured slope s for LINPACK BAND is about 8.3. Using LINPACK BAND it would require 3 years of VAX 11/780 time to solve these equations for $h = 0.03$ and 300 centuries for $h = 0.01$. The best iteration module (SYMMETRIC SOR SI) has slope of about 3.5, which is even better than the conjecture made above. This module could solve these equations for $h = 0.01$ in less than 100 hours (4 days). The module RED-BLACK + REDUCED SYSTEM CG is more efficient for smaller values of N, say N less than 30.

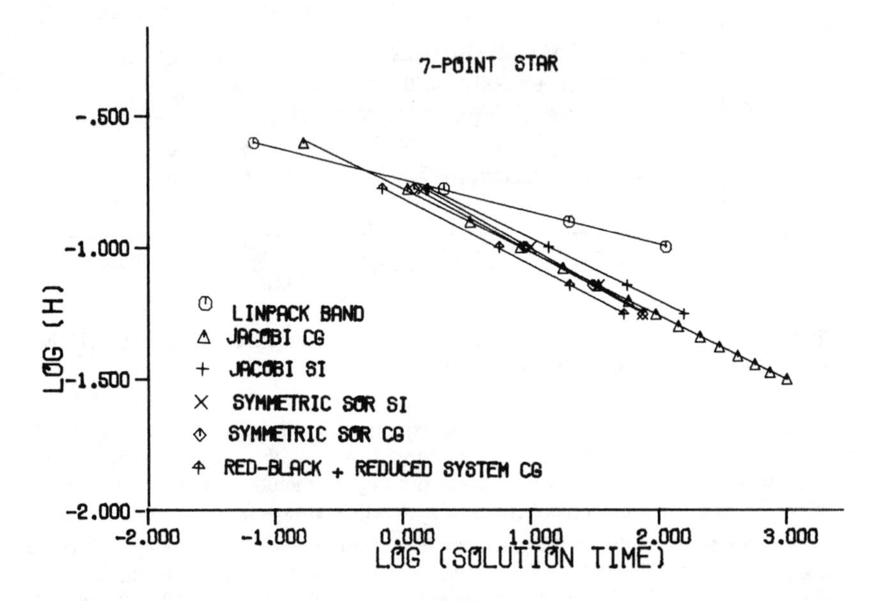

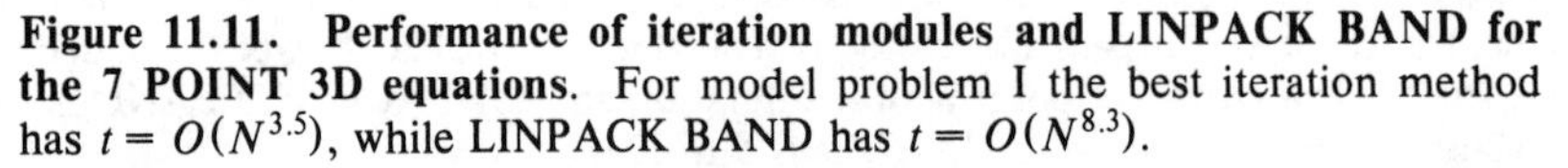
Figure 11.11. Performance of iteration modules and LINPACK BAND for the 7 POINT 3D equations. For model problem I the best iteration method has $t = O(N^{3.5})$, while LINPACK BAND has $t = O(N^{8.3})$.

Figure 11.12 shows the performance of two **sparse matrix** modules for these equations. The efficiency is better than for the band elimination modules, but only by a constant factor of about 10 or 15 percent. These data support the belief that iterative methods are superior to sparse matrix methods for the equations that arise from 3-dimensional elliptic problems.

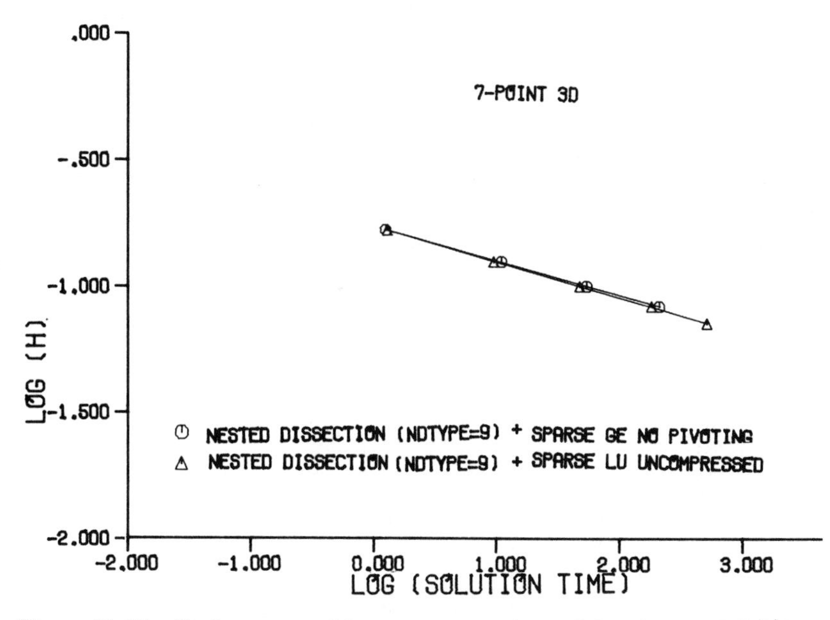

Figure 11.12. Performance of two sparse matrix modules for the 7 POINT 3D equations. These modules have $t = O(N^8)$, approximately.

11.F REFERENCES

Dyksen, W. R., E. N. Houstis, R. E. Lynch and J. R. Rice [1984], "The performance of the collocation and Galerkin methods with Hermite bicubics", SIAM J. Numer. Anal. 21, 695–715.

Dyksen, W. R. and J. R. Rice [1984], "A new ordering scheme for the Hermite bicubic collocation equations", in *Elliptic Problem Solvers* II, (G. Birkhoff and A. Schoenstadt, eds.), Academic Press, New York, pp. 467–480.

Eisenstat, S., A. George, R. Grimes, D. Kincaid and A. Sherman [1979], "Some comparisons of software packages for large sparse linear systems", in *Advances in Computer Methods for Partial Differential Equations* III, (R. Vichnevetsky and R. S. Stepleman, eds.), IMACS, Rutgers Univ., New Brunswick, N. J., pp. 98–106.

Rice, J. R. [1983], "Performance analysis of 13 methods to solve the Galerkin method equations", Linear Alg. Applic. 53, pp. 533–546.

Varga, R. S., [1962], *Matrix Iterative Analysis*, Prentice-Hall, Englewood Cliffs, N. J.

PART 4

CHAPTERS 12–16

CONTRIBUTOR'S GUIDE

Part 4 describes the ELLPACK system organization so that a new problem solving module may be contributed to it. This organization is based on the principle of software parts technology.

Chapter 12

SOFTWARE PARTS FOR ELLIPTIC PROBLEMS

12.A OVERVIEW

The ELLPACK system is founded on the principle of **software parts technology** [Batz et al., 1983; Rice, 1982]. A **software part** is a program for performing a well-defined computational task. The form of input and output of a part is rigidly defined in such a way that various software parts can be joined together to perform more complex tasks. That is, the output of one software part is the input to a second software part, and so on. Thus, the specification of a set of **interfaces** defines an environment in which software parts can be easily intermixed. The purpose of the **ELLPACK Contributor's Guide** is to describe a specific environment for software parts that can be used to solve elliptic boundary value problems.

In order to use a software parts technology, the overall computational problem must be decomposable into subproblems. In ELLPACK the solution of an elliptic problem is decomposed into the following tasks.

DOMAIN PROCESSING
: The relationship of each grid point to the boundary is determined, and the location of each boundary/grid intersection is located.

DISCRETIZATION
: The partial differential equation and its boundary conditions are converted to a system of linear algebraic equations.

INDEXING
: The equations and/or unknowns in the linear algebraic system are reordered to increase the efficiency of a solution method.

SOLUTION
: The linear algebraic system is solved to produce an approximation to the solution of the partial differential equation.

OUTPUT
: The solution and/or its derivatives are tabulated or plotted.

Each of these tasks corresponds to a software part type. The **interfaces** between software parts have the names

PROBLEM DEFINITION INTERFACE
DISCRETE DOMAIN INTERFACE
DISCRETE OPERATOR INTERFACE
EQUATION/UNKNOWN INDEXING INTERFACE
LINEAR ALGEBRAIC SOLUTION INTERFACE

Thus, a domain processor produces output at the discrete domain interface, while an operator discretization module produces output at the discrete operator interface, and so on. In this way the standard input to each module is the set of interfaces which precede it, while its standard output is the interface which follows it. The situation is illustrated in Figure 12.1 where the organization of a typical ELLPACK run is displayed. Note that a main program, called the **ELLPACK control program**, invokes each of the selected software parts in turn. This program also sets up the initial interface (problem definition).

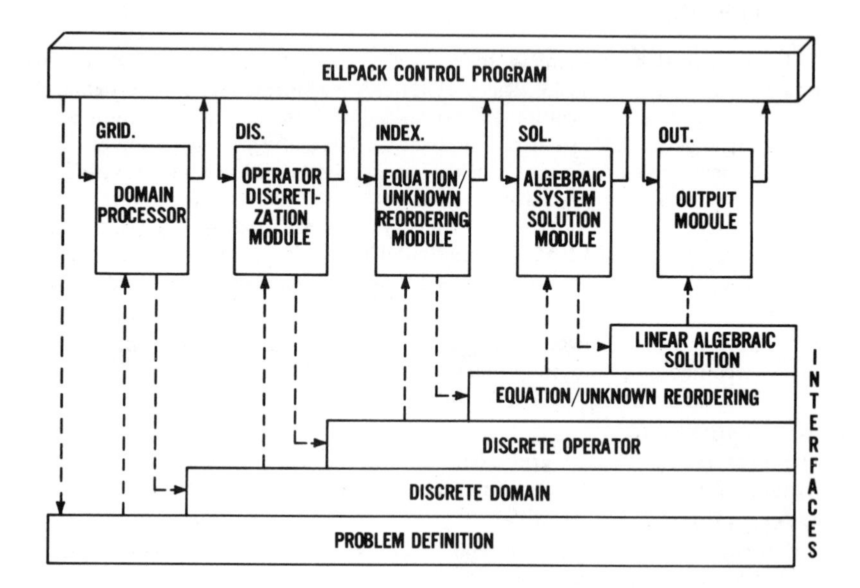

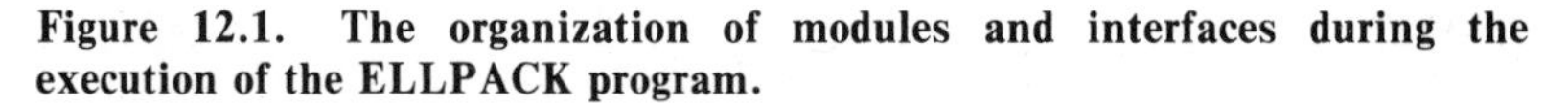
Figure 12.1. The organization of modules and interfaces during the execution of the ELLPACK program.

The ELLPACK interfaces are implemented as collections of Fortran variables and arrays; a detailed description of these is given in Chapter 13. Chapter 14 provides information on how problem-solving modules may gain access to them.

In addition to interface input, a software part may require other special input parameters. An individual module may require access to workspace whose size is problem-dependent. The use of such additional input is described in Chapter 14.

There are two other types of software parts in ELLPACK. A TRIPLE module is one that combines the operations of discretization, indexing and solution. Such parts are used to implement algorithms that do not neatly decompose into the form given above. A TRIPLE module does not explicitly assign values to variables at the discrete operator interface or the equation/unknown indexing interface. PROCEDURE modules are not associated with any particular ELLPACK interface. Instead, they implement various operations such as displaying the pattern of nonzeros in the discretization matrix or computing its eigenvalues. Modules of this type may access the information at any interface; there are usually no interface variables to which they are required to assign values.

ELLPACK contributors write subprograms that implement these software parts. Each part has the following basic structure.

PROLOGUE
: Each part must first determine if its input is admissible. For example, an operator discretization module may require that the problem be self-adjoint, or an algebraic equation solution module may require that the system be symmetric. All such restrictions must be checked before the module is executed. Then, it may be necessary to convert its input to a form more suitable for the algorithm to be employed. For example, a band matrix solver may need to create a band matrix in working storage.

NUCLEUS
: This performs the computations necessary to complete the task associated with the software part.

EPILOGUE
: Finally, the module must assign values to all output variables associated with the part. This may require that internal data structures are converted to a standard output interface data structure.

Contributed modules are also expected to adhere to a uniform set of programming standards for such things as printed output, error handling, portability, and documentation. These are described in Chapter 15. Several utility programs available to module authors to implement these standards are also described there.

The desire to provide modules with complete and easy to use information about the problem to be solved has led to quite complicated interfaces, and hence setting up these interfaces in Fortran programs is quite tedious, even when the PDE being solved is simple. In order to ease the burden on users, an **ELLPACK preprocessor** has been developed which reads a simple, high-level description of the problem and produces an appropriate **ELLPACK control program** which invokes ELLPACK software parts. Thus, ELLPACK users describe their problem in (nearly) mathematical notation, and select problem-solving modules by simply giving their names. The preprocessor analyzes the PDE, sets up the initial (problem definition) interface as well as that part of the discrete domain interface common to both rectangular and nonrectangular domains, allocates interface and workspace arrays, and sets up calls to requested modules. A diagram of an ELLPACK run as perceived by users is displayed in Figure 12.2. Note that it is a two stage process. The preprocessor reads an ELLPACK input program and creates a Fortran program which performs all the requested computations. This program is then compiled and executed.

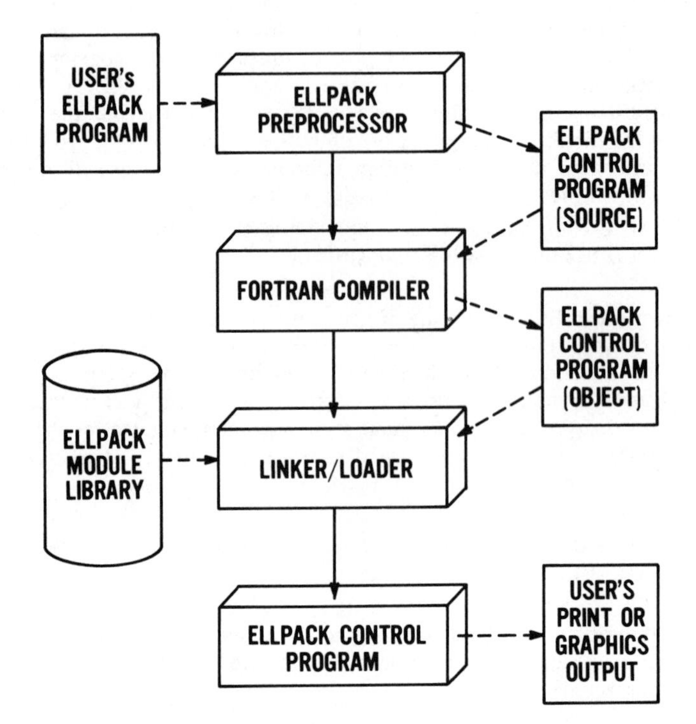

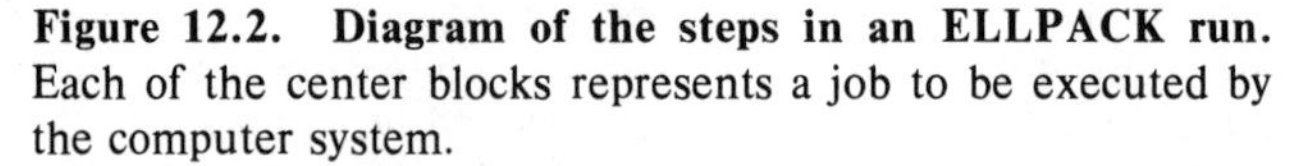

Figure 12.2. Diagram of the steps in an ELLPACK run. Each of the center blocks represents a job to be executed by the computer system.

Details about the structure of the ELLPACK preprocessor, itself a Fortran program, are given in Part 5. It should be clear, however, that the preprocessor must know many specific details about the contributed modules in order to carry out its function (e.g., their names, their parameters and defaults, the size of interface and workspace arrays as a function of problem parameters, etc.). The preprocessor gathers this information from a file called the **ELLPACK definition file**. Contributors must provide entries in this file for their modules. Information on how this is done is contained in Chapter 16.

12.B CONTRIBUTOR'S CHECKLIST

In this section we summarize the steps that must be taken in order to contribute a module to ELLPACK. The remainder of the Contributor's Guide describes these steps in detail. The numbers in brackets refer to sections in which further information can be found.

A. Obtain module name and module identifier characters from the ELLPACK coordinator. [12.C]

B. Prepare module.

   1. Observe naming conventions for subprograms and common blocks. [12.C]

   2. Set all output interface variables:

      a. I1NEQN, I1NCOE, R1COEF, I1IDCO, L1SYMM for DISCRETIZATION modules. [13.C]

      b. I1ENDX, I1UNDX for INDEXING modules. [13.D]

      c. R1UNKN for SOLUTION modules. [13.E]

      d. I1NEQN, I1NCOE, and R1UNKN for TRIPLE modules. [13.F]

      e. L1UINI for PROCEDURE modules that initialize the vector of unknowns R1UNKN. [13.F]

   3. Be sensitive to the global control variable I1LEVL. Provide summary output in standard format for I1LEVL=1. [15.A]

   4. Check initial interface variables for consistency. Is the problem posed one that your module is designed to solve? [15.B]

   5. Use the subprograms Q1ERRH and Q1ERRT to report fatal errors. Be sure that your error messages will be understandable to users. [15.B]

   6. Strive for portability in your code. Avoid nonstandard Fortran constructs. Obtain machine-dependent constants from the utility programs I1MACH and R1MACH. [15.C]

   7. Provide meaningful comments in your code. Identify the module to which each subprogram belongs. [15.D]

C. Prepare a solution evaluation module (DISCRETIZATION and TRIPLE modules only). [13.G]

D. Prepare entry for ELLPACK definition file. [16.C]

   1. Prepare an "ELLPACK call" segment. All module private parameters must have defaults.

   2. Prepare a FORTRAN segment to give the calling sequence of your main subprogram.

   3. Prepare an ANSWER segment to give the call sequence of your solution evaluation module (DISCRETIZATION and TRIPLE modules only). [13.G]

   4. Prepare a DC segment to give information on your module's storage requirements.

   5. Prepare the optional segments SETUP and DECLARATIONS if necessary.

E. Prepare documentation in standard form for inclusion in the *ELLPACK Users Guide.* [15.D]

12.C CONVENTIONS

PDE AND BOUNDARY CONDITION COEFFICIENTS

ELLPACK users specify their partial differential equation (PDE) in one of two forms:

$$a*\mathrm{UXX} + b*\mathrm{UYY} + c*\mathrm{UZZ} + d*\mathrm{UXY} + e*\mathrm{UXZ} + f*\mathrm{UYZ} \\ + g*\mathrm{UX} + h*\mathrm{UY} + i*\mathrm{UZ} + j*\mathrm{U} = k,$$

$$(a*\mathrm{UX})\mathrm{X} + (b*\mathrm{UY})\mathrm{Y} + (c*\mathrm{UZ})\mathrm{Z} + j*\mathrm{U} = k.$$

The first case is called **standard form** while the second is called **self-adjoint form.** Although the order in which the terms of the PDE may be written by ELLPACK users is arbitrary, there is a standard ordering in which these coefficients are stored internally. It is

1. coefficient of UXX,
2. coefficient of UXY,
3. coefficient of UYY,
4. coefficient of UX,
5. coefficient of UY,
6. coefficient of U,
7. coefficient of UZZ,
8. coefficient of UXZ,
9. coefficient of UYZ,
10. coefficient of UZ.

Note that the order of the coefficients that are independent of z remains the same in both two dimensional (2D) and three dimensional (3D) problems. When the user writes the equation in self-adjoint form then the coefficient a is associated with UXX, the coefficient b with UYY, and the coefficient c with UZZ.

The most general form of boundary condition specified by ELLPACK users is

$$p*\mathrm{UX} + q*\mathrm{UY} + r*\mathrm{UZ} + s*\mathrm{U} = t.$$

The standard ordering for boundary condition coefficients is

1. coefficient of U
2. coefficient of UX
3. coefficient of UY
4. coefficient of UZ

Note again that the order of coefficients that are independent of z remains the same for both 2D and 3D problems.

RECTANGULAR DOMAINS

Rectangular domains are treated as a special case in ELLPACK, and in this case a standard numbering of sides (or faces in 3D) is assumed. For the rectangle (AX,BX) × (AY,BY) × (AZ,BZ) this is

1. right side (x = BX).
2. bottom side (y = AY).
3. left side (x = AX).

4. top side (y = BY).
5. front side (z = BZ).
6. back side (z = AZ).

In the nonrectangular case boundary pieces are numbered in the same order as they are given by the user.

Internal Ellpack Names

The ELLPACK language gives its users considerable flexibility in the construction of programs that use the ELLPACK modules. They may declare their own variables and insert their own FORTRAN code anywhere within the ELLPACK Control Program. PDE and boundary condition coefficients may be given in terms of user-defined functions. Thus, there is considerable opportunity for ELLPACK users to generate variables and subprograms with the same names as those used internal to the ELLPACK system. It is also quite possible for two contributed modules to contain different subprograms with the same name.

In order to avoid conflicts among ELLPACK, its contributors, and its users, ELLPACK interface variables, common blocks, and subprograms use a systematic naming convention. All such objects have six-character names with first character C, I, L, R, or Q and second character a digit: for example, C200PP, Q1PCOE, R4SSMN, and I1MACH. The exact form of internal object names is

ABCCDD

where

A = **module type**. Possible values are

- Q → subroutine
- R → real variable or function
- I → integer variable or function
- L → logical variable or function
- C → common block

B = **owner type**. Possible values are

- 0 → internal object (for use by preprocessor or ELLPACK control program)
- 1 → interface object or utility program (available for use by all modules)
- 2 → domain processor object
- 3 → DISCRETIZATION module object
- 4 → INDEXING module object
- 5 → SOLUTION module object
- 6 → TRIPLE module object
- 7 → PROCEDURE module object
- 8 → OUTPUT module object
- 9 → other

CC = **module identifier**. If $2 < B < 8$, then this is two characters assigned to the module by the ELLPACK coordinator; otherwise they are any two alphanumeric characters.

DD = any two alphanumeric characters chosen by the module contributor

For example, a discretization module assigned the identifier 5P would have subprograms named Q35Pxx, and common blocks named C35Pxx, where xx are any characters. The main subprogram of a module should use the characters MN for xx.

12.D REFERENCES

Batz, J. C., P. M. Cohen, S. T. Redwine and J. R. Rice [1983], "The application-specific task area", IEEE Computer 16, pp. 78–85.

Comer, D. E., J. R. Rice, H. D. Schwetman, and L. Snyder [1981], Project Quanta, Report CSD-TR 366, Department of Computer Sciences, Purdue University, West Lafayette, IN.

Rice, J. R. [1982a], "Software for scientific computation", in *Computer Aided Design of Technological Systems* (G. Leininger, ed.), IFAC, Pergamon Press, pp. 339–344.

Chapter 13

INTERFACE SPECIFICATIONS

The following is a detailed description of the module interfaces in ELLPACK. A summary of the variables at each interface is given in Table 13.1, and a list of all ELLPACK variables in alphabetical order with short descriptions is given in Table 13.2. Information on how modules may gain access to them is given in Chapter 14.

13.A PROBLEM INTERFACE

DOMAIN DEFINITION

Two logical variables indicate the basic form of the geometry:

L1TWOD = .TRUE. if the problem domain is two-dimensional,
.FALSE. if the problem domain is three-dimensional,

L1RECT = .TRUE. if the domain is rectangular,
.FALSE. otherwise.

Only rectangular domains are allowed for three-dimensional problems and hence L1RECT=.TRUE. whenever L1TWOD=.FALSE..

Nonrectangular domains are defined as the intersection of the interior of a simple closed curve in the plane (the "boundary") and the exteriors of a set of simple closed curves contained wholly within the boundary ("holes"). ELLPACK users may also define curves that are to be excluded from the domain ("arcs") on which additional side conditions are prescribed. All curves are taken to be piecewise smooth. The boundary, holes, and arcs are each described parametrically by ELLPACK users. Thus, the number of boundary pieces, the range of the parameter along each piece, and a routine that returns (x,y) for each value of the parameter are available. The number of boundary pieces is given by

I1NBND = the number of pieces defining the boundary of the domain. (I1NBND = 4 for 2D rectangular domains, and 6 for 3D rectangular domains.)

The following additional objects are available only in the case L1RECT=.FALSE.:

R1BRNG = array giving initial and final parameter values for each boundary piece, that is, R1BRNG$(1,k)$ = initial value and R1BRNG$(2,k)$ = final value for the kth side, k = 1, ..., I1NBND.

```
SUBROUTINE Q1BDRY (P,X,Y,K)
INTEGER K
REAL P,X,Y
  input:    K,P
  output:   returns coordinates X and Y associated with the parameter
              value P on boundary piece K.
```

The pieces describing each closed loop are numbered in order so that piece k joins with pieces $k-1$ and $k+1$. One variable indicates the orientation with which the boundary is described. (holes have the opposite orientation):

L1CLKW = .TRUE. if the boundary is parametrized with a clockwise orientation,
.FALSE. if the boundary is parametrized with a counterclockwise orientation.

Finally, two switches are available to indicate whether the domain specified by the user has holes or arcs:

L1HOLE = .TRUE. if the domain has holes,
.FALSE. otherwise,

L1ARCC = .TRUE. if the domain has internal arcs,
.FALSE. otherwise.

PDE DEFINITION

One logical variable indicates the basic form in which the user has written the partial differential equation: standard form or self-adjoint form. It is

L1SELF = .TRUE. if the user has written the PDE in the form
$(a*\text{UX})\text{X} + (b*\text{UY})\text{Y} + ...,$
.FALSE. otherwise.

Two subprograms are used to define the PDE. In the two-dimensional case (L1TWOD=.TRUE.) they are

```
SUBROUTINE Q1PCOE (X,Y,CPDE)
REAL X,Y,CPDE(6)
  input:    X,Y
  output:   CPDE(i) = value of ith PDE coefficient at (X,Y)
REAL FUNCTION R1PRHS (X,Y)
REAL X,Y
  input:    X,Y
  output:   Returns value of right hand side of the PDE at (X,Y)
```

Two subprograms are used to define the boundary conditions. They are

```
SUBROUTINE Q1BCOE (K,X,Y,CBC)
INTEGER K
REAL X,Y,CBC(3)
  input:    K,X,Y
  output:   CBC(i) = value of the ith boundary condition coefficient
              on side K at (X,Y)
```

```
REAL FUNCTION R1BRHS (K,X,Y)
INTEGER K
REAL X,Y
  input:   K,X,Y
  output:  Returns the value of the right side of the Kth boundary
             condition at the point (X,Y)
```

Note: If the problem is three dimensional (L1TWOD=.FALSE.), then the subprograms Q1PCOE, R1PRHS, Q1BCOE, and R1BRHS each have an additional argument, Z, which follows Y in the parameter list. Also, the arrays CPDE and CBC are expanded to lengths 10 and 4 in this case. See Section 12.C for the coefficient ordering used in Q1PCOE and Q1BCOE.

Two arrays of switches are available to help modules determine the status of the PDE and boundary condition coefficients:

I1CFST = PDE coefficient status vector. I1CFST(i) gives the status of the ith coefficient of the partial differential equation for $i = 1, ..., 10$. The coefficient ordering of Section 12.C is used.

I1BCST = boundary coefficient status array. I1BCST(i,k) gives the status of the ith coefficient in the kth boundary condition, $i = 1, ..., 4$, $k = 1, ...,$ I1NBND. The coefficient ordering of Section 12.C is used.

The possible values for coefficient status are

0 → coefficient is identically zero (0),
1 → coefficient is identically one (1),
2 → coefficient is independent of x, y, and z (constant),
3 → coefficient is dependent upon x, y, or z (variable).

For example, I1CFST can be used to determine whether an equation that was written in standard form is, in fact, trivially self-adjoint. In two dimensions this is true when I1CFST(i) < 3, $i = 1, ..., 10$ and I1CFST(i) $= 0$, $i = 2, 4, 5, 8, 9, 10$. If periodic boundary conditions are specified on the kth boundary piece, then I1BCST(i,k) $= 0$ for $i = 1, 2, 3, 4$.

Several sets of logical variables are available that summarize the state of the coefficient status arrays and extend the information in them in certain cases. The following variables give summary information about the PDE.

L1CRST = .FALSE. if no cross derivative terms are present [i.e., I1CFST(2) = I1CFST(8) = I1CFST(9) = 0],
.TRUE. otherwise,

L1CSTC = .TRUE. if the coefficients of the PDE are all constant [i.e., I1CFST(i) < 3, $i = 1, ..., 10$],
.FALSE. otherwise,

L1POIS = .TRUE. if the PDE is the Poisson equation [i.e., if L1TWOD and I1CFST(k) $= 0$, $k = 2, 4, 5, 6$ and I1CFST(k) $= 1$, $k = 1, 3$ else if I1CFST(k) $= 0$, $k = 2, 4, 5, 6, 8, 9, 10$ and I1CFST(k) $= 1$, $k = 1, 3, 7$],
.FALSE. otherwise,

L1HMEQ = .TRUE. if the PDE is homogeneous [i.e., R1PRHS(X,Y) ≡ 0.0],
.FALSE. otherwise,

L1LAPL = .TRUE. if the PDE is the Laplace equation [i.e., L1POIS and L1HMEQ],
.FALSE. otherwise.

When L1CSTC=.TRUE. variables become available (they are undefined otherwise) which provide the constant values of the PDE coefficients. They are

R1CUXX = coefficient of UXX,

R1CUXY = coefficient of UXY,

R1CUYY = coefficient of UYY,

R1CCUX = coefficient of UX,

R1CCUY = coefficient of UY,

R1CCCU = coefficient of U,

R1CUZZ = coefficient of UZZ,

R1CUXZ = coefficient of UXZ,

R1CUYZ = coefficient of UYZ,

R1CCUZ = coefficient of UZ.

The following variables summarize boundary condition information:

L1CSTB = .TRUE. if the coefficients of the boundary conditions are all constants [i.e., I1BCST(i,k) < 3, i = 1, ..., 4, k = 1, ..., I1NBND],
.FALSE. otherwise,

L1HMBC = .TRUE. if the boundary conditions are homogeneous [i.e., R1BRHS(k,X,Y)=0.0, k = 1, ..., I1NBND],
.FALSE. otherwise,

I1BCTY = vector of boundary condition types. I1BCTY(k) gives the type of boundary condition on the kth boundary piece, k = 1, ..., I1NBND. Possible choices are

1 → condition of form U = R1BRHS(k,X,Y); i.e. I1BCST(1,k) = 1, I1BCST(2,k) = I1BCST(3,k) = I1BCST(4,k) = 0,
2 → pure derivative condition; i.e. I1BCST(1,k) = 0,
3 → nonperiodic, but not in form 1 or 2,
4 → periodic boundary condition.

Note that the boundary condition U = 0.2*SIN(X) is of type 1, while 5.0*U = SIN(X) is of type 3. I1BCST can be inspected to determine the exact nature of the boundary condition in either case. Note also that periodic boundary conditions are only allowed on rectangular domains (L1RECT=.TRUE.), and if a periodic condition is specified on one side of the rectangle it must also be specified along the opposite side. Several switches provide even further summary of this information. They are

L1DRCH = .TRUE. if all boundary conditions are of form U = R1BRHS(k,X,Y) [i.e., I1BCTY(k) = 1 for k = 1, ..., I1NBND],
.FALSE. otherwise,

L1NEUM = .TRUE. if all boundary conditions are of purely derivative type [i.e., I1BCTY(k) = 2 for k = 1, ..., I1NBND],
.FALSE. otherwise,

L1PRDX = .TRUE. if the boundary conditions are periodic in x [i.e., L1RECT and I1BCTY(1) = I1BCTY(3) = 4],
.FALSE. otherwise,

L1PRDY = .TRUE. if the boundary conditions are periodic in y [i.e., L1RECT and I1BCTY(2) = I1BCTY(4) = 4],
.FALSE. otherwise,

L1PRDZ = .TRUE. if the boundary conditions are periodic in z [i.e., (not L1TWOD) and L1RECT and I1BCTY(5) = I1BCTY(6) = 4],
.FALSE. otherwise,

L1PRDC = .TRUE. if all boundary conditions are periodic [i.e., L1PRDX and L1PRDY and (L1TWOD or L1PRDZ)],
.FALSE. otherwise,

L1MIXD = .TRUE. if not (L1DRCH or L1NEUM or L1PRDC),
.FALSE. otherwise.

Note that L1MIXD does not mean mixed boundary conditions in the usual sense. L1MIXD is .TRUE., for instance, if the Dirichlet condition U = 0.0 is prescribed on all sides except one where 2.0*U = 0.0 is given.

When the specified boundary value problem has a unique solution only up to an arbitrary additive constant (as in the Poisson equation with Neumann and/or periodic boundary conditions) then the ELLPACK user may specify this using the procedure NON-UNIQUE. This affects several problem definition variables.

L1NUNQ = .TRUE. if the solution is non-unique,
.FALSE. otherwise.

If L1NUNQ=.TRUE. the user indicates the desired solution by giving its value at a single point in the closure of the domain. This information is given by

R1UNQX = x coordinate of the selected point,

R1UNQY = y coordinate of the selected point,

R1UNQZ = z coordinate of the selected point (R1UNQZ = 0.0 in 2D problems),

R1UNQU = value of solution at (R1UNQX,R1UNQY,R1UNQZ).

Modules may place restrictions on the location of this point. If this is done the restrictions should be stated in the module documentation.

13.B DISCRETE DOMAIN INTERFACE

Variables Defining the Rectangular Grid

The following variables provide information about the rectangular grid specified by the user:

R1AXGR, R1BXGR = range of x grid,

R1AYGR, R1BYGR = range of y grid,

R1AZGR, R1BZGR = range of z grid (R1AZGR = R1BZGR = 0.0 for 2D problems),

I1NGRX = number of grid lines in x,

I1NGRY = number of grid lines in y,

I1NGRZ = number of grid lines in z (I1NGRZ = 1 for 2D problems),

R1GRDX(i) = location of ith grid line in x, i = 1, ..., I1NGRX [R1GRDX(1) = R1AXGR, R1GRDX(I1NGRX) = R1BXGR],

R1GRDY(j) = location of jth grid line in y, j = 1, ..., I1NGRY [R1GRDY(1) = R1AYGR, R1GRDY(I1NGRY) = R1BYGR],

R1GRDZ(k) = location of kth grid line in z, k = 1, ..., I1NGRZ [R1GRDZ(1) = R1AZGR, R1GRDZ(I1NGRZ) = R1BZGR].

The entries in the arrays R1GRDX, R1GRDY, and R1GRDZ are strictly increasing.

The grid is defined by the lines x = R1GRDX(i), i = 1, ..., I1NGRX; y = R1GRDY(j), j = 1, ..., I1NGRY; and z = R1GRDZ(k), k = 1, ..., I1NGRZ. The points where these lines intersect are called grid points. Thus, in 2D the (i,j)th grid point is (R1GRDX(i),R1GRDY(j)). Most grid points have four **neighbors**. For example, the neighbors of the (i,j)th grid point are $(i,j+1)$, $(i,j-1)$, $(i+1,j)$, $(i-1,j)$ and they are called the north, south, east, and west neighbors, respectively. When i = 1 there is no west neighbor, when i = I1NGRX there is no east neighbor, and so on. In 3D each grid point has at most six neighbors.

Several quantities are available that summarize various properties of the grid. They are

R1HXGR = (R1BXGR − R1AXGR)/(I1NGRX − 1),

R1HYGR = (R1BYGR − R1AYGR)/(I1NGRY − 1),

R1HZGR = (R1BZGR − R1AZGR)/(I1NGRZ − 1) [R1HZGR = 0.0 for 2D problems],

L1UNFX = .TRUE. if the grid is uniformly spaced in the x direction [i.e., R1GRDX(i+1) − R1GRDX(i) = R1HXGR, i = 1, ..., I1NGRX − 1],
.FALSE. otherwise,

L1UNFY = .TRUE. if the grid is uniformly spaced in the y direction [i.e., R1GRDY(j+1) − R1GRDY(j)=R1HYGR, j = 1, ..., I1NGRY − 1],
.FALSE. otherwise,

L1UNFZ = .TRUE. if the grid is uniformly spaced in the z direction [i.e., R1GRDZ(k+1) − R1GRDZ(k) = R1HZGR, k = 1, ..., I1NGRZ − 1],
.FALSE. otherwise,

L1UNFG = .TRUE. if the grid is uniformly spaced in every variable [i.e., L1UNFX and L1UNFY and L1UNFZ],
.FALSE. otherwise,

R1EPSG = real variable giving the uncertainty in the grid. For example, x is considered equal to R1GRDX(i) if

$$\text{R1GRDX}(i) - \text{R1EPSG} < x < \text{R1GRDX}(i) + \text{R1EPSG}.$$

The machine-dependent constant R1EPSG is used principally by the domain processor. Ideally, its value is small enough so that uncertainties in the grid of size R1EPSG do not affect the accuracy of the PDE solution, and yet large enough so that roundoff errors do not affect the domain processing.

Variables Defining the Relationship Between Domain and Grid

When the domain is rectangular (L1RECT = .TRUE.), the relationship between the grid and the domain is very simple: the boundary of the domain coincides with the extreme grid lines, and all grid points (R1GRDX(i),R1GRDY(j)) are either inside the domain ($1 < i <$ I1NGRX, $1 < j <$ I1NGRY) or on its boundary (i = 1, i = I1NGRX, j = 1, j = I1NGRY). Similar properties hold in three dimensions. When the domain is nonrectangular (L1RECT = .FALSE.) the situation is much more complex. In this case the ELLPACK domain processor is invoked to determine the nature of each grid point and the location of all points where the boundary intersects grid lines. It then provides this information in various forms so that later processing by ELLPACK modules requires a minimum of geometric analysis.

The points where the boundary (including holes and arcs) intersects grid lines are called **boundary points**. The domain processor numbers the boundary points beginning with 1 while traversing the boundary in the order specified by the user. All points where boundary pieces join or arcs begin and end are considered boundary points, even when they do not lie on grid lines. Before points on a new hole or arc are numbered the first point of the previous closed loop or arc is repeated in the list. The following variables give basic information about the boundary points:

I1NBPT = number of boundary points.

R1XBND = vector of x coordinates of boundary points. R1XBND(i) is the x coordinate of the ith point, i = 1, ..., I1NBPT.

R1YBND = vector of y coordinates of boundary points. R1YBND(i) is the y coordinate of the ith point, i = 1, ..., I1NBPT.

I1PECE = vector of boundary piece indices. I1PECE(i) is the number of the boundary piece on which the ith boundary point lies, i = 1, ..., I1NBPT. If the ith point is where two pieces meet, then I1PECE(i) gives the smaller index.

R1BPAR = vector of boundary point parameter values. R1BPAR(i) gives the value of the parameter that determines the location of the ith boundary point. That is,

CALL Q1BDRY(R1BPAR(i),X,Y,I1PECE(i))

will return X = R1XBND(i), Y = R1YBND(i), i = 1, ..., I1NBPT.

I1BPTY = vector of boundary point types. I1BPTY(i) is the type of the ith boundary point. Possible values are

1 → on a y grid line (horizontal),
2 → on an x grid line (vertical),
3 → at a grid point (both),
4 → corner point not on grid line (interior corner),
5 → at end of a boundary piece that does not join continuously with next piece (jump).

If more than one case applies, the largest value is selected.

I1BGRD = vector of boundary point grid square locations. I1BGRD(i) gives the grid square in which the ith boundary point is located, i = 1, ..., I1NBPT. The value is encoded as $j + 1000k$, which implies

$$\text{R1GRDX}(j) \leqslant \text{R1XBND}(i) < \text{R1GRDX}(j+1)$$

and

$$\text{R1GRDY}(k) \leqslant \text{R1YBND}(i) < \text{R1GRDY}(k+1).$$

The status of each grid point (R1GRDX(i),R1GRDY(j)), i = 1, ..., I1NGRX, j = 1, ..., I1NGRY, is given in the array I1GRTY.

I1GRTY = array of grid point types. I1GRTY(i,j) gives the type of the grid point (R1GRDX(i),R1GRDY(j)) for i = 1, ..., I1NGRX and j = 1, ..., I1NGRY. Possible values of the type m are:

$m = 0$ → outside domain, away from boundary,
m = I1PACK − 1 → inside domain, away from boundary,
$0 < m <$ I1PACK − 1 → on the boundary at boundary point m,
I1PACK $\leqslant m$ → inside domain, near boundary,
$m \leqslant 1 -$ I1PACK → outside domain, near boundary.

The constant I1PACK is available at the discrete domain interface (its value is 1000, unless there are more than 998 boundary points). A grid point is considered near the boundary if any of its four neighbors is on the boundary or if a boundary curve must be crossed when travelling along grid lines from the grid point to any of its four neighbors. The associated boundary points are called the **boundary neighbors** of the grid point in question. When $|\text{I1GRTY}(i,j)| \geqslant \text{I1PACK}$ then information about the boundary neighbors of the (i,j)th grid point is encoded in the I1GRTY entry. In these cases, $|\text{I1GRTY}(i,j)| = k + \text{I1PACK}n$, where

k = the index of the lowest numbered boundary neighbor,
n = integer whose four binary digits give the directions in which a boundary neighbor can be found:

0001 → boundary neighbor to north,
0010 → boundary neighbor to east,
0100 → boundary neighbor to south,
1000 → boundary neighbor to west.

Thus, $n = 9$ (1001 binary) implies that there are boundary neighbors to the north and west.

For example, if I1PACK = 1000, I1GRTY(i,j) = −3017 implies that the grid point (R1GRDX(i),R1GRDY(j)) is outside the domain, but has boundary neighbors to the north and east, one of which is boundary point 17.

The complement of I1GRTY's boundary neighbor information is available in the array I1BNGH. For each boundary point, I1BNGH gives the directions in which neighboring interior grid points can be found:

I1BNGH = vector of neighboring interior grid points. I1BNGH(i) gives pointers to grid points inside the domain which are adjacent to the ith boundary point. Specifically, I1BNGH(i) is an integer whose four binary digits give the directions in which interior neighbors are located:

0001 → north,
0010 → east,
0100 → south,
1000 → west.

Thus, I1BNGH(i) = 3 (0011 binary) implies that the ith boundary point has interior neighbors to the north and east of it. Note that interior corners (I1BPTY(i) = 4) never have neighboring grid points.

The variable I1NBPT, which gives the number of boundary points, is determined by the ELLPACK domain processor, which executes only when the ELLPACK control program is run. The preprocessor must estimate this number so that it can allocate storage for the arrays giving information about the boundary points. The preprocessor's estimate is available in the following variable:

I1MBPT = storage allocated for arrays requiring length I1NBPT.

SAMPLE OF DOMAIN PROCESSOR OUTPUT

We now give a sample of domain processor output. The domain we choose is the semicircular region with a semicircular hole displayed in Figure 13.1. The outer boundary is parametrized by (using ELLPACK notation)

```
LINE 0.0,4.0 TO 0.0,-4.0
X=4.0*COS((T-0.5)*PI), Y=4.0*SIN((T-0.5)*PI)    FOR T=0.0 TO 1.0
```

while the hole is parametrized by

```
LINE 1.5,-1.0 TO 1.5,1.0
X=1.5+COS((0.5-T)*PI), Y=SIN((0.5-T)*PI)         FOR T=0.0 TO 1.0
```

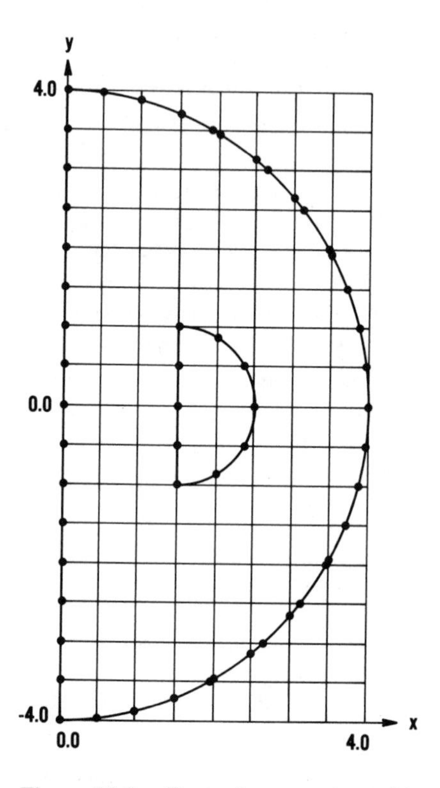

Figure 13.1. Example nonrectangular domain for the discussion of the discrete domain interface.

We choose a uniform grid of size 9 by 17 in the range 0.0 to 4.0 in x and -4.0 to 4.0 in y. The data produced for this problem follow. Here I1PACK = 1000, so interior grid points have I1GRTY=999, grid points near the boundary have $|$I1GRTY$|$ > 1000, and grid points on the boundary have 0 < I1GRTY < 999.

```
+++++++++++++++++++++++++++++++++++++++++++++++++++++++++
+                                                       +
+    TABLE  OF  GRID POINT  TYPES  ON   9 X  17  GRID   +
+                                                       +
+++++++++++++++++++++++++++++++++++++++++++++++++++++++++
```

THE POINT XGRID(1), YGRID(1) IS AT THE LOWER LEFT.

17	*	1	-12001	-4045	-4044	0	0	0	0	0
16	*	2	9002	1045	3043	-12042	-4041	0	0	0
15	*	3	8003	999	999	1042	3040	-12039	0	0
14	*	4	8004	999	999	999	999	3038	-8038	0
13	*	5	8005	999	999	999	999	2037	-12036	0
12	*	6	8006	999	4052	999	999	999	3035	-8035
11	*	7	8007	2052	52	12052	999	999	2034	-8034
10	*	8	8008	2051	51	-11051	12054	999	2035	-12032
9	*	9	8009	2050	50	-10050	55	8055	2032	32
8	*	10	8010	2049	49	-14049	9055	999	2031	-9031
7	*	11	8011	2048	48	9048	999	999	2030	-8030
6	*	12	8012	999	1048	999	999	999	6028	-8029
5	*	13	8013	999	999	999	999	2027	-9027	0
4	*	14	8014	999	999	999	999	6025	-8026	0
3	*	15	8015	999	999	4022	6023	-9024	0	0
2	*	16	12016	4019	6020	-9021	-1023	0	0	0
1	*	17	-9017	-1019	-1020	0	0	0	0	0
	*									
	***	1	2	3	4	5	6	7	8	9

```
+++++++++++++++++++++++++++++++++++++++++++++++++++++++++++++++
+                                                             +
+  TABLE  OF  THE  BOUNDARY  POINT  TYPES ON   9 X  17  GRID  +
+                                                             +
+++++++++++++++++++++++++++++++++++++++++++++++++++++++++++++++
```

NUMBER	R1XBND	R1YBND	R1BPAK	R1PECE	R1BPTY	R1BGRD	R1BNGH
1	.0000	4.0000E+00	.0000	1	BOTH	17001	0
2	.0000	3.5000E+00	6.2500E-02	1	BOTH	18001	2
3	.0000	3.0000E+00	1.2500E-01	1	BOTH	15001	2
4	.0000	2.5000E+00	1.8750E-01	1	BOTH	14001	2
5	.0000	2.0000E+00	2.5000E-01	1	BOTH	13001	2
6	.0000	1.5000E+00	3.1250E-01	1	BOTH	12001	2
7	.0000	1.0000E+00	3.7500E-01	1	BOTH	11001	2
8	.0000	5.0000E-01	4.3750E-01	1	BOTH	10001	2
9	.0000	.0000	5.0000E-01	1	BOTH	9001	2
10	.0000	-5.0000E-01	5.6250E-01	1	BOTH	8001	2
11	.0000	-1.0000E+00	6.2500E-01	1	BOTH	7001	2
12	.0000	-1.5000E+00	6.8750E-01	1	BOTH	6001	2
13	.0000	-2.0000E+00	7.5000E-01	1	BOTH	5001	2
14	.0000	-2.5000E+00	8.1250E-01	1	BOTH	4001	2
15	.0000	-3.0000E+00	8.7500E-01	1	BOTH	3001	2
16	.0000	-3.5000E+00	9.3750E-01	1	BOTH	2001	2
17	3.9684E-09	-4.0000E+00	1.0000E+00	1	BOTH	1001	0
18	5.0000E-01	-3.9686E+00	3.9893E-02	2	VERT	1002	1
19	1.0000E+00	-3.8730E+00	8.0431E-02	2	VERT	1003	1
20	1.5000E+00	-3.7081E+00	1.2236E-01	2	VERT	1004	1
21	1.9365E+00	-3.5000E+00	1.6086E-01	2	HORZ	2004	8
22	2.0000E+00	-3.4641E+00	1.6667E-01	2	VERT	2005	1
23	2.5000E+00	-3.1225E+00	2.1490E-01	2	VERT	2006	1
24	2.6458E+00	-3.0000E+00	2.3005E-01	2	HORZ	3006	8
25	3.0000E+00	-2.6458E+00	2.6995E-01	2	VERT	3007	1
26	3.1225E+00	-2.5000E+00	2.8510E-01	2	HORZ	4007	8
27	3.4641E+00	-2.0000E+00	3.3333E-01	2	HORZ	5007	8
28	3.5000E+00	-1.9365E+00	3.3914E-01	2	VERT	5008	1

29	3.7081E+00	-1.5000E+00	3.7764E-01	2	HORZ	6008	8
30	3.8730E+00	-1.0000E+00	4.1957E-01	2	HORZ	7008	8
31	3.9686E+00	-5.0000E-01	4.6011E-01	2	HORZ	8008	8
32	4.0000E+00	.0000	5.0000E-01	2	BOTH	9009	8
33	3.9686E+00	5.0000E-01	5.3989E-01	2	HORZ	10008	8
34	3.8730E+00	1.0000E+00	5.8043E-01	2	HORZ	11008	8
35	3.7081E+00	1.5000E+00	6.2236E-01	2	HORZ	12008	8
36	3.5000E+00	1.9365E+00	6.6086E-01	2	VERT	12008	4
37	3.4641E+00	2.0000E+00	6.6667E-01	2	HORZ	13007	8
38	3.1225E+00	2.5000E+00	7.1490E-01	2	HORZ	14007	8
39	3.0000E+00	2.6458E+00	7.3005E-01	2	VERT	14007	4
40	2.6458E+00	3.0000E+00	7.6995E-01	2	HORZ	15006	8
41	2.5000E+00	3.1225E+00	7.8510E-01	2	VERT	15006	4
42	2.0000E+00	3.4641E+00	8.3333E-01	2	VERT	15005	4
43	1.9365E+00	3.5000E+00	8.3914E-01	2	HORZ	16004	8
44	1.5000E+00	3.7081E+00	8.7764E-01	2	VERT	16004	4
45	1.0000E+00	3.8730E+00	9.1957E-01	2	VERT	16003	4
46	5.0000E-01	3.9686E+00	9.6011E-01	2	VERT	16002	4
47	.0000	4.0000E+00	1.0000E+00	2	JUMP	17001	0
48	1.5000E+00	-1.0000E+00	.0000	3	BOTH	7004	14
49	1.5000E+00	-5.0000E-01	2.5000E-01	3	BOTH	8004	8
50	1.5000E+00	.0000	5.0000E-01	3	BOTH	9004	8
51	1.5000E+00	5.0000E-01	7.5000E-01	3	BOTH	10004	8
52	1.5000E+00	1.0000E+00	1.0000E+00	3	BOTH	11004	11
53	2.0000E+00	8.6603E-01	1.6667E-01	4	VERT	10005	1
54	2.3660E+00	5.0000E-01	3.3333E-01	4	HORZ	10005	2
55	2.5000E+00	.0000	5.0000E-01	4	BOTH	9006	7
56	2.3660E+00	-5.0000E-01	6.6667E-01	4	HORZ	8005	2
57	2.0000E+00	-8.6603E-01	8.3333E-01	4	VERT	7005	4

13.C DISCRETIZED PDE INTERFACE

Discretization modules construct a system of algebraic equations whose solution provides an approximation to the unknown function u. The following variables must be set by these modules:

I1NEQN = the number of equations generated,

I1NCOE = maximum number of unknowns per equation,

R1COEF = two-dimensional array containing nonzero elements of the matrix generated by the discretization module. The ith row of R1COEF contains the nonzero elements in the ith row of the matrix. The entries need not be stored in order of increasing column number.

I1IDCO = two-dimensional array of column indices corresponding to the nonzero matrix entries stored in the array R1COEF. I1IDCO(i,j) is the column index of the matrix element stored in R1COEF(i,j). That is, R1COEF(i,j) is the $(i,$I1IDCO$(i,j))$th element of the matrix. Each element of I1IDCO [that is, I1IDCO(i,j), $i = 1, ...,$ I1NEQN, $j = 1, ...,$ I1MNCO] must contain a value. If the element R1COEF(i,j) does not contain a matrix entry, then I1IDCO$(i,j) = 0$.

R1BBBB = right side vector. R1BBBB(i) = right side of the ith equation.

L1SYMM = .TRUE. if the matrix is symmetric,
.FALSE. otherwise.

The storage allocated for these arrays are given by

I1MNEQ = row dimension of R1COEF and I1IDCO; length of R1BBBB,

I1MNCO = column dimension of R1COEF and I1IDCO.

The variables I1MNEQ and I1MNCO are set by the ELLPACK control program and should not be changed by modules.

Authors of discretization modules are also encouraged to contribute procedure modules which initialize the vector of unknowns of their linear system given a function which estimates the solution. Further information about such modules is given in Section 13.F.

13.D EQUATION/UNKNOWN INDEXING INTERFACE

Equation/unknown renumbering modules generate a reordering for the equations and unknowns in the linear system produced by the discretization module in order to increase the efficiency of the solution process. Their output is two permutation vectors.

I1ENDX = equation reordering permutation vector. The ith row in the permuted matrix is stored in row I1ENDX(i) of R1COEF.

I1UNDX = unknown reordering permutation vector. The column index of the element R1COEF(i,j) in the permuted matrix is I1UNDX(I1IDCO(i,j)). Thus the kth unknown in the original system occupies the I1UNDX(k)th position in the reordered system.

If $A(i,j)$ is the original coefficient matrix, then, after the indexing, the reordered coefficient matrix is $B(i,j) = A(\text{I1ENDX}(i), \text{I1UNDX}^{-1}(j))$, where I1UNDX^{-1} is the inverse of the permutation vector I1UNDX.

The storage allocated for these arrays is given by the variables

I1MEND = size of the array I1ENDX
I1MUND = size of the array I1UNDX.

The variables I1MEND and I1MUND are set by the ELLPACK control program and should not be changed by modules.

It is sometimes necessary for modules to use the inverses of the permutations given here. Constructing the inverse, IP, of the permutation P can be done as follows.

```
      DO 10 K=1,I1NEQN
         IP(P(K)) = K
   10 CONTINUE
```

Modules requiring an inverse permutation should construct it in reusable workspace (see Section 14.C).

It is often useful to compose indexing modules, that is, apply two indexing modules in sequence with the second working on the matrix reordered by the first. This is fairly easy to do in most cases, and authors of indexing modules are encouraged to allow for it whenever possible. In order to effect the composition, instead of saying the ith equation gets mapped to the kth equation (I1ENDX(i) = k), one should say the equation where the ith equation came from gets mapped to the kth equation. This is represented by the assignment I1ENDX(INVERS(i)) = k, where INVERS is the inverse of the previously existing permutation I1ENDX. The reordering of the unknowns works similarly. The ELLPACK preprocessor automatically sets the arrays I1ENDX and I1UNDX to the identity permutation following each call on a discretization module to initialize the indexing arrays.

13.E ALGEBRAIC EQUATION SOLUTION INTERFACE

Solution modules solve the linear system of equations created by the discretization module and reordered by the permutations I1ENDX and I1UNDX. [The sparse matrix format used by ELLPACK for the discretized PDE interface may be inappropriate for some solution algorithms (e.g., band solvers), and hence an initial storage reformatting phase may be required.] Solution modules have only one array of output:

R1UNKN = vector of length I1NEQN containing the solution to the linear algebraic system.

The storage allocated to R1UNKN is given by

I1MUNK = size of the array R1UNKN.

This variable is set by the ELLPACK control program and should not be changed by modules.

Solution modules must initially use the ordering for equations and unknowns given by the permutation vectors I1ENDX and I1UNDX respectively. (The ordering may subsequently be changed by operations such as pivoting). On output the array R1UNKN must be in the order specified by I1UNDX. That is, R1UNKN(I1UNDX(j)) is the value of the discretization module's jth unknown. After the solution module is invoked the ELLPACK control program unscrambles the unknowns so that R1UNKN contains the unknowns in the order originally generated by the discretization module.

Several logical variables are available which solution modules may be used to determine properties of the indexing which has been selected:

L1ASIS = .TRUE. if I1ENDX(i) = I1UNDX(i) = i, i = 1, ..., I1NEQN,
.FALSE. otherwise,

L1RDBL = .TRUE. if the indexing is of red-black type,
.FALSE. otherwise.

After a discretization module is called the control program automatically sets L1ASIS=.TRUE. and L1RDBL=.FALSE.. The indexing vectors I1ENDX and I1UNDX are also set to the identity permutation at that time. Whenever an indexing module is called, the control program sets L1ASIS = .FALSE.. The indexing module AS IS sets L1ASIS = .TRUE., while the indexing module RED BLACK sets L1RDBL = .TRUE..

An additional logical variable is available to signal whether the solution vector R1UNKN has been set to an initial value in preparation for a module implementing an iterative algorithm.

L1UINI	= .TRUE.	if R1UNKN has been given an initial value prior to the execution of the solution module,
	.FALSE.	otherwise.

This variable is always set to .FALSE. by the ELLPACK control program immediately after the execution of a discretization module (the array R1UNKN is not disturbed, however). The initial value may be specified by ELLPACK users by way of various procedure modules which have been contributed for this purpose. When a nonlinear or time-dependent problem is being solved then this parameter can be reset directly by users to indicate that the value remaining from the previous solution should be used. The ordering of R1UNKN is initially the same as used by the discretization module, i.e., the indexing vector I1UNDX has not been applied.

13.F TRIPLE AND PROCEDURE MODULES

Triple modules combine the operations of discretization, indexing, and solution. The output of these modules are the scalar variables I1NEQN and I1NCOE (see Section 13.C) and the vector R1UNKN (see Section 12.E). The variable I1NCOE should be set to zero. These modules must also provide a subprogram to evaluate the solution. (see Section 13.G).

PROCEDURE modules perform auxiliary computations which do not fit precisely into the categories of DISCRETIZATION, INDEXING, SOLUTION, or TRIPLE. Examples of such operations include the computation of eigenvalues of the discretization matrix or a display of the nonzero structure of the matrix. There are no predefined input or output interfaces for procedures. They may read input from any interface. The module documentation should be very precise about the circumstances wherein the module is applicable, and note changes made to ELLPACK interface variables, if any.

One particularly useful type of PROCEDURE module is that which initializes the vector R1UNKN for use by an iterative solution algorithm. Such modules are usually contributed by authors of DISCRETIZATION modules, since the meaning of the elements of the solution vector is a function of the particular discretization. In this case the module is given the name "SET UNKNOWNS FOR module-name." Typically, such a module requires that the user specify a function (a "module parameter" — see Section 14.B) which estimates the solution in the domain. Such modules must set the interface variable L1UINI to .TRUE. to alert SOLUTION modules that R1UNKN has been initialized. The ordering of unknowns in R1UNKN is the same as that of the DISCRETIZATION module.

13.G SOLUTION EVALUATION

In ELLPACK the solution to the PDE problem is a function which, along with its first and second order partial derivatives, can be evaluated anywhere in the domain. The output of the numerical method, however, is simply the vector

R1UNKN, and only the discretization module knows how to map this information into a solution of the PDE problem. Thus, a contributor of a DISCRETIZATION or TRIPLE module must provide a Fortran function for evaluating the computed solution and its derivatives anywhere in the domain. This function takes the form

R0SOLV = R*k*xxEV(X,Y,R1UNKN,IDERIV)

or

R0SOLV = R*k*xxEV(X,Y,Z,R1UNKN,IDERIV)

This function returns the value of the computed solution or one of its derivatives at an arbitrary point (X,Y) or (X,Y,Z) in the domain. The name of the function must follow the conventions of Section 12.C (k is a digit that depends on the type of module, and the characters xx are assigned by the ELLPACK coordinator). IDERIV uses the ordering of Section 12.C to specify the function whose value is to be returned. Thus, for example, IDERIV=1 selects UXX and IDERIV=6 selects U. An ELLPACK variable may be tested to check whether the solution evaluation module has been called since the last solution or triple module executed. This is useful for solution evaluation modules which must do special processing the first time they are called.

L1NEQD = .TRUE. on the first call of a solution evaluation module after execution of a discretization or triple module,
.FALSE. otherwise.

A function of the form given above is readily produced by finite-element modules. It is much more difficult for finite-difference modules, since they typically only produce estimates of the solution at grid points. ELLPACK provides interpolation utilities to aid modules of this type. To use these programs, discretization or triple modules need only provide a subroutine which loads an array, R1TABL, with values of the solution at all grid points in the domain or on the boundary.

R1TABL = table of solution values at grid points. R1TABL(i,j,k) is the value of the computed solution at (R1GRDX(i), R1GRDY(j), R1GRDZ(k)), $i = 1, ...,$ I1NGRX, $j = 1, ...,$ I1NGRY, $k = 1, ...,$ I1NGRZ. When the domain is nonrectangular a value need only be provided for those (i,j) such that I1GRTY$(i,j)>0$.

ELLPACK interpolation utilities extend this to all points in the domain. The interpolant's derivatives are used for UX, UY, UXX, etc.

ELLPACK users select the type of interpolation used. This is either local quadratic interpolation (the default, see Chapter 6) or tensor products of one-dimensional B-splines. The spline interpolation procedure requires substantially more computation than local quadratic interpolation, but is much more accurate when solutions resulting from high-order-accurate finite-difference discretizations must be evaluated at points not on the grid. The order of accuracy of the spline interpolant is selected by the module contributor by setting the variable I1KORD.

I1KORD = order to be used in spline interpolation, if the user selects it. The resulting interpolant will be a piecewise polynomial of degree I1KORD − 1 in x with I1KORD − 2 continuous derivatives along each line x = constant, and a piecewise polynomial of degree I1KORD − 1 in y with I1KORD − 2 continuous derivatives along each line y = constant. 1 < I1KORD < 20.

It is important that I1KORD be set only just before the end of a discretization or triple module's execution. This is because the PDE coefficients might be a function of the previously computed solution, which, in turn, is a function of the previous value of I1KORD.

Neither interpolation technique applies to arbitrary nonrectangular domains, although the use of local quadratics requires only that the boundary coincide with the grid lines. Thus, in the nonrectangular case a simple low-order extrapolation is used to extend the solution to grid points outside the domain. The result is that the interpolant is less accurate near the boundary of nonrectangular domains. See Chapter 6 for further details.

Contributors should be aware of a potential conflict when their modules are used to solve quasilinear problems by iteration. Suppose the coefficient of UX in the PDE is specified as U(X,Y). Suppose further that the DISCRETIZATION module has a private common block in which it stores information to help it compute U when requested to do so by an OUTPUT module. When the DISCRETIZATION module is recalled, it will store new information in this common block, thus destroying data required for the computation of the PDE coefficients using the previously computed solution. Similar problems can occur when users attempt the solution of time-dependent problems as in Chapter 5. There is no trouble in most cases, since data passed to the solution evaluation module is typically dependant only upon the domain and the grid, which is not changed in these iterations. Modules with potential conflicts should describe them in their documentation.

13.H CONTROL VARIABLES

A number of other variables are used to control various actions of contributed modules. It is crucial that modules be sensitive to these if the system is to behave coherently. Variables which control the transfer of input and output include:

I1INPT = logical unit number for standard input [same as I1MACH(1)],

I1OUTP = logical unit number for standard output [same as I1MACH(2)],

I1SCRA = logical unit number for a sequential scratch file.

Contributors are discouraged from using Fortran READs to obtain auxiliary input from users. There are several alternative methods to get such information from the user to individual modules. The use of module parameters (described in Section 14.B) is the prime example. Note also that since Fortran code can be embedded in ELLPACK programs, users can declare arrays and assign them values for passing to modules.

The amount of printed output is controlled by the variable I1LEVL. This is set directly by users through the LEVEL phrase in the OPTIONS statement:

I1LEVL = level of printed output on unit I1OUTP. Possible values are

- 0 → no printed output,
- 1 → minimal printed output (default case),
- 2 → more detailed output,
- 3, 4, 5 → debug modes (detail increases with I1LEVL).

Users select the type of pagination to be used on the standard output file in the PAGE phrase of the OPTIONS statement. The user's selection is stored in the variable I1PAGE, which is primarily used by the control program to determine when to put page ejects between the output of two different modules:

I1PAGE = type of pagination to use on the standard output unit, I1OUTP. Possible values are

- 0 → no page ejects,
- 1 → page eject before DISCRETIZATION, TRIPLE, TABLE, or SUMMARY (default),
- 2 → page eject before each module and OUTPUT item.

More detailed information about the printed output used by ELLPACK modules is given in Chapter 15.

Two other important global variables are

L1TIME = .TRUE. if the user has asked for module timings,
.FALSE. otherwise,

L1FATL = .TRUE. if a module has detected a fatal error and requests termination of the run,
.FALSE. otherwise.

These two variables are usually of concern only to the ELLPACK control program. L1FATL is usually set only by the ELLPACK utility program Q1ERRT, which is called by modules to request error termination (see Section 15.B). L1TIME is set by the user through the TIME phrase of the OPTIONS statement. When module timing has been selected the ELLPACK control program times each module execution and prints a summary at the end of the run. Some modules may also print the timing of individual portions of their code if this is of interest, and there are ELLPACK utilities to assist modules to do this in a transportable way (see Section 15.E). However, this should only be done if the user has requested such timings (i.e., L1TIME=.TRUE.).

Finally, two real-valued machine-dependent constants are available as standard ELLPACK variables:

PI = the mathematical constant π,

R1EPSM = the machine epsilon. This is the largest real number ϵ such that $1+\epsilon = 1$ on this machine. (Same as R1MACH(3)).

Several other machine dependent constants are returned by the standard ELLPACK functions I1MACH and R1MACH (see Section 15.E).

Table 13.1. ELLPACK Module Interfaces. The following variables and subprograms are available for use by all ELLPACK modules. Only authorized modules may change their values. Name is the name of the variable used in the ELLPACK control program. Common is the name of the COMMON block where this variable is stored. Dimension gives the constants or variables which must be used to dimension arrays (an entry of 1 denotes a scalar). The last column gives conditions that must be true in order for the variable to be defined. Table 13.2 gives a brief description of each variable.

PROBLEM DEFINITION INTERFACE

Name	*Common*	*Type*	*Dimension*	*Restrictions*
L1ARCC	C1LVPR	logical	1	.NOT.L1RECT
Q1BCOE		subroutine		
I1BCST	C1BCST	integer	4,I1NBND	
I1BCTY	C1BCTY	integer	I1NBND	
Q1BDRY		subroutine		.NOT.L1RECT
R1BRHS		real function		
R1BRNG	C1BRNG	real	2,I1NBND	.NOT.L1RECT
R1CCCU	C1RVPR	real	1	L1CSTC
R1CCUX	C1RVPR	real	1	L1CSTC
R1CCUY	C1RVPR	real	1	L1CSTC
R1CCUZ	C1RVTR	real	1	L1CCTC
I1CFST	C1CFST	integer	10	
L1CLKW	C1LVPR	logical	1	.NOT.L1RECT
L1CRST	C1LVPR	logical	1	
L1CSTB	C1LVPR	logical	1	
L1CSTC	C1LVPR	logical	1	
R1CUXX	C1RVPR	real	1	L1CSTC
R1CUXY	C1RVPR	real	1	L1CSTC
R1CUXZ	C1RVPR	real	1	L1CSTC
R1CUYY	C1RVPR	real	1	L1CSTC
R1CUYZ	C1RVPR	real	1	L1CSTC
R1CUZZ	C1RVPR	real	1	L1CSTC
L1DRCH	C1LVPR	logical	1	
L1HMBC	C1LVPR	logical	1	
L1HMEQ	C1LVPR	logical	1	
L1HOLE	C1LVPR	logical	1	.NOT.L1RECT
L1LAPL	C1LVPR	logical	1	
L1MIXD	C1LVPR	logical	1	
I1NBND	C1IVPR	integer	1	
L1NEUM	C1LVPR	logical	1	
L1NUNQ	C1LVPR	logical	1	
Q1PCOE		subroutine		
L1POIS	C1LVPR	logical	1	
L1PRDC	C1LVPR	logical	1	
L1PRDX	C1LVPR	logical	1	
L1PRDY	C1LVPR	logical	1	
L1PRDZ	C1LVPR	logical	1	
R1PRHS		real function		
L1RECT	C1LVPR	logical	1	
L1SELF	C1LVPR	logical	1	
L1TWOD	C1LVPR	logical	1	

PROBLEM DEFINITION INTERFACE

Name	Common	Type	Dimension	Restrictions
R1UNQU	C1RVPR	real	1	L1NUNQ
R1UNQX	C1RVPR	real	1	L1NUNQ
R1UNQY	C1RVPR	real	1	L1NUNQ
R1UNQZ	C1RVPR	real	1	L1NUNQ

DISCRETE DOMAIN INTERFACE

Name	Common	Type	Dimension	Restrictions
R1AXGR	C1RVGR	real	1	
R1AYGR	C1RVGR	real	1	
R1AZGR	C1RVGR	real	1	
I1BGRD	C1BGRD	integer	I1MBPT	.NOT.L1RECT
I1BNGH	C1BNGH	integer	I1MBPT	.NOT.L1RECT
R1BPAR	C1BPAR	real	I1MBPT	.NOT.L1RECT
I1BPTY	C1BPTY	integer	I1MBPT	.NOT.L1RECT
R1BXGR	C1RVGR	real	1	
R1BYGR	C1RVGR	real	1	
R1BZGR	C1RVGR	real	1	
R1GRDX	C1GRDX	real	I1NGRX	
R1GRDY	C1GRDY	real	I1NGRY	
R1GRDZ	C1GRDZ	real	I1NGRZ	
I1GRTY	C1GRTY	integer	I1NGRX,I1NGRY	.NOT.L1RECT
R1HXGR	C1RVGR	real	1	
R1HYGR	C1RVGR	real	1	
R1HZGR	C1RVGR	real	1	
I1MBPT	C1IVGR	integer	1	.NOT.L1RECT
I1NBPT	C1IVGR	integer	1	.NOT.L1RECT
I1NGRX	C1IVGR	integer	1	
I1NGRY	C1IVGR	integer	1	
I1NGRZ	C1IVGR	integer	1	
I1PACK	C1LVGR	logical	1	.NOT.L1RECT
I1PECE	C1PECE	integer	I1MBPT	.NOT.L1RECT
L1UNFG	C1LVGR	logical	1	
L1UNFX	C1LVGR	logical	1	
L1UNFY	C1LVGR	logical	1	
L1UNFZ	C1LVGR	logical	1	
R1XBND	C1XBND	real	I1MBPT	.NOT.L1RECT
R1YBND	C1YBND	real	I1MBPT	.NOT.L1RECT

DISCRETE OPERATOR INTERFACE

Name	Common	Type	Dimension	Restrictions
R1BBBB	C1BBBB	real	I1MNEQ	
R1COEF	C1COEF	real	I1MNEQ,I1MNCO	
I1IDCO	C1IDCO	integer	I1MNEQ,I1MNCO	
I1MNEQ	C1IVDI	integer	1	
I1MNCO	C1IVDI	integer	1	
I1NCOE	C1IVDI	integer	1	
I1NEQN	C1IVDI	integer	1	
L1SYMM	C1LVDI	logical	1	

EQUATION/ UNKNOWN INDEXNG INTERFACE				
Name	*Common*	*Type*	*Dimension*	*Restrictions*
L1ASIS	C1LVIN	logical	1	
I1ENDX	C1ENDX	integer	I1MEND	
I1MEND	C1IVIN	integer	1	
I1MUND	C1IVIN	integer	1	
L1RDBL	C1LVIN	logival	1	
I1UNDX	C1UNDX	integer	I1MUND	

ALGEBRAIC EQUATION SOLUTION INTERFACE				
Name	*Common*	*Type*	*Dimension*	*Restrictions*
I1MUNK	C1IVSO	integer	1	
L1UINI	C1LVSO	logical	1	
R1UNKN	C1UNKN	real	I1MUNK	

OTHER GLOBAL VARIABLES				
Name	*Common*	*Type*	*Dimension*	*Restrictions*
R1EPSG	C1RVGL	real	1	
R1EPSM	C1RVGL	real	1	
L1FATL	C1LVCN	logical	1	
I1KORD	C1IVCN	integer	1	
I1KWRK	C1IVCN	integer	1	
I1INPT	C1IVCN	integer	1	
I1LEVL	C1IVCN	integer	1	
L1NEWD	C1LVCN	logical	1	
R1NRM1	C1RNRM	real	1	
R1NRM2	C1RNRM	real	1	
R1NRMI	C1RNRM	real	1	
I1OUTP	C1IVCN	integer	1	
I1PAGE	C1IVCN	integer	1	
I1SCRA	C1IVCN	integer	1	
R1TABL	C1TABL	real	I1NGRX,I1NGRY,I1NGRZ	
L1TIME	C1LVCN	logical	1	
PI	C1RVGL	real	1	
R1WORK	blank	real	I1KWRK	

Table 13.2. Summary of Ellpack Objects All ELLPACK variables of interest to module contributors are given below in alphabetical order along with a brief description. Column (a) gives the name of the ELLPACK variable, and column (b) is the interface name. Interface names are PR (problem definition), GR (discrete domain), DI (operator discretization), IN (equation/unknown indexing), SO (algebraic equation solution), and GL (other global variables).

(*a*)	(*b*)	*Description*
L1ARCC	PR	Switch for domain with arc
L1ASIS	IN	Switch for as-is ordering
R1AXGR	GR	Initial *x* grid point
R1AYGR	GR	Initial *y* grid point
R1AZGR	GR	Initial *z* grid point
R1BBBB	DI	Matrix right side vector
Q1BCOE	PR	Subroutine for boundary conditions
I1BCST	PR	Boundary condition coefficient status vector
I1BCTY	PR	Boundary condition type vector
Q1BDRY	PR	Subroutine which defines the boundary
I1BGRD	GR	Boundary point grid square vector
R1BNGH	GR	Boundary point neighbors vector
R1BPAR	GR	Boundary point parameter value vector
I1BPTY	GR	Boundary point type vector
R1BRHS	PR	Function for right side of boundary conditions
R1BRNG	PR	Parameter range array for boundary pieces
R1BXGR	GR	Final *x* grid point
R1BYGR	GR	Final *y* grid point
R1BZGR	GR	Final *z* grid point
R1CCCU	PR	Constant coefficient of U
R1CCUX	PR	Constant coefficient of UX
R1CCUY	PR	Constant coefficient of UY
R1CCUZ	PR	Constant coefficient of UZ
I1CFST	PR	PDE coefficient status vector
L1CLKW	PR	Switch for clockwise boundary definition
R1COEF	DI	Array of nonzero matrix elements
L1CRST	PR	Switch for cross term (UXY, UXZ, UYZ)
L1CSTB	PR	Switch for constant coefficient boundary condition
L1CSTC	PR	Switch for constant coefficient PDE
R1CUXX	PR	Constant coefficient of UXX
R1CUXY	PR	Constant coefficient of UXY
R1CUXZ	PR	Constant coefficient of UXZ
R1CUYY	PR	Constant coefficient of UYY
R1CUYZ	PR	Constant coefficient of UYZ
R1CUZZ	PR	Constant coefficient of UZZ
L1DRCH	PR	Switch for Dirichlet boundary conditions
I1ENDX	IN	Equation reordering permutation vector
R1EPSG	GL	Grid uncertainty tolerance
R1EPSM	GL	Machine epsilon. Same as R1MACH(3).
L1FATL	GL	Fatal error switch
R1GRDX	GR	Grid point vector in *x*
R1GRDY	GR	Grid point vector in *y*
R1GRDZ	GR	Grid point vector in *z*
I1GRTY	GR	Array of grid point types

(*a*)	(*b*)	*Description*
L1HMBC	PR	Switch for homogeneous boundary condition
L1HMEQ	PR	Switch for homogeneous PDE
L1HOLE	PR	Switch for domain with hole
R1HXGR	GR	Average grid spacing in x
R1HYGR	GR	Average grid spacing in y
R1HZGR	GR	Average grid spacing in z
I1IDCO	DI	Column indices for matrix elements in R1COEF
I1INPT	GL	Unit number for standard input. Same as I1MACH(1).
I1KWRK	GL	Storage allocated to (blank) common workspace
L1LAPL	PR	Switch for Laplace's equation
I1LEVL	GL	Level of printed output
I1MBPT	GR	Storage allocated per boundary point vector
I1MEND	IN	Storage allocated for vector I1ENDX
L1MIXD	PR	Switch for .NOT.L1DRCH .AND. .NOT.L1NEUM
I1MNCO	DI	Column dimension of R1COEF, I1IDCO
I1MNEQ	DI	Row dimension of R1COEF, I1IDCO
I1MUND	IN	Storage allocated for vector I1UNDX
I1MUNK	SO	Storage allocated for vector R1UNKN
I1NBND	PR	Number of boundary pieces
I1NBPT	GR	Number of boundary points
I1NCOE	DI	Maximum number of nonzeros per generated equation
I1NEQN	DI	Number of equations generated
L1NEUM	PR	Switch for normal derivative boundary conditions
L1NEWD	GL	Switch for new discretization
I1NGRX	GR	Number of grid points in x
I1NGRY	GR	Number of grid points in y
I1NGRZ	GR	Number of grid points in z
L1NUNQ	PR	Switch for nonunique solution
I1OUTP	GL	Unit number for standard output. Same as I1MACH(2).
I1PACK	GR	Packing factor for domain processor data
Q1PCOE	PR	Subroutine for PDE coefficients
I1PECE	GR	Boundary point boundary piece vector
PI	GL	The mathematical constant π
L1POIS	PR	Switch for Poisson's equation
L1PRDC	PR	Switch for all periodic boundary conditions
L1PRDX	PR	Switch for periodic boundary conditions x
L1PRDY	PR	Switch for periodic boundary conditions y
L1PRDZ	PR	Switch for periodic boundary conditions z
R1PRHS	PR	Function for right-hand side of the PDE
L1RDBL	IN	Switch for red-black ordering
L1RECT	PR	Switch for rectangular domain
I1SCRA	GL	Logical unit number for scratch storage
L1SELF	PR	Switch for PDE written in self-adjoint form
L1SYMM	DI	Switch for Symmetric matrix
R1TABL	GL	Table of solution on grid for interpolation
L1TIME	GL	Switch for module timing
L1TWOD	PR	Switch for two-dimensional domain
L1UINI	SO	Switch for initialized solution vector
I1UNDX	IN	Unknowns reordering permutation vector
L1UNFG	GR	Switch for uniform grid
L1UNFX	GR	Switch for uniform x grid

(*a*)	(*b*)	*Description*
L1UNFY	GR	Switch for uniform y grid
L1UNFZ	GR	Switch for uniform z grid
R1UNKN	SO	Vector for solution of matrix problem
R1UNQU	PR	Known value of u to fix nonunique solutions
R1UNQX	PR	Coordinate in x at which R1UNQU applies
R1UNQY	PR	Coordinate in y at which R1UNQU applies
R1UNQZ	PR	Coordinate in z at which R1UNQU applies
R1WORK	GL	Common workspace (blank common)
R1XBND	GR	Vector of boundary point x coordinates
R1YBND	GR	Vector of boundary point y coordinates

Chapter 14

MODULE INTERFACE ACCESS

14.A STORAGE SCHEME FOR INTERFACE VARIABLES

There are two ways that modules may gain access to interface variables: through the module's calling sequence or through COMMON blocks. Each variable is available by each method; however, module contributors are urged to access interface variables through COMMON blocks because this shortens the symbol table required by the ELLPACK preprocessor (see Section 16.A). Since the former method is straightforward we only describe the arrangement of interface variables in COMMON blocks.

Scalars are arranged according to their interface and type. Thus, for example, all logical variables in the problem definition interface are collected together. This arrangement is essential for portability. The naming convention for these COMMON blocks is

C1aaxx

where:

aa = variable type. Possible values are
- RV = real variables,
- IV = integer variables,
- LV = logical variables.

xx = interface indicator. Possible values are
- PR = problem definition,
- GR = discrete domain,
- DI = discrete operator,
- IN = equation/unknown indexing,
- SO = linear algebraic equation solution,
- CN = control variables
- GL = other global variables.

The lengths of array interface variables depend upon problem size, so each array is placed in a separate COMMON block to assure proper alignment. The form of these COMMON block names is simply

C1xxxx

where xxxx is the last four characters in the array name.

The exact declaration of the COMMON blocks used by the ELLPACK control program is given below. Arrays are dimensioned using variables of the form \$I1xxxx. This denotes that the ELLPACK variable I1xxxx contains the constant which the preprocessor will write in place of the character string "\$I1xxxx" when it generates the ELLPACK control program.

```
C
C===========    PROBLEM DEFINITION INTERFACE
C
      LOGICAL                L1ARCC,L1CLKW,L1CRST,L1CSTB,L1CSTC,
     A                       L1DRCH,L1HMBC,L1HMEQ,L1HOLE,L1LAPL,
     B                       L1MIXD,L1NEUM,L1NUNQ,L1POIS,L1PRDC,
     C                       L1PRDX,L1PRDY,L1PRDZ,L1RECT,L1SELF,
     D                       L1TWOD
      COMMON / C1LVPR /  L1ARCC,L1CLKW,L1CRST,L1CSTB,L1CSTC,
     A                       L1DRCH,L1HMBC,L1HMEQ,L1HOLE,L1LAPL,
     B                       L1MIXD,L1NEUM,L1NUNQ,L1POIS,L1PRDC,
     C                       L1PRDX,L1PRDY,L1PRDZ,L1RECT,L1SELF,
     D                       L1TWOD
      COMMON / C1IVPR /  I1NBND
      COMMON / C1RVPR /  R1CUXX,R1CUXY,R1CUYY,R1CCUX,R1CCUY,
     A                       R1CCCU,R1CUZZ,R1CUXZ,R1CUYZ,R1CCUZ,
     B                       R1UNQX,R1UNQY,R1UNQZ,R1UNQU
C
      COMMON / C1BCST /  I1BCST(4,$I1NBND)
      COMMON / C1BCTY /  I1BCTY($I1NBND)
      COMMON / C1CFST /  I1CFST(10)
C
C     THE FOLLOWING ARE ACTIVE ONLY WHEN L1RECT=.FALSE.
C
      EXTERNAL               Q1BDRY
      COMMON / C1BRNG /  R1BRNG(2,$I1NBND)
C
C===========    DISCRETE DOMAIN INTERFACE
C
      LOGICAL                L1UNFG,L1UNFX,L1UNFY,L1UNFZ
      COMMON / C1LVGR /  L1UNFG,L1UNFX,L1UNFY,L1UNFZ
      COMMON / C1RVGR /  I1NGRX,I1NGRY,I1NGRZ,I1NBPT,I1MBPT,
     A                       I1PACK
      COMMON / C1IVGR /  R1AXGR,R1AYGR,R1AZGR,R1BXGR,R1BYGR,
     A                       R1BZGR,R1HXGR,R1HYGR,R1HZGR
C
      COMMON / C1GRDX /  R1GRDX($I1NGRX)
      COMMON / C1GRDY /  R1GRDY($I1NGRY)
      COMMON / C1GRDZ /  R1GRDZ($I1NGRZ)
C
C     THE FOLLOWING ARE ACTIVE ONLY WHEN L1RECT=.FALSE.
C
      COMMON / C1GRTY /  I1GRTY($I1NGRX,$I1NGRY)
      COMMON / C1BNGH /  I1BNGH($I1MBPT)
      COMMON / C1BGRD /  I1BGRD($I1MBPT)
      COMMON / C1BPTY /  I1BPTY($I1MBPT)
      COMMON / C1PECE /  I1PECE($I1MBPT)
      COMMON / C1BPAR /  R1BPAR($I1MBPT)
      COMMON / C1XBND /  R1XBND($I1MBPT)
      COMMON / C1YBND /  R1YBND($I1MBPT)
C
C===========    DISCRETE OPERATOR INTERFACE
C
      LOGICAL                L1SYMM
      COMMON / C1LVDI /  L1SYMM
      COMMON / C1IVDI /  I1NEQN,I1MNEQ,I1NCOE,I1MNCO
C
```

```
      COMMON / C1COEF / R1COEF($I1MNEQ,$I1MNCO)
      COMMON / C1IDCO / I1IDCO($I1MNEQ,$I1MNCO)
      COMMON / C1BBBB / R1BBBB($I1MNEQ)
C
C===========      EQUATION/UNKNOWN INDEXING INTERFACE
C
      LOGICAL                  L1ASIS,L1RDBL
      COMMON / C1LVIN / L1ASIS,L1RDBL
      COMMON / C1IVIN / I1MEND,I1MUND
C
      COMMON / C1ENDX / I1ENDX($I1MEND)
      COMMON / C1UNDX / I1UNDX($I1MUND)
C
C===========      ALGEBRAIC EQUATION SOLUTION INTERFACE
C
      LOGICAL                  L1UINI
      COMMON / C1LVSO / L1UINI
      COMMON / C1IVSO / I1MUNK
C
      COMMON / C1UNKN / R1UNKN($I1MUNK)
C
C===========      OTHER GLOBAL VARIABLES
C
      LOGICAL                  L1TIME,L1FATL,L1NEWD
      COMMON / C1LVCN / L1TIME,L1FATL,L1NEWD
      COMMON / C1IVCN / I1LEVL,I1INPT,I1OUTP,I1SCRA
     A                         I1KWRK,I1KORD
      COMMON / C1RVGL / R1EPSG,R1EPSM,PI
      COMMON / C1RNRM / R1NRM1,R1NRM2,R1NRMI
C
      COMMON / C1TABL / R1TABL($I1NGRX,$INGRY,$INGRZ)
      COMMON                   R1WORK($IKWRK)
```

These COMMON blocks may be used within modules as needed. It is appropriate to mark the collection used by comments (e.g. ELLPACK INTERFACE).

14.B MODULE-PARAMETER ACCESS

ELLPACK modules may obtain additional input directly from users through the use of module parameters. For example, the solution module SOR needs to know the maximum number of iterations that it should attempt in solving the linear system of equations. It also accepts a value for the overrelaxation parameter OMEGA. This information is not part of the interface for solution modules; instead ELLPACK users specify these quantities in the module invocation as

```
SOLUTION.      SOR (ITMAX=175,OMEGA=1.6)
```

The ELLPACK preprocessor inserts assignment statements for all module parameters before the module CALL statement, and thus one way to obtain the parameter value is to include it as an argument in the module's subroutine call. The preprocessor can also write any declaration statements required for the invocation of a module, and hence parameters can also be passed through labelled COMMON. All module parameters must have default values (for SOR, ITMAX=100 is the default and OMEGA is chosen adaptively). In addition, names of Fortran function subprograms may also be passed as arguments. (The preprocessor will automatically generate the necessary Fortran EXTERNAL statements in this case.) Chapter 16 provides details on how to provide information about module parameters to the ELLPACK preprocessor.

14.C WORKSPACE ACESS

ELLPACK modules have access to two types of workspace: reusable and private. The reusable workspace is blank COMMON and is called R1WORK in the ELLPACK control program. This workspace is used by all modules and hence can only be used for temporary storage. The preprocessor allocates enough space to satisfy the largest workspace requirement for the set of modules invoked by the user's ELLPACK program. The ELLPACK variable I1KWRK gives the number of words that has been allocated to the array R1WORK.

Caution: ELLPACK modules must check that the size of reusable workspace is large enough to fulfill their requirements. This is especially important in light of the fact that ELLPACK users may override the estimates of workspace size specified by module authors (the estimate may be too large or too small in some cases).

Occasionally it is necessary for modules to store information for later use in arrays whose lengths are problem-dependent, and the preprocessor can allocate such private COMMON blocks for modules. Operator discretization modules are typical examples. They must provide routines to evaluate the solution after the linear system is solved, and hence they may need to retain some information about how the solution vector relates to the domain. This information cannot be put in blank COMMON because intervening modules might destroy it. It is important that modules use blank COMMON whenever possible, since this is an important means of reducing the overall storage requirements of ELLPACK runs. Information on how to specify workspace requirements to the preprocessor is given in Chapter 16.

Finally, one sequential file is also available for scratch storage by ELLPACK modules. The unit number of this file is given by the interface variable I1SCRA.

Chapter 15

PROGRAMMING STANDARDS

One of the goals of ELLPACK is to provide a substantial collection of software parts for elliptic PDEs that is easy to use and understand. The rigid interface definitions allow the easy combination of software parts, but several other considerations are also important if this software is to be truly usable by others.

1. The printed output should be intelligible and easy to relate to the user's ELLPACK program.

2. The system should be easy to implement on a wide range of computers.

3. The system should be easy to maintain.

These considerations have led to the establishment of a number of conventions or standards for printed output, error handling, portability, and documentation which must also be respected by module authors. These conventions insure that individual ELLPACK modules present themselves in a coherent manner to users, thus enhancing the integrity and friendliness of the system as a whole.

15.A PRINTED OUTPUT

It is important that the form of output of each module appear consistent to the user. To accomplish this, contributors are asked to adhere to the following rules.

1. Use the problem definition interface variable I1OUTP as the logical unit number for all printed output.

2. Print no more than 80 characters per line. (This makes output easier to read on most computer terminals.)

3. Be sensitive to the users output level request (the variable I1LEVL), especially the requests for I1LEVL = 0 and I1LEVL = 1.

Specific guidelines for printed output at the different output levels follow.

I1LEVEL = 0 — No printed output except fatal error messages (see Section 15.B).

I1LEVEL = 1 — Print a heading indicating that the module has started execution. After a successful execution, print a short summary of the output (e.g., number of equations generated for a discretization module, or bandwidth for a banded linear system solver). Finally, print a message saying that the execution was successful. The standard form for each of these items is illustrated in Figure 15.1. Note that the module should skip three lines before the first line printed and skip no lines after the last. A utility subprogram, Q1BANR, is available to print the initial heading describing the module type (i.e., DISCRETIZATION, INDEXING, SOLUTION, TRIPLE, OUTPUT). See Section 15.E.

I1LEVEL = 2 — Provide all I1LEVL = 1 output, plus a more detailed listing of module parameters, assumed operating conditions, and output quantities. This simply gives the user a more complete record of the computation that was performed.

I1LEVEL = 3, 4, 5 — Provide all I1LEVL = 2 output, increasing the detail as I1LEVL increases. This allows the user to request output for purposes of analysis or debugging. For example, I1LEVL = 3 could yield a listing of all generated interface variables, I1LEVL = 4 an execution trace, and I1LEVL = 5 a detailed execution trace with listing of important intermediate quantities. I1LEVL = 3 should make a discretization module produce a detailed listing of the equations, the unknowns, and their numbering.

Finally, contributors are encouraged to use the 1P format specification when using E format printing of real numbers to insure that at least one significant digit is printed before the decimal point. In particular, the format 1PE15.8 is suggested. The use of this convention adds further consistency to module output.

These guidelines are clearly not appropriate for every module. Although only the items at I1LEVL = 0 and I1LEVL = 1 are mandatory, contributors are strongly encouraged to provide users with the means to debug ELLPACK programs suggested here.

15.B ERROR HANDLING

No ELLPACK module should begin processing a problem that it is obviously not designed to solve. Due to the large variety of module characteristics, the ELLPACK preprocessor does not verify whether each module has been presented with legal input. Thus, each module must exhaustively check the validity of its initial interface. This is usually easy to do. For example, such questions as "Is the mesh fine enough?", "Is the operator written in self-adjoint form?", "Are the boundary conditions homogeneous?", and "Is the matrix symmetric?" can all be

```
-------------------------
 DISCRETIZATION MODULE
-------------------------

    5 - P O I N T   S T A R

    DOMAIN                                RECTANGLE
    DISCRETIZATION                          UNIFORM
    NUMBER OF EQUATIONS                          99
    MAX NUMBER OF UNKNOWNS PER EQ.                5
    MATRIX IS                             SYMMETRIC

    EXECUTION SUCCESSFUL

-----------------
INDEXING MODULE
-----------------

    R E V E R S E   C U T H I L L   M C K E E

    EXECUTION SUCCESSFUL

-----------------
 SOLUTION MODULE
-----------------

    L I N P A C K   S P D   B A N D

    NUMBER OF EQUATIONS          99
    BANDWIDTH                     9
    REQUIRED WORKSPACE         1435
    EXECUTION SUCCESSFUL
```

Figure 15.1. Sample format of I1LEVL = 1 module output.

answered by simple checks on interface variables. Contributors should use caution when using logical variables from the problem definition interface to make such tests, however. For example, modules that only discretize self-adjoint equations should not immediately initiate a fatal error when they discover that L1SELF = .FALSE. The problem may be trivially self-adjoint, i.e., L1CSTC = .TRUE. and I1CFST(2) = I1CFST(4) = I1CFST(5) = 0. In this case L1SELF = .FALSE. merely signifies that the user wrote Poisson's equation as UXX + UYY = F(X,Y) instead of (UX)X + (UY)Y = F(X,Y). Other types of fatal errors might not be practical to detect until processing has begun. For example, a routine for computing a Cholesky decomposition may discover that the matrix is not positive definite or that working storage is not large enough.

When a fatal error has occurred, processing cannot reasonably continue. This must be reported to the user and ELLPACK provides a standard mechanism to do this. Modules should use the ELLPACK utility programs Q1ERRH and Q1ERRT ("error header" and "error trailer" respectively) as follows:

```
CALL Q1ERRH(nH<module-name>,n)
     Write out an error message (single-spaced)
CALL Q1ERRT
RETURN
```

where n is the number of characters in "module-name". This results in a message of the form

```
$$$$$Q1ERRH:
**************************************************
**********    F A T A L    E R R O R    **********
**************************************************
IN ELLPACK MODULE ⟨module-name⟩
--------------------------------------------------
   ... module's error message appears here ...
--------------------------------------------------
$$$$$Q1ERRT:  ELLPACK RUN ABORTED
```

The call to Q1ERRH prints the first six lines; the call to Q1ERRT prints the last two. In some implementations Q1ERRT might stop the program; however, contributors should not rely on this and should return to the ELLPACK control program. Q1ERRT will set the ELLPACK variable L1FATL to .TRUE. to signal that a fatal error has occurred.

In some implementations the second call might also set operating system flags to signal that further processing of the run should stop. These facilities also allow postprocessors to easily assess whether an ELLPACK program has successfully run to completion.

It is very important that error messages be meaningful to the user. For example, the message

```
ERROR 352 IN Q506FS
```

is not nearly as useful as

```
STORAGE OVERFLOW IN PROCESSING EQUATION 52.
INCREASE WORKSPACE AND RESTART.
```

15.C PORTABILITY

ELLPACK is targetted for a very wide class of machines: those with at least 32 bit floating-point arithmetic, "standard" Fortran (1966) compilers, and operating system facilities to support Fortran preprocessor systems. It is the responsibility of module authors to insure that their programs will run on any such system with as few changes as possible. We outline some important considerations in this section.

First, the module must compile successfully. Although ANSI Fortran 77 is the current industry standard, some target machines do not yet fully support it. Thus, module authors are cautioned against the use of Fortran 77 constructs that are not already available as extensions on most Fortran 66 compilers. Several specific suggestions are

1. Do not rely on the fact that DO loops are always executed at least once in Fortran 66. They may not be in Fortran 77.

2. Do not use Hollerith strings except in format statements. They are not allowed in Fortran 77.

3. Do not expect the operating system to initialize all variables to zero for you.

4. Be aware of the fact that integer, real and logical variables may be represented by storage units of different lengths on different machines. Hence, when using private COMMON blocks, write a separate one for each type of variable.

As far as possible, modules should be written in a way that is independent of the particular arithmetic environment of the host computer. To help achieve this, the PORT library utility subroutines I1MACH and R1MACH [Fox, Hall, and Schryer, 1978] are distributed with ELLPACK. These routine should be used to obtain information about the local arithmetic environment. They are described in detail in Section 15.E. Many mathematical constants can be obtained in a machine-independent way also, e.g. e = EXP(1.0E0). (Note that the mathematical constants PI and R1EPSM, the machine epsilon, are both also available in common block R1RVGL.)

Nonstandard operating system requests should also be avoided. The exceptions are the subroutines Q1TIME (returns elapsed CP time since start of job) and Q1DATE (returns the current date) which must be provided by a local systems programmer upon installation of ELLPACK. These utilities are described further in Section 15.E.

15.D DOCUMENTATION

ELLPACK contributors provide documentation for their modules in two forms: a hard-copy summary for inclusion in the **ELLPACK User's Guide** (Chapter 6), and comments in the codes themselves. The necessity for the first type is obvious. The need for good comments in modules arises from the fact that the code will undoubtedly be scrutinized by humans at some point in its lifetime. An error in the module may need to be located, for instance. In some exceptional cases a generated ELLPACK program and modules may be captured and modified for a particular applications problem that does not exactly fit within the framework of ELLPACK. As a result, certain guidelines for internal documentation of ELLPACK modules have been established.

1. The subprogram representing the main entry point into the module should contain a detailed description of the module's function. This should include a detailed specification of the assumed input, a description of the algorithms implemented (with references), the name and address of the author, and a history of modifications. The suggested form for this information is illustrated in Figure 15.2.

2. Each subprogram must identify the ELLPACK module to which it belongs.

3. The comments in the body of the code should be sufficient for a person who is generally familiar with the implemented algorithm to be able to follow the logic of the program easily. At the very least, the comments should agree with the code.

The required form for hard-copy documentation is illustrated in Part 2 of this book.

```
      SUBROUTINE QKxxMN (parameter-list)
C
C-----------------------------------------------------------------
C  E L L P A C K   type-of-module   M O D U L E   module-name
C-----------------------------------------------------------------
C
C  P U R P O S E
C
C    BRIEF STATEMENT OF THE MODULE'S FUNCTION.
C
C  D E S C R I P T I O N
C
C    DETAILED EXPLANATION OF THE PROBLEMS TREATED AND ALGORITHMS
C    USED.
C
C  R E S T R I C T I O N S
C
C    SPECIFIC LIMITATIONS ON ACCEPTABLE PROBLEMS.
C
C  P A R A M E T E R S
C
C    A DESCRIPTION OF EACH OF THE MODULE PARAMETERS.
C
C  A R R A Y   D I M E N S I O N S
C
C    ESTIMATES OF THE REQUIRED SIZES OF ALL PARAMETER AND
C    INTERFACE ARRAYS IN TERMS OF PROBLEM PARAMETERS.
C
C  R E F E R E N C E S
C
C    REFERENCES CONTAINING INFORMATION ABOUT ALGORITHMS USED.
C
C  A U T H O R
C
C    NAME, ADDRESS AND TELEPHONE NUMBER OF THE MODULE AUTHOR(S).
C
C  V E R S I O N
C
C    DATE(S) WRITTEN AND MODIFIED.
C
C-----------------------------------------------------------------
```

Figure 15.2. Suggested form for internal module documentation.

15.E UTILITY PROGRAMS

Several utility programs are available in the ELLPACK module library to aid contributors in performing various routine tasks. The types of routines are:

* standardized output,
* error handling,
* machine-dependent constants,
* time and date.

Contributors are strongly encouraged to use the supplied utilities whenever appropriate. This will help reduce the size of the module library and avoid some duplication of effort. A complete description of each routine follows. There is a utility routine D1MACH that is essentially identical to R1MACH except for double precision.

```
      SUBROUTINE Q1BANR (MODTYP)
C
C  TYPE     : STANDARDIZED OUTPUT
C  PURPOSE  : PRINTS BANNER DENOTING MODULE TYPE
C
C  PRINTS A BANNER INDICATING THE TYPE OF MODULE.
C  THE BANNER PRINTED DEPENDS UPON THE INTEGER ARGUMENT
C  MODTYP AS FOLLOWS
C
C      MODTYP=2   ==>   DOMAIN PROCESSOR
C      MODTYP=3   ==>   DISCRETIZATION MODULE
C      MODTYP=4   ==>   INDEXING MODULE
C      MODTYP=5   ==>   SOLUTION MODULE
C      MODTYP=6   ==>   TRIPLE MODULE
C      MODTYP=7   ==>   PROCEDURE MODULE
C      MODTYP=8   ==>   OUTPUT MODULE
C
C  THIS NUMBERING CORRESPONDS TO THE MODULE TYPE INDEX
C  IN THE SUBPROGRAM NAMING CONVENTION (SECTION 12.C).
C
C  THIS SUBPROGRAM SHOULD BE CALLED BEFORE ANY WRITE
C  STATEMENT IN THE CASE I1LEVL>0.
C
```

```
      SUBROUTINE Q1ERRH (MODNAM,N)
C
C   TYPE     : ERROR HANDLING
C   PURPOSE  : FATAL ERROR MESSAGE HEADING
C
C   PRINTS A FATAL ERROR MESSAGE HEADING IN STANDARD FORM.
C   USED TO INITIATE A FATAL ERROR SEQUENCE AS FOLLOWS
C
C         CALL Q1ERRH(NHMODULE-NAME,N)
C         WRITE THE ERROR MESSAGE (PRINT NO BLANK LINES)
C         CALL Q1ERRT
C         RETURN
C
C   THE INPUT ARRAY MODNAM IDENTIFIES THE MODULE. IT IS A
C   HOLLERITH STRING WHOSE LENGTH IS PASSED AS THE SECOND
C   ARGUMENT.
C
```

```
      SUBROUTINE Q1ERRT
C
C   TYPE     : ERROR HANDLING
C   PURPOSE  : FATAL ERROR MESSAGE TRAILER
C
C   PRINTS A FATAL ERROR MESSAGE TRAILER IN STANDARD FORM
C   AND SETS A FATAL ERROR FLAG FOR THE ELLPACK CONTROL PROGRAM.
C   USED TO TERMINATE A FATAL ERROR SEQUENCE AS FOLLOWS
C
C         CALL Q1ERRH (NHMODULE-NAME,N)
C         WRITE THE ERROR MESSAGE (PRINT NO BLANK LINES)
C         CALL Q1ERRT
C         RETURN
C
```

```
      INTEGER FUNCTION I1MACH (N)
C
C    TYPE      : MACHINE-DEPENDENT CONSTANTS
C    PURPOSE  : INTEGER-VALUED MACHINE CONSTANTS
C
C    RETURNS THE NTH INTEGER MACHINE-DEPENDENT CONSTANT FROM
C    THE LIST BELOW.
C
C    THIS ROUTINE IS BASED UPON THE FOLLOWING MODEL FOR
C    THE REPRESENTATION OF NUMBERS:
C
C
C    INTEGERS :  S-DIGIT, BASE A FORM
C
C         SIGN [ X(S-1)*A**(S-1) + ... + X(1)*A + X(0) ]
C
C      WHERE  0 .LE. X(I) .LE. A  FOR I=0,...,S-1
C
C
C    FLOATING-POINT NUMBERS :  T-DIGIT, BASE B FORM
C
C         SIGN (B**E)*[ X(1)/B + ... + X(T)/B**T ]
C
C      WHERE  0 .LT. X(1)
C             0 .LE. X(I) .LE. B  FOR I=1,..,T
C             EMIN .LE. E .LE. EMAX
C
C
C      N                             I1MACH(N)
C    ---------------------------------------------------------
C
C     I/O UNIT NUMBERS
C
C      1      THE STANDARD INPUT UNIT.
C      2      THE STANDARD OUTPUT UNIT.
C      3      THE STANDARD PUNCH UNIT.
C      4      THE STANDARD ERROR MESSAGE UNIT.
C
C     WORDS
C
C      5      THE NUMBER OF BITS PER INTEGER STORAGE UNIT.
C      6      THE NUMBER OF CHARACTERS PER INTEGER STORAGE UNIT.
C
C     INTEGERS
C
C      7      A (THE BASE)
C      8      S (THE NUMBER OF BASE A DIGITS)
C      9      A**S-1 (THE LARGEST MAGNITUDE)
C
C     FLOATING-POINT NUMBERS
C
C     10      B (THE BASE)
C
C     SINGLE-PRECISION
C
C     11      T (THE NUMBER OF BASE B DIGITS)
C     12      EMIN (THE SMALLEST EXPONENT E)
C     13      EMAX (THE LARGEST EXPONENT E)
C
C     DOUBLE-PRECISION
C
C     14      T (THE NUMBER OF BASE B DIGITS)
C     15      EMIN (THE SMALLEST EXPONENT E)
C     16      EMAX (THE LARGEST EXPONENT E)
C
```

```
C  THIS SUBPROGRAM IS THE SAME AS THE ROUTINE I1MACH OF
C  [FOX, HALL, AND SCHRYER, 1978].
C
```

```
      REAL FUNCTION R1MACH (N)
C
C   TYPE    : MACHINE-DEPENDENT CONSTANTS
C   PURPOSE : REAL-VALUED MACHINE CONSTANTS
C
C   RETURNS THE NTH REAL MACHINE-DEPENDENT CONSTANT FROM
C   THE LIST BELOW.
C
C   THIS ROUTINE IS BASED UPON THE FOLLOWING MODEL FOR
C   THE REPRESENTATION OF NUMBERS:
C
C
C   INTEGERS :  S-DIGIT, BASE A FORM
C
C        SIGN [ X(S-1)*A**(S-1) + ... + X(1)*A + X(0) ]
C
C     WHERE  0 .LE. X(I) .LE. A  FOR I=0,...,S-1
C
C
C   FLOATING-POINT NUMBERS :  T-DIGIT, BASE B FORM
C
C        SIGN (B**E)*[ X(1)/B + ... + X(T)/B**T ]
C
C     WHERE  0 .LT. X(1)
C            0 .LE. X(I) .LE. B  FOR I=1,...,T
C            EMIN .LE. E .LE. EMAX
C
C
C     N                          R1MACH(N)
C   ----------------------------------------------------------
C
C     1     B**(EMIN-1) (THE SMALLEST POSITIVE MAGNITUDE)
C     2     B**EMAX*(1-B**(-T)) (THE LARGEST MAGNITUDE)
C     3     B**(-T) (THE SMALLEST RELATIVE SPACING)
C     4     B**(1-T) (THE LARGEST RELATIVE SPACING)
C     5     LOG10(B)
C
C  THIS SUBPROGRAM IS THE SAME AS THE ROUTINE R1MACH OF
C  [FOX, HALL, AND SCHRYER, 1978].
C
```

```
      SUBROUTINE Q1TIME (T)
C
C   TYPE    : TIME AND DATE
C   PURPOSE : CPU USAGE
C
C   SETS THE REAL PARAMETER T TO THE ELAPSED CPU TIME SINCE
C   THE START OF THE RUN (IN SECONDS).
C
C   THIS CAN BE USED TO TIME SECTIONS OF CODE AS FOLLOWS
C
C        CALL Q1TIME(T0)
C        ... CODE TO BE TIMED ...
C        CALL Q1TIME(T1)
C        TIME = T1-T0
C
```

```
C     THE VARIABLE TIME THEN CONTAINS THE REQUIRED CPU TIME.
C     THE TIMES RETURNED WILL USUALLY ONLY BE ACCURATE UP TO
C     A MILLISECOND, SO SHORT PIECES OF CODE CANNOT BE TIMED
C     ACCURATELY.
C
```

```
      SUBROUTINE Q1DATE (IMONTH,IDAY,IYEAR)
C
C     TYPE    : TIME AND DATE
C     PURPOSE : CURRENT DATE
C
C     SETS THE THREE INTEGER PARAMETERS IMONTH, IDAY AND IYEAR
C     TO THE CURRENT MONTH (1-12), DAY (1-31) AND YEAR (LAST
C     TWO DIGITS ONLY).
C
```

15.F REFERENCES

Fox, P. A., A. D. Hall and N. L. Schryer [1978], "Algorithm 528: Framework for a portable library", ACM Trans. Math. Software 4, pp. 177–188.

Chapter 16

PREPROCESSOR DATA

The purpose of the ELLPACK preprocessor is to read a user's ELLPACK program, recognize it, and generate an appropriate ELLPACK control program to carry out the user's requests. In order to accomplish this the preprocessor must know certain information about each of the ELLPACK problem solving modules. In this chapter we describe briefly the way in which this information is supplied to the preprocessor by contributors. We begin with a general orientation to the workings of the preprocessor.

16.A INTRODUCTION TO THE PREPROCESSOR

The ELLPACK preprocessor's job is done in three phases (see Figure 16.1).

* **System Definition**
  Read a standard file to determine the characteristics of the current system (including its problem solving modules).
* **Analysis**
  Parse the user's ELLPACK program to determine characteristics of the problem to be solved and the sequence of modules to be executed.
* **Code Generation**
  Generate the ELLPACK control program.

Much of the preprocessor's work is performed by the ELLPACK **template processor**. The template processor is a form of macro processor. Its basic operation is to copy input to output, replacing all occurrences of **template variables** by values which have been specified for them. The input file, which is called a **template**, may contain meta-statements which perform operations such as assigning values to template variables and conditionally processing input text based upon the current value of template variables. A complete description of the ELLPACK template processor may be found in Appendix C.

The form of the control program generated by the preprocessor is specified in a two part ELLPACK template file. During its system definition phase, the preprocessor instructs the template processor to read the first part of the ELLPACK template file. This initializes many template variables and defines sub-templates giving Fortran code which may be used to invoke each of the problem solving modules in the current system; no output is produced, however. The user's ELLPACK program is read during the analysis phase, and the template processor's symbol table is updated to reflect the user's PDE problem and module selection.

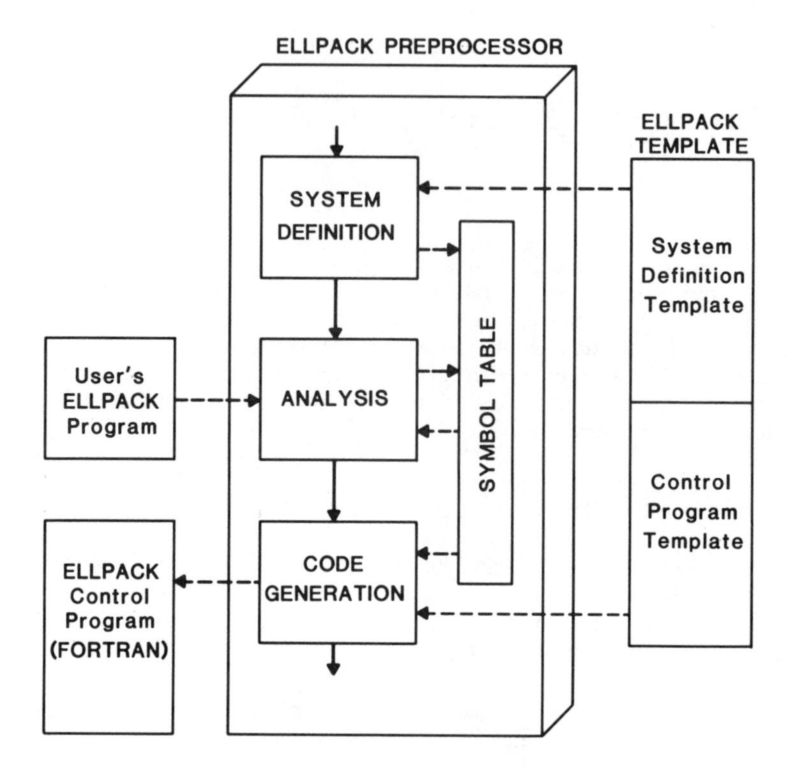

Figure 16.1. Organization of the ELLPACK preprocessor. The ELLPACK template is read in two major pieces.

In the final phase the template processor reads a template for the ELLPACK control program and processes it, replacing all template variables by their actual values in the symbol table. Code templates for selected problem solving modules are merged into the control program template and processed also at that time. The resulting output is the ELLPACK control program in Fortran.

The major advantage of this organization is that substantial changes to the form of the control program that the preprocessor generates can be effected with fairly simple changes to the ELLPACK template file. Similar changes must be made when new modules are added to the system. In this case modifications must also be made to a preprocessor subroutine called **DIMCAL**. DIMCAL is called by the preprocessor whenever it recognizes a call to a module in the user's ELLPACK program. DIMCAL updates template variables which save information about the module's interface requirements (e.g. the maximum number of equations that a discretization module will generate given the selected grid size).

The format of the ELLPACK template and the DIMCAL routine are quite complex. As a result, an additional processor called **DC** has been written to convert a very high level specification of module data into an ELLPACK template and a DIMCAL routine. In this way, all information which affects the generation of the ELLPACK control program is stored in one place, making it quite easy to personalize the system to satisfy local requirements without wholesale rewriting of the preprocessor itself. This process is illustrated in Figure 16.2. The module data read by DC is called the **ELLPACK definition file**. ELLPACK contributors are expected to provide records for this file which describe their modules. The format of the ELLPACK definition file, which is very similar to that of ELLPACK programs, is described in detail in the next section.

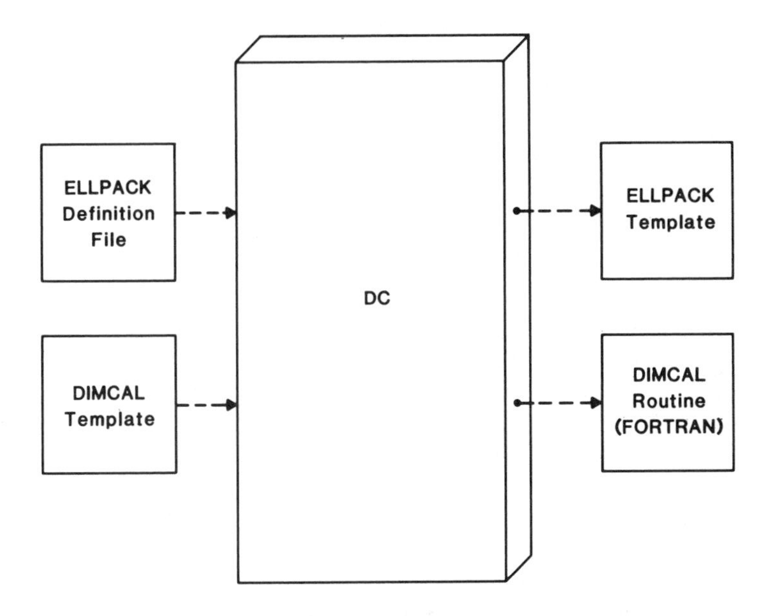

Figure 16.2. Schematic diagram of the use of DC. The program DC must be run when modules are added or deleted from the system. DC reads the ELLPACK definition file and produces a new ELLPACK template as well as a Fortran subprogram (DIMCAL) to be linked to the preprocessor.

The actual syntax of the ELLPACK language may also be changed with a concomitant increase in effort. This syntax is defined by a grammar which is input to a compiler-compiler system (called **PG**), which generates the preprocessor's analyzer. This is shown in Figure 16.3.

All programs required to implement the preprocessor are coded in standard Fortran. A further description of the preprocessor and its installation is available in Part 5 of this book. The ELLPACK template processor and PG system are described in Appendices B and C.

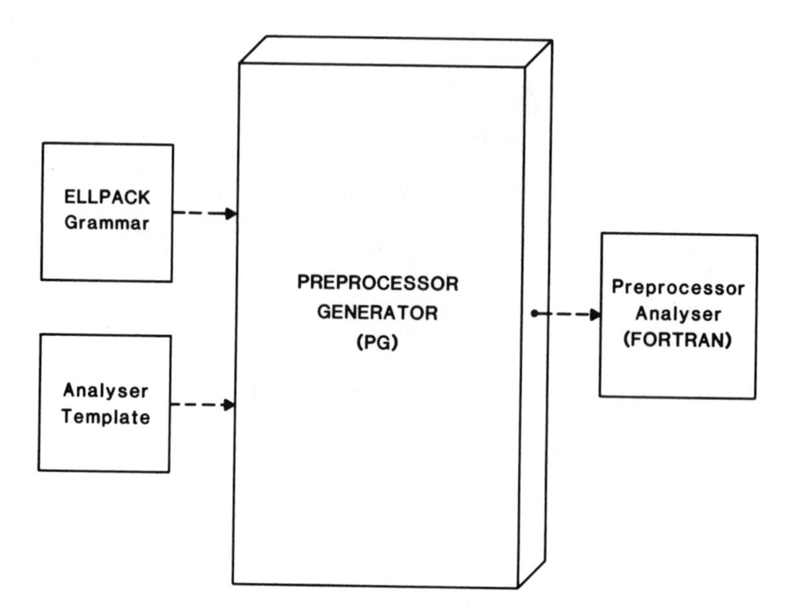

Figure 16.3. Schematic diagram of the use of PG. The program PG must be run when the syntax of the ELLPACK language is changed. PG reads a grammar for the ELLPACK language and produces a new parser for the preprocessor.

16.B THE ELLPACK DEFINITION FILE

In order to properly invoke problem solving modules, the ELLPACK preprocessor must know the following information about each module.

1. Module name and type.
2. Module parameters and their default values.
3. Fortran statements that call the module.
4. Required size of all ELLPACK interface arrays as a function of problem parameters.
5. Special Fortran declaration statements required by the module, if any.
6. Fortran statements that evaluate the computed solution (required from modules that discretize).

The executable Fortran statements generated by the preprocessor in order to to invoke a module can be divided up into three parts. These are

* module parameter assignments,
* setup statements,
* module call.

First a set of simple assignment statements give values to each of the module parameters. If the ELLPACK user has not specified values for some of these parameters, then defaults supplied by the module contributor are used. Next, a block of code that performs module-specific setup operations is inserted. This code performs the function of the module prologue discussed in Section 12.A, and it is timed separately from the rest of the module. Finally, the code that calls the requested module is inserted.

All this information is encoded into the ELLPACK definition file. Entries in this file are written in a high-level specifications language similar to the ELLPACK language itself. That is, they are composed of several segments (some optional) whose names begin in column 1 (only the first two characters are significant). After the first line of a segment, column 1 must be left blank. If the last nonblank character on a line is an ampersand (&), then the next line is a continuation of the current line. Blank lines and lines with the character + in column 1 are treated as comments. The SETUP, FORTRAN, ANSWER, and DECLARATIONS segments are composed of Fortran statements to be copied into the ELLPACK control program; these must follow the syntax of Fortran. Each of these code segments is actually a template, and hence template variables are available for substitution into the code (useful in declaration statements, for example), and for conditional code generation (see Section 16.C). An informal description of each segment of the ELLPACK definition file follows (these must appear in the order presented below).

ELLPACK Call (Mandatory)

This segment gives the module name, its type, and the defaults for *all* module parameters. This information is given exactly the way it would appear in an ELLPACK program. Some examples are

```
DIS.            SPLINE GALERKIN (KGO=3,NCDGO=1)
IND.            RED BLACK
SOLUTION.       SOR (IADAPT=1,ITMAX=100,CME=0.0,OMEGA=1.0 &
                     ZETA=AMAX1(0.5E-5,R1EPSM))
```

The default parameter assignments are copied into the Control Program exactly as they appear here. As a result, default parameter values may be arbitrary Fortran expressions and may use any interface variables that are active when the module is called. If names of module parameters do not follow Fortran's default naming conventions (initial character A–H,0–Z for reals, I-N for integers), then these variables must be declared in the ELLPACK control program. Such declaration statements are given in the DECLARATIONS segment of the ELLPACK definition file as described below.

Modules may also have parameters that are names of Fortran function subprograms. These are denoted by a double equal sign as in the following example:

```
PROCEDURE.   INITIALIZE UNKNOWNS FOR 5-POINT STAR (U==ZERO)
```

Here the formal parameter name is U and the default is ZERO. (ZERO is a standard ELLPACK function of two real arguments that returns 0.0 for all input.) All function names that are used as formal parameters must also be declared in the DC segment (described below) using the statement

```
        EXTERNAL <formal-parameter-name>
```

The preprocessor will then automatically declare the actual function name specified by the user as EXTERNAL in the ELLPACK control program.

SETUP (OPTIONAL)

This gives a Fortran CALL statement to be copied into the ELLPACK control program immediately before the module call. The called subprogram performs preprocessing to transform the initial ELLPACK interface into one natural for the method. This code is timed separately from the module call itself. For example, the SETUP segment for a band solver might look like

```
SETUP.
      CALL Q5BSSU(I5BSLB,I5BSUB)
```

The routine in this example takes the matrix at the equation/unknown indexing interface and reformats it, generating a band matrix in the reusable workspace. None of the ELLPACK interface variables need appear as arguments, since they are all global variables available in some named COMMON block. It explicitly returns the lower and upper bandwidths (I5BSLB and I5BSUB), which will be passed on to the module proper when it is called. Note that the word SETUP is on a line by itself and the remaining lines follow the rules of Fortran. All ELLPACK interface variables and module parameter values are available at this point in the control program. If module parameters that are function names must be passed to the subprogram, they should appear as $P(⟨name⟩) in the calling sequence, where ⟨name⟩ is the formal parameter name specified in the ELLPACK call segment. The preprocessor will substitute the actual function name for this string when the ELLPACK control program is generated.

FORTRAN (MANDATORY)

This segment gives the Fortran CALL statement for the contributed module. The word FORTRAN must appear on a line by itself and the remaining lines follow the rules of Fortran. The subprogram name should have MN ("main") as its last two characters. All ELLPACK interface variables and current module parameter values are available at this point in the control program. A simple example is a call to the band solver described in the SETUP segment above.

```
FORTRAN.
        CALL Q5BSMN(R1WORK,I1NEQN,I5BSLB,I5BSUB,R1BBBB,I1MNQ,
     *              R1UNKN)
```

The module in this example obtains some ELLPACK interface variables by way of its calling sequence. This example also shows how variables created in the SETUP phase (I5BSLB and I5BSUB) may be passed along to the module. Alternatively, these variables could be passed through a named COMMON block declared using the DECLARATIONS segment (described below). If module parameters that are function names must be passed, they should appear as $P(⟨name⟩) in the calling sequence, where ⟨name⟩ is the formal parameter name specified in the ELLPACK call segment. The preprocessor will substitute the actual function name for this string when the ELLPACK control program is generated.

ANSWER (MANDATORY FOR DISCRETIZATION AND TRIPLE MODULES)

This segment gives a Fortran assignment statement (usually a function call) that returns the value of the computed solution or one of its derivatives at a given point in the domain. This code is required because only modules that discretize the PDE know how to relate the interface vector R1UNKN to a solution of the problem. The ordering of unknowns in R1UNKN is the same as that generated by the DISCRETIZATION or TRIPLE module.

The requested function value should be placed in the real variable R0SOLV. The Fortran statements given here are not placed in the ELLPACK control program. Instead, they are inserted in the function R0SOLV written by the preprocessor. A simple example is

```
ANSWER.
        R0SOLV = R3FEEV(X,Y,R1UNKN,IDERIV,L1NEWD)
```

X and Y (and Z for three-dimensional problems) give the point at which the function value is desired and IDERIV is an integer that indicates which function is desired. (The canonical ordering of function and derivative values given in Section 12.B is used; thus IDERIV=6 denotes the solution function and IDERIV=4 denotes its first derivative with respect to x.) The logical variable L1NEWD is .TRUE. if the solution has not been evaluated since the last solution or triple module; some solution evaluation schemes may require preprocessing in this case.

A different procedure is required if ELLPACK interpolation utilities are to be used. In this case modules need only provide a table of solution values at grid points, i.e. R1TABL(i,j,k) = U(R1GRDX(i), R1GRDY(j), R1GRDZ(k)), i=1,...,I1NGRX, j=1,...,I1NGRY, k=1,...,I1NGRZ. The array R1TABL is the COMMON block C1TABL. This table must be loaded before the first call on the interpolation utilities. A typical ANSWER segment in this case is

```
ANSWER.
        IF (L1NEWD) CALL Q35PVL
        INTERPOLATE (R1TABL,I1KORD)
```

The subprogram Q35PVL loads the table on first call (i.e. when L1NEWD = .TRUE.). It must be provided by the DISCRETIZATION or TRIPLE module. The second statement is interpreted as "compute the requested function by interpolation of order I1KORD on the data in the array R1TABL". The variable I1KORD is set by the DISCRETIZATION or TRIPLE module at run time (see Section 13.G). Note that this allows the order of the piecewise polynomial interpolant to depend upon module parameters. If I1KORD is not dynamic, then it may be replaced by an integer constant in the INTERPOLATE function of the ANSWER segment. In this case, it need not be assigned a value by the DISCRETIZATION or TRIPLE module.

Two assignment statements must be added to the DC segment (described below) if ELLPACK interpolation utilities are used. First, the statement

```
$L(NEEDR1TABL) = .TRUE.
```

alerts the preprocessor that R1TABL is needed, so that storage will be allocated for it. Second, the B-spline interpolation utilities require a workspace array whose size depends upon the largest value of I1KORD which will be required (1 < I1KORD < 20). The variable I1MXKD must be set to this value.

Declarations (Optional)

This segment gives module-specific declaration statements to be included in the ELLPACK control program. These are most often type declarations or COMMON declarations for module parameters. It is also possible to declare arrays whose sizes are problem dependent. For example, the following declarations segment will generate an array of size I1NGRX by I1NGRY in the common block C35PNN.

```
DECLARATIONS.
      INTEGER I35PNN($I1NGRX,$I1NGRY)
      COMMON /C35PNN/ I35PNN
```

The strings $I1NGRX and $I1NGRY denote template variables which will be replaced by constants which are the values of the interface variables I1NGRX and I1NGRY when the ELLPACK control program is generated. Any integer-valued template variable may be used in this way. These include $I1NBND, $I1MBPT, $I1NGRX, $I1NGRY, $I1NGRZ, $I1MNEQ, $I1MNCO, $I1MEND, $I1MUND, $I1MUNK. Their meanings correspond exactly to the variables with similar names described in Chapter 13. A list of all predefined template variables is given in Table 16.1.

Table 16.1. Summary of Preprocessor Variables. ELLPACK preprocessor variables of interest to module contributors. The first column is the name of the template variable used by the preprocessor. These variables have the same meaning as their interface variable counterparts (except I1KBAN). The second column gives the name used to obtain the value of the template variable in expressions in the DC segment. Template variables which are assigned values in DC segments are designated by a "*" in column three.

Name	DC name	*	Name	DC name	*
L1ARCC	$L(L1ARCC)		I1MUND	$I(I1MUND)	*
L1ASIS	$L(L1ASIS)		I1MUNK	$I(I1MUNK)	*
L1CLKW	$L(L1CLKW)		I1KWRK	$I(I1KWRK)	*
L1CRST	$L(L1CRST)		I1NBND	$I(I1NBND)	
L1CSTB	$L(L1CSTB)		I1NGRX	$I(I1NGRX)	
L1CSTC	$L(L1CSTC)		I1NGRY	$I(I1NGRY)	
L1DRCH	$L(L1DRCH)		I1NGRZ	$I(I1NGRZ)	
L1HMBC	$L(L1HMBC)		L1NUNQ	$L(L1NUNQ)	
L1HMEQ	$L(L1HMEQ)		L1POIS	$L(L1POIS)	
L1HOLE	$L(L1HOLE)		L1PRDC	$L(L1PRDC)	
I1KBAN	$I(I1KBAN)	*	L1PRDX	$L(L1PRDX)	
I1MXKO	$I(I1MXKO)	*	L1PRDY	$L(L1PRDY)	
L1LAPL	$L(L1LAPL)		L1PRDZ	$L(L1PRDZ)	
I1MBPT	$I(I1MBPT)		L1RECT	$L(L1RECT)	
I1MEND	$I(I1MEND)	*	L1SELF	$L(L1SELF)	
L1MIXD	$L(L1MIXD)		L1TIME	$L(L1TIME)	
I1MNCO	$I(I1MNCO)	*	L1TWOD	$L(L1TWOD)	
I1MNEQ	$I(I1MNEQ)	*			

Occasionally an array is required whose size is a function of the predefined template variables or of a module's parameters. To declare an array of this type, one must define a new template variable and use it as a dimension. Section 16.C describes how template variables may be defined. When such variables are used to dimension arrays, they must be defined in the DC segment rather than in a code template. (Otherwise the template variable might not be defined in the template processor until after the dimension statement is generated.)

A potential problem exists when two distinct modules have declarations for the same variable. If an ELLPACK program invokes each of these modules, then both sets of declarations will be copied into the ELLPACK control program; this results in multiple declarations for these variable, a fatal error for many compilers. To remedy this, there is a facility which allows modules to specify the same set of declarations as another module. To do this one specifies

```
DECLARATIONS. ⟨module name⟩
```

where ⟨module name⟩ is the name of the module whose declarations should be copied.

DC (MANDATORY)

This segment specifies a set of extended Fortran statements which can be used to determine problem dependent array sizes. The DC processor will place these statements in the subroutine DIMCAL which then becomes part of the preprocessor. When the preprocessor's analyzer recognizes a module call in a user's ELLPACK program, DIMCAL is called to determine how that module affects the size of ELLPACK interface arrays. Modules report the sizes required for these arrays by providing code which set Fortran variables. The variables which must be set depend upon the type of module.

Module type	*Must set*
DISCRETIZATION	I1MNEQ,I1MNCO,I1MXKO,I1KBAN,I1KWRK
INDEXING	I1MEND,I1MUND,I1KWRK
SOLUTION	I1MUNK,I1KWRK
TRIPLE	I1MUNK,I1MXKO,I1KWRK
PROCEDURE	I1KWRK
OUTPUT	I1KWRK

The meanings of these variables are

I1MNEQ = estimated number of equations to be generated by discretization,

I1MNCO = estimated maximum number of nonzero coefficients in any equation,

I1MXKO = mximum order (degree+1) of polynomials to be used in piecewise polynomial interpolation of the PDE solution (this is required only of modules which use the interpolation option),

I1KBAN = estimated half bandwidth of generated equations [maximum $|i-j|$ for all i,j such that the (i,j) entry of the matrix is nonzero],

I1KWRK = required reusable workspace size,

I1MEND = size of interface array I1ENDX,

I1MUND = size of interface array I1UNDX,

I1MUNK = size of interface array R1UNKN.

All but I1KBAN and I1MXKO correspond exactly to variables available in the ELLPACK control program.

Any executable Fortran statement not requiring a statement number may be used in DC. In particular, this code may assign values to any number of intermediate integer variables with arbitrary Fortran names. The code may obtain the current value of the template variable ⟨name⟩ in Table 16.1 by referring to $I(⟨name⟩) or $L(⟨name⟩) for integers and logicals respectively. (When the DC processor creates the DIMCAL routine, these references are replaced with calls to subprograms which retrieve the current value of the template variable from the symbol table.) A simple example of a DC segment for a discretization module is

```
DC.
        NX = $I(I1NGRX)
        NY = $I(I1NGRY)
        I1MNEQ = NX*NY
        I1MNCO = 5
        I1KBAN = NX
        IF($L(L1DRCH)) I1MNEQ = (NX-2)*(NY-2)
        IF($L(L1DRCH)) I1KBAN = NX-2
        I1MWRK = NX*NY + 42
        I1MXKO = 4
        $L(NEEDR1TABL) = .TRUE.
```

The current value of any integer or logical module parameter can be obtained by referring to $IP(⟨name⟩) or $LP(⟨name⟩) respectively. Thus, for example, if the size of the reusable workspace required by the module is I1NGRX*I1NGRY + ITMAX + 2, where ITMAX is a module parameter, then the assignment

```
        I1KWRK = $I(I1NGRX)**$I(I1NGRY) + $IP(ITMAX) + 2
```

would be appropriate.

DC can also be used to define new template variables to be used as dimensions in the DECLARATIONS segment. This can be done simply by assigning a value to an extended variable of the form $I⟨name⟩. For example, suppose that two arrays are to be declared, one of size ITMAX (a module parameter) and the other of size I1NGRX+2. This is accomplished by

```
DECLARATIONS.
        REAL Q5SRRT($I5SRRT)
        INTEGER I5SRIT($I5SRIT)

DC.
        $I(I5SRRT) = $I(I1NGRX) + 2
        $I(I5SRIT) = $IP(ITMAX)
```

If an extended variable is to the left of an equal sign, then the DIMCAL statements generated by the DC processor will store the computed value in the preprocessor's symbol table by the name given in parentheses. To avoid conflicts the newly defined template variables should use the naming convention of Section 12.C.

16.C EXAMPLES OF FILE ENTRIES

The following are samples of entries in the ELLPACK definition file for modules of various types.

```
+     ================================================================
+             EXAMPLE 1   :   A FINITE DIFFERENCE MODULE
+     ================================================================
+
+     NOTE THAT DIFFERENT SUBPROGRAMS ARE CALLED WHEN THE DOMAIN
+     IS NONRECTANGULAR.  CONDITIONAL CODE GENERATION STATEMENTS
+     ARE USED TO ACCOMPLISH THIS; SEE SECTON 16.D (THIS AVOIDS
+     HAVING BOTH SUBPROGRAMS LOADED INTO MEMORY WHEN ONLY ONE
+     IS TO BE USED.)

DISCRETIZATION.    5 POINT STAR

FORTRAN.
*IF(L1RECT)
      CALL Q35PMN
*ELSE
      CALL Q35GMN
*ENDIF

DECLARATIONS.
      COMMON / C35PNU / I35PNU($I1NGRX,$I1NGRY)
*IF(L1RECT)
*ELSE
      COMMON / C35GBN / I35GBN($I1MBPT)
*ENDIF

ANSWER.
*IF(L1RECT)
      IF (L1NEWD)  CALL Q35PVL
*ELSE
      IF (L1NEWD)  CALL Q35GVL
*ENDIF
      INTERPOLATE( R1TABL, 4 )

DC.
      NX = $I(I1NGRX)
      NY = $I(I1NGRY)
      I1KWRK = 35
      I1MXKO = 4
      I1MNCO = 5
      IF ((.NOT.$L(L1RECT))  .AND.  (NOT.$L(L1DRCH)))
     A    I1MNCO = 7
      I1MNEQ     = (NX-2)*(NY-2)
      IF (.NOT. $L(L1DRCH)) I1MNEQ = NX*NY
      IF (.NOT. $L(L1RECT)) I1MNEQ = I1MNEQ+NY+NY
      I1KBAN     = NX-2
      IF  (.NOT.  $L(L1DRCH)) I1KBAN = NX
      IF ((.NOT.  $L(L1RECT))  .AND.  (.NOT.  $L(L1DRCH)))
     A    I1KBAN = 2*NX+NX/2
      $L(NEEDR1TABL) = .TRUE.
```

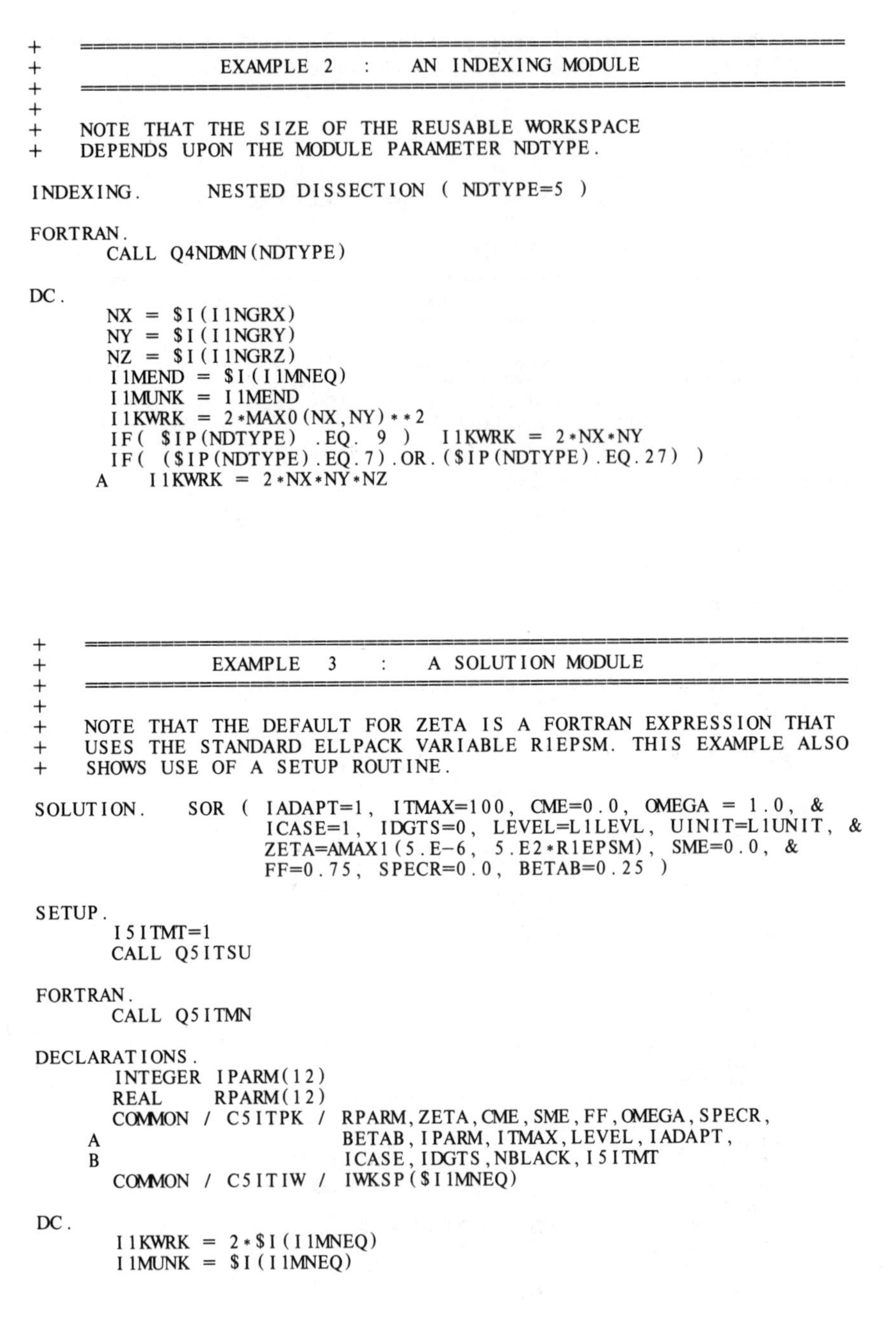

```
+   ================================================================
+              EXAMPLE 2  :   AN INDEXING MODULE
+   ================================================================
+
+   NOTE THAT THE SIZE OF THE REUSABLE WORKSPACE
+   DEPENDS UPON THE MODULE PARAMETER NDTYPE.

INDEXING.      NESTED DISSECTION ( NDTYPE=5 )

FORTRAN.
      CALL Q4NDMN(NDTYPE)

DC.
      NX = $I(I1NGRX)
      NY = $I(I1NGRY)
      NZ = $I(I1NGRZ)
      I1MEND = $I(I1MNEQ)
      I1MUNK = I1MEND
      I1KWRK = 2*MAX0(NX,NY)**2
      IF( $IP(NDTYPE) .EQ. 9 )  I1KWRK = 2*NX*NY
      IF( ($IP(NDTYPE).EQ.7).OR.($IP(NDTYPE).EQ.27) )
     A   I1KWRK = 2*NX*NY*NZ




+   ================================================================
+             EXAMPLE 3   :   A SOLUTION MODULE
+   ================================================================
+
+   NOTE THAT THE DEFAULT FOR ZETA IS A FORTRAN EXPRESSION THAT
+   USES THE STANDARD ELLPACK VARIABLE R1EPSM. THIS EXAMPLE ALSO
+   SHOWS USE OF A SETUP ROUTINE.

SOLUTION.    SOR ( IADAPT=1, ITMAX=100, CME=0.0, OMEGA = 1.0, &
                   ICASE=1, IDGTS=0, LEVEL=L1LEVL, UINIT=L1UNIT, &
                   ZETA=AMAX1(5.E-6, 5.E2*R1EPSM), SME=0.0, &
                   FF=0.75, SPECR=0.0, BETAB=0.25 )

SETUP.
      I5ITMT=1
      CALL Q5ITSU

FORTRAN.
      CALL Q5ITMN

DECLARATIONS.
      INTEGER IPARM(12)
      REAL    RPARM(12)
      COMMON / C5ITPK / RPARM,ZETA,CME,SME,FF,OMEGA,SPECR,
     A                  BETAB,IPARM,ITMAX,LEVEL,IADAPT,
     B                  ICASE,IDGTS,NBLACK,I5ITMT
      COMMON / C5ITIW / IWKSP($I1MNEQ)

DC.
      I1KWRK = 2*$I(I1MNEQ)
      I1MUNK = $I(I1MNEQ)
```

```
+    ================================================================
+                   EXAMPLE  4    :    A TRIPLE MODULE
+    ================================================================
+
+    BOTH THE SIZE OF THE REUSABLE WORKSPACE AND THE ORDER
+    OF PIECEWISE POLYNOMIAL INTERPOLATION DEPEND UPON THE
+    VALUE OF THE MODULE PARAMETER IORDER HERE.

TRIPLE.      HODIE FFT (IORDER = 4)

FORTRAN.
      CALL Q6H2MN (IORDER)
      I1KORD = IORDER + 1

ANSWER.
      IF (L1NEWD) CALL Q6H2VL
      INTERPOLATE (R1TABL,I1KORD)

DC.
      NX = $I(N1NGRX)
      NY = $I(I1NGRY)
      I1MUNK = NX*NY
      I1KWRK = NX*($IP(IORDER)*NY/2-1) + 2*(NX + NY)
      I1MXKO = 7
      $L(NEEDR1TABL) = .TRUE.



+    ================================================================
+                   EXAMPLE  5    :    A PROCEDURE MODULE
+    ================================================================
+
+    THIS MODULE INITIALIZES THE COMPUTED SOLUTION
+    VECTOR R1UNKN USING A FUNCTION PROVIDED BY THE
+    USER WHICH ESTIMATES THE SOLUTION ON THE DOMAIN.

PROCEDURE.   SET UNKNOWNS FOR 5 POINT STAR (U==ZERO)

FORTRAN.
      CALL Q75PIN ($P(U))

DC.
      EXTERNAL(U)
```

16.D CONDITIONAL CODE GENERATION

In addition to its string substitution facilities, the ELLPACK template processor has a simple facility which allows the selection of Fortran statements to be included in the ELLPACK control program based upon the values of template variables. This facility may be used in the SETUP, FORTRAN, ANSWER, and DECLARATIONS segments of the ELLPACK definition file. We provide a brief outline of this facility here; details may be found in Appendix C.

The basic conditional code generation statement takes the form

```
*IF(⟨name⟩)
     ... FORTRAN STATEMENTS ...
    *ELSE
     ... FORTRAN STATEMENTS ...
    *ENDIF
```

where ⟨name⟩ is the name of a logical template variable (see Table 16.1). If the current value of the template variable is .TRUE., then the first block of statements is generated; otherwise the second is generated. The *IFs can be nested and the *ELSE clause can be omitted if it is not required. A simple example is

```
FORTRAN.
*IF(L1RECT)
      CALL Q35PMN
*ELSE
      CALL Q35GMN
*ENDIF
```

which generates the first call for problems on rectangular domains and the second for problems on general domains.

Conditional expressions are not supported in the *IF statement, but fairly complex conditions can be tested using nested *IFs. For example, code to be generated for problems with nonrectangular domains and homogeneous boundary conditions can be specified by

```
*IF(L1RECT)
*ELSE
*IF(L1HMBC)
        ... FORTRAN STATEMENTS ...
*ENDIF
*ENDIF
```

Sometimes the use of this technique requires the duplication of blocks of code in the definition file. To avoid this, one may define new template variables. The extended statements

```
*SET(NAME=.TRUE.)
*SET(NAME=.FALSE.)
```

set the template variable ⟨name⟩ to .TRUE. and .FALSE. respectively. The variable name is an arbitrary string, but it is wise to adhere to the interface variable name conventions to avoid conflicts. The following example shows a more complex conditional code generation.

```
ANSWER.
*SET(L3COHM=.FALSE.)
*IF(L1HMBC)
*IF(L1MIXD)
*ELSE
*SET(L3COHM=.TRUE.)
*ENDIF
*ENDIF
*IF(L3COHM)
+         USE IF L1HMBC .AND. .NOT. L1MIXD
      R0SOLV = R3COAA(X,Y,R1UNKN,IDERIV,L1NEWD)
*ELSE
+         USE OTHERWISE
      R0SOLV = R3COBB(X,Y,R1UNKN,IDERIV,L1NEWD)
*ENDIF
```

PART 5

CHAPTERS 17–18

SYSTEM PROGRAMMING GUIDE

Chapter 17 gives explicit instructions for installing the ELLPACK system. The ELLPACK system is constructed so that it can be changed in many ways. Problem solving modules can be added or deleted; the ELLPACK language itself can be changed. Chapter 18 outlines how these changes are made.

Chapter 17

INSTALLING ELLPACK

In order to install ELLPACK one need not know anything about elliptic boundary value problems; one should, however, be reasonably familiar with the local operating system, especially the facilities for compiling large collections of Fortran programs and collecting them into libraries. One should also be familiar with the construction of control card procedures involving several job steps, and with the allocation and reference of sequential files. Knowledge of local graphics facilities is also helpful. All this is necessary because the installation procedure is described in a machine-independent manner. This chapter is self-contained, so a system programmer need not read the entire ELLPACK guide in order to get the system operational.

17.A OVERVIEW OF THE SYSTEM

ELLPACK is a software system for solving partial differential equations; the answers obtained are functions of two or three variables. To solve such problems ELLPACK users write programs in a very high level language. This language has declaration statements which allow users to define the partial differential equation, the region on which it is defined, and boundary conditions that are to be satisfied. Executable statements are available to perform the various steps required to solve the problem. The ELLPACK language is an extension of Fortran, and Fortran statements may be freely interspersed with ELLPACK statements. A preprocessor translates the user's ELLPACK program to ordinary Fortran. The resulting main program, called the **ELLPACK control program**, is then compiled, linked to a library of problem-solving modules, and executed. The user's output may include both tabulated and plotted results. This process is illustrated in Figure 12.2.

The main components of the system are thus the **ELLPACK preprocessor** and the **ELLPACK module library**. The entire system is written in Fortran 66, although one may also obtain a version with some limited Fortran 77 facilities, such as CHARACTER*1. The preprocessor is simply a Fortran main program and the module library a collection of Fortran subprograms. Outside of optional (and limited) graphics facilities, very little machine-dependent code exists in the system.

The ELLPACK system is designed to be easily extendable. The preprocessor, for instance, obtains all information about the form of the ELLPACK control program from a file, called the **ELLPACK template**, which it reads each time it executes. (A template is a form of macro.) The ELLPACK template also contains most of the information required to generate code to invoke individual problem-solving modules. The remainder of this information is isolated in several small subprograms (collectively called **DIMCAL**) which initialize the preprocessor's symbol table and describe module-dependent array sizes. A high level description of all this information is contained in the **ELLPACK definition file**. When new modules are added or the form of the control program is to be altered, one simply modifies the ELLPACK definition file appropriately. A program called **DC** is then run to produce a new ELLPACK template as well as the revised DIMCAL routines. The form of the ELLPACK language itself can be modified in a similar manner. This is done by modifying the grammar which is input to the program **PG**, a compiler-compiler system. The output of PG is a collection of subprograms which form the part of the preprocessor that recognizes the ELLPACK language. These processes are illustrated in Figures 16.2 and 16.3 respectively and described in more detail in Chapter 18. **It is not necessary to perform any of these operations to install the basic ELLPACK system.**

The preprocessor and numerous problem-solving modules (including the domain processor and all output modules) were written at Purdue University, and responsibility for their correctness lies with the ELLPACK project there. The majority of problem-solving modules were contributed to ELLPACK by other authors, and in these cases responsibility for correctness lies with the contributing author. The ELLPACK project will pass along user error reports to module authors, and make simple changes where appropriate, but prompt response to problems requiring substantial rewriting of these modules cannot be guaranteed.

17.B INSTALLATION PROCEDURE SUMMARY

The steps required to install the basic ELLPACK system on any computer are given below.

1. **Read the ELLPACK Tape**. The ELLPACK tape contains the Fortran source for the ELLPACK preprocessor and each of the ELLPACK problem solving modules. Also included is the standard ELLPACK template which is read by the preprocessor each time it executes. A sample ELLPACK program is included, along with the ELLPACK control program generated by the preprocessor, and the output produced on the Purdue system. Also distributed are the sources for the programs DC and PG, which can be used to rebuild the preprocessor. The inputs to these programs which define the standard preprocessor are also included. The general format of the ELLPACK tape is given in Section 17.C.

2. **Prepare the ELLPACK Modules**. Several routines which return machine-dependent constants must be modified, and routines returning the current date and elapsed CPU time must be provided. Graphics routines must be interfaced to a local graphics library (ELLPACK may be run without graphics if desired). These items are described in detail in Section 17.D.

3. **Create the ELLPACK Module Library**. All ELLPACK modules must be compiled and collected together in a library which may be used by the local linker utility.

4. **Prepare the ELLPACK Preprocessor**. Routines which return machine-dependent constants must be modified, and routines returning the current date and elapsed CPU time must be provided (these are the same routines as those in the ELLPACK module library). Several statements which open sequential files and assign logical unit numbers may need to be adjusted. A **program** statement may need to be added (CDC Fortran). On CDC machines the ELLPACK template must be modified to add a **program** statement. These items are described in detail in Section 17.E.

5. **Create the ELLPACK Preprocessor**. Compile the preprocessor and link it, creating an executable ELLPACK preprocessor.

6. **Develop Control Card Procedures to Run the System**. A control card procedure which performs all the functions illustrated in Figure 12.2 must be prepared. Several of these are illustrated in Section 17.F.

7. **Checkout the System**. Run the sample ELLPACK program and compare its results with the output file found on the ELLPACK tape. Note that the floating-point values produced on your computer might not exactly match those computed on the Purdue system.

17.C THE ELLPACK TAPE

The physical format of the ELLPACK tape (e.g., density, blocksize, character set, etc.) is selected when ELLPACK is ordered and hence will vary from site to site. The description of the files written on the ELLPACK tape given here might also not correspond exactly to what is distributed to each site. A complete inventory of files will be distributed with each tape. The list here is given to facilitate the discussion in Sections 17.D and 17.E. Subscribers may also choose between single and double precision versions as well as between Fortran 66 and Fortran 77 versions. The only non-Fortran 66 facilities used in the Fortran 77 version are CHARACTER*1 and quoted strings. These facilities greatly reduce the memory requirements of the preprocessor.

The ELLPACK tape is logically divided into six parts with one to many files in each part. The following is a brief description of each of these parts; there are about 100 files on the tape.

Installation Instructions

A single file contains a machine-readable version of an addendum to the installation instructions given in this chapter.

Part A.0 (Text) Installation instructions.

ELLPACK Modules

The source for each ELLPACK module occupies a separate file. Low-level ELLPACK utilities are collected together as if they were a module. All subprograms which may be system dependent or machine dependent are collected into the first two files.

Part B.0 (Fortran source) System dependent subprograms (two files).

Part B.k (Fortran source) The kth ELLPACK module, for $k=1,...,n$, where n is the number of modules distributed.

ELLPACK Preprocessor

The source for the ELLPACK preprocessor occupies five separate subparts: the main program, system dependent subprograms, the DIMCAL subprograms (which are produced by the DC processor), and two other fairly large collections of subprograms. Also included is the standard ELLPACK template which is read by the preprocessor each time it executes.

Part C.0 (Fortran source) The preprocessor main program.

Part C.1 (Fortran source) System dependent subprograms.

Part C.2 (Fortran source) The DIMCAL subroutines.

Part C.3 (Fortran source) Additional preprocessor subprograms.

Part C.4 (Fortran source) Additional preprocessor subprograms.

Part C.5 (Text) The ELLPACK template.

Sample Program

A sample ELLPACK program is distributed along with the resulting ELLPACK control program which should be generated by the preprocessor. A third file contains the printed output of the ELLPACK execution produced on the Purdue VAX/UNIX system.

Part D.0 (Text) An ELLPACK input program.

Part D.1 (Fortran source) The ELLPACK control program which should be produced by the preprocessor given the input on file D.0. Compiling this program, linking it to the ELLPACK module library, and executing it should produce the output in file D.2.

Part D.2 (Text) The output produced by executing the ELLPACK control program in file D.1 on the Purdue VAX/UNIX system

DC Processor

This part contains the source and input data required to run the DC processor. DC is run only when new modules are added to the system or the form of the ELLPACK control program is to be changed. This is described further in Chapter 18. The DC processor is not required for the installation of basic ELLPACK.

Part E.0 (Fortran source) The DC processor main program.

Part E.1 (Fortran source) System dependent subprograms.

Part E.2 (Fortran source) The remaining DC subprograms.

Part E.3 (Text) DIMCAL template. This is the input to DC that defines the structure of the DIMCAL routines which it generates.

Part E.4 (Text) The ELLPACK definition file. This is the input to DC which defines the current ELLPACK modules and the form of the ELLPACK control program.

THE PG SYSTEM

This part contains the source and input data for the **preprocessor generator system** (PG). PG is a compiler-compiler system which is used to build the syntactic analyzers used by the ELLPACK preprocessor and the DC processor. It must be used to change the form of the ELLPACK language or the form of the ELLPACK definition file. This is described further in Chapter 18 and Appendix B. **The PG system is not required for the installation of basic ELLPACK.**

Part F.0 (Fortran source) The PG system main program.

Part F.1 (Fortran source) System dependent subprograms.

Part F.2 (Fortran source) The remaining PG subprograms.

Part F.3 (Text) Symbols A standard input to PG. This is a template giving the form of the syntactic analysis routines generated by PG.

Part F.4 (Text) The input grammar defining the ELLPACK language.

Part F.5 (Text) The input grammar defining the ELLPACK definition file.

Part F.6 (Text) The input grammar defining PG grammars.

17.D PREPARING THE ELLPACK MODULES

The ELLPACK modules are sets of Fortran subprograms which must be compiled and collected together into a relocatable library which is compatible with the local linker/loader utility. There are no name conflicts among these routines. The control program written by the preprocessor is linked to this library in each ELLPACK run. The required subprograms are found in parts B.1 through B.n on the ELLPACK tape, where n is the number of modules distributed.

Module contributors have been asked to code their programs in standard Fortran 66. The only non-Fortran 66 construct that one should expect to find is the occasional use of quoted strings. If a Fortran 77 compiler is to be used one should be aware that (a) Hollerith strings may be used as actual arguments in calling sequences, (b) integer variables may be used to contain character data (one character per word), and (c) some local variables may be assumed to retain their values after a RETURN (no SAVE statements are used). These should not cause problems in most Fortran compilers.

There are several genuine system dependences in the ELLPACK modules, but the number of such items is small, and they have been localized to a few routines. In the following, we describe four distinct types of dependences: machine constants, time and date routines, error handling, and graphics. Each of the subprograms discussed in this section is found in part B.0 of the ELLPACK tape.

Machine Constants

All machine-dependent constants required by ELLPACK modules are localized in three small function subprograms, I1MACH, R1MACH and D1MACH. The use of these routines is described in Section 15.E. The machine constants returned by these routines are stored in local tables (using Fortran DATA statements), and tables appropriate to many machine types are included as comments. To specialize these routines for your computer simply comment the current table (put C's in column 1) and activate the desired table (remove C's from column 1). If the constants for your machine are not provided, then you must provide them. A machine constant of particular interest is I1MACH(2), which gives the logical unit number of the standard output file. This is the unit to which all ELLPACK printed output is written.

Time and Date Routines

The ELLPACK module library is expected to contain two subroutines, Q1TIME and Q1DATE, which return the elapsed CPU time and the current date, respectively. A system programmer must provide these routines (or suitable dummies) when the system is installed. The required calling sequences of these routines in given in Section 15.E.

Error Handling

To report fatal errors to users ELLPACK modules use a sequence of the form

```
CALL Q1ERRH(nH<module name>,n)
     ... write error message ...
CALL Q1ERRT
```

Q1ERRH and Q1ERRT are ELLPACK utilities which print standard error message headers and trailers, respectively. The latter routine also sets a fatal error flag which the ELLPACK control program checks when the module finishes execution (see Section 15.B). If the flag is set, the control program stops. A system programmer could optionally insert code into these routines to perform such tasks as writing a message into the user's "dayfile", or setting an operating system switch to indicate a fatal error in the job step.

Graphics

It is possible to run ELLPACK without graphics, but that is ill advised, since this is a very useful facility. To run ELLPACK without graphics one simply advises users to avoid the PLOT verb in the OUTPUT segment. If they request graphical output they will obtain unsatisfied external references when the ELLPACK control program is linked to the ELLPACK module library. Since implementing ELLPACK graphics requires only the addition of several graphics primitives to the ELLPACK module library, it is sometimes easier to ignore graphics initially and then add it after the basic system is operational.

ELLPACK users may request plots of the domain on which the problem is defined (with the computational grid superimposed) as well as contour plots of the solution and related functions. These plots are available only when the domain is two-dimensional. Five routines are included in the ELLPACK module library to do these things:

Q8PDR2 Plot of rectangular domain and grid

Q8PDNR Plot of nonrectangular domain and grid

Q8PLR2 Contour plot of function defined on a rectangular domain

Q8PLNR Contour plot of function defined on a nonrectangular domain

Q8PLDR Low-level routine used by contour plotter

The contour plots are controlled by a device-independent contour plotter [Snyder, 1978]. Sample plots produced by the routines Q8PDR2, Q8PLR2, Q8PDNR, and Q8PLNR are displayed in Figure 17.1.

To draw lines these routines actually call on a set of low-level system-independent graphics primitives. To implement ELLPACK graphics one simply provides versions of these primitives which interface with local graphics facilities. The ELLPACK graphics primitives have the following names and functions:

Q0GP00 Begin plotting (open plot file)

Q0GP99 Finish plotting (close plot file)

Q0GPBG Begin plot frame

Q0GPEN End plot frame

Q0GPAX Draw axes

Q0GPMU Pen up move

Q0GPMD Pen down move

Q0GPLN Draw polygonal line

Q0GPWC Draw text

Q0GPWI Draw integer

Q0GPWR Draw real number

The ELLPACK preprocessor inserts calls to Q0GP00 and Q0GP99 at the beginning and end of the ELLPACK control program respectively whenever plotting is to be done. These are called exactly once. ELLPACK output modules call Q0GPBG and Q0GPAX to set up each plot frame and Q0GPEN to end the plot frame.

A version of the plotting primitives is available on part B.0 of the ELLPACK tape. A sample collection of ELLPACK graphics primitives based on a particular plotting package [DISSPLA, 1981] follows. All parameters are input variables.

```
      SUBROUTINE Q0GP00
C
C-----------------------------------------------------------------
C  ELLPACK GRAPHICS PRIMITIVES  /  DISSPLA VERSION
C-----------------------------------------------------------------
C
C   Q0GP00 / INITIALIZE PLOTTING (OPEN PLOT FILE).
C
      CALL UNIPLT
      RETURN
      END
```

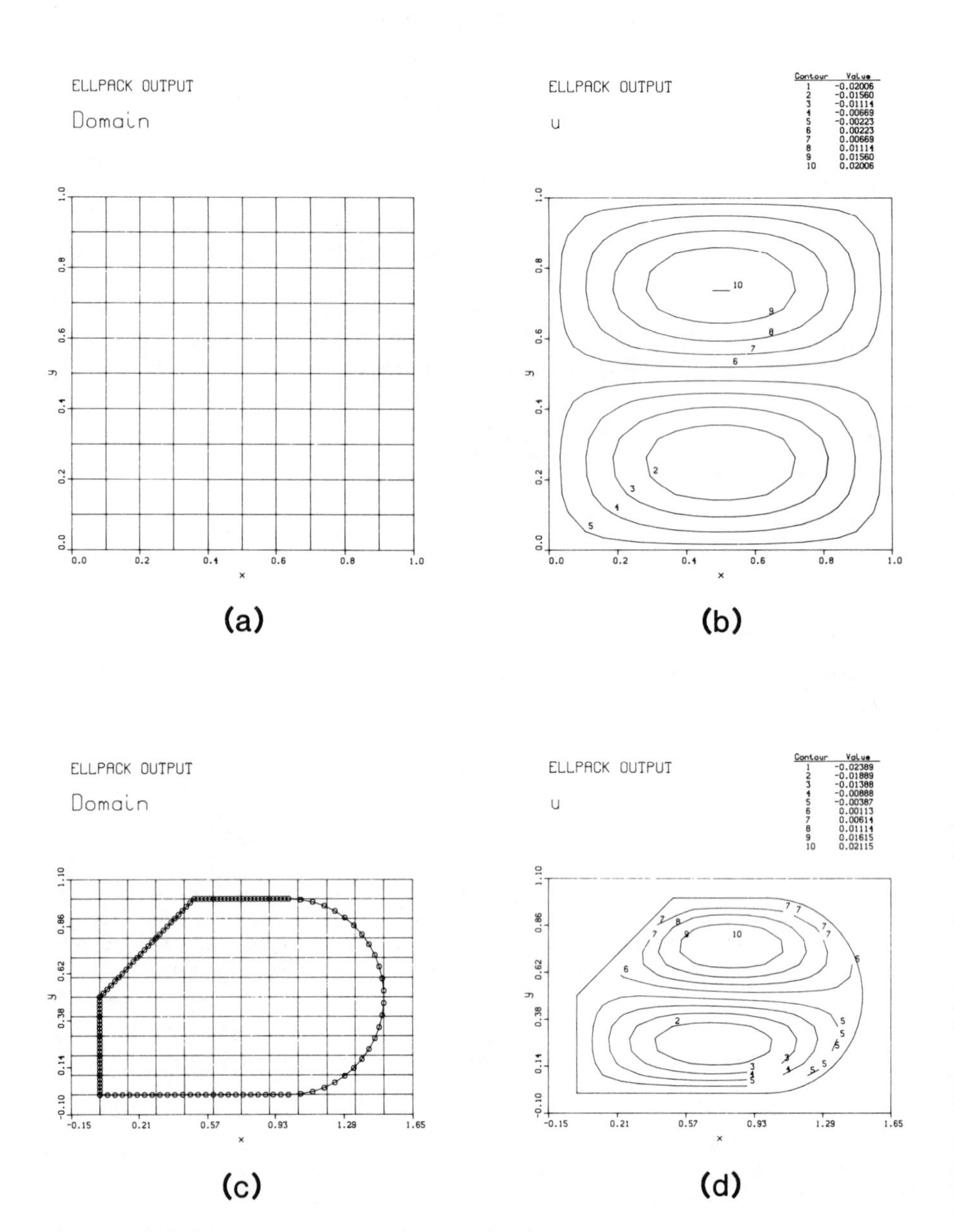

Figure 17.1. Sample plots produced by ELLPACK. (a) A simple rectangular grid on a rectangular domain. (b) Contour plot of u(x,y). (c) Rectangular grid over a nonrectangular domain. (d) Contour plot of u(x,y); the contours are truncated near curved boundaries.

```
      SUBROUTINE Q0GP99
C
C-----------------------------------------------------------------------
C  ELLPACK GRAPHICS PRIMITIVES   /   DISSPLA VERSION
C-----------------------------------------------------------------------
C
C   Q0GP99 / TERMINATE PLOTTING (CLOSE PLOT FILE).
C
      CALL DONEPL
      RETURN
      END

      SUBROUTINE Q0GPBG (PAGEX,PAGEY)
C
C-----------------------------------------------------------------------
C  ELLPACK GRAPHICS PRIMITIVES   /   DISSPLA VERSION
C-----------------------------------------------------------------------
C
C   Q8PGBG  /  BEGIN PLOT FRAME.
C
C     PAGEX   =  HORIZONTAL LENGTH (IN INCHES)
C     PAGEY   =  VERTICAL LENGTH   (IN INCHES)
C
      CALL RESET('ALL')
      CALL PAGE(PAGEX,PAGEY)
      CALL NOBRDR
      RETURN
      END

      SUBROUTINE Q0GPEN
C
C-----------------------------------------------------------------------
C  ELLPACK GRAPHICS PRIMITIVES   /   DISSPLA VERSION
C-----------------------------------------------------------------------
C
C   Q0GPEN  /  END PLOT FRAME.
C
      CALL ENDPL(0)
      RETURN
      END

      SUBROUTINE Q0GPAX(XORIGN,XLEN,XFIRST,DX,XLAST,LABLX,NLABLX,
     *                  YORIGN,YLEN,YFIRST,DY,YLAST,LABLY,NLABLY)
C
C-----------------------------------------------------------------------
C  ELLPACK GRAPHICS PRIMITIVES   /   DISSPLA VERSION
C-----------------------------------------------------------------------
C
C Q0GPAX  /  DRAW AXES.
C
C   XORIGN  =  PHYSICAL X-COORDINATE OF AXES ORIGIN  (IN INCHES)
C   XLEN    =  LENGTH OF X AXIS                      (IN INCHES)
C   XFIRST  =  X-COORDINATE OF FIRST POINT ON X-AXIS (USER UNITS)
C   DX      =  X-AXIS TICK MARK INCREMENT            (USER UNITS)
C   XLAST   =  X-COORDINATE OF LAST  POINT ON X-AXIS (USER UNITS)
C   LABLX   =  LABEL FOR X-AXIS, PACKED
C   NLABLX  =  NUMBER OF CHARACTERS IN LABLX
C   YORIGN  =  PHYSICAL Y-COORDINATE OF AXES ORIGIN  (IN INCHES)
C   YLEN    =  LENGTH OF Y AXIS                      (IN INCHES)
C   YFIRST  =  Y-COORDINATE OF FIRST POINT ON Y-AXIS (USER UNITS)
C   DY      =  Y-AXIS TICK MARK INCREMENT            (USER UNITS)
C   YLAST   =  Y-COORDINATE OF LAST  POINT ON Y-AXIS (USER UNITS)
C   LABLY   =  LABEL FOR Y-AXIS, PACKED
C   NLABLY  =  NUMBER OF CHARACTERS IN LABLY
C
```

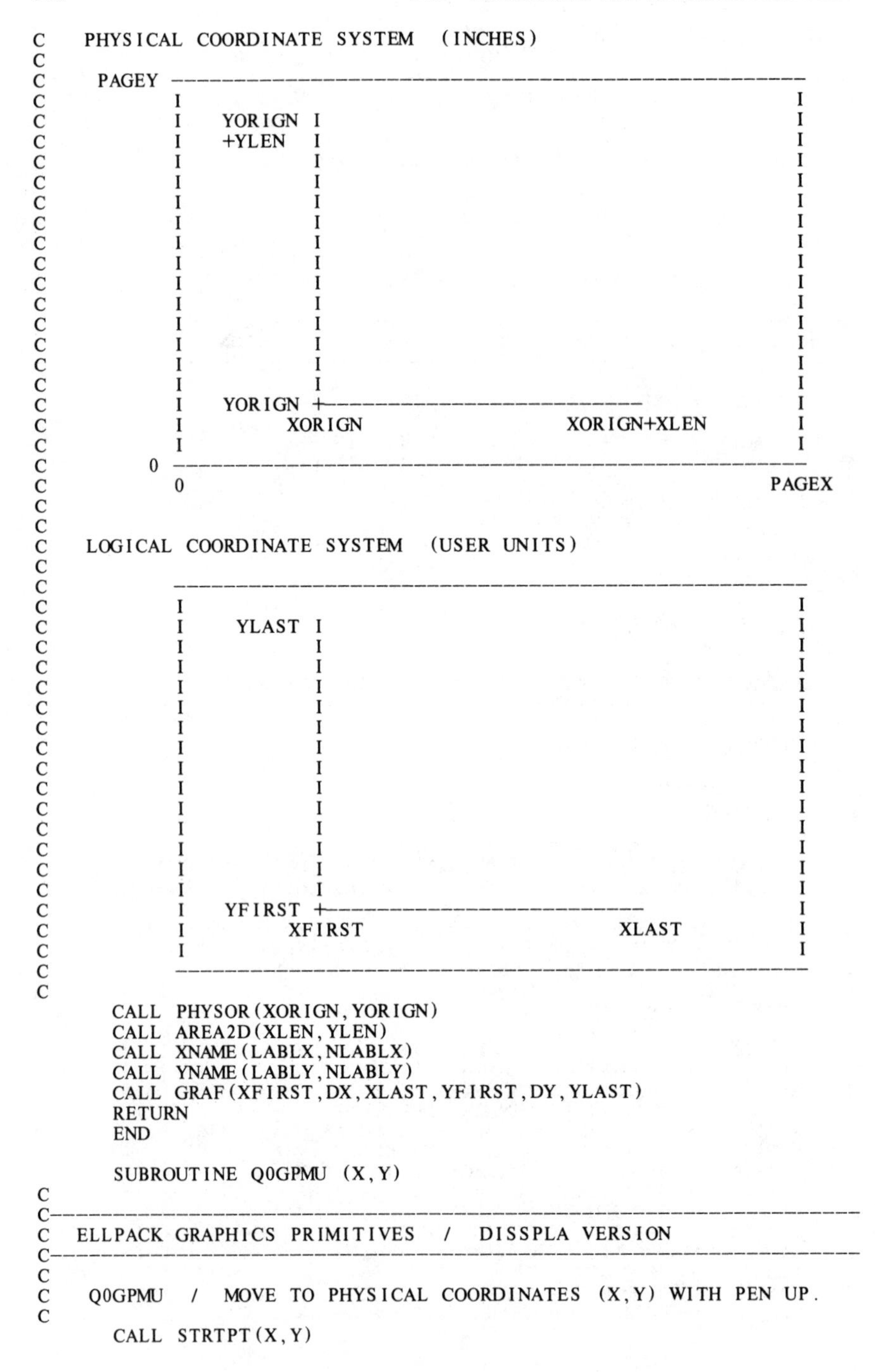

```
C   PHYSICAL COORDINATE SYSTEM  (INCHES)
C
C    PAGEY ------------------------------------------------------------
C          I                                                          I
C          I    YORIGN I                                              I
C          I    +YLEN  I                                              I
C          I           I                                              I
C          I           I                                              I
C          I           I                                              I
C          I           I                                              I
C          I           I                                              I
C          I           I                                              I
C          I           I                                              I
C          I           I                                              I
C          I           I                                              I
C          I           I                                              I
C          I           I                                              I
C          I           I                                              I
C          I    YORIGN +------------------------------                I
C          I        XORIGN                     XORIGN+XLEN            I
C          I                                                          I
C        0 ------------------------------------------------------------
C          0                                                         PAGEX
C
C
C   LOGICAL COORDINATE SYSTEM  (USER UNITS)
C
C          ------------------------------------------------------------
C          I                                                          I
C          I     YLAST I                                              I
C          I           I                                              I
C          I           I                                              I
C          I           I                                              I
C          I           I                                              I
C          I           I                                              I
C          I           I                                              I
C          I           I                                              I
C          I           I                                              I
C          I           I                                              I
C          I           I                                              I
C          I           I                                              I
C          I           I                                              I
C          I           I                                              I
C          I    YFIRST +------------------------------                I
C          I        XFIRST                          XLAST             I
C          I                                                          I
C          ------------------------------------------------------------
C
      CALL  PHYSOR(XORIGN,YORIGN)
      CALL  AREA2D(XLEN,YLEN)
      CALL  XNAME(LABLX,NLABLX)
      CALL  YNAME(LABLY,NLABLY)
      CALL  GRAF(XFIRST,DX,XLAST,YFIRST,DY,YLAST)
      RETURN
      END

      SUBROUTINE Q0GPMU (X,Y)
C
C-----------------------------------------------------------------------
C   ELLPACK GRAPHICS PRIMITIVES   /   DISSPLA VERSION
C-----------------------------------------------------------------------
C
C    Q0GPMU  /  MOVE TO PHYSICAL COORDINATES (X,Y) WITH PEN UP.
C
      CALL STRTPT(X,Y)
```

```
      RETURN
      END

      SUBROUTINE Q0GPMD (X,Y)
C
C-----------------------------------------------------------------------
C  ELLPACK GRAPHICS PRIMITIVES  /  DISSPLA VERSION
C-----------------------------------------------------------------------
C
C   Q0GPMD  /  MOVE TO PHYSICAL COORDINATES (X,Y) WITH PEN DOWN.
C
      CALL CONNPT(X,Y)
      RETURN
      END

      SUBROUTINE Q0GPLN (XVEC,YVEC,N,IMARK)
C
C-----------------------------------------------------------------------
C  ELLPACK GRAPHICS PRIMITIVES  /  DISSPLA VERSION
C-----------------------------------------------------------------------
C
C  Q0GPLN  /  DRAW A POLYGONAL LINE CONNECTING THE POINTS WITH
C             LOGICAL COORDINATES (XVEC(K),YVEC(K)), K=1,...,N.
C             IF IMARK.NE.0 AN ON-CENTER PLOT SYMBOL SHOULD BE
C             DRAWN AT EACH OF THE POINTS.
C
      REAL XVEC(N), YVEC(N)
C
      IF (IMARK .NE. 0)  GO TO 50
         CALL CURVE(XVEC,YVEC,N,0)
         GO TO 100
   50    CONTINUE
         CALL MARKER(1)
         CALL CURVE(XVEC,YVEC,N,1)
  100 CONTINUE
      RETURN
      END

      SUBROUTINE Q0GPWC (TEXT,NTEXT,X,Y,H)
C
C-----------------------------------------------------------------------
C  ELLPACK GRAPHICS PRIMITIVES  /  DISSPLA VERSION
C-----------------------------------------------------------------------
C
C   Q0GPWC  /  PLOT THE NTEXT CHARACTERS GIVEN IN THE PACKED
C              ARRAY TEXT STARTING AT PHYSICAL COORDINATES (X,Y)
C              USING CHARACTERS OF HEIGHT H.
C
      CHARACTER*(*) TEXT
C
      CALL HEIGHT(H)
      CALL MESSAG(TEXT,NTEXT,X,Y)
      RETURN
      END

      SUBROUTINE Q0GPWI (I,X,Y,H)
C
C-----------------------------------------------------------------------
C  ELLPACK GRAPHICS PRIMITIVES  /  DISSPLA VERSION
C-----------------------------------------------------------------------
C
C   Q0GPWI  /  PLOT THE INTEGER I BEGINNING AT PHYSICAL
C              COORDINATES (X,Y) USING CHARACTERS OF HEIGHT H.
C
      CALL HEIGHT(H)
```

```
      CALL INTNO(I,X,Y)
      RETURN
      END

      SUBROUTINE Q0GPWR (R,X,Y,H)
C
C-----------------------------------------------------------------------
C  ELLPACK GRAPHICS PRIMITIVES   /   DISSPLA VERSION
C-----------------------------------------------------------------------
C
C   Q0GPWR  /  PLOT THE REAL NUMBER R BEGINNING AT PHYSICAL
C              COORDINATES (X,Y) USING CHARACTERS OF HEIGHT H.
C
      CALL HEIGHT(H)
      CALL REALNO(R,5,X,Y)
      RETURN
      END
```

17.E PREPARING THE ELLPACK PREPROCESSOR

The ELLPACK preprocessor consists of a Fortran main program and a large collection of Fortran subprograms. These must be compiled and linked to form an executable preprocessor. It is desirable, but not necessary, to organize the preprocessor subprograms into a library compatible with the system linker utility. This is especially useful if the preprocessor is to be frequently updated with new problem-solving modules. The preprocessor library must be distinct from the ELLPACK module library. The required programs are found in parts C.0 through C.4 on the ELLPACK tape.

The preprocessor's job is nonnumeric in nature. Each time it executes, two files are read — the ELLPACK template and the user's ELLPACK program — and two output files are produced, one containing a listing and the other containing the ELLPACK control program. The entire preprocessor is coded in standard Fortran 66 with characters stored one per integer word, although a version in which INTEGER is replaced by CHARACTER*1 where appropriate is distributed on request. It is preferable to use character variables if possible, since this greatly reduces the memory requirements of the preprocessor.

The machine dependences in the preprocessor are of three types, which will be described separately: utility routines, input/output, and PROGRAM statements. All subprograms which may require change are found in the part C.1 on the ELLPACK tape.

Utility Routines

Copies of the routines Q1TIME, Q1DATE, and I1MACH which were prepared for the ELLPACK module library (see Section 17.D) must also be included in the preprocessor library.

INPUT/OUTPUT

The preprocessor reads two files and writes two files each time it executes. The way in which these files are opened, closed, and assigned unit numbers may need to be changed on a given computer system. These files and their default logical unit numbers are summarized below:

Unit	*Type*	*Purpose*
IQTMPF	read	ELLPACK template (part C.5 of ELLPACK tape)
IQINPF	read	User's ELLPACK program
IQLSTF	write	Listing and error messages
IQPROG	write	ELLPACK control program

Units IQTMFP, I1INPF, and IQPROG represent simple sequential card image files. Unit IQLSTF is a standard print file.

The initialization of these files is performed in the preprocessor subroutine PPDRIV. In this subroutine logical unit numbers are assigned by the following assignment statements

```
IQTMPF = 4
IQINPF = 5
IQLSTF = 6
IQPROG = 8
```

which may be modified as required for the local system. Explicit OPEN and REWIND statements for these files may also be included here. If necessary, CLOSE statements may also be inserted at the end of PPDRIV to close any of these files.

All preprocessor I/O on units IQTMPF, IQINPF, and IQPROG is done through two low-level subroutines called IORDLN and IOWRLN, for reading and writing, respectively. The versions of these routines distributed with ELLPACK perform simple formatted (80A1 format) I/O. There is substantial overhead in most Fortran run-time systems to do such formatted input and output. **Significant savings in the running time of the preprocessor can be attained if these routines are replaced with machine-specific versions which avoid the use of Fortran formatted I/O.** This, of course, is optional.

The preprocessor's job is the generation of the ELLPACK control program which is itself a Fortran main program that must assign logical unit numbers and open and close files. The ELLPACK Control Program generated by the default ELLPACK system uses the following files:

Unit	*Type*	*Purpose*
I1INPT	read	Standard Fortran input file
I1OUTP	write	Standard Fortran output file
I1SCRA	read/write	Module scratch file
I0TIME	read/write	Module timing file

Units I1INPT and I0TIME are simple sequential card image files. Unit I1OUTP is a standard print file. It should be the same as IQLSTF above. Unit I1SCRA may be used for unformatted sequential I/O.

Fortran statements are included in the ELLPACK template (part C.5) and the ELLPACK definition file (part E.4) which assign values to these unit numbers. The assignments in the default ELLPACK system are

```
I1INPT = I1MACH(1)
I1OUTP = I1MACH(2)
I1SCRA = 1
I0TIME = 2
```

These values may be changed simply by editing the ELLPACK template. The same changes should be made in the ELLPACK definition file (otherwise the defaults will be in effect again if the preprocessor is ever regenerated by the DC processor). If explicit OPEN or REWIND statements are required for any of these files, they may be included immediately after the logical unit assignments. If any of these files needs to be explicitly closed, CLOSE statements may be inserted immediately before the STOP statement in these files.

PROGRAM STATEMENTS

Certain Fortran compilers, notably those from CDC, require a PROGRAM statement at the beginning of main programs. Both the ELLPACK preprocessor and the ELLPACK control program will require PROGRAM statements on these systems. A PROGRAM statement is included as a comment in the preprocessor main program (part C.0). The ELLPACK template and ELLPACK definition files also contain comment cards with PROGRAM statements. These should be modified as required if a PROGRAM statement is needed.

17.F SAMPLE CONTROL CARD PROCEDURES

In this section we give sample control card procedures which run an ELLPACK job. The basic procedure is illustrated in Figure 12.2. We first describe the steps in a machine independent way and then give examples using the terminology of specific operating systems. The files we require are:

TEMPLATE = the ELLPACK template (permanent). This is part C.5 from the ELLPACK tape.
USERPROG = the user's ELLPACK input program.
CTRLPROG = the ELLPACK control program (temporary).
LISTING = the standard output file for printed output.
EMLIB = the ELLPACK module library (permanent).
STD-IN = the standard input file.
SCRATCH1 = scratch file used by the control program (temporary).
SCRATCH2 = scratch file used by the control program (temporary).

The job steps which must be performed are

1. Assign: TEMPLATE for input (unit IQTMPF of Section 17.E).
USERPROG for input (unit IQINPT of Section 17.E).
LISTING for output (unit IQLSTF of Section 17.E).
CTRLPROG for output (unit IQPROG of Section 17.E).

2. Execute: PREPROCESSOR
3. Compile: CTRLPROG
4. Link: CTRLPROG to EMLIB
5. Assign: STD-IN for input (unit I1INPT of Section 17.E)
   LISTING for output (unit I1OUTP of Section 17.E)
   SCRATCH1 for input/output (unit I1SCRA of Section 17.E)
   SCRATCH2 for input/output (unit I0TIME of Section 17.E)
6. Execute: CTRLPROG

SPERRY 1100 EXEC SYSTEM

Sample control cards used to run ELLPACK on the Sperry 1100 EXEC System [Sperry, 1979] are given below. Note that logical unit 1 is the card punch in Sperry EXEC and hence default logical unit assignment for the module scratch file must be changed. We assume that the assignment I1SCRA = 1 is changed to I1SCRA = 9 in the ELLPACK Template (and in the ELLPACK definition file). In addition, we assume that the following files exist:

ELLPACK*TEMPLATE:
A catalogued data file containing the ELLPACK template.

ELLPACK*SYSTEM:
A catalogued program file containing an element, called PREPROCESSOR, which is the absolute version of the ELLPACK preprocessor.

ELLPACK*MODULES:
A catalogued program file containing relocatable versions of all subprograms in the ELLPACK modules. This is the ELLPACK module library. There should be one Fortran subprogram per element. The PREP processor must have been applied to this file.

SYSTEM*GRAPHICS:
The local system graphics library.

An EXEC control card procedure which uses these files to run an ELLPACK job follows:

```
@MSG,N
@MSG,N   EXECUTE ELLPACK PREPROCESSOR
@MSG,N
@ASG,A ELLPACK*TEMPLATE.
@ASG,T ELLPACK*CTRLPROG.
@USE 4,ELLPACK*TEMPLATE
@USE 8,ELLPACK*CTRLPROG
@XQT ELLPACK*SYSTEM.PREPROCESSOR
@ADD ... USER'S ELLPACK PROGRAM ...
@FREE ELLPACK*TEMPLATE.
@MSG,N
@MSG,N   COMPILE ELLPACK CONTROL PROGRAM
@MSG,N
@COPY,I ELLPACK*CTRLPROG.,TPF$.ECP$
@FREE ELLPACK*CTRLPROG.
@FTN,NOZ TPF$.ECP$,TPF$.ECP$
@MSG,N
```

```
@MSG,N  LINK TO ELLPACK MODULE LIBRARY
@MSG,N
@MAP,EIX ,TPF$.ECP$
 IN TPF$.ECP$
LIB ELLPACK*MODULES
LIB SYSTEM*GRAPHICS
END
@MSG,N
@MSG,N  EXECUTE ELLPACK CONTROL PROGRAM
@MSG,N
@ASG,T ELLPACK*MODSCRA.
@ASG,T ELLPACK*TIMINGS.
@USE 9,ELLPACK*SCRATCH
@USE 2,ELLPACK*TIMINGS
@XQT TPF$.ECP$
@ADD  ... USER DATA, IF ANY ...
@FREE ELLPACK*MODSCRA.
@FREE ELLPACK*TIMINGS.
```

IBM JOB CONTROL LANGUAGE

Here we give a rough set of IBM OS/360 JCL which may be used to run ELLPACK. The following data sets are assumed to exist:

PPLOAD:
The ELLPACK preprocessor in load module form, with main program called ELLPRE.

TEMPLATE:
The ELLPACK template.

EMODLIB:
The ELLPACK module library with routines in object module form.

The JCL statements follow. Such statements should be embodied in a catalogued procedure.

```
/*
/*  EXECUTE ELLPACK PREPROCESSOR
/*
//PREPROS   EXEC  PGM=ELLPRE
//STEPLIB         DD DSN=PPLOAD,DISP=SHR
//FT04F001        DD DSN=TEMPLATE,DISP=SHR
//FT08F001        DD DSN=ECP,DISP=(NEW,PASS),UNIT=SYSDA,
//              SPACE=(TRK(5,5)),
//              DCB=(RECFM=FB,LRECL=80,BLKSIZE=800)
//FT06F001        DD SYSOUT=A
//FT05F001        DD *
      ... USER'S ELLPACK PROGRAM ...
/*
/*  COMPILE-LOAD-GO OF ELLPACK CONTROL PROGRAM
/*
//ELLRUN     EXEC  FORTHCLG
//FORT.SYSIN     DD DSN=ECP,DISP=OLD
//LKED.SYSLIB    DD DSN=EMODLIB,DISP=SHR
//               DD DSN=SYS1.FORTLIB,DISP=SHR
//GO.FT01F001    DD DSN=MODSCRA,DISP=NEW,UNIT=SYSDA,
//            SPACE=(TRK,(5,5)),
//            DCB=(RECFM=VB,LRECL=512,BLKSIZE=512)
//GO.FT02F001    DD DSN=ECPSCRA,DISP=NEW,UNIT=SYSDA,
//            SPACE=(TRK,(5,5)),
//            DCB=(RECFM=FB,LRECL=80,BLKSIZE=800)
```

```
//GO.FT05F001   DD *
      ... USER'S DATA, IF ANY ...
```

UNIX Shell Script

For a UNIX system we assume that all files relating to the ELLPACK system are found in the directory /usr/ellpack/lib. These include the executable preprocessor (pp), the ELLPACK template file (template), the library of problem-solving modules (emlib.a), and the local graphics library (plotlib.a). The following simple shell script can be used to run ELLPACK. It is based upon the Berkeley Unix system [Ritchie, 1974].

```
#   UNIX Shell Script --- Execution of the ELLPACK System
#
#  Usage :   ellpack userprog
#
# where userprog is the name of a file containing the
# user's ELLPACK program
#
# The listing produced by the ELLPACK preprocessor is left in the
# file userprog.lst, while the output produced by the execution
# of the ELLPACK control program is left in the file userprog.out
#
ELLPACK=/usr/ellpack/lib
PP=$ELLPACK/pp
TEMPLATE=$ELLPACK/template
EMLIB=$ELLPACK/emlib.a
PLOTLIB=$ELLPACK/plotlib.a
#
#   Run the preprocessor
#
rm -f ctrlprog.f $1.lst
$PP $1 ctrlprog.f $TEMPLATE $1.lst
#
#   Compile the control program and link to modules and graphics
#
f77 -o ctrlprog ctrlprog.f $EMLIB $PLOTLIB
#
#   Execute the control program
#
ctrlprog > $1.out
```

The following file assignments are appropriate for the preprocessor routine PPDRIV.

```
IQTMPF = 4
IQINPF = 5
IQLSTF = 6
IQPROG = 8
CALL GETARG(1,CINPF)
CALL GETARG(2,CPROG)
CALL GETARG(3,CTMPF)
CALL GETARG(4,CLSTF)
OPEN(UNIT=IQINPF,FILE=CINPF,STATUS='OLD')
OPEN(UNIT=IQTMPF,FILE=CTMPF,STATUS='OLD')
OPEN(UNIT=IQPROG,FILE=CPROG,STATUS='NEW')
OPEN(UNIT=IQLSTF,FILE=CLSTF,STATUS='NEW')
```

GETARG is a Unix Fortran subprogram which allows Fortran programs to access the command line arguments. CINPF, CPROG, CTMPF, and CLSTF are declared as CHARACTER*100. With these statements in PPDRIV, the script statement "pp $1 ctrlprog.f $TEMPLATE $1.lst" dictates that $1 is the user's ELLPACK program, ctrlprog.f is the control program produced by the preprocessor, $TEMPLATE is the ELLPACK template file, and $1.lst is the preprocessor output listing file. Clearly, this script can (and should) be embellished to detect and respond to errors of various kinds; we have not done so here so that the script would be easy to understand.

CDC NOS Control Cards

We assume that the ELLPACK template file (TEMPLT) and the ELLPACK module library (EMLIB) are permanent files cataloged under user number ELLPACK. The PROGRAM statement in the preprocessor is taken to be

```
  PROGRAM MAIN  (INPUT=102B,OUTPUT=102B,TEMPLT=102B,CTRLPG=102B,
 *   TAPE1=TEMPLT,TAPE5=INPUT,TAPE6=OUTPUT,TAPE8=CTRLPG)
```

while the generated control program should contain

```
  PROGRAM ELPK  (INPUT,OUTPUT,TIMING,SCRA,TAPE1=SCRA,TAPE2=TIMING,
 *   TAPE5=INPUT,TAPE6=OUTPUT)
```

If the user's ELLPACK program is on the local file USERPG, the following NOS control cards [Control Data, 1979] will process it.

```
REWIND,CTRLPG,SCRA,TIMING,ELLGO.
ATTACH,PP,TEMPLT,EMLIB/UN=ELLPACK.
REWIND,PP,TEMPLT,EMLIB.
PP,USERPG,,TEMPLT,CTRLPG.
REWIND,CTRLPG.
FTN5,I=CTRLPG,B=ELLGO.
LIBRARY,EMLIB.
ELLGO.
```

CDC CYBER 205 VSOS Control Cards

We assume that the ELLPACK preprocessor (PP), template (TEMPLATE), and module library (EMLIB) are permanent files owned by pool ELLPACK. The main program for the preprocessor contains the PROGRAM statement

```
  PROGRAM MAIN  (UNIT4=TEMPLATE,UNIT5=INPUT,
 *                UNIT6=OUTPUT,UNIT8=FORT)
```

The generated control program contains the statement

```
  PROGRAM ELPK  (UNIT5=INPUT,UNIT6=OUTPUT,
 *                UNIT1=SCRATCH,UNIT2=TIMING)
```

If the user's ELLPACK program is on local file USERPG, then the following VSOS control cards [Control Data, 1982] will process it.

```
PATTACH,ELLPACK.
PP,UNIT5=USERPG,UNIT8=CTRLPG.
FORTRAN,I=CTRLPG,O=B.
PATTACH,FORTPOOL.
LOAD,LIB=EMLIB,FORTLIB.
GO.
```

17.G REFERENCES

Control Data [1979], NOS Version 1 Reference Manual, Control Data Corp. Minneapolis, MN.

Control Data [1982], CDC VSOS Verson 2 Reference Manual, Control Data Corp. Minneapolis, MN.

DISSPLA [1981], User's Manual, Version 9.0, ISSCO, 4186 Sorrento Valley Blvd., San Diego, CA.

Ritchie, D. M. and K. Thompson [1974], "The UNIX time-sharing system", Comm. ACM 17, pp. 365-375.

Snyder, W. V. [1978], "Algorithm 531, Contour plotting", ACM Trans. on Math. Software 4, pp. 290–294.

Sperry [1979], *Sperry Univac Series 1100 Executive System Programmer Reference*, vols. 1 and 2, Sperry Corporation Computer Systems, St. Paul, MN.

Chapter 18

TAILORING ELLPACK

The ELLPACK system is designed to be easily modified and extended. One can modify it to

1. change the form of the ELLPACK control program,
2. add new problem solving modules to the ELLPACK module library, or
3. change the syntax of the ELLPACK language.

These capabilities give the system great flexibility. For example, one might want to add extra code to the ELLPACK control program to monitor the use of ELLPACK. This is very easy to do. With only slightly more effort one could add a new PROCEDURE module to do postprocessing of the solution. With more work one could transform the ELLPACK language to use the terminology of a particular application area. One could even arrange for ELLPACK to explicitly recognize time-dependent problems and call specialized problem solving modules for their solution.

Our purpose in this chapter is to provide some guidance as to how ELLPACK can be tailored to suit local needs. The first section gives a brief overview of the operation of the preprocessor; the second describes how the ELLPACK control program is organized. In the third section one can find detailed instructions on how to modify the control program and add new modules. Finally, the last section describes how to change the form of the ELLPACK language itself. Note that in order to fully understand this material, one must be reasonably familiar with the entire ELLPACK system; in particular, one should have read Chapter's 2 and 4 of the User's Guide, Chapters 12–16 of the Contributor's Guide, and Chapter 17 of the System Programmer's Guide.

18.A ORGANIZATION OF THE PREPROCESSOR

To use ELLPACK one writes a program in the **ELLPACK language** which defines the elliptic boundary value problem to be solved and indicates the solution method that is to be used. This language is an extension to Fortran with very high level statements for defining partial differential equations, their domains, and boundary conditions; in addition, there are powerful problem solving statements which perform the basic tasks necessary to solve such problems. The **ELLPACK preprocessor** reads this program and produces the **ELLPACK control program**, a Fortran program which performs all the necessary operations using subprograms in the **ELLPACK module library**.

The ELLPACK preprocessor is constructed using a lexical/syntactic analyzer and a template processor. The **template processor** is a type of macro processor especially designed for manipulating Fortran text. The input to the template processor is called a **template**. It is simply a text file which contains references to template variables. When the template processor is executed it writes out its input, substituting values from its symbol table wherever template variables are referenced. The value of a template variable may be an arbitrary text string, possibly containing several lines and references to other template variables. Directives in the input file may be used to set the value of template variables, to conditionally process text based upon the current value of template variables, or to process text repetitively (looping).

The following is a simple example of a template:

```
*IF(DOUBLE)
*SET(TYPE = 'DOUBLE PRECISION')
*SET(PREFIX = 'D')
*ELSE
*SET(TYPE = 'REAL')
*SET(PREFIX = 'S')
*ENDIF
      $(TYPE) A($(NDIM),$(NDIM)), B($(NDIM))
      READ(5,*) N, ((A(I,J),J=1,N),I=1,N), (B(I),I=1,N)
      CALL $(PREFIX)POFA(A,$(NDIM),N,INFO)
      CALL $(PREFIX)POSL(A,$(NDIM),N,B)
      WRITE(6,*) (B(I),I=1,N)
      STOP
      END
```

Lines beginning with an asterisk (*) are directives to the template processor. Items beginning with a dollar sign ($) are references to template variables. If the template processor's symbol table initially contained the values DOUBLE='.TRUE.' and DNIM='100', then the output of the template processor for this example would be the following:

```
      DOUBLE PRECISION A(100,100), B(100)
      READ(5,*) N, ((A(I,J),J=1,N),I=1,N), B(I),I=1,N)
      CALL DPOFA(A,100,N,INFO)
      CALL DPOSL(A,100,N,B)
      WRITE(6,*) (B(I),I=1,N)
      STOP
      END
```

The ELLPACK control program is generated by running a template of the control program through the template processor. Before this is done the symbol table must contain information about each of the problem solving modules in ELLPACK, such as their names and the Fortran code required to invoke them. This information is obtained by initially running the template processor with an input file containing directives assigning values to many template variables. The form of the ELLPACK control program also depends upon the partial differential equation being solved and the names of the problem solving modules actually requested. The lexical/syntactic analyzer updates the template processor's symbol table with this information based upon the results of parsing the user's ELLPACK program.

The ELLPACK processor, then, does its job in three phases:

Phase I: System Definition.
The first part of the ELLPACK template is read, initializing the symbol table with many template variables, including those defining the problem solving modules currently available, their properties, and the Fortran code required to invoke them.

Phase II: Analysis.
The User's ELLPACK program is read and parsed. The problem to be solved and the modules to be invoked are determined, and the symbol table is updated to reflect these things.

Phase III: Code Generation.
The remainder of the ELLPACK template, containing the template of the ELLPACK control program, is read and processed. References to template variables are replaced with their actual values from the symbol table.

The operation of the preprocessor is illustrated in Figure 18.1. The entire system is coded in standard Fortran. Phases I and III are implemented by separate calls of the subroutine **TPDRV** which is the template processor. Control of the three-phase execution of the preprocessor is found in the subroutine **PPDRIV**.

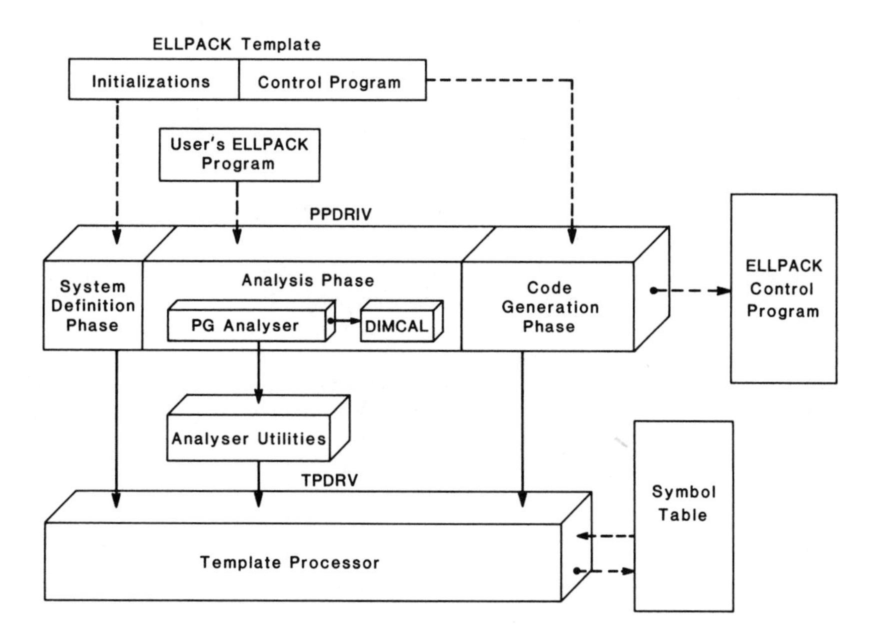

Figure 18.1. A detailed look at the preprocessor. Solid arrows indicate control flow; dashed lines indicate data flow.

The phase II analyzer uses many of the low-level routines of the template processor to do such things as reading, writing, and interacting with the symbol table. The analyzer utilities provide an interface to these programs. **DIMCAL** is a set of two small subprograms (DIMCL1 and DIMCL2) which return information about the storage requirements of each problem solving modle as a function of problem parameters. This is the only place in which information about problem solving modules explicitly appears in the code of the preprocessor (all other information is obtained via the ELLPACK template). The dimension calculation code is supplied by module conributors, and hence these routines must be modified, recompiled, and the preprocessor relinked each time a module is added to the system. Further information about DIMCAl may be found in Section 18.C.

The major part of the ELLPACK lexical/syntactic analyzer is generated automatically using the **PG system**, a compiler-compiler specifically designed for the generation of Fortran preprocessors. PG reads a grammar defining the language constructs to be recognized and indicating actions which are to be taken when they are recognized. PG then generates a set of Fortran subprograms which parses the desired language and takes the desired actions when language elements are recognized. These subprograms make up the PG analyzer displayed in Figure 18.1. The PG system is described further in Section 18.C and in Appendix B.

18.B ORGANIZATION OF THE ELLPACK CONTROL PROGRAM

The ELLPACK control program is the principle output of the ELLPACK preprocessor. It is a Fortran program which performs the tasks specified in the user's ELLPACK program. This program is not normally seen by ELLPACK users. However, its structure must be understood if changes are to be made to it. We briefly describe the format of the control program in the system originally distributed by Purdue University. In general, control programs generated at local sites may have some changes required in their implementations.

With some minor exceptions, the control program produced by the ELLPACK preprocessor will be compatible with ANSI standard Fortran (1966). The only exception may be in the use of quoted string constants. The correctness of the control program also depends upon the correctness of user-supplied Fortran code in the DECLARATIONS, GLOBAL, FORTRAN, and SUBPROGRAMS segments. These segments are copied verbatim into the control program. This is also true of various user-supplied arithmetic expressions given as part of the EQUATION and BOUNDARY segments.

The basic components of the Control Program are summarized below.

Standard Declarations

The control program begins with a set of declaration statements defining all internal ELLPACK variables. Variables are organized into groups called **interfaces**. Each interface (except the first and last) is the input to a module of one type and the output of a module of another type. All variables are allocated in named COMMON blocks. Arrays occupy individual blocks and scalars are organized into blocks based upon their type (INTEGER, REAL, LOGICAL). Blank COMMON is available for working storage. The storage allocated to each array variable depends upon the user's problem, the grid, and the problem solving modules invoked. A complete description of ELLPACK interface variables is contained in Chapter 13.

SPECIAL DECLARATIONS

User-supplied declaration statements (from the DECLARATIONS and GLOBAL segments) are inserted next, followed by additional declarations required by the problem solving modules that have been invoked.

INITIALIZATIONS

The first block of executable Fortran code generated initializes variables at the initial module interface. A number of global constants are also defined here.

MAIN PROGRAM

A block of executable Fortran code is inserted for each executable ELLPACK segment in the order encountered in the user's ELLPACK program. These segments are GRID, DISCRETIZATION, INDEXING, SOLUTION, TRIPLE, PROCEDURE, OUTPUT, and FORTRAN. The specific code inserted depends upon the problem and the specific modules invoked. Code in FORTRAN segments is copied verbatim.

STANDARD SUBPROGRAMS

A set of standard Fortran subprograms used by the system to define various aspects of the PDE problem follows next. These subprograms are:

Q1PCOE = Coefficients of the partial differential equation.

R1PRHS = Right side of the partial differential equation.

Q1BCOE = Coefficients of the boundary conditions.

R1BRHS = Right sides of the boundary conditions.

Q1BDRY = Defines the boundary of the domain, (present only when problems on non rectangular domains are solved).

The operation of these routines is described in detail in Chapter 13. They are written explicitly by the preprocessor because they return values computed by Fortran expressions written by the user in the EQUATION and BOUNDARY segments of the ELLPACK program. User-specified declarations given in the GLOBAL segment are also inserted into these routines. An additional function, called R0SOLV, is also written by the preprocessor. This function returns the value (or derivative) of the computed solution function at a given point. The way this is computed depends upon the DISCRETIZATION or TRIPLE modules specified by the user.

USER SUBPROGRAMS

The contents of the SUBPROGRAMS segment is copied verbatim at the end of the control program.

EXAMPLE 18.1 The ELLPACK control program for Example 4.D.1

The remainder of this section gives a guided tour through a sample ELLPACK control program. We use the program produced by the preprocessor given Example 4.D1 as input. This example solves a problem on a nonrectangular domain. User code is included to perform some special pre- and postprocessing. A detailed description of the code follows. Note that variables and subprograms follow a specific naming convention. This convention is described in Section 12.C.

```
 1: C      PROGRAM ELPK
 2: C
 3: C==========      PROBLEM DEFINITION INTERFACE
 4: C
 5:        LOGICAL                  L1ARCC, L1CLKW, L1CRST, L1CSTB, L1CSTC,
 6:       A                         L1DRCH, L1HMBC, L1HMEQ, L1HOLE, L1LAPL,
 7:       B                         L1MIXD, L1NEUM, L1NUNQ, L1POIS, L1PRDC,
 8:       C                         L1PRDX, L1PRDY, L1PRDZ, L1RECT, L1SELF,
 9:       D                         L1TWOD
10:        COMMON / C1LVPR /        L1ARCC, L1CLKW, L1CRST, L1CSTB, L1CSTC,
11:       A                         L1DRCH, L1HMBC, L1HMEQ, L1HOLE, L1LAPL,
12:       B                         L1MIXD, L1NEUM, L1NUNQ, L1POIS, L1PRDC,
13:       C                         L1PRDX, L1PRDY, L1PRDZ, L1RECT, L1SELF,
14:       D                         L1TWOD
15:        COMMON / C1IVPR /        I1NBND
16:        COMMIN / C1RVPR /        R1CUXX, R1CUXY, R1CUYY, R1CCUX, R1CCUY,
17:       A                         R1CCCU, R1CUZZ, R1CUXZ, R1CUYZ, R1CCUZ,
18:       B                         R1UNQX, R1UNQY, R1UNQZ, R1UNQU,
19:        COMMON   / C1BCST /      I1BCST(4,4)
20:        COMMON   / C1BCTY /      I1BCTY(4)
21:        COMMON   / C1CFST /      I1CFST(10)
22:        EXTERNAL                 Q1BDRY
23:        COMMON / C1BRNG /        R1BRNG(2,4)
24: C
25: C==========      DISCRETE DOMAIN INTERFACE
26: C
27:        LOGICAL                  L1UNFG, L1UNFX, L1UNFY, L1UNFZ
28:        COMMON   / C1LVGR /      L1UNFG, L1UNFX, L1UNFY, L1UNFZ
29:        COMMON   / C1IVGR /      I1NGRX, I1NGRY, I1NGRZ, I1NBPT, I1MBPT,
30:       A                         I1PACK
31:        COMMON   / C1RVGR /      R1AXGR, R1AYGR, R1BZGR, R1BXGR, R1BYGR,
32:       A                         R1BZGR, R1HXGR, R1HYGR, R1HZGR
33:        COMMON   / C1GRDX /      R1GRDX(11)
34:        COMMON   / C1GRDY /      R1GRDY(16)
35:        COMMON   / C1GRDZ /      R1GRDZ(1)
36:        COMMON   / C1GRTY /      I1GRTY(11,16)
37:        COMMON   / C1BNGH /      I1BNGH(148)
38:        COMMON   / C1BGRD /      I1BGRD(148)
39:        COMMON   / C1BPTY /      I1BPTY(148)
40:        COMMON   / C1PECE /      I1PECE(148)
41:        COMMON   / C1BPAR /      R1BPAR(148)
42:        COMMON   / C1XBND /      R1XBND(148)
43:        COMMON   / C1YBND /      R1BYND(148)
44: C
45: C==========      DISCRETE OPERATOR INTERFACE
46: C
47:        LOGICAL                  L1SYMM
48:        COMMON   / C1LVDI /      L1SYMM
49:        COMMON   / C1IVDI /      I1NEQN, I1MNEQ, I1NCOE, I1MNCO
50:        COMMON   / C1COEF /      R1COEF(203,7)
51:        COMMON   / C1IDCO /      I1IDCO(203,7)
52:        COMMON   / C1BBBB /      R1BBBB(203)
53: C
55: C==========      EQUATION/UNKNOWN REORDERING INTERFACE
55: C
56:        LOGICAL                  L1ASIS, L1RDBL
57:        COMMON   / C1LVIN /      L1ASIS, L1RDBL
58:        COMMON   / C1IVIN /      L1MEND, L1MUND
59:        COMMON   / C1ENDX /      I1ENDX(203)
60:        COMMON   / C1UNDX /      I1UNDX(203)
61: C
62: C==========      ALGEBRAIC EQUATION SOLUTION INTERFACE
63: C
64:        LOGICAL                  L1UINI
65:        COMMON   / C1LVSO /      L1UINI
66:        COMMON   / C1LVSO /      I1MUNK
67:        COMMON   / C1UNKN /      R1UNKN(203)
68: C
69: C==========      OTHER GLOBAL CONTROL VARIABLES
70: C
71:        LOGICAL                  L1TIME, L1FATL, L1NEWD
```

```
 72:        COMMON   / C1IVCN /   I1TIME, L1FATL, L1NEWD
 73:        COMMON   / C1IVCN /   I1LEVL, L1PAGE, L1INPT, L1OUTP, L1SCRA,
 74:       A                      I1KWRK, I1KORD
 75:        COMMON   / C1RVGL /   R1EPSG, R1EPSM, PI
 76:        COMMON   / C1RNRM /   R1NRM1, R1NRM2, R1NRMI
 77:        COMMON   / C1RTABL/   R1TABL(11,16,1)
 78:        COMMON                R1WORK(16849)
 79: C
 80:        COMMON   / C1RVBS /   R1BSTP(1)
 81:        COMMON   / C0IVCN /   I0GROT, I0MODN, I0TIME
 82:        INTEGER               I0GROT(6)
 83:        COMMON   / C0LVCN /   L0HVDI, L0HVAN
 84:        LOGICAL               L0HVDI, L0HVAN
 85: C
 86:        COMMON / PARAMS / BETA,BOV2,BOV2SQ,YINF,DELTA,TWOPI,TWOPID
 87:        EXTERNAL              C
 88: C
 89:        COMMON   / C35PNU /   I35PNU(11,16)
 90:        COMMON   / C35GBN /   I35GBN(148)
 91: C
 92:        I1INPT = I1MACH(1)
 93:        I1OUTP = I1MACH(2)
 94:        I1SCRA = 1
 95:        I0TIME = 2
 96:        PI     = 4.*ATAN(1.)
 97:        R1EPSM = R1MACH(3)
 98:        I1BCST(1,1) = 4
 99:        I1BCST(1,2) = 1
100:        I1BCST(1,3) = 4
101:        I1BCST(1,4) = 31
102:        I1MBPT      = 148
103:        I1BCTY(1)   = 2
104:        I1BCTY(2)   = 1
105:        I1BCTY(3)   = 2
106:        I1BCTY(4)   = 3
107:        CALL Q0INIT( .TRUE., .FALSE., .FALSE., .TRUE.,
108:       A             .FALSE., .FALSE., .TRUE., .FALSE., .FALSE.,
109:       B             .TRUE., .FALSE., .FALSE., .FALSE., .FALSE.,
110:       C             .TRUE., 4, .FALSE., 203,
111:       D             203, 7, 203, 203,
112:       E             16849, 2065, 0, .TRUE., .TRUE.,
113:       F             .TRUE., .TRUE. )
114:        CALL Q1PCOE(0.0, 0.0, R1CUXX)
115: C
116:        CALL Q8GP00
117: C
118:        YINF  = 0.5
119:        BETA  = 2.0
120:        DELTA = 0.2
121:        WRITE(6,2000) YINF,BETA,DELTA
122:   2000 FORMAT('1 PARAMETERS FOR THIS RUN ARE ' /
123:       A       ' ---------------------------- ' //
124:       B       5X,'YINF  = ',1PE12.5 / 5X,'BETA  = ',1PE12.5 /
125:       C       5X,'DELTA = ',1PE12.5 /)
126:        BOV2 = 0.5*BETA
127:        BOV2SQ = BOV2*BOV2
128:        TWOPI = 2.0*PI
129:        TWOPID = TWOPI*DELTA
130: C
131: C============     SETUP GRIDX
132: C
133:        I1NGRX = 11
134:        L1UNFX = .TRUE.
135:        R1AXGR = 0.0
136:        R1BXGR = 0.5
```

```
137:        CALL Q0GRUF(R1AXGR, R1BXGR, R1GRDX, I1NGRX, R1HXGR, 11, 'X')
138: C
139: C===========    SETUP GRIDY
140: C
141:        I1NGRY = 16
142:        L1UNFY = .TRUE.
143:        R1AYGR = W(0.5)
144:        R1BYGR = YINF
145:        CALL Q0GRUF(R1AYGR, R1BYGR, R1GRDY, I1NGRY, R1HYGR, 16, 'Y')
146:        L1UNFG = L1UNFX .AND. L1UNFY .AND. L1UNFZ
147: C
148:        R1BRNG(1,1) = W(0.0)
149:        R1BRNG(2,1) = YINF
150:        R1BRNG(1,2) = 0.0
151:        R1BRNG(2,2) = 0.5
152:        R1BRNG(1,3) = 0.0
153:        R1BRNG(2,3) = YINF-W(0.5)
154:        R1BRNG(1,4) = 0.0
155:        R1BRNG(2,4) = 0.5
156: C
157: C===========    DOMAIN
158: C
159:        L1CLKW = .TRUE.
160:        I1NBND = 4
161:        CALL Q2DPMN
162:        IF (L1FATL)  CALL Q1FATL
163: C
164: C===========    5-POINT STAR
165: C
166:        CALL Q35GMN
167:        I0DISM = 1
168:        L0HVDI = .TRUE.
169:        L1UINI = .FALSE.
170:        CALL Q0ASIS
171:        IF (L1FATL)  CALL Q1FATL
172:        IF (.NOT. L0HVDI)  CALL Q0ERPP(2)
173: C
174: C===========    LINPACK BAND
175: C
176:        CALL Q5LBSU(I5BDLW, I5BDUP)
177:        IF (L1FATL)  CALL Q1FATL
178:        CALL Q5LBMN  (I5BDLW, I5BDUP)
179:        I0MODN = I0DISM
180:        L0HVAN = .TRUE.
181:        L1NEWD = .TRUE.
182:        CALL Q0UNDX
183:        IF (L1FATL)  CALL Q1FATL
184: C
185: C===========    PLOT
186: C
187:        CALL Q8PLNR(C, 'C      ', -1, -1 )
188:        IF (L1FATL)  CALL Q1FATL
189:        NPTS = 21
190:        WRITE(6,2001)
191:   2001 FORMAT('      TABLE OF CONCENTRATION ALONG INTERFACE' /
192:      A      '      ----------------------------------------' //
193:      B      '             X                  C(X,W(X))         '/ )
194:        DX = 0.5/FLOAT(NPTS-1)
195:        DO 100 I = 1,NPTS
196:           X = FLOAT(I-1)*DX
197:           CVAL = C(X,W(X))
198:           WRITE(6,2002) X,CVAL
199:    100 CONTINUE
200:   2002 FORMAT(1X,1P,2E18.6)
201:        CALL Q8GP99
```

```
202:         STOP
203:         END
204:         SUBROUTINE Q1PCOE(X, Y, R0CPDE)
205: C
206: C ======= DEFINE EQUATION COEFFICIENTS
207: C
208:         COMMON  / C1RVGL /  R1EPSG, R1EPSM, PI
209:         COMMON / PARAMS / BETA,BOV2,BOV2SQ,YINF,DELTA,TWOPI,TWOPID
210:         REAL               R0CPDE(6)
211: C
212:         R0CPDE( 1) = 1.0
213:         R0CPDE( 2) = 0.0
214:         R0CPDE( 3) = 1.0
215:         R0CPDE( 4) = 0.0
216:         R0CPDE( 5) = 0.0
217:         R0CPDE( 6) = -BOV2SQ
218: C
219:         RETURN
220:         END
221:         REAL FUNCTION R1PRHS(X, Y)
222: C
223: C ======= DEFINE THE RIGHT SIDE OF THE EQUATION
224: C
225:         COMMON  / C1RVGL /  R1EPSG, R1EPSM, PI
226:         COMMON / PARAMS / BETA,BOV2,BOV2SQ,YINF,DELTA,TWOPI,TWOPID
227:         R1PRHS = 0.0
228: C
229:         RETURN
230:         END
231:         SUBROUTINE Q1BCOE(I0SIDE, X, Y, R0CBC)
232: C
233: C ======= DEFINE THE BOUNDARY CONDITIONS
234: C
235:         COMMON  / C1RVGL /  R1EPSG, R1EPSM, PI
236:         COMMON / PARAMS / BETA,BOV2,BOV2SQ,YINF,DELTA,TWOPI,TWOPID
237:         REAL R0CBC(3)
238:         COMMON  / C0IVCN /  I0GROT, I0MODN, I0TIME
239:         INTEGER             I0GROT(6)
240: C
241:         I0BCND = I0SIDE
212:         GO TO (10001,10002,10003,10004), I0BCND
243: 10001 CONTINUE
244:         R0CBC(1) = 0.0
245:         R0CBC(2) = 1.0
246:         R0CBC(3) = 0.0
247:         GO TO 9999
248: C
249: 10002 CONTINUE
250:         R0CBC(1) = 1.0
251:         R0CBC(2) = 0.0
252:         R0CBC(3) = 0.0
253:         GO TO 9999
254: C
255: 10003 CONTINUE
256:         R0CBC(1) = 0.0
257:         R0CBC(2) = 1.0
258:         R0CBC(3) = 0.0
259:         GO TO 9999
260: C
261: 10004 CONTINUE
262:         R0CBC(1) = +BOV2
263:         R0CBC(2) = -DW(X)
264:         R0CBC(3) = 1.0
265:         GO TO 9999
266: C
```

```
267: C
268:  9999 CONTINUE
269:       RETURN
270:       END
271:       REAL FUNCTION R1BRHS(I0SIDE, X, Y)
272: C
273: C ======= DEFINE THE RIGHT SIDES OF THE BOUNDARY CONDITIONS
274: C
275:       COMMON  / C1RVGL /  R1EPSG, R1EPSM, PI
276:       COMMON / PARAMS / BETA,BOV2,BOV2SQ,YINF,DELTA,TWOPI,TWOPID
277:       COMMON  / C0IVCN /  I0GROT, I0MODN, I0TIME
278:       INTEGER             I0GROT(6)
279: C
280:       I0BCND = I0SIDE
281:       GO TO (10001,10002,10003,10004), I0BCND
282: 10001 CONTINUE
283:       R1BRHS = 0.0
284:       GO TO 9999
285: C
286: 10002 CONTINUE
287:       R1BRHS = 0.0
288:       GO TO 9999
289: C
290: 10003 CONTINUE
291:       R1BRHS = 0.0
292:       GO TO 9999
293: C
294: 10004 CONTINUE
295:       R1BRHS = -BETA*SINH(BOV2*Y)
296:       GO TO 9999
297: C
298: C
299:  9999 CONTINUE
300:       RETURN
301:       END
302:       SUBROUTINE Q1BDRY(R0PARM, X, Y, I0PECE)
303: C
304: C ======= DEFINE THE BOUNDARY PIECES
305: C
306:       COMMON  / C1RVGL /  R1EPSG, R1EPSM, PI
307:       COMMON / PARAMS / BETA,BOV2,BOV2SQ,YINF,DELTA,TWOPI,TWOPID
308:       GO TO (10001,10002,10003,10004), I0PECE
309: 10001 CONTINUE
310:       T = R0PARM
311:       X=0.0
312:       Y=T
313:       GO TO 9999
314: C
315: 10002 CONTINUE
316:       T = R0PARM
317:       X=T
318:       Y=YINF
319:       GO TO 9999
320: C
321: 10003 CONTINUE
322:       T = R0PARM
323:       X=0.5
324:       Y=YINF-T
325:       GO TO 9999
326: C
327: 10004 CONTINUE
328:       T = R0PARM
329:       X=0.5-T
330:       Y=W(0.5-T)
331:       GO TO 9999
```

```
332: C
333:  9999 RETURN
334:       END
335:       REAL FUNCTION R0SOLV(IDERIV, X, Y)
336: C
337: C ======= RETURN THE SOLUTION AT THE SPECIFIED POINT
338: C
339:       COMMON  / C1UNKN /  R1UNKN(203)
340:       COMMON  / C1TABL /  R1TABL(11, 16, 1)
341:       COMMON  / C1LVCN /  I1LEVL, I1PAGE, I1INPT, I1OUTP, I1SCRA,
342        A                     I1KWRK, I1KORD
343:       COMMON  / C0IVCN /  L1TIME, L1FATL, L1NEWD
344:       LOGICAL             L1TIME, L1FATL, L1NEWD
345:       COMMON  / C0LVCN /  I0GROT, I0MODN, I0TIME
346:       INTEGER             I0GROT(6)
347:       COMMON  / C0LVCN /  L0HVDI, L0HVAN
348:       LOGICAL             L0HVDI, L0HVAN
349: C
350:       IF (.NOT. L0HVAN)  CALL Q0ERPP(3)
351:       GO TO (10001), I0MODN
352: 10001 CONTINUE
353: C
354: C===========        5-POINT STAR
355: C
356:       IF (L1NEWD)  CALL Q35GVL
357:       IF (L1NEWD)  CALL Q2XTMN(R1TABL)
358:       R0SOLV = R1QD2I(X, Y, R1TABL, IDERIV)
359:       GO TO 9999
360: C
361: C
362:  9999 CONTINUE
363:       L1NEWD = .FALSE.
364:       RETURN
365:       END
366:       REAL FUNCTION W(X)
367: C                SHAPE OF THE SOLID-LIQUID INTERFACE
368:       COMMON / PARAMS / BETA,BOV2,BOV2SQ,YINF,DELTA,TWOPI,TWOPID
369:       W = DELTA*COS(TWOPI*X)
370:       RETURN
371:       END
372:       REAL FUNCTION DW(X)
373: C                 DERIVATIVE OF SOLID-LIQUID INTERFACE SHAPE
374:       COMMON / PARAMS / BETA,BOV2,BOV2SQ,YINF,DELTA,TWOPI,TWOPID
375:       DW = -TWOPID*SIN(TWOPI*X)
376:       RETURN
377:       END
378:       REAL FUNCTION C(X,Y)
379: C                 COMPUTES CONCENTRATION FROM PDE SOLUTION
380:       COMMON / PARAMS / BETA,BOV2,BOV2SQ,YINF,DELTA,TWOPI,TWOPID
381:       C = 1.0 + EXP(-BETA*Y) + U(X,Y)*EXP(-BOV2*Y)
382:       RETURN
383:       END
```

The following notes describe the items in the above program in detail.

Lines	Notes
1:	PROGRAM statement. This comment may be replaced by a PROGRAM card in some implementations.
2– 84:	ELLPACK interface variables. These standard declarations define the interfaces between ELLPACK problem solving modules. They are copied in this form

Lines	Notes
	into every ELLPACK program. Lines 22–23 and 36–43 only appear when problems on nonrectangular domains are solved (in which case the ELLPACK domain processor is called). Array sizes depend upon the problem being solved and the problem solving modules invoked. Note the blank common workspace R1WORK declared in line 78. This is available for use as scratch storage by all modules; its size is the maximum workspace required by all the modules invoked in this program.
86:	GLOBAL declaration. This statement is copied from the GLOBAL segment in the user's ELLPACK program.
87:	External declaration. This appears as a result of the PLOT(C) directive in the user's ELLPACK program. The user function C will be passed to a standard ELLPACK function which makes a contour plot of a function of two variables.
89– 90:	Module-specific declarations. These are additional arrays required by the module 5-POINT STAR; their sizes depend upon the grid size.
92–114:	General initializations. These statements set up the initial interface in ELLPACK which defines the problem that is to be solved. Much of this work is done inside the standard routine Q0INIT. The first four assignments set the value of logical unit numbers for files used for standard input, standard output, scratch storage, and timing information, respectively.
116:	Graphics initialization. This call initializes the local graphics system, and appears only when plotting is requested.
118–129:	User code. These statements are copied verbatim from the first FORTRAN segment in the user's ELLPACK program.
130–162:	Domain processing. These statements are written wherever a GRID segment occurs in the user's ELLPACK program. The first two parts, lines 130–137 and lines 138–146 set up uniform grids in x and y, respectively. The third part, lines 148–162, appears only when the problem involves a nonrectangular domain. The subprogram Q2DPMN is the ELLPACK domain processor.
163–172:	DISCRETIZATION module. The first statement invokes the 5-POINT STAR module. The remaining statements are found in each discretization module invocation. The first assigns a unique identifier for this module in

Lines	Notes
	this run, the second signals that a discretization has been done, and the third indicates that the solution vector has not yet been initialized. The subprogram Q0ASIS sets permutation vectors for equations and unknowns to the identity. The final two statements perform error checking.
173–183:	SOLUTION module. These statements invoke the LINPACK BAND module. This module requires a setup phase, and hence there are two separate call statements required (these would be timed separately if the user requested module timing statistics). The remaining statements appear whenever a solution module is called. The first three assignment statements indicate that the system of equations generated by the 5-POINT STAR module (I0MODN = I0DISM) has now been solved (L0HVAN = .TRUE.), and that the solution has not yet been evaluated (L1NEWD = .TRUE.). The subroutine Q0UNDX applies the inverse unknown permutation to the solution vector. This is required if an INDEXING module was invoked in the user's ELLPACK program; it has no effect here.
184–188:	OUTPUT module. Here a subprogram which produces a contour plot of a function of two variables on a nonrectangular domain is called. The name of the function to be plotted, as well as a character string containing its name, is passed to the routine.
189–200:	User code. Statements from the second FORTRAN segment in the user's ELLPACK program are copied here.
201:	Graphics finalization. A statement which signals the local graphics system that plotting has been completed is inserted here whenever plotting has been requested.
204–334:	Problem definition subprograms. These are subprograms called by ELLPACK modules to evaluate the coefficients and right sides of the partial differential equation and its boundary conditions. Fortran expressions from the EQUATION and BOUNDARY segments are copied directly into these routines without syntax checking. (See lines 217, 295, and 329–330, for example.) Note also that the Fortran declarations from the user's GLOBAL segment are copied into each of these routines.
335–365:	Solution evaluation. After a solution has been computed, ELLPACK users may freely use the functions U(X,Y), UX(X,Y), etc. to evaluate the computed solution and its derivatives. These functions, in turn, call R0SOLV to perform the evaluation. One section of code is included in this routine for each DISCRETIZATION and TRIPLE module invoked. A computed GOTO statement (line 351) selects the "current"

Lines	Notes
	evaluation procedure. In this example there is only one choice (lines 353–359). The 5-POINT STAR module sets up a standard formatted table of the solution on first call (line 356–357), and then calls an ELLPACK utility to interpolate into this table thereafter.
366–383:	User subprograms. The contents of the SUBPROGRAMS segment in the user's ELLPACK program is copied at the end of the Control Program.

18.C HOW TO ADD MODULES TO ELLPACK

Chapters 12–16 describe the standard interfaces between modules of various types in the ELLPACK system. This description should be sufficient to allow sophisticated users to actually prepare new problem-solving modules for the system. In this section we describe the next step in the process, that is, how to define a new module to the preprocessor once it has been written.

As indicated in Section 18.A, two parts of the system must be updated to define a new module, the ELLPACK template and the DIMCAL subprograms. The template contains the Fortran code segments that must be inserted into the ELLPACK control program to invoke the module, and the DIMCAL routines compute dimensions of standard ELLPACK interface arrays whose sizes are problem-dependent. The syntax and conventions used in the template and DIMCAL are quite complex, much more suited to machine interpretation than to human interpretation. Because of this, a very high level description of the form of the control program, and the ELLPACK modules in particular, is kept in a standard file called the **ELLPACK definition file**. Another processor, called **DC**, reads the ELLPACK definition file and a template for the DIMCAL subprograms, and produces a correctly formatted ELLPACK template and DIMCAL. This process is illustrated in Figure 16.2. Although many changes to the system can be effected by modifying the ELLPACK template directly, it is recommended that all changes to the form of the control program be made by modifying the ELLPACK definition file and running DC (it will probably never be necessary to modify the DIMCAL template). In this way the ELLPACK definition file will remain a readable, up-to-date description of the state of the ELLPACK system. The ELLPACK definition file which defines the standard ELLPACK system is found in file E.4 of the ELLPACK tape (see Section 17.C).

The Ellpack Definition File

The ELLPACK definition file (EDF) can be considered as a "program" in a very high-level language similar in style to the ELLPACK language itself. The statements in this language are called segments. Each segment begins with a segment name starting in column 1 and ending with a period (only three characters are significant). The initial line of a segment may be followed by 1 or more additional lines with a blank in column 1. If the last nonblank character in a line is an ampersand (&), then the next line is considered part of the current line. The last line in the EDF is signaled by the characters "END." in columns 1–4.

Comments, indicated by a plus (+) in column one may appear anywhere; the comment line +1+ causes a page eject in the listing produced by the DC processor. Blank lines are also considered as comments. Many segments contain Fortran statements or expressions; these must follow the rules of Fortran. Most of the segments are, in fact, Fortran code templates. Thus, it is useful to be familiar with the syntax required by the ELLPACK template processor. This is described in detail in Appendix C. We do not present an exhaustive description of EDF syntax in this section. Rather, we intend to describe enough so that one can easily read the existing EDF and understand what it does.

The ELLPACK definition file can informally be divided into four parts:

1. Initializations.
2. Definition of OPTIONS.
3. Definition of modules.
4. The control program template.

Normally, only the third part need by changed when adding new modules. When making other changes to the form of the control program, the fourth part might also need changing. The first part consists of three segments: ELLPACK, SYMBOLS, and DC.

ELLPACK.
This segment appears first. It specifies default values for many of the template variables used to determine the form of the control program.

SYMBOLS.
This segment, which appears second, is copied directly into the ELLPACK template. It defines many subtemplates which may be used repeatedly in the control program. Examples of these templates include the code written to set up a grid, standard code written before each module invocation (i.e., a Fortran comment, and possibly a call on the timing routine), and the standard code written after each DISCRETIZATION module.

DC.
This segment specifies Fortran code that is to be copied into DIMCAL rather than into the ELLPACK template. It appears several times after the SYMBOLS segment to define code to initialize and finalize the dimension calculations required by the preprocessor. It also appears later in the file to define the size requirements of interface arrays required by each ELLPACK module. This is done by simply assigning values to preprocessor variables which are used to store the sizes of interface arrays. The constructs $I(⟨name⟩) and $L(⟨name⟩) are used in this segment to refer to the current value of the template variable ⟨name⟩ in the template processor symbol table. These are converted to calls to the symbol table manager. Further information about DC can be found in Section 16.B.

The second part of the ELLPACK definition file defines the constructs allowed in the OPTIONS segment of a user's ELLPACK program. In the ELLPACK language, OPTION statements take the form ⟨variable⟩ = ⟨value⟩.

Some options are "executable" in the sense that they may be varied during the execution of the ELLPACK run, while others are "nonexecutable" in that they only affect how the preprocessor works. In the former case one wants an assignment written into the ELLPACK control program, while in the latter case one wants a preprocessor template variable to be modified. Up to three EDF segments may be required to fully specify an OPTION.

OPTION.
This specifies the name of the option and its default value. This will also be the name of the preprocessor template variable or the ELLPACK control program variable that the option affects.

ALIAS.
This specifies an alternate name for the option.

TYPE.
If this segment appears, then the option is executable. A TYPE of FORTRAN specifies that the option assignment should only be copied into the ELLPACK control program. A TYPE of BOTH specifies that the corresponding template variable should also be updated. As an example, consider the following sequence of EDF segments:

```
OPTION.     I1LEVL = 1
ALIAS.      LEVEL
TYPE.       FORTRAN
```

This specifies that OPTIONS of the forms "I1LEVL = ⟨value⟩" and "LEVEL = ⟨value⟩" will be accepted in ELLPACK programs, and that when they occur the former should be copied directly into the control program. The default value (if the user does not specify this option) is 1.

The third part of the ELLPACK definition file contains a complete description of each of the problem-solving modules in the system. An entry for a particular module may require up to six different segments: an ELLPACK call (one of DISCRETIZATION, INDEXING, SOLUTION, TRIPLE, PROCEDURE, or OUTPUT), DECLARATION, SETUP, FORTRAN, ANSWER, and DC.

ELLPACK CALL
This gives the type of module, its name, its parameters, and their defaults. The segment name used indicates the type of module; in fact, the format of this segment shows exactly how the module must be invoked in ELLPACK programs.

DECLARATION.
This segment gives any special declaration statements which must be written into the control program for this module.

SETUP.
This segment gives Fortran statements which perform any interface setup operations which must be timed separately.

FORTRAN.

This segment gives Fortran statements which call the subprogram which executes this module from the ELLPACK module library.

ANSWER.

This segment gives Fortran statements which evaluate the computed solution function (only required of DISCRETIZATION and TRIPLE modules).

DC.

This segment gives Fortran statements which give the sizes of interface arrays as a function of problem parameters.

The DC processor will write information into DIMCAL when processing the DC segment; all other segments will cause information to be added to the ELLPACK template. Each of the DECLARATION, SETUP, FORTRAN, and ANSWER segments is, in fact, a Fortran code template. Module contributors must prepare entries of this type when submitting modules for inclusion in ELLPACK, and so a detailed description of these segments is given in the Contributor's Guide, Section 16.B.

The final part of the EDF gives the control program template; it is copied directly into the ELLPACK template by the DC processor.

Installing the DC Processor

The DC processor consists of a Fortran main program and a collection of Fortran subprograms. Like the preprocessor, the DC processor's job is nonnumeric in nature. When it executes, two files are read — the ELLPACK definition file and the DIMCAL template — and three output files are produced, one containing a listing, one containing the ELLPACK template, and the other containing DIMCAL routines for use by the preprocessor. Like the preprocessor, DC does its job in three steps.

1. Initialization.
2. Analysis.
3. Code generation.

In the second phase the ELLPACK definition file is read and digested, and in the third phase the output files are produced. DC has been constructed using many of the same tools used to build the preprocessor. In particular, its lexical and syntactic analysis routines are generated automatically using the PG system (see Section 18.D and Appendix B). The template processor (see Appendix C) is used to maintain the DC symbol table and to perform much of the output generation.

The DC processor occupies part E of the ELLPACK tape. file E.0 is the main program, file E.1 contains system dependent routines, file E.2 contains the DC subprograms, and files E.3 and E.4 contain the ELLPACK definition file and DIMCAL template which define the standard ELLPACK system. Since DC also requires many of the same underlying utility programs as the preprocessor, it is necessary to have the preprocessor subprograms (those on files C.1, C.3, and C.4 on the ELLPACK tape) organized as a library on your system so that the DC programs can be linked to them. If this was not originally done, then it must be done before proceeding.

Once the preprocessor library is prepared, the installation procedure is quite simple, and can be summarized as follows.

1. Adjust the file specifications in the routine DCDRIV (file E.1 of the ELLPACK tape), if necessary.
2. Compile the the DC main program (file E.0 of the ELLPACK tape), the routine DCDRIV, and the DC subprograms (file E.2 of the ELLPACK tape).
3. Link these programs to the preprocessor library.
4. Prepare a command procedure which will execute DC on your system.

The only changes which may be required in the DC code are in the routine DCDRIV and involve the use of external files. The way in which these files are opened, closed, and assigned unit numbers may need to be changed on a given computer system. The required files are given below.

Unit	*Type*	*Purpose*
ITEMPF	write	ELLPACK template
IDMCLF	write	DIMCAL routines
ISCRA3	read/write	Scratch file
ISCRA4	read/write	Scratch file
IQINPF	read	ELLPACK definition file
IQLSTF	write	Listing and error messages
IDCTPF	read	DIMCAL template

Units ITEMPF, IDMCLF, ISCRA3, ISCRA4, IQINPF, and IDCTPF are simple sequential card image files, while IQLSTF is a standard print file.

Unit numbers are assigned in DCDRIV by the following statements:

```
ITEMPF = 1
IDMCLF = 2
ISCRA3 = 3
ISCRA4 = 4
IQINPF = 5
IQLSTF = 6
IDCTPF = 7
```

Explicit OPEN and REWIND statements for these files may be included here. These statements should be modified as required. If necessary, CLOSE statements may be inserted at the end of DCDRIV to close any of these files.

18.D HOW TO CHANGE THE ELLPACK LANGUAGE

Changing the ELLPACK language can mean any number of things. This ranges from simple syntax changes to the addition of new segments. The former is straightforward, while the latter may be quite difficult, requiring changes to the preprocessor, the DC processor, the control program, and the module interfaces.

Thus, there is no simple cookbook recipe for changing the ELLPACK language. This is a job for ELLPACK "experts", and if one intends to do it, then it will be necessary to spend the time to understand all the parts of the system and how they fit together.

Any change in the ELLPACK language will certainly require changes in the portion of the preprocessor which parses the ELLPACK language. This is the PG analyzer shown in Figure 18.1. These subprograms were automatically generated by the PG (Preprocessor Generator) system. This system takes a formal definition of the syntax of the ELLPACK language as input, and produces a set of subprograms which parse ELLPACK programs. This same system is used to generate the analyzer for the DC processor. In this section we we briefly describe how the PG system can be used to build preprocessors.

PG GRAMMARS

The syntax of the ELLPACK language can be described succinctly by a formal grammar. A formal grammar may be specified by a set of production rules which show how high level constructs in the language are defined in terms of lower level constructs. For example, a simple IF statement in Fortran might be defined by the production rule

```
SIMPLE-IF  ->  'IF'  '('  LOGICAL-EXPRESSION  ')'  SIMPLE-STATEMENT
```

This says that a SIMPLE-IF is made up of the characters "IF", followed by the character "(", followed by a LOGICAL-EXPRESSION, followed by the character ")", followed by a SIMPLE-STATEMENT. Items such as SIMPLE-STATEMENT are called variables, while characters such as I, F, (, and) are called terminal symbols. The variables LOGICAL-EXPRESSION and SIMPLE-STATEMENT are defined by other production rules. A grammar for Fortran, then, would be a complete set of production rules which shows how to get from a starting variable such as FORTRAN-PROGRAM to any syntacticly correct program made up of only terminal symbols. The problem of parsing a computer program is the reverse of this; given a particular program, one wants to determine whether it is legal. This can be done by working backwards in the production rules. If a path from the particular program to the starting symbol can be found, then the program is syntactically correct.

Armed with a formal system for defining languages, one may then attempt to automatically generate a parser for a language given its grammar. programs which do this are called **compiler-compilers**. In order to process its input language, a compiler must take certain actions when elements of the language are recognized. Thus, the input to a compiler-compiler is not only a set of production rules defining the syntax of the language, but also a specification of what actions must be performed when the parser recognizes that a given rule has been applied.

The PG system, constructed at Purdue University as part of the ELLPACK project, is such a compiler-compiler. It was especially designed for the automatic generation of Fortran preprocessors. A detailed description of PG is given in Appendix B. PG was used to generate parsers for the ELLPACK preprocessor, and the DC processor, as well as for the PG system itself. Thus, formal grammars have been written to define the input languages for each of these processors.

PG grammars can be considered as programs in a very high level language similar in style to the ELLPACK language itself. The statements in this language are called segments. Each segment begins with a segment name starting in column 1 and ending with a period (only three characters are significant). The initial line of a segment may be followed by one or more additional lines with a blank in column 1. If the last nonblank character in a line is an ampersand (&), then the next line is considered part of the current line. The last line in the grammar is signaled by the characters "END." in columns 1–4. Comments, indicated by a star (*) in column 1, may appear anywhere. Blank lines are also considered as comments. We do not present an exhaustive description of of the syntax of PG grammars in this section. Instead we refer readers to the description found in Appendix B.

PG grammars are made up of three separate parts.

1. Global declarations.
2. Driver program definition.
3. Production rules.

The production rules are all found in the MAIN segment of the PG grammar. The PG system generates one Fortran subroutine for each production rule to performs the requested actions. Also in this segment is the specification of a Fortran CALL statement for a user-supplied driver routine for the analyzer. The GLOBAL segment defines a set of Fortran declarations to be included in each of the rule-processing routines.

The main part of the grammar is the set of production rules. Each rule begins with the name of a variable, followed by the characters "->", followed by the definition of the variable in terms of other variables and terminal symbols. More than one "->" may appear to indicate alternate definitions. As a simple example, consider the following set of production rules:

```
GRDPTS    ->    GRDSIZ XYZ '-'? 'POINTS'
GRDSIZ    ->    NAME
          ->    DIGIT+
XYZ       ->    'X'
          ->    'Y'
          ->    'Z'
```

The character "?" after an item means that it is optional, and the character "+" after an item indicates that one or more occurrences of the item should be matched. The first rule defines GRDPTS as a GRDSIZ, followed by an XYZ, followed optionally by the character "–", followed by the characters "POINTS". A GRDSIZ is defined as either a NAME or one or more DIGITs. NAME and DIGIT are predefined variables which match Fortran variable names and the digits 0–9, respectively. Finally, an XYZ is either the character "X", the character "Y", or the character "Z". Thus, this set of rules will match any of the following strings:

```
10 X POINTS
5263 Z-POINTS
```

A3 Y POINTS

Each production rule may be followed by Fortran-like statements which show what actions are to be taken when the rule is applied. The form of these statements is described in Appendix B. Typical actions might involve updating a global variable in the analyzer, or updating the (template processor) symbol table to reflect the item matched.

Installing the PG System

The PG system consists of a Fortran main program and a collection of Fortran subprograms. Like the preprocessor, the PG system's job is nonnumeric in nature. When it executes, two files are read, the input grammar and the analyzer template, and two output files are produced, one containing a listing, and the other containing analyzer routines which parse the language defined by the input grammar. Like the preprocessor, PG does its job in three steps.

1. Initialization.
2. Analysis.
3. Code Generation.

In the second phase the input grammar is read and digested, and in the third phase the output files are produced. PG has been constructed using many of the same tools used to build the preprocessor. In particular, the template processor (see Appendix C) is used to maintain the PG symbol table and to perform the output generation. In fact, the lexical and syntactic analysis routines in PG are also defined by a PG grammar, and hence are generated automatically by PG itself.

The PG system is not normally installed when ELLPACK is first implemented on a particular system. In the remainder of this section we describe the installation procedure. The PG system occupies part F of the ELLPACK tape. File F.0 is the main program, file F.1 contains system-dependent routines, file F.2 contains the PG subprograms, and file F.3 contains the analyzer template. The next two files (F.4 and F.5) contain the grammars which define the standard ELLPACK language and the standard DC input language (i.e., the syntax of the ELLPACK definition file). The final file (F.6) contains the grammar which defines the PG language itself. Since PG also requires many of the same underlying utility programs as the preprocessor, it is necessary to have the preprocessor subprograms (those on files C.1, C.3, and C.4 on the ELLPACK tape) organized as a library on your system so that the PG programs can be linked to them.

Once the preprocessor library is prepared the installation procedure is quite simple, and can be summarized as follows.

1. Adjust the file specifications in the routine PGCDRV (file F.1 of the ELLPACK tape), if necessary.
2. Compile the PG main program (file F.0 of the ELLPACK tape), the routine PGCDRV, and the PG subprograms (file F.2 of the ELLPACK tape).
3. Link these programs to the preprocessor library.

4. Prepare a system command procedure which will execute PG on your system.

The only changes which may be required in the PG code are in the routine PGCDRV and involve the use of external files. The way in which these files are opened, closed, and assigned unit numbers may need to be changed on a given computer system. The required files are given below:

Unit	*Type*	*Purpose*
IQSYMF	read	Analyzer template
IQINPF	read	Input grammar
IQLSTF	write	Listing and error messages
IQOUTF	read/write	Scratch file
IQPROG	write	Generated analyzer program

Units IQSYMF, IQINPF, IQOUTF, and IQPROG are simple sequential card image files, while IQLSTF is a standard print file.

Unit numbers are assigned in PGCDRV by the following statements:

```
IQSYMF = 4
IQINPF = 5
IQLSTF = 6
IQOUTF = 7
IQPROG = 8
```

Explicit OPEN and REWIND statements for these files may also be included here. These statements should be modified as required. If necessary, CLOSE statements may be inserted at the end of PGCDRV to close any of these files.

APPENDICES

A. THE PDE POPULATION
B. THE PG (PREPROCESSOR GENERATOR) SYSTEM
C. THE TEMPLATE PROCESSOR
D. EXAMPLES OF SOME ELLPACK STATEMENTS

Appendix A

THE PDE POPULATION

Wayne R. Dyksen
Purdue University

Elias N. Houstis
University of Thessaloniki

John R. Rice
Purdue University

A.1 INTRODUCTION

The motivation for creating this PDE population is for use in the evaluation of numerical methods and PDE software. The need and rationale for a systematic approach to evaluation is given in Chapter 8. A properly chosen problem population is an essential ingredient for a sound evaluation of numerical methods and software.

It is important that one be able to create relevant subpopulations as one inevitably wants to evaluate methods for particular subclasses of PDEs (e.g., ones that are separable, have singularities or have mixed boundary conditions). Experience shows that no one method is best for all elliptic PDEs and one of the important tasks of research is to create and/or identify methods that are especially efficient for particular classes of PDEs. Once one embarks on such a task one sees that this population, which originally might seem large and bulky, is actually rather small for the uses to be made of it. It is only the fact that it can be substantially expanded in various directions through parametrization that gives one hope that it is adequate for a wide variety of evaluations.

The use of this population for experimental performance evaluation is illustrated in Chapters 10 and 11. Those results are very interesting and suggestive, but they do not provide conclusive scientific evidence about the relative merits of different methods. Many of the conclusions suggested by the results in Chapters 10 and 11 can be scientifically established with more extensive experiments. Indeed, one of the purposes of the ELLPACK system is to allow people to make small or large experiments with methods and problems particularly relevant to their applications.

We first discuss the sources of the PDEs, how they are described and how a structure has been created in the population through the use of quantitative (but subjective) measures of features.

A.2 CHARACTERISTICS OF THE PROBLEMS

SOURCES

A source parameter is assigned to each PDE which ranges from 0 (artificial problem) to 100 (actual real world problem). This feature, as the others introduced later, is subjective in nature and the values given must be taken as approximate indications of our intuitive feelings. The PDE $u_{xx} + u_{yy} = 1$ might be completely artificial for one person and be the actual application PDE for another. We have tried to be consistent in these values.

Many problems have been normalized so the maximum value of the solution is 1.0 and almost all have this value between .1 and 100. Many of the domains have been standardized to the unit square, $0 \leqslant x, y \leqslant 1$. The sources of the PDE are:

A. **Problems used in previous studies**. Nine problems are included which were used by [Eisenstat and Schultz, 1973] or [Houstis et al., 1975 and 1978]. Some of these PDEs have had parameters added and most have been normalized so the maximum value of the solution is about 1.0.

B. **Artificial problems**. Many problems have been created just to exhibit various mathematical behaviors of interest (e.g. singularities, oscillations or wave fronts). Such behaviors are important for theory or application (or both) and one needs to have them present in the population in an easily identifiable manner.

C. **Problems adapted from the "real world"**. A persistent conflict is the desire to have PDEs which represent the "real world" and the necessity to know their true solutions. The following strategies to adapt real world problems have been used:

1. choosing explicit functions which model the physical solutions and then determining appropriate boundary conditions and/or right side to make this the true solution;

2. using using truncated series expansions (of high accuracy) with appropriate small modifications in the boundary conditions or right side;

3. solving solving nonlinear problems approximately, then substituting the tabulated numerical solution into the operator (using quadratic interpolation from a 10 by 10 grid) to obtain a linear problem which in turn, is solved approximately. In these cases the true solution is not known, but the machine-readable population contains tabulated values of a hopefully accurate numerical solution.

PROBLEM FEATURES AND COMPLEXITY CLASSIFICATION

We identify as problem features the *smoothness* and *local variation* of the operator, the boundary conditions and the solution. These features are quantified on a one-dimensional scale of 0 to 100 even though there are rather independent properties associated with the terms smoothness or local variation. These features are measured subjectively from the following descriptions of the scale.

A. **Smoothness.** This refers to the mathematical properties of the functions or operators involved. Key points on the scale are:

00	=	entire functions or constants
10	=	analytic; very well behaved
30	=	very smooth, discontinuity possible in some higher derivative (5 or so)
50	=	still smooth, third derivative discontinuity possible
70	=	not rough to the eye, but possibly only 1 continuous derivative
80	=	continuous, functions might be theoretically smooth but rough on a gross scale,
90	=	possibly discontinuous, nearly singular functions or operators
100	=	strong singularities like $1/x$ or $1/x^2$.

B. **Local variation.** This refers to how much a function changes (relative to its size) in a small part of its domain. These variations might be oscillations, wave fronts, peaks or boundary layers. Key points on the scale are

00	=	very smooth, uniform,
10	=	mild variation, probably convex, some nonuniformity, e.g. $sin(2x)$, e^{3x} on [0,1],
25	=	modest variation or oscillation; mild wave front or peak, e.g. $\sin(6x)$, $1/(1+100x^4)$ on [0,1],
40	=	considerable peak or oscillation; order of magnitude change occurs within 10 – 15 percent of domain,
60	=	sharp peaks, wave fronts, boundary layers or oscillations; 100 percent change in magnitude occurs within 5 percent of domain,
75	=	practically a discontinuity in magnitude; continuity observable only with a fine scale examination,
90	=	actual discontinuity in magnitude; extreme oscillation, step functions, e.g. $\sin(300x)$ on [0,1].

The overall problem complexity is represented by the average of the above six feature measures (i.e., smoothness and local variation for each of the operator, the boundary conditions and the solution). The problems in this population do not have complexities exceeding 58 (only one exceeds 50), a level which might be interpreted as "rather messy with one or two substantial complications". The problem feature measures are included in the descriptions along with the source parameter.

The following tables are given:

Table A1: The PDEs grouped according to types of the operator and boundary conditions (e.g. Helmholtz and Dirichlet or constant coefficients and mixed),

Table A2: The PDEs listed with abbreviated feature descriptions,

Table A3: The PDEs grouped according to the smoothness of the operator and right side,

Table A4: The PDEs grouped according to the smoothness of the solution.

Table A.1. Classification of Problems According to Operator and Boundary Conditions. Note that problems 1, 4 and 35 appear in two places in the table since they have boundary conditions of the form $u + au_n = g$ and hence have Dirichlet boundary conditions for $\alpha = 0$. Problem 24 appears in two places since it has boundary conditions of the form $\beta u + u_x + u_y = 0$.

Operator	*Constant Coefficients*			*Nonconstant Coefficients*		
	Dirichlet	*Neumann*	*Mixed*	*Dirichlet*	*Neumann*	*Mixed*
Laplace	3, 4, 7 8, 10, 11, 17, 33, 34, 35, 47, 50		4, 31, 35, 38 55			
Helmholtz Type	9, 41, 53			6, 20, 39, 44, 45, 48, 49		
Self-Adjoint	5			1, 13, 22, 25, 28, 54,		1, 19, 23, 52
General	14, 46	42	43	12, 15, 16, 18, 21, 26, 27, 29, 30, 32, 36, 37, 56	24	2, 23, 24, 40 51

Table A.2. Problem Characteristics. The principal characteristics are tabulated below using the following encodings:

A	Analytic	N	Neumann Boundary Condition
BL	Boundary Layer	NS	Nearly Singular
C	Constant (coefficients)	O	Oscillatory
CC	Computationally Complex	P	Parameterized or Peaked
D	Dirichlet Boundary Condition	S	Singular (infinite)
E	Entire	SD	Singular Derivative
H	Homogeneous	U	Unknown
J	Jump Discontinuity	VS	Variable Smoothness
M	Mixed Boundary Condition	WF	Wave Front

Characteristics in parentheses are possible for certain special parameter values.

Problem Number	*Operator*	*Right Side*	*Solution*	*Boundary Conditions*	*Domain*
1P	A	E	E	M	Unit Square
2	E	A	A	M,H	Unit Square
3P	C	S,SD	S,SD	D,H	Unit Square
4P	C	E	E	M	Unit Square
5P	C	E	E	D,H	Unit Square
6	E,NS	A	A,O	D,H	Unit Square
7	C	C	SD	D,H	Unit Square
8P	C	SD	SD,WF	D	Unit Square
9P	C,NS	E,NS	E,BL	D	Unit Square
10P	C	E,P	E,P	D,H	Unit Square

Problem Number	*Operator*	*Right Side*	*Solution*	*Boundary Conditions*	*Domain*
11P	C	A,O	A,O	D	Unit Square
12P	E,O	E,O	E,O	D	Unit Square
13	J	S	SD	D	Unit Square
14P	C	S	S	D	Unit Square
15P	A,NS	S	SD	D	Unit Square
16P	A,NS	C	U,BL	D,H	Variable Square
17P	C	A,NS	A,NS,WF	D	Unit Square
18P	E	A,NS	A,NS,WF	D	Unit Square
19P	S	S	E	M,H	Square
20P	NS,P,CC	P	E,P	D	Rectangle
21	E	E	E	D	Unit Square
22	SD	S	E	D	Unit Square
23P	SD	SD	SD,WF	M,H	Unit Square
24P	S,NS	S,NS	U,P	M,H	Square
25P	SD	S	E	D,H	Unit Square
26P	A	A	U,SD	D,H	Variable Square
27	A,NS	C	U,BL	D,H	Square
28P	J	C	U,WF	D,H	Square
29P	S	H	U,VS,BL	D	Unit Square
30P	A,CC	A,CC	A,NS	D	Unit Square
31	C	C	E,(SD)	M	Square
32	A	A	E	D,H	Rectangle
33	C	E	E,O	D	Rectangle
34	C	C	E,(SD)	D	Square
35P	C	H	E,O,BL	M	Square
36P	S	S	A,BL	D	Unit Square
37	E	E	E	D	Unit Square
38P	C	H	E,O,VS	D	Rectangle
39P	CC,S	CC,S	U,BL	D,C	Unit Square
40P	E	A	A	M	Unit Square
41P	C,NS	SD,NS	SD	D,H	Square
42P	C	H	A,O	N	Variable Rectangle
43	C	H	E	M	Square
44P	CC	CC	U,BL	D,H	Unit Square
45P	C,NS	H	U,BL	D	Unit Square
46P	C,NS	H	U,BL	D	Variable Rectangle
47P	C	S	SD,VS	D	Unit Square
48P	CC	CC	U	U	Unit Square
49P	CC	CC	U,SD,BL	D,C	Unit Square
50	C	H	E,O	D	Rectangle
51P	S	C	U,SD,WF	M,H	Unit Square
52P	CC	H	U,O	M,C	Unit Square
53P	C,NS	E,O	E,O	D	Unit Square
54P	E,CC	S,CC	SD,VS	D	Unit Square
55P	C	H	S,VS,BL	M	Rectangle
56P	S	CC	U,O(SD)	M	Rectangle

Table A.3. Classification of Problems According to Smoothness of the Operator and Right Side. The following codes are used:

A	=	Analytic;	C	=	Constants;
CC	=	Computationally Complicated;	DD	=	Discontinuous Derivatives;
E	=	Entire;	O	=	Oscillatory;
P	=	Peak;	S	=	Singular

Smoothness		*Problem Numbers*
Operator	*Right Side*	
C	C	7, 31, 34, 35, 38, 42, 43, 45, 46, 50, 55
C	E	4, 5, 9, 10, 33, 53
C	A	11, 17
C	DD	3, 8, 41
C	S	3, 14, 47
C	O	6, 11, 53
C	P	10
E	E	12, 21, 37
E	A	2, 6, 18, 40
E	S	54
A	C	16, 27
A	E	1
A	A	26, 30, 32
A	S	15
DD	C	28
DD	DD	13, 23, 25
DD	S	22, 25
S	C	29, 51, 56
S	S	19, 24, 36
O	A	6
O	O	12
C	CC	17, 18
S	CC	56
CC	C	52
CC	P	20
CC	CC	30, 39, 44, 48, 49, 54

Table A.4. Classification of Problems According to Smoothness of the Solution

Solution Smoothness	*Problem Numbers*
Entire	1, 4, 5, 9, 10, 12, 19, 20, 21, 22, 25, 31, 32, 33, 34, 35, 37, 38, 43, 50, 53
Analytic	2, 6, 11, 17, 18, 30, 36, 40, 42
Singular Derivatives	3, 7, 13, 14, 15, 41, 47, 51, 54, 56
Oscillatory	6, 11, 12, 33, 35, 38, 42, 50, 53, 56
Wave Front	8, 17, 18, 23, 28, 51
Discontinuous Derivatives	8, 23
Singular	54, 55
Boundary Layer	7, 9, 15, 16, 27, 29, 44, 45, 46, 49
Peak	10, 20, 24
Tabulated Solution	16, 24, 26, 27, 28, 29, 39, 44, 45, 46 48, 49, 51, 52

A.3 FORMAT OF PDE PROBLEM DESCRIPTIONS

Section A.5 contains a mathematical description of each PDE along with contour plots for each specific parameter set included in the PDE population. The description begins with the problem number and source followed by a mathematical description of the PDE. Then brief comments are given for the operator, right side, boundary conditions, solution and parameters (if any). Sometimes functions appearing in the mathematical description are defined in these comments.

Four **generic functions** are used:

$f(x,y)$ = right side of PDE determined so that the given true solution is correct.

$f(x)$, $g(y)$ = right sides of boundary conditions determined so that the given true solution is correct.

$T(x,y)$ = an approximate solution used in some PDEs whose true solution is unknown.

Contour plots are given for one or more particular PDEs for each problem. The border of the plots contains the following information:

1. values of the parameters (the variables A, B, ..., denote α, β, ...);

2. maximum and minimum values of the solution; the contours are equispaced between these values;

3. the classification parameters in the form

S.P O1.O2 B1.B2 S1.S2

where

S	=	source parameter	P	=	problem complexity
$\alpha 1$	=	smoothness feature	α	=	local variation

α = O for the operator
= B for the boundary conditions
= S for the solution

There is a machine-readable description of the PDE population which consists of two files: EQNFIL and MACFIL. EQNFIL has entries which are either complete statements of the PDE in the ELLPACK language or a reference to an entry in MACFIL with values given for parameters. See Figure A.1 for a short example. The information given starts with the problem number, feature parameter values and a code for various attributes of the PDE which are used within the ELLPACK system. The machine-readable description uses A, B, C, etc. for α, β, γ, etc. Then ELLPACK language code is given for the operator and boundary conditions. Finally, there is Fortran code for any functions that appear in the operator, right side or boundary conditions. The latter code can be very long.

MACFIL entries are just like EQNFIL descriptions of a PDE except that the places where parameter values are to be substituted are indicated by &A, &B, etc. A refers to the first parameter, B the second, and so on. There are somewhat more than 8500 lines in these two files.

A.4 ACKNOWLEDGEMENTS

The creation of this PDE population has been supported in part as follows: National Science Foundation grant MCS77-01408, United States Army Contract DAAG29-75-C-0024, National Science Foundation grant MCS79-01437, and Department of Energy contract DE-AC02-81ER-10997.

A.5 PROBLEM DEFINITIONS AND PLOTS

The 56 problems of the PDE population are defined and contour plots given for many values of the problem parameters. References for the sources of the problems are given in Section A.6 and indicated on the problem description by numbers in square brackets.

```
'EOR
'EOF
***************
* PROBLEM       2*
***************
'EOR
1            000.04        000.00        004.05        010.02
1            2000200000020
1            TWO  DIMENSIONS
1            UXX  +  (1.+Y*Y)*UYY  -  UX  -  (1.+Y*Y)*UY  =  F(X,Y)
2            MIXED
2            U  +  UX  =  0.27*EXP(Y)                          ON  X=0.
2            U  -  UX  =  0.                                     ON  X=1.
2            U  +  UX  =  0.27*EXP(X)                          ON  Y=0.
2            U  -  UX  =  0.135*(ALOG(2.)-1.)*(X*X-X)**2  ON  Y=1.
3        FUNCTION  TRUE(X,Y)
3        TRUE  =  0.135*(EXP(X+Y)+(X*X-X)**2*ALOG(1.+Y*Y))
3        RETURN
3        END
3        FUNCTION  F(X,Y)
3        F  =  0.135*(  (-4.*X*X*X+18.*X*X-14.*X+2.)*ALOG(1.+Y*Y)
3       $       -  2.*((X*X-X)**2)*(Y*Y+Y**3+Y-1)/(1.+Y*Y)  )
3        RETURN
3        END
--------------------------------------------------------------------
'EOR
'EOF
***************
* PROBLEM     3 *
***************
'EOR
*PARAMETER  SET        1(A=1.5)
*             000.43     090.60        000.00        070.40
EXPAND  3/1.5/
'EOR
*PARAMETER  SET        2(A=2.5)
*             000.35     080.50        000.00        060.20
EXPAND  3/2.5/
'EOR
*PARAMETER  SET        3(A=3.5)
*             000.28     070.30        000.00        050.15
EXPAND  3/3.5/
'EOR
*PARAMETER  SET        4(A=4.5)
*             000.23     055.20        000.00        040.20
EXPAND  3/4.5/
```

Figure A.1 A sample from EQNFIL. The PDE description in machine-readable form is shown for problem 2, the references to MACFIL are shown for problem 3.

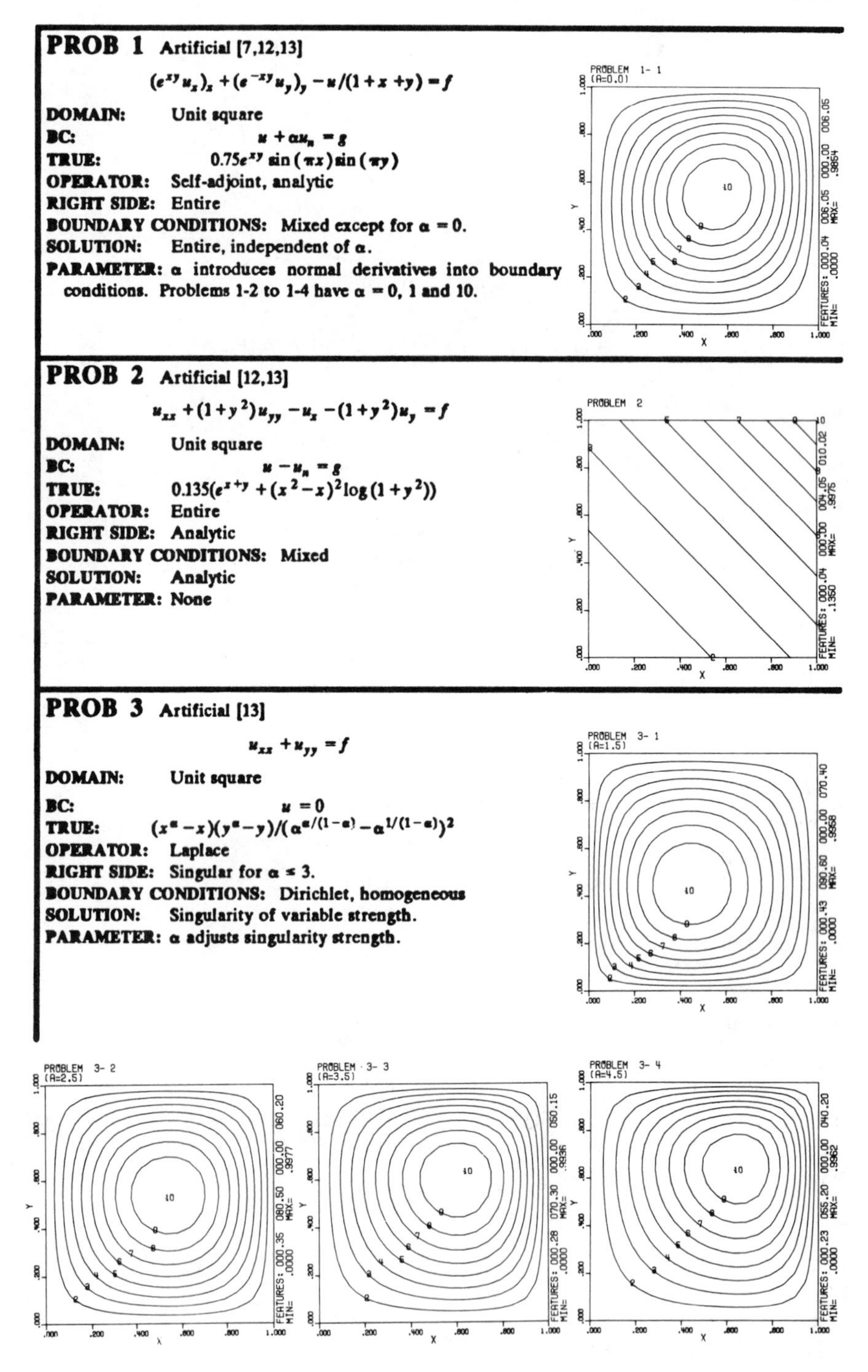

PROB 1 Artificial [7,12,13]

$$(e^{xy}u_x)_x + (e^{-xy}u_y)_y - u/(1+x+y) = f$$

DOMAIN: Unit square
BC: $u + \alpha u_n = g$
TRUE: $0.75e^{xy}\sin(\pi x)\sin(\pi y)$
OPERATOR: Self-adjoint, analytic
RIGHT SIDE: Entire
BOUNDARY CONDITIONS: Mixed except for $\alpha = 0$.
SOLUTION: Entire, independent of α.
PARAMETER: α introduces normal derivatives into boundary conditions. Problems 1-2 to 1-4 have $\alpha = 0$, 1 and 10.

PROB 2 Artificial [12,13]

$$u_{xx} + (1+y^2)u_{yy} - u_x - (1+y^2)u_y = f$$

DOMAIN: Unit square
BC: $u - u_n = g$
TRUE: $0.135(e^{x+y} + (x^2-x)^2\log(1+y^2))$
OPERATOR: Entire
RIGHT SIDE: Analytic
BOUNDARY CONDITIONS: Mixed
SOLUTION: Analytic
PARAMETER: None

PROB 3 Artificial [13]

$$u_{xx} + u_{yy} = f$$

DOMAIN: Unit square
BC: $u = 0$
TRUE: $(x^\alpha - x)(y^\alpha - y)/(\alpha^{\alpha/(1-\alpha)} - \alpha^{1/(1-\alpha)})^2$
OPERATOR: Laplace
RIGHT SIDE: Singular for $\alpha \le 3$.
BOUNDARY CONDITIONS: Dirichlet, homogeneous
SOLUTION: Singularity of variable strength.
PARAMETER: α adjusts singularity strength.

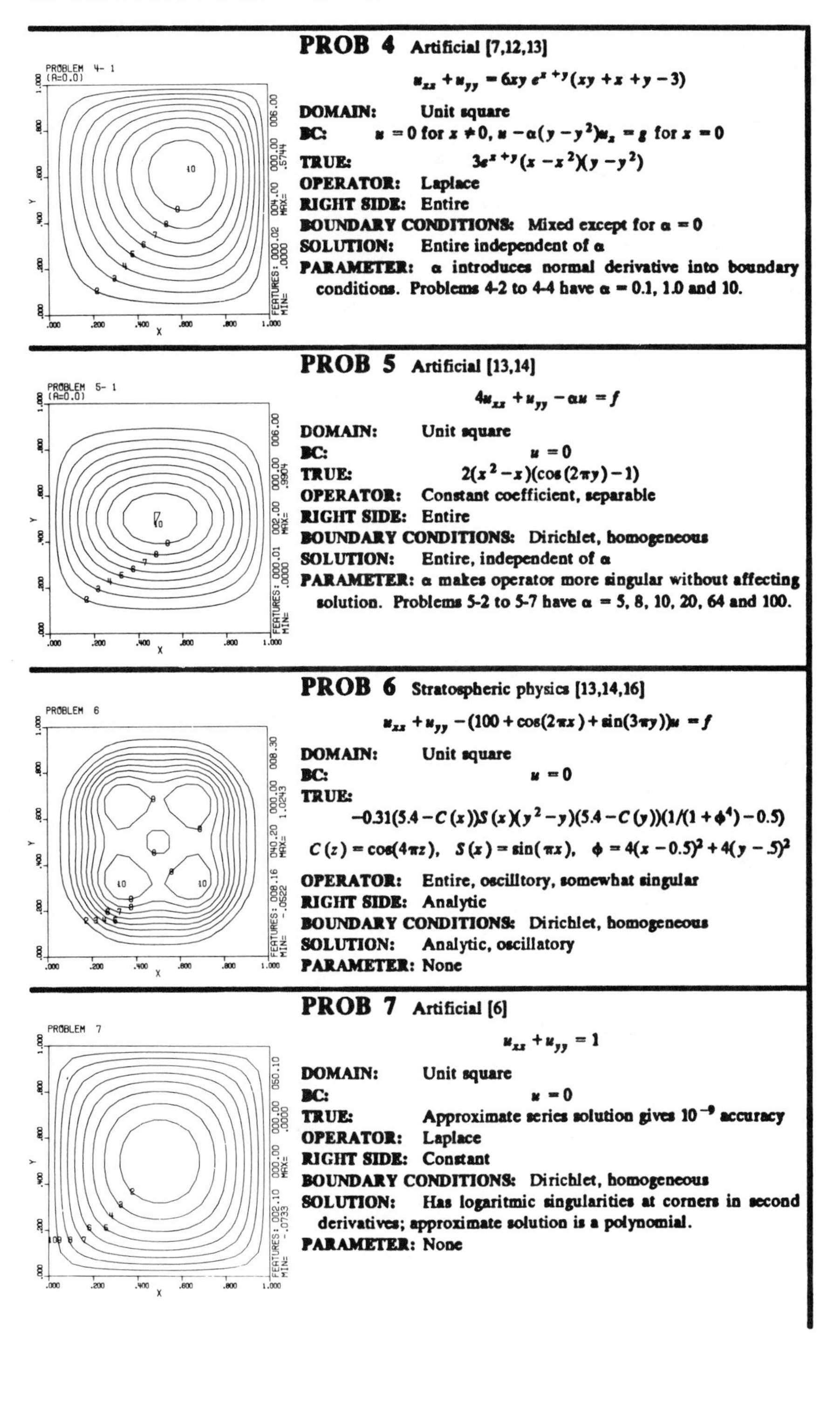

PROB 4 Artificial [7,12,13]

$$u_{xx} + u_{yy} = 6xy\,e^{x+y}(xy + x + y - 3)$$

DOMAIN: Unit square

BC: $u = 0$ for $x \neq 0$, $u - \alpha(y - y^2)u_x = g$ for $x = 0$

TRUE: $3e^{x+y}(x - x^2)(y - y^2)$

OPERATOR: Laplace

RIGHT SIDE: Entire

BOUNDARY CONDITIONS: Mixed except for $\alpha = 0$

SOLUTION: Entire independent of α

PARAMETER: α introduces normal derivative into boundary conditions. Problems 4-2 to 4-4 have $\alpha = 0.1$, 1.0 and 10.

PROB 5 Artificial [13,14]

$$4u_{xx} + u_{yy} - \alpha u = f$$

DOMAIN: Unit square

BC: $u = 0$

TRUE: $2(x^2 - x)(\cos(2\pi y) - 1)$

OPERATOR: Constant coefficient, separable

RIGHT SIDE: Entire

BOUNDARY CONDITIONS: Dirichlet, homogeneous

SOLUTION: Entire, independent of α

PARAMETER: α makes operator more singular without affecting solution. Problems 5-2 to 5-7 have $\alpha = 5, 8, 10, 20, 64$ and 100.

PROB 6 Stratospheric physics [13,14,16]

$$u_{xx} + u_{yy} - (100 + \cos(2\pi x) + \sin(3\pi y))u = f$$

DOMAIN: Unit square

BC: $u = 0$

TRUE:

$$-0.31(5.4 - C(x))S(x)(y^2 - y)(5.4 - C(y))(1/(1 + \phi^4) - 0.5)$$

$$C(z) = \cos(4\pi z), \quad S(x) = \sin(\pi x), \quad \phi = 4(x - 0.5)^2 + 4(y - .5)^2$$

OPERATOR: Entire, oscilltory, somewhat singular

RIGHT SIDE: Analytic

BOUNDARY CONDITIONS: Dirichlet, homogeneous

SOLUTION: Analytic, oscillatory

PARAMETER: None

PROB 7 Artificial [6]

$$u_{xx} + u_{yy} = 1$$

DOMAIN: Unit square

BC: $u = 0$

TRUE: Approximate series solution gives 10^{-9} accuracy

OPERATOR: Laplace

RIGHT SIDE: Constant

BOUNDARY CONDITIONS: Dirichlet, homogeneous

SOLUTION: Has logaritmic singularities at corners in second derivatives; approximate solution is a polynomial.

PARAMETER: None

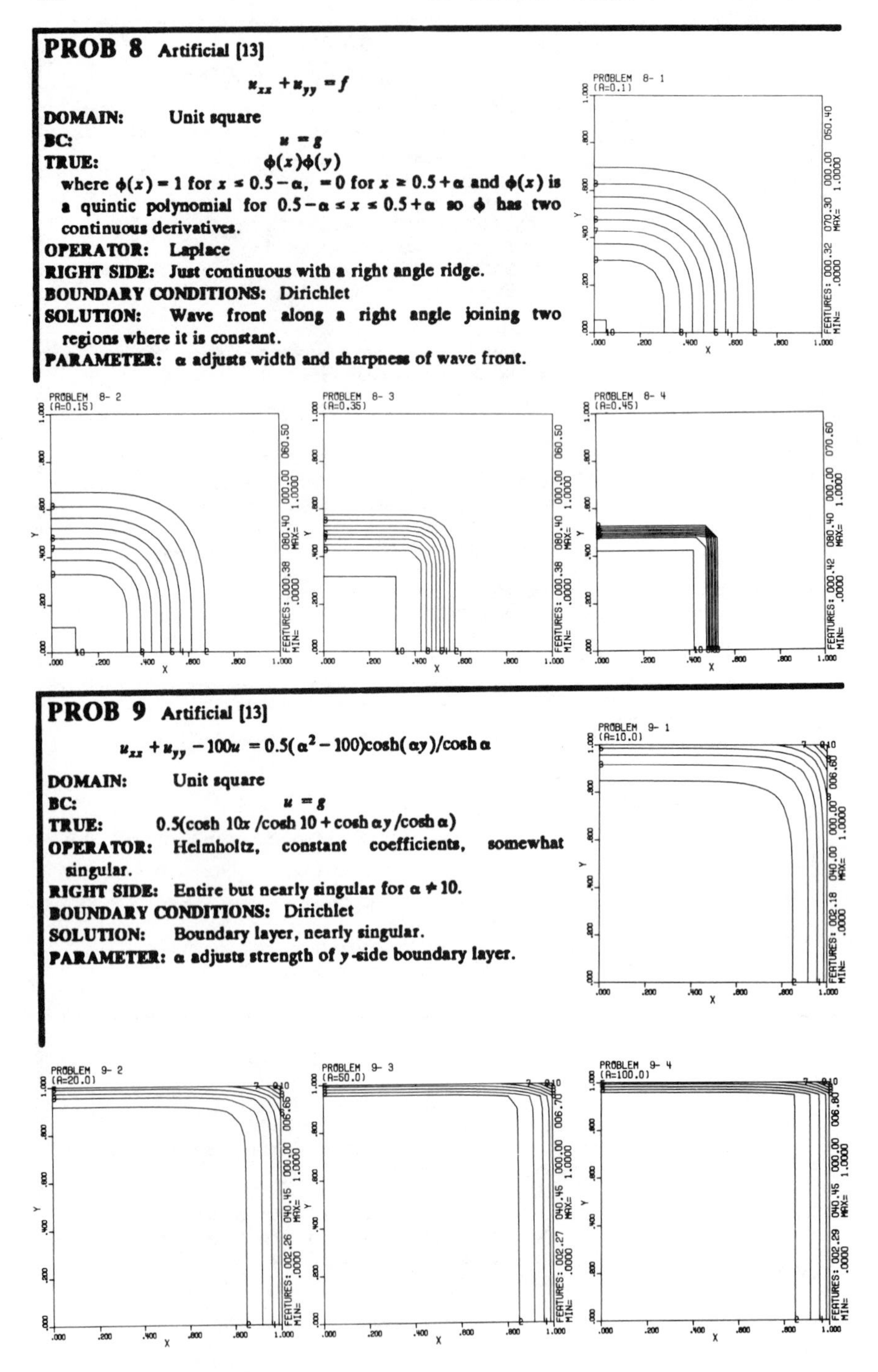

PROB 8 Artificial [13]

$$u_{xx} + u_{yy} = f$$

DOMAIN: Unit square

BC: $u = g$

TRUE: $\phi(x)\phi(y)$
where $\phi(x) = 1$ for $x \le 0.5-\alpha$, $= 0$ for $x \ge 0.5+\alpha$ and $\phi(x)$ is a quintic polynomial for $0.5-\alpha \le x \le 0.5+\alpha$ so ϕ has two continuous derivatives.

OPERATOR: Laplace

RIGHT SIDE: Just continuous with a right angle ridge.

BOUNDARY CONDITIONS: Dirichlet

SOLUTION: Wave front along a right angle joining two regions where it is constant.

PARAMETER: α adjusts width and sharpness of wave front.

PROB 9 Artificial [13]

$$u_{xx} + u_{yy} - 100u = 0.5(\alpha^2 - 100)\cosh(\alpha y)/\cosh \alpha$$

DOMAIN: Unit square

BC: $u = g$

TRUE: $0.5(\cosh 10x/\cosh 10 + \cosh \alpha y/\cosh \alpha)$

OPERATOR: Helmholtz, constant coefficients, somewhat singular.

RIGHT SIDE: Entire but nearly singular for $\alpha \neq 10$.

BOUNDARY CONDITIONS: Dirichlet

SOLUTION: Boundary layer, nearly singular.

PARAMETER: α adjusts strength of y-side boundary layer.

PROB 10 Artificial [13]

$$u_{xx} + u_{yy} = f$$

DOMAIN: Unit square

BC: $u = 0$

TRUE: $e^{-\alpha[(x-0.5)^2+(y-\beta)^2]}(x^2-x)(y^2-y)$

OPERATOR: Laplace

RIGHT SIDE: Strongly peaked if α large, but entire.

BOUNDARY CONDITIONS: Dirichlet, homogeneous

SOLUTION: Strongly peaked for large α.

PARAMETER: α adjusts strength of the peak, β moves it in the y-direction.

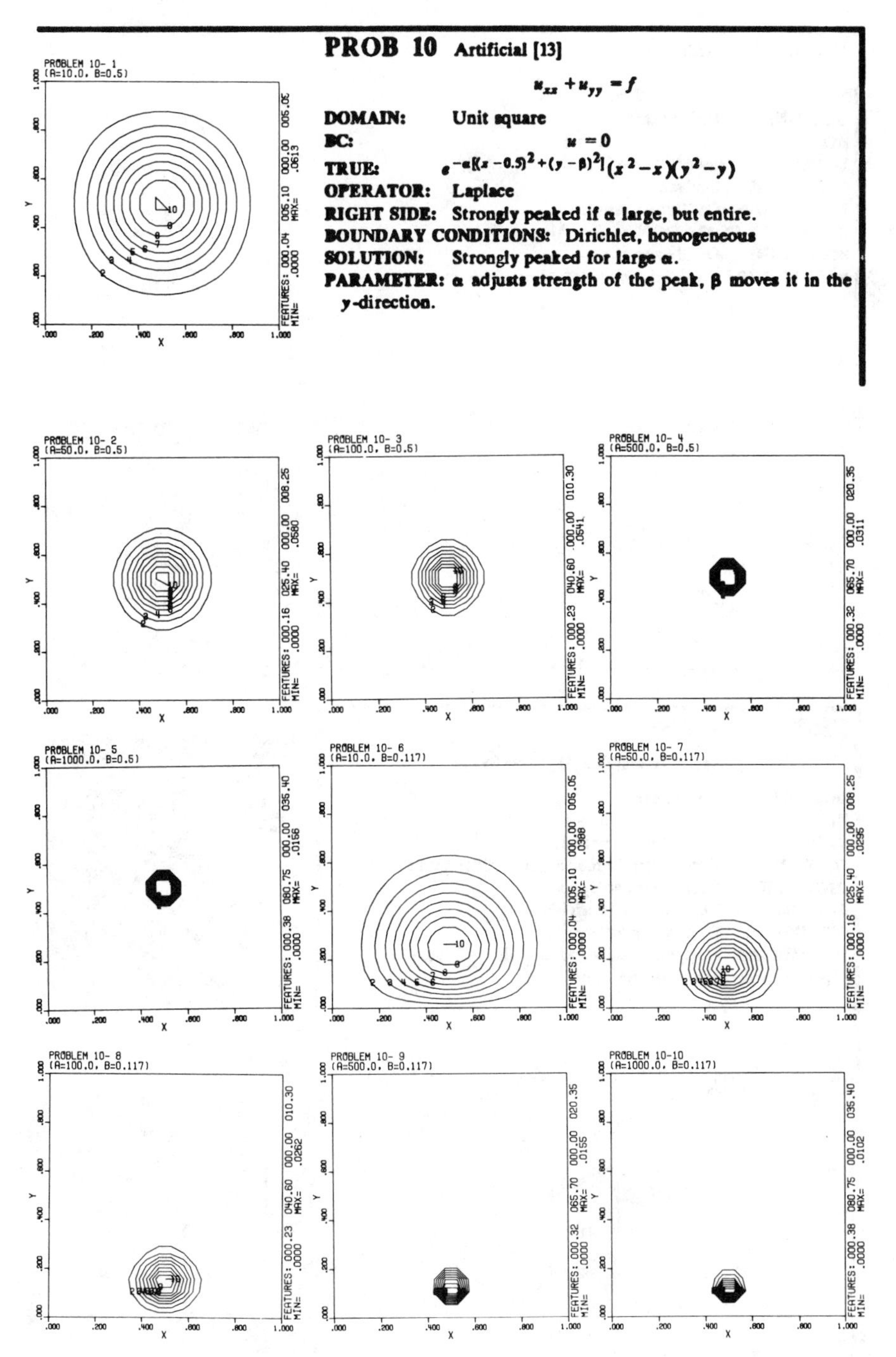

PROB 11 Artificial

$$u_{xx} + u_{yy} = f$$

DOMAIN: Unit square
BC: $u = g$
TRUE: $\sin[\alpha(x-y+2)^5/(1+(x-y+2)^4)]$
OPERATOR: Laplace
RIGHT SIDE: Oscillatory, analytic
BOUNDARY CONDITIONS: Dirichlet
SOLUTION: Oscillatory
PARAMETER: α adjusts frequency of oscillations.

PROB 12 Artificial

$$u_{xx} + u_{yy} + [1+\sin(\alpha x)]u_x - \cos(\alpha y)u = f$$

DOMAIN: Unit square
BC: $u = g$
TRUE: $\cos\beta y + \sin\beta(x-y)$
OPERATOR: Oscillatory, Laplacian plus lower terms
RIGHT SIDE: Oscillatory, analytic
BOUNDARY CONDITIONS: Dirichlet
SOLUTION: Oscillatory, entire
PARAMETER: α adjusts oscillation in PDE coefficients.
β adjusts oscillation in the solution.

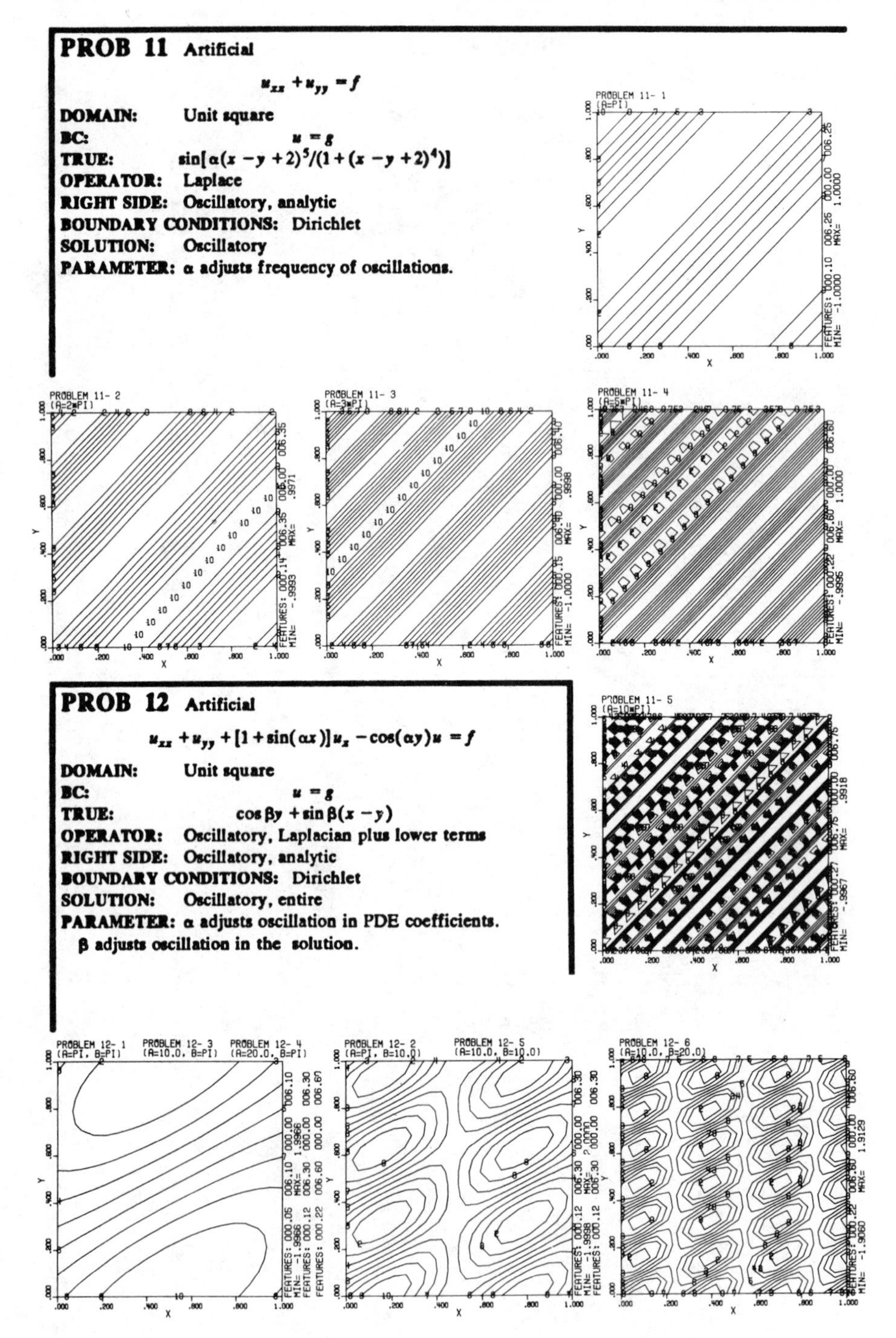

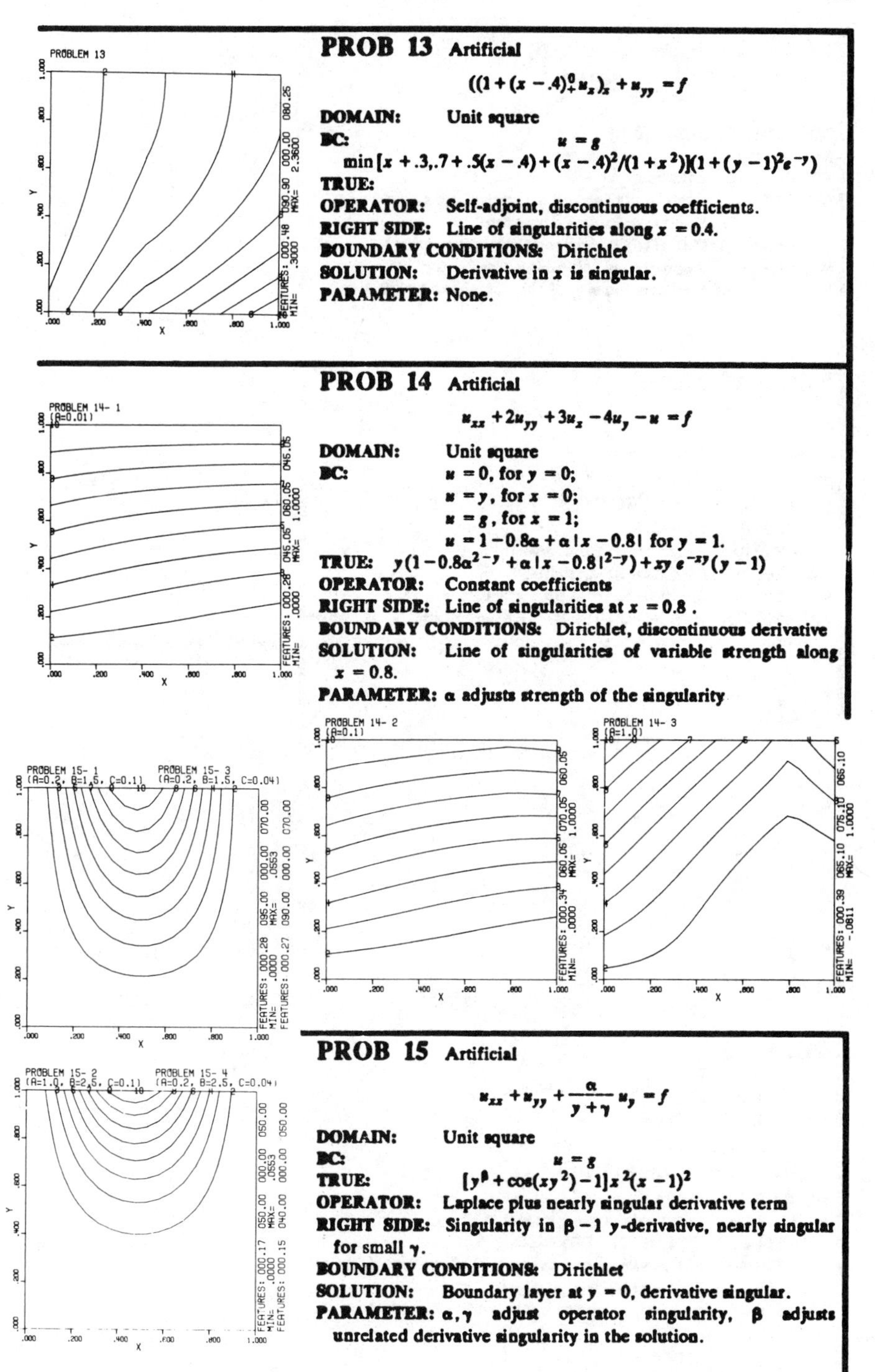

PROB 13 Artificial

$$((1+(x-.4)_+^0 u_x)_x + u_{yy} = f$$

DOMAIN: Unit square

BC: $u = g$

$\min[x+.3,.7+.5(x-.4)+(x-.4)^2/(1+x^2)](1+(y-1)^2 e^{-y})$

TRUE:

OPERATOR: Self-adjoint, discontinuous coefficients.

RIGHT SIDE: Line of singularities along $x = 0.4$.

BOUNDARY CONDITIONS: Dirichlet

SOLUTION: Derivative in x is singular.

PARAMETER: None.

PROB 14 Artificial

$$u_{xx} + 2u_{yy} + 3u_x - 4u_y - u = f$$

DOMAIN: Unit square

BC: $u = 0$, for $y = 0$;
$u = y$, for $x = 0$;
$u = g$, for $x = 1$;
$u = 1 - 0.8\alpha + \alpha|x - 0.8|$ for $y = 1$.

TRUE: $y(1 - 0.8\alpha^{2-y} + \alpha|x-0.8|^{2-y}) + xy\,e^{-xy}(y-1)$

OPERATOR: Constant coefficients

RIGHT SIDE: Line of singularities at $x = 0.8$.

BOUNDARY CONDITIONS: Dirichlet, discontinuous derivative

SOLUTION: Line of singularities of variable strength along $x = 0.8$.

PARAMETER: α adjusts strength of the singularity

PROB 15 Artificial

$$u_{xx} + u_{yy} + \frac{\alpha}{y+\gamma} u_y = f$$

DOMAIN: Unit square

BC: $u = g$

TRUE: $[y^\beta + \cos(xy^2) - 1]x^2(x-1)^2$

OPERATOR: Laplace plus nearly singular derivative term

RIGHT SIDE: Singularity in $\beta - 1$ y-derivative, nearly singular for small γ.

BOUNDARY CONDITIONS: Dirichlet

SOLUTION: Boundary layer at $y = 0$, derivative singular.

PARAMETER: α, γ adjust operator singularity, β adjusts unrelated derivative singularity in the solution.

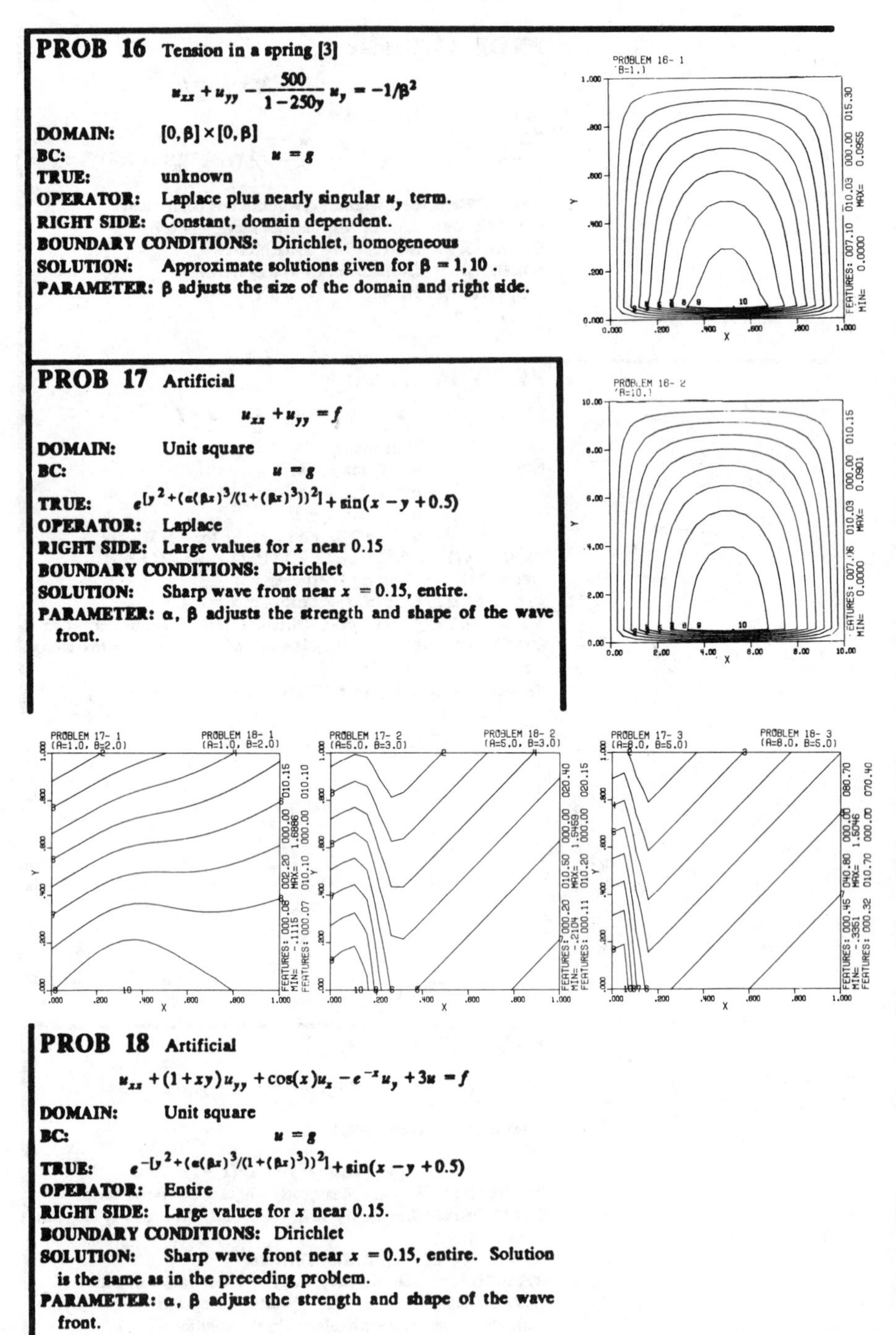

PROB 16 Tension in a spring [3]

$$u_{xx} + u_{yy} - \frac{500}{1-250y} u_y = -1/\beta^2$$

DOMAIN: $[0,\beta] \times [0,\beta]$

BC: $u = g$

TRUE: unknown

OPERATOR: Laplace plus nearly singular u_y term.

RIGHT SIDE: Constant, domain dependent.

BOUNDARY CONDITIONS: Dirichlet, homogeneous

SOLUTION: Approximate solutions given for $\beta = 1, 10$.

PARAMETER: β adjusts the size of the domain and right side.

PROB 17 Artificial

$$u_{xx} + u_{yy} = f$$

DOMAIN: Unit square

BC: $u = g$

TRUE: $e^{[y^2+(\alpha(\beta x)^3/(1+(\beta x)^3))^2]} + \sin(x-y+0.5)$

OPERATOR: Laplace

RIGHT SIDE: Large values for x near 0.15

BOUNDARY CONDITIONS: Dirichlet

SOLUTION: Sharp wave front near $x = 0.15$, entire.

PARAMETER: α, β adjusts the strength and shape of the wave front.

PROB 18 Artificial

$$u_{xx} + (1+xy)u_{yy} + \cos(x)u_x - e^{-x}u_y + 3u = f$$

DOMAIN: Unit square

BC: $u = g$

TRUE: $e^{-[y^2+(\alpha(\beta x)^3/(1+(\beta x)^3))^2]} + \sin(x-y+0.5)$

OPERATOR: Entire

RIGHT SIDE: Large values for x near 0.15.

BOUNDARY CONDITIONS: Dirichlet

SOLUTION: Sharp wave front near $x = 0.15$, entire. Solution is the same as in the preceding problem.

PARAMETER: α, β adjust the strength and shape of the wave front.

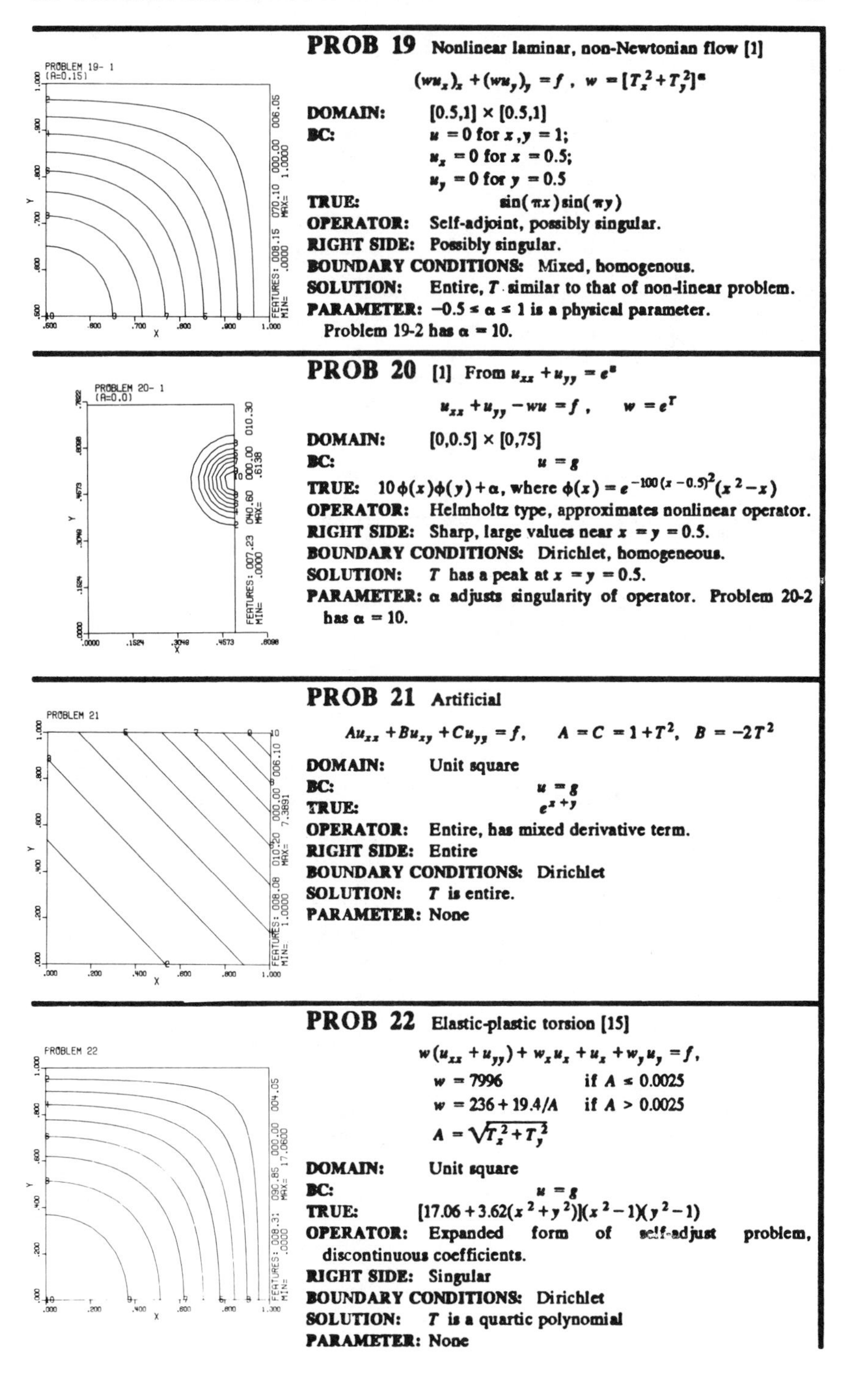

PROB 19 Nonlinear laminar, non-Newtonian flow [1]

$$(wu_x)_x + (wu_y)_y = f\,, \quad w = [T_x^2 + T_y^2]^\alpha$$

DOMAIN: $[0.5,1] \times [0.5,1]$

BC: $u = 0$ for $x, y = 1$;
$u_x = 0$ for $x = 0.5$;
$u_y = 0$ for $y = 0.5$

TRUE: $\sin(\pi x)\sin(\pi y)$

OPERATOR: Self-adjoint, possibly singular.

RIGHT SIDE: Possibly singular.

BOUNDARY CONDITIONS: Mixed, homogenous.

SOLUTION: Entire, T similar to that of non-linear problem.

PARAMETER: $-0.5 \le \alpha \le 1$ is a physical parameter. Problem 19-2 has $\alpha = 10$.

PROB 20 [1] From $u_{xx} + u_{yy} = e^u$

$$u_{xx} + u_{yy} - wu = f\,, \qquad w = e^T$$

DOMAIN: $[0,0.5] \times [0,75]$

BC: $u = g$

TRUE: $10\phi(x)\phi(y) + \alpha$, where $\phi(x) = e^{-100(x-0.5)^2}(x^2 - x)$

OPERATOR: Helmholtz type, approximates nonlinear operator.

RIGHT SIDE: Sharp, large values near $x = y = 0.5$.

BOUNDARY CONDITIONS: Dirichlet, homogeneous.

SOLUTION: T has a peak at $x = y = 0.5$.

PARAMETER: α adjusts singularity of operator. Problem 20-2 has $\alpha = 10$.

PROB 21 Artificial

$$Au_{xx} + Bu_{xy} + Cu_{yy} = f, \qquad A = C = 1 + T^2, \; B = -2T^2$$

DOMAIN: Unit square

BC: $u = g$

TRUE: e^{x+y}

OPERATOR: Entire, has mixed derivative term.

RIGHT SIDE: Entire

BOUNDARY CONDITIONS: Dirichlet

SOLUTION: T is entire.

PARAMETER: None

PROB 22 Elastic-plastic torsion [15]

$$w(u_{xx} + u_{yy}) + w_x u_x + u_x + w_y u_y = f,$$
$$w = 7996 \qquad \text{if } A \le 0.0025$$
$$w = 236 + 19.4/A \qquad \text{if } A > 0.0025$$
$$A = \sqrt{T_x^2 + T_y^2}$$

DOMAIN: Unit square

BC: $u = g$

TRUE: $[17.06 + 3.62(x^2 + y^2)](x^2 - 1)(y^2 - 1)$

OPERATOR: Expanded form of self-adjust problem, discontinuous coefficients.

RIGHT SIDE: Singular

BOUNDARY CONDITIONS: Dirichlet

SOLUTION: T is a quartic polynomial

PARAMETER: None

PROB 23 Nonlinear laminar, non-Newtonian flow [1]

$$w(u_{xx}+u_{yy})+w_x u_x+w_y u_y=f, \quad \text{see below for } w$$

DOMAIN: Unit square

BC: $u_x=0$, for $x=0,1$;
$u=2\cos(\pi x)$ for $y=0$;
$u=\cos(\pi x)$ for $y=1$.

TRUE: $(\phi(y)+1)\cos(\pi x)$

where $\phi(y)=1$ for $y\le 0.5-\delta$, $=0$ for $y\ge 0.5+\delta$ and $\phi(y)$ is a quintic polynomial for $0.5-\delta\le y\le 0.5+\delta$ so ϕ has two continuous derivatives.

OPERATOR: Expanded from self-adjoint problem, analytic.

RIGHT SIDE: Analytic

BOUNDARY CONDITIONS: Mixed

SOLUTION: Has jumps in third y-derivatives.

PARAMETER: Three cases for w given in terms of

$$A=\sqrt{T_x^2+T_y^2}, \quad c=1, \quad w=1/(\alpha+\beta A)$$

$$c=2, \quad w=e^{[A/(\alpha+\beta A)]}/A$$

$$c=3, \quad w=\alpha\tanh(\beta A)/A$$

Physical parameters α,β of (387.75,50) and (554.5,0.544) have been used in practice.

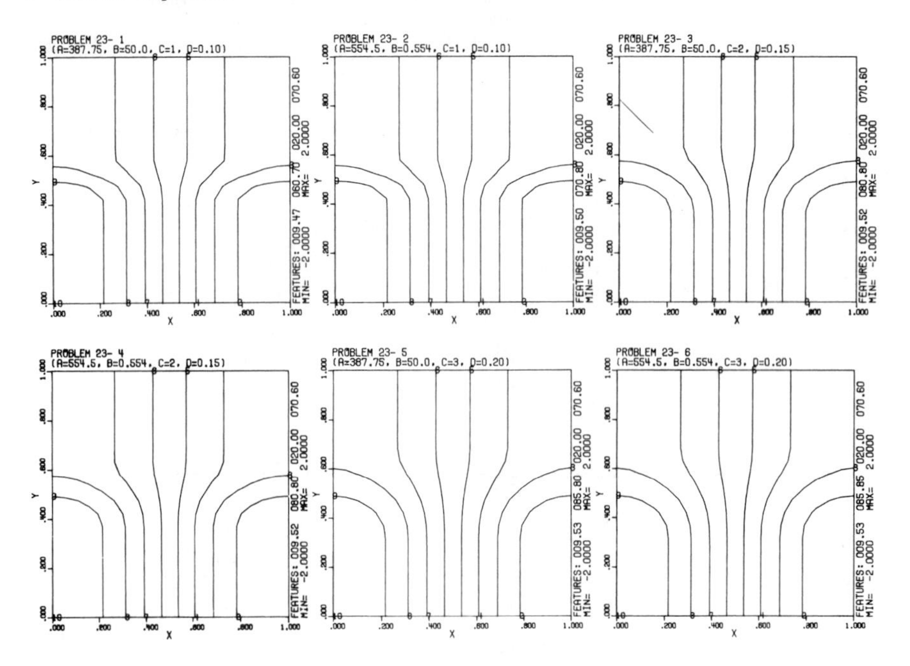

PROB 24 Friction in a brake shoe [3]

$$u_{xx} + u_{yy} + \frac{3h_x}{h} u_x = \frac{\gamma h_x}{h^3}, \quad h = \sin(\alpha \pi x y)$$

DOMAIN: $[.1,1] \times [.1,1]$

BC: $\beta u + u_x + u_y = 0$

TRUE: Unknown

OPERATOR: Laplace plus u_x term which is possibly singular.

RIGHT SIDE: Analytic, possibly nearly singular.

BOUNDARY CONDITIONS: Mixed, homogeneous.

SOLUTION: Approximate solutions given for 8 cases.

PARAMETER: α, β and γ are physical parameters,

1. $\alpha = 0.25$ $\beta = 100$ $\gamma = -0.1$
2. $\alpha = 0.25$ $\beta = 1000$ $\gamma = 0.1$
3. $\alpha = 0.5$ $\beta = 1$ $\gamma = -0.01$
4. $\alpha = 0.5$ $\beta = 10$ $\gamma = -0.1$
5. $\alpha = 1.$ $\beta = 1$ $\gamma = -0.1$
6. $\alpha = 1.$ $\beta = 10$ $\gamma = -0.1$
7. $\alpha = 1.$ $\beta = 100$ $\gamma = -1.$
8. $\alpha = 1.$ $\beta = 1000$ $\gamma = -1.$

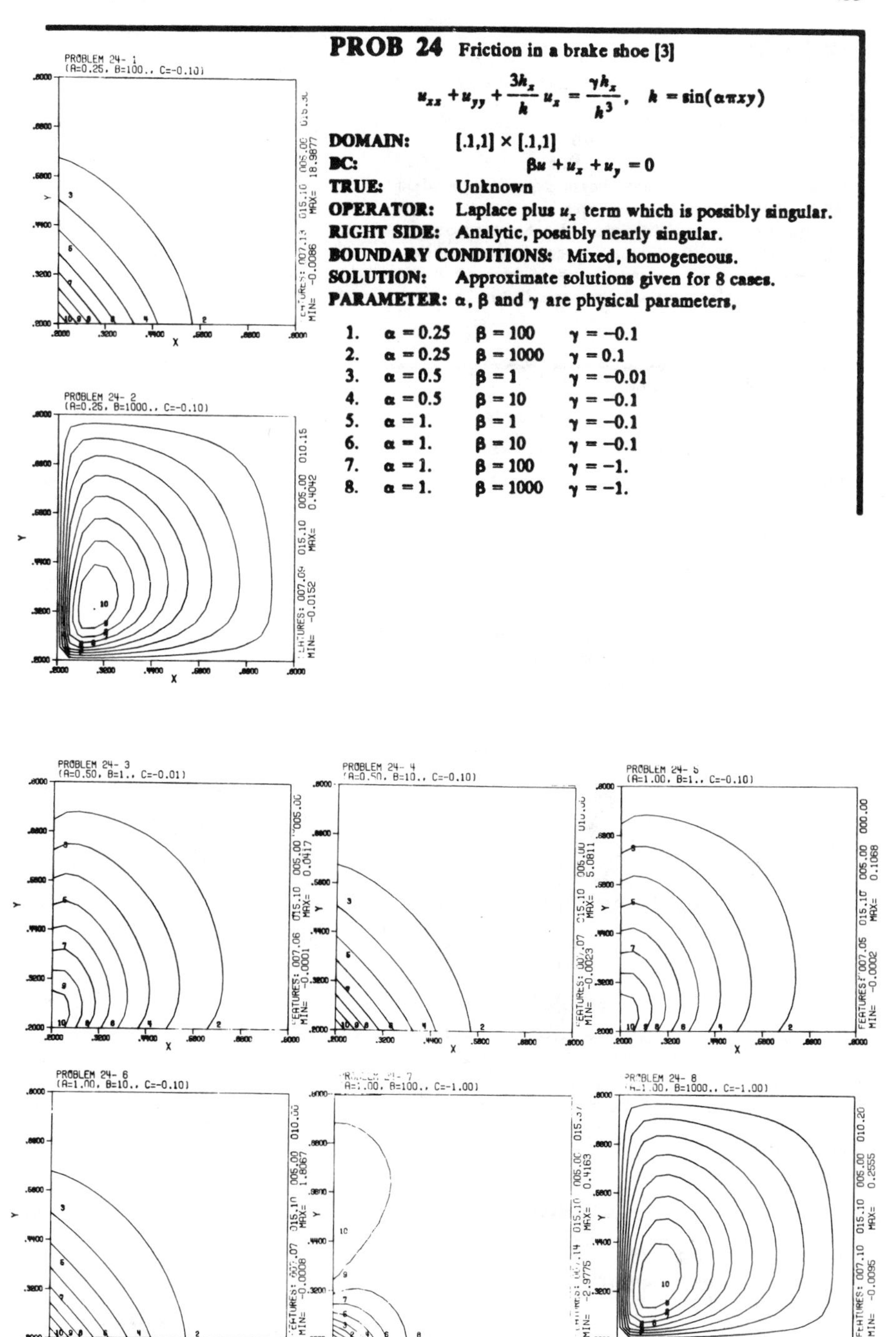

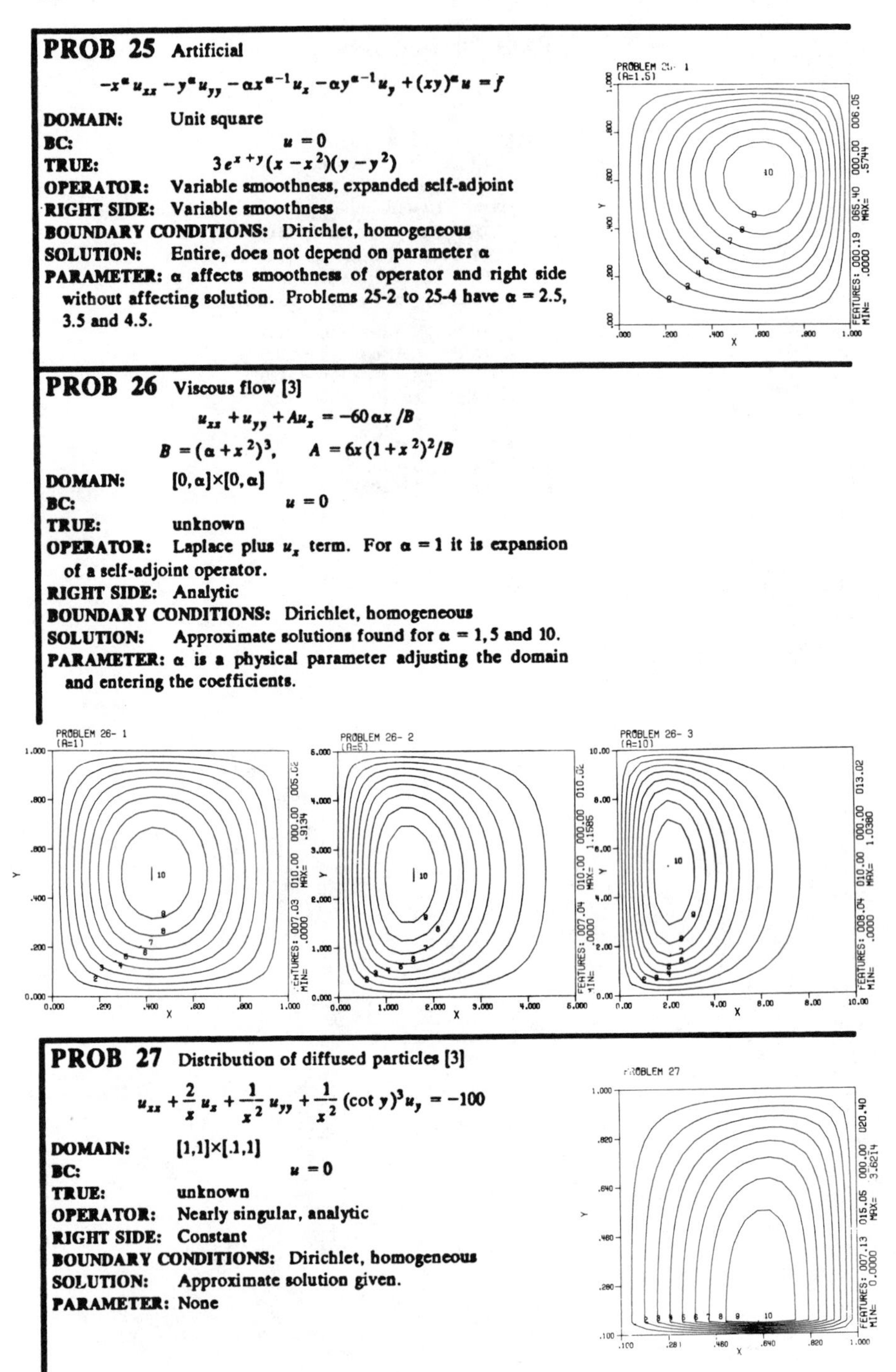

PROB 25 Artificial

$$-x^{\alpha} u_{xx} - y^{\alpha} u_{yy} - \alpha x^{\alpha-1} u_x - \alpha y^{\alpha-1} u_y + (xy)^{\alpha} u = f$$

DOMAIN: Unit square

BC: $u = 0$

TRUE: $3e^{x+y}(x-x^2)(y-y^2)$

OPERATOR: Variable smoothness, expanded self-adjoint

RIGHT SIDE: Variable smoothness

BOUNDARY CONDITIONS: Dirichlet, homogeneous

SOLUTION: Entire, does not depend on parameter α

PARAMETER: α affects smoothness of operator and right side without affecting solution. Problems 25-2 to 25-4 have $\alpha = 2.5$, 3.5 and 4.5.

PROB 26 Viscous flow [3]

$$u_{xx} + u_{yy} + A u_x = -60 \alpha x / B$$

$$B = (\alpha + x^2)^3, \quad A = 6x(1+x^2)^2/B$$

DOMAIN: $[0,\alpha]\times[0,\alpha]$

BC: $u = 0$

TRUE: unknown

OPERATOR: Laplace plus u_x term. For $\alpha = 1$ it is expansion of a self-adjoint operator.

RIGHT SIDE: Analytic

BOUNDARY CONDITIONS: Dirichlet, homogeneous

SOLUTION: Approximate solutions found for $\alpha = 1, 5$ and 10.

PARAMETER: α is a physical parameter adjusting the domain and entering the coefficients.

PROB 27 Distribution of diffused particles [3]

$$u_{xx} + \frac{2}{x} u_x + \frac{1}{x^2} u_{yy} + \frac{1}{x^2} (\cot y)^3 u_y = -100$$

DOMAIN: $[.1,1]\times[.1,1]$

BC: $u = 0$

TRUE: unknown

OPERATOR: Nearly singular, analytic

RIGHT SIDE: Constant

BOUNDARY CONDITIONS: Dirichlet, homogeneous

SOLUTION: Approximate solution given.

PARAMETER: None

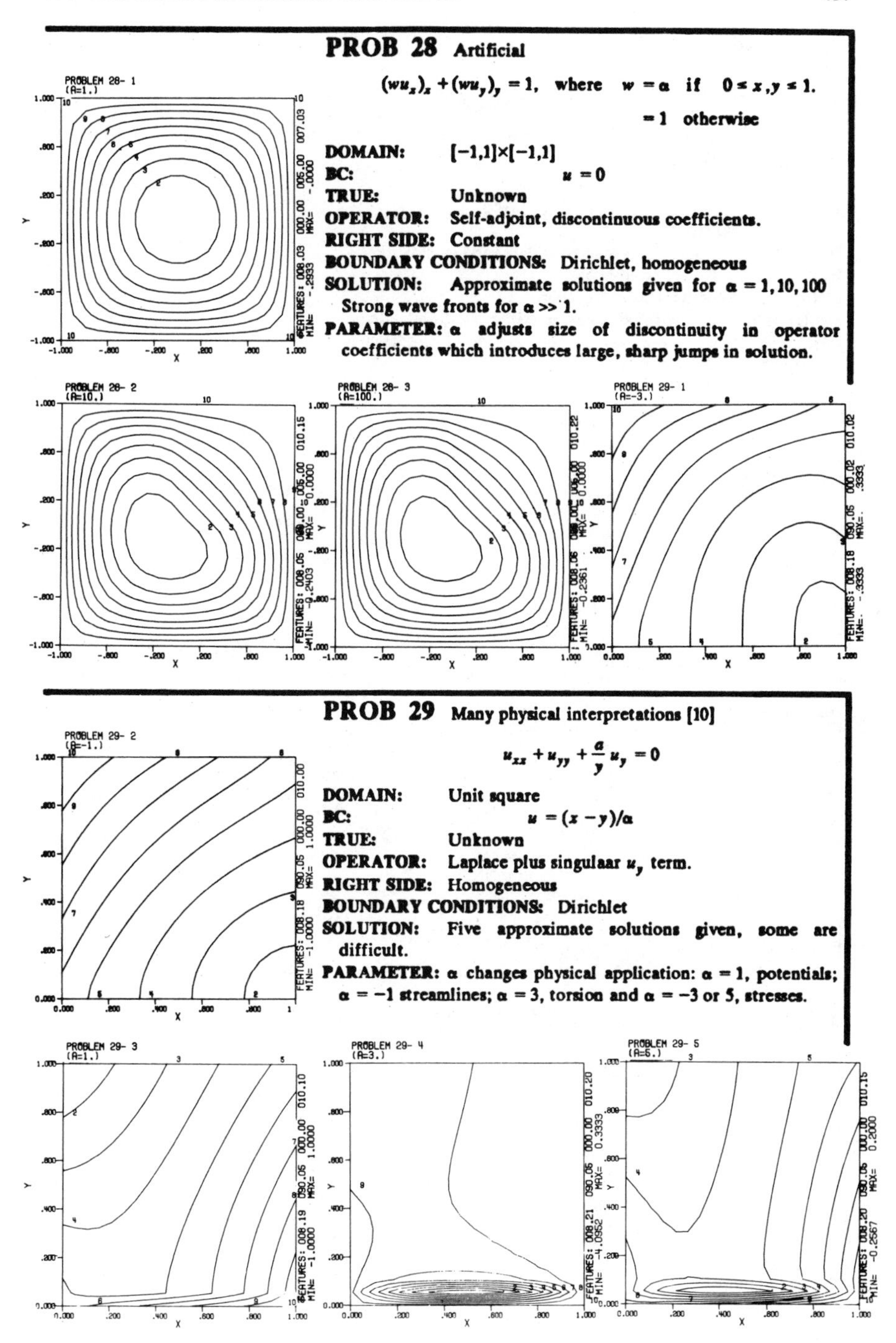

PROB 28 Artificial

$$(wu_x)_x + (wu_y)_y = 1, \quad \text{where} \quad w = \alpha \ \text{if} \ 0 \le x, y \le 1.$$
$$= 1 \ \text{otherwise}$$

DOMAIN: $[-1,1]\times[-1,1]$

BC: $u = 0$

TRUE: Unknown

OPERATOR: Self-adjoint, discontinuous coefficients.

RIGHT SIDE: Constant

BOUNDARY CONDITIONS: Dirichlet, homogeneous

SOLUTION: Approximate solutions given for $\alpha = 1, 10, 100$ Strong wave fronts for $\alpha >> 1$.

PARAMETER: α adjusts size of discontinuity in operator coefficients which introduces large, sharp jumps in solution.

PROB 29 Many physical interpretations [10]

$$u_{xx} + u_{yy} + \frac{a}{y} u_y = 0$$

DOMAIN: Unit square

BC: $u = (x - y)/\alpha$

TRUE: Unknown

OPERATOR: Laplace plus singulaar u_y term.

RIGHT SIDE: Homogeneous

BOUNDARY CONDITIONS: Dirichlet

SOLUTION: Five approximate solutions given, some are difficult.

PARAMETER: α changes physical application: $\alpha = 1$, potentials; $\alpha = -1$ streamlines; $\alpha = 3$, torsion and $\alpha = -3$ or 5, stresses.

PROB 30 Artificial

$$[2+(y-1)e^{-\alpha y^4}]u_{xx}+[1+\frac{1}{1+(2x)^{\beta}}]u_{yy}$$
$$+\gamma[x(x-1)+(y-0.3)(6-0.7)]u=f$$

DOMAIN: Unit square

BC: $u=g$

TRUE: $\frac{x+y^2}{1+(2x)^{\beta-1}}+(y-1)(1+x)e^{\alpha y^4}+\gamma(x+y)\cos(xy)$

OPERATOR: Coefficients may be widely varying, singular.

RIGHT SIDE: Complicated behavior

BOUNDARY CONDITIONS: Dirichlet

SOLUTION: Complicated behavior, with wave fronts, etc.

PARAMETER: α, β, γ adjust the contribution of 3 independent complexities of the problem.

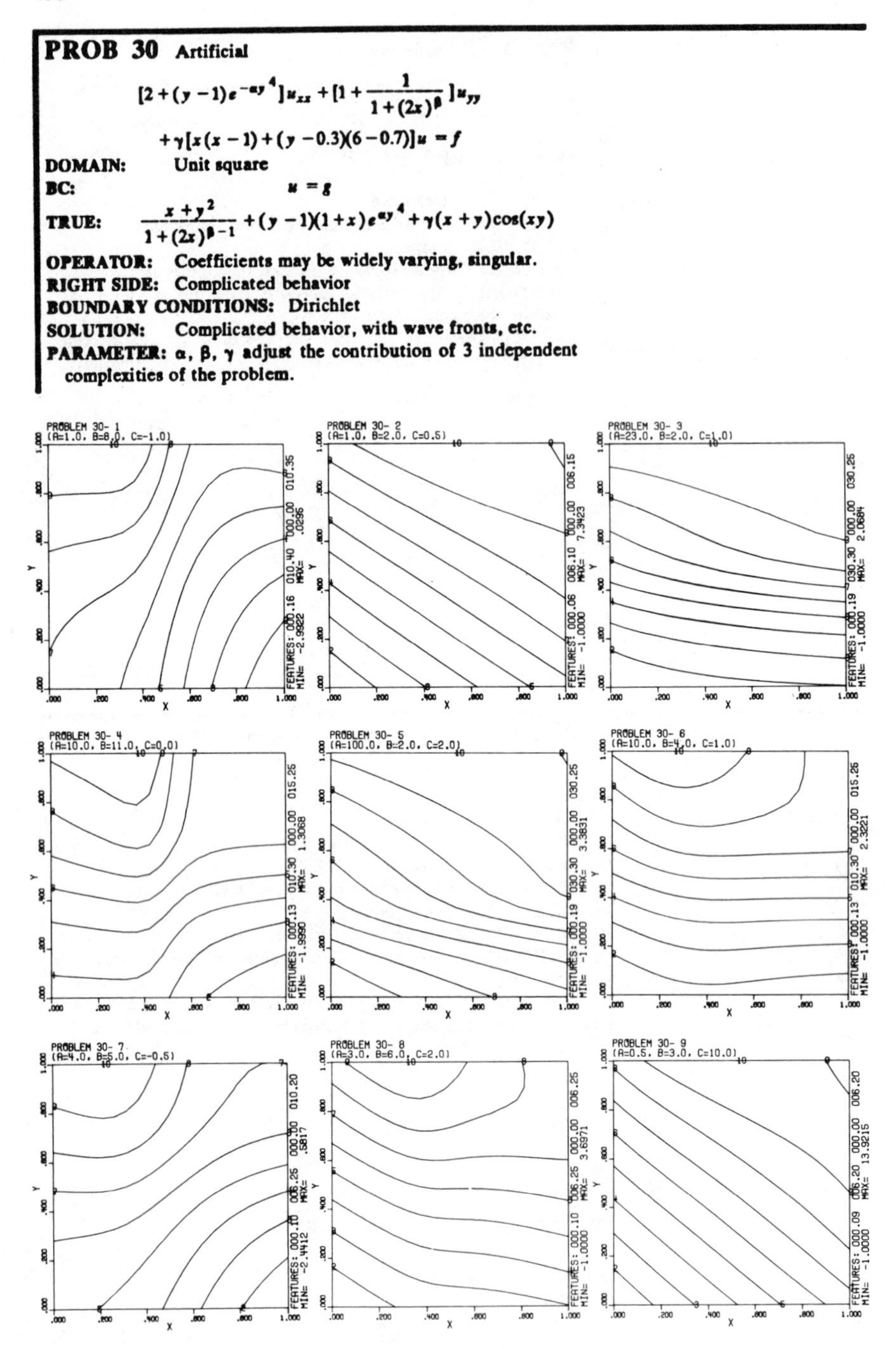

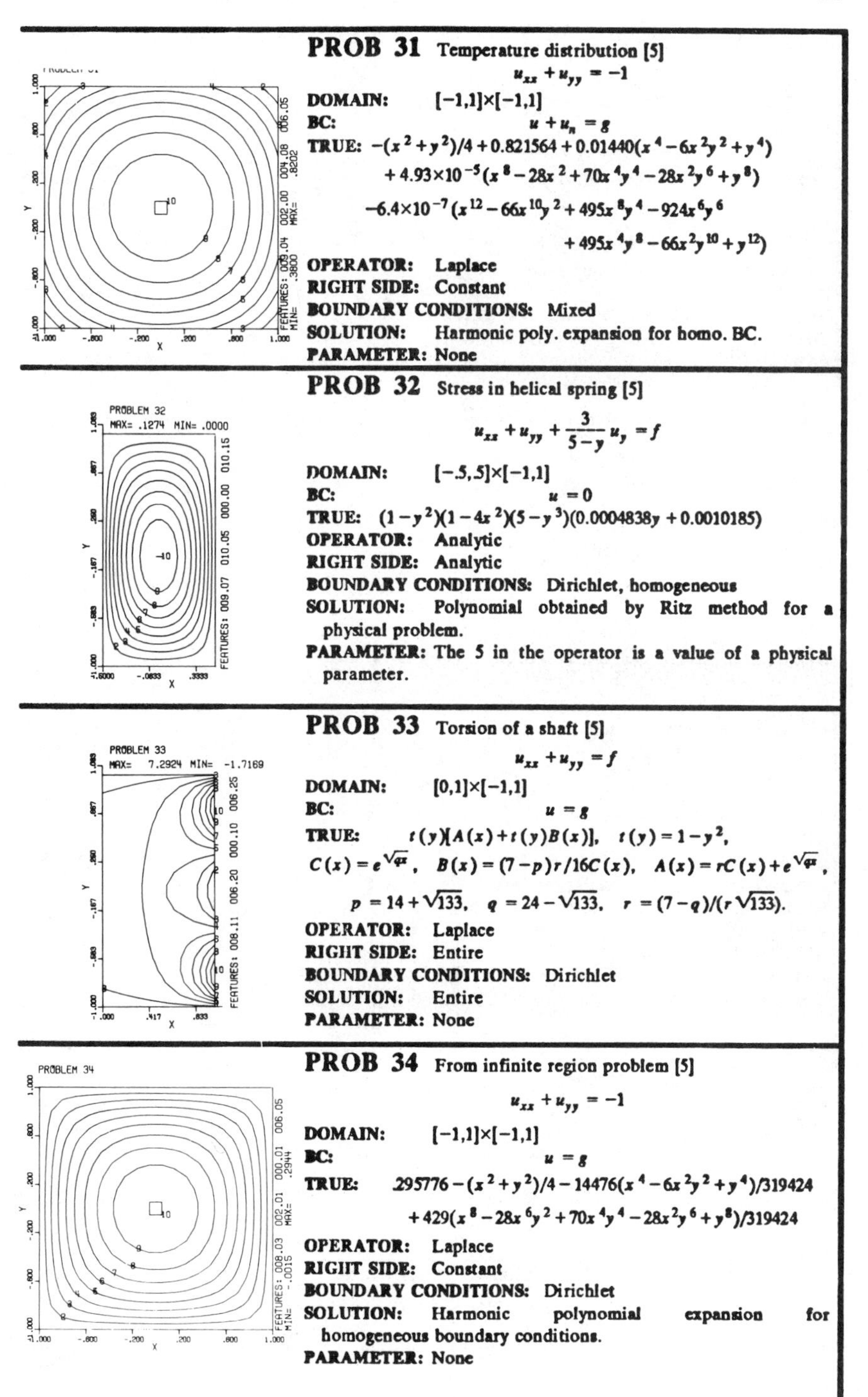

PROB 31 Temperature distribution [5]

$$u_{xx} + u_{yy} = -1$$

DOMAIN: $[-1,1]\times[-1,1]$

BC: $u + u_n = g$

TRUE: $-(x^2+y^2)/4 + 0.821564 + 0.01440(x^4 - 6x^2y^2 + y^4)$
$+ 4.93\times10^{-5}(x^8 - 28x^2 + 70x^4y^4 - 28x^2y^6 + y^8)$
$-6.4\times10^{-7}(x^{12} - 66x^{10}y^2 + 495x^8y^4 - 924x^6y^6$
$+ 495x^4y^8 - 66x^2y^{10} + y^{12})$

OPERATOR: Laplace
RIGHT SIDE: Constant
BOUNDARY CONDITIONS: Mixed
SOLUTION: Harmonic poly. expansion for homo. BC.
PARAMETER: None

PROB 32 Stress in helical spring [5]

$$u_{xx} + u_{yy} + \frac{3}{5-y}u_y = f$$

DOMAIN: $[-.5,.5]\times[-1,1]$

BC: $u = 0$

TRUE: $(1-y^2)(1-4x^2)(5-y^3)(0.0004838y + 0.0010185)$

OPERATOR: Analytic
RIGHT SIDE: Analytic
BOUNDARY CONDITIONS: Dirichlet, homogeneous
SOLUTION: Polynomial obtained by Ritz method for a physical problem.
PARAMETER: The 5 in the operator is a value of a physical parameter.

PROB 33 Torsion of a shaft [5]

$$u_{xx} + u_{yy} = f$$

DOMAIN: $[0,1]\times[-1,1]$

BC: $u = g$

TRUE: $t(y)[A(x) + t(y)B(x)]$, $t(y) = 1 - y^2$,
$C(x) = e^{\sqrt{qx}}$, $B(x) = (7-p)r/16C(x)$, $A(x) = rC(x) + e^{\sqrt{qx}}$,
$p = 14 + \sqrt{133}$, $q = 24 - \sqrt{133}$, $r = (7-q)/(r\sqrt{133})$.

OPERATOR: Laplace
RIGHT SIDE: Entire
BOUNDARY CONDITIONS: Dirichlet
SOLUTION: Entire
PARAMETER: None

PROB 34 From infinite region problem [5]

$$u_{xx} + u_{yy} = -1$$

DOMAIN: $[-1,1]\times[-1,1]$

BC: $u = g$

TRUE: $.295776 - (x^2+y^2)/4 - 14476(x^4 - 6x^2y^2 + y^4)/319424$
$+ 429(x^8 - 28x^6y^2 + 70x^4y^4 - 28x^2y^6 + y^8)/319424$

OPERATOR: Laplace
RIGHT SIDE: Constant
BOUNDARY CONDITIONS: Dirichlet
SOLUTION: Harmonic polynomial expansion for homogeneous boundary conditions.
PARAMETER: None

PROB 35 Torsion for a beam [5]

$$u_{xx} + u_{yy} = 0$$

DOMAIN: $[-1,1]\times[-1,1]$

BC: $u = g$ for $y = \pm 1$,

$(1+\alpha)u + \alpha u_n = g$ for $x = \pm 1$.

TRUE: $1.1786 - 0.1801p + 0.006q$

$p(x,y) = x^4 - 6x^2y^2 + y^4$,

$q(x,y) = x^8 - 28x^6y^2 + 70x^4y^4 - 28x^2y^6 + y^8$

OPERATOR: Laplace, homogeneous

RIGHT SIDE: Zero

BOUNDARY CONDITIONS: Mixed, Dirichlet for $\alpha = 0$.

SOLUTION: Harmonic polynomial combination.

PARAMETER: α adjusts contribution of mixed boundary condition; $\alpha = 0$ is the physical problem.

PROBLEM 35- 1 (A=0.01)

FEATURES: 009.11 002.00 000.10 006.50 MIN= 1.0070 MAX= 1.9950

PROBLEM 35- 2 (A=0.1)

PROBLEM 35- 3 (A=1.0)

PROBLEM 35- 4 (A=10.)

PROB 36 Adapted from Problem 27

$$(1+\beta)u_{xx} + \frac{3}{x+\alpha}u_x + \frac{1}{(x+\alpha)^2}u_{yy} + \frac{\cot y}{x+\alpha}u_y = f$$

DOMAIN: Unit square

BC: $u = g$

TRUE: $(1-\beta)e^{x+y} + \beta \log_e(x+\alpha)$

OPERATOR: Possibly singular coefficients for $\alpha = 0$.

RIGHT SIDE: Analytic except for $\alpha = 0$; then singular.

BOUNDARY CONDITIONS: Dirichlet

SOLUTION: Logarithmic singularity for $\alpha = 0$.

PARAMETER: α adjusts distance of singularity from domain, β adjusts relative size of exponential and logarithmic terms in solution.

PROBLEM 36- 1 (A=0.0, B=0.0)

FEATURES: 008.28 098.50 000.00 006.05 MIN= 1.0000 MAX= 7.3891

PROBLEM 36- 2 (A=0.1, B=-0.5)

FEATURES: 006.23 090.25 000.00 010.15 MIN= 2.4342 MAX= 11.0369

PROBLEM 36- 3 (A=0.0, B=1.0)

FEATURES: 007.49 100.45 000.00 100.50 MIN= -2.9444 MAX= -.0000

PROBLEM 36- 4 (A=0.5, B=0.0)

FEATURES: 006.20 090.25 000.00 005.00 MIN= 1.0000 MAX= 7.3891

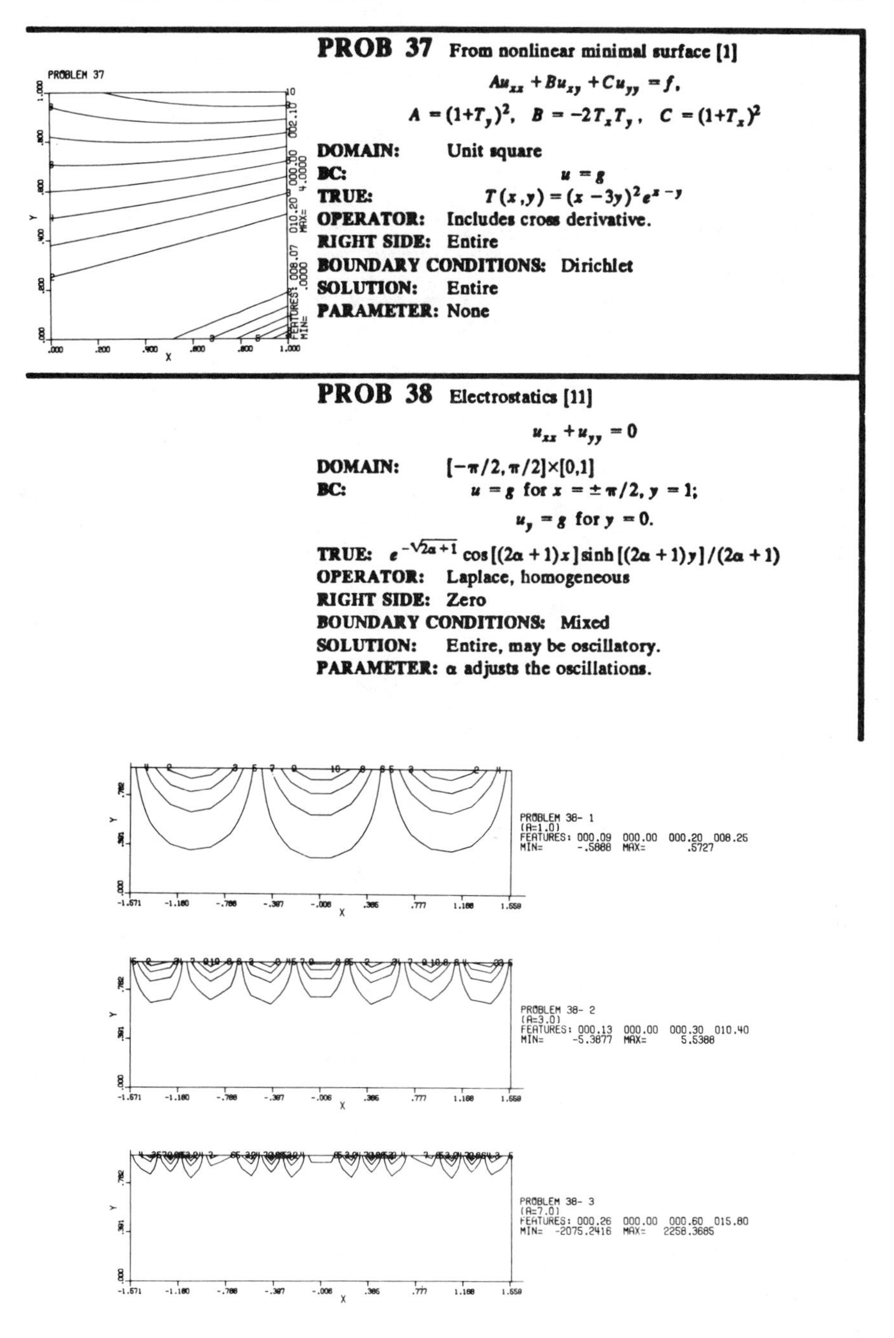

PROB 37 From nonlinear minimal surface [1]

$$Au_{xx} + Bu_{xy} + Cu_{yy} = f,$$

$$A = (1+T_y)^2, \quad B = -2T_xT_y, \quad C = (1+T_x)^2$$

DOMAIN: Unit square
BC: $u = g$
TRUE: $T(x,y) = (x-3y)^2e^{x-y}$
OPERATOR: Includes cross derivative.
RIGHT SIDE: Entire
BOUNDARY CONDITIONS: Dirichlet
SOLUTION: Entire
PARAMETER: None

PROB 38 Electrostatics [11]

$$u_{xx} + u_{yy} = 0$$

DOMAIN: $[-\pi/2, \pi/2] \times [0,1]$
BC: $u = g$ for $x = \pm\pi/2, y = 1$;
$u_y = g$ for $y = 0$.

TRUE: $e^{-\sqrt{2\alpha+1}} \cos[(2\alpha+1)x] \sinh[(2\alpha+1)y]/(2\alpha+1)$
OPERATOR: Laplace, homogeneous
RIGHT SIDE: Zero
BOUNDARY CONDITIONS: Mixed
SOLUTION: Entire, may be oscillatory.
PARAMETER: α adjusts the oscillations.

PROB 39 From nonlinear problem [4]

$$u_{xx} + u_{yy} + [1 - h(x)^2 w(x,y)^2]/\beta u = 0$$

DOMAIN: Unit square
BC: $u = 1$
TRUE: unknown
OPERATOR: Helmholtz type, homogeneous
RIGHT SIDE: Zero
BOUNDARY CONDITIONS: Dirichlet, constant
SOLUTION: Approximate solution $w(x,y)$ calculated and tabulated for 5 cases.
PARAMETER: $h(x) = 1/x$ for $B = 0.5, 1$ (Cases 1 and 2) $h(x) = e^x$ $\beta = 0.25, 0.5, 1$ (Cases 3, 4 and 5).

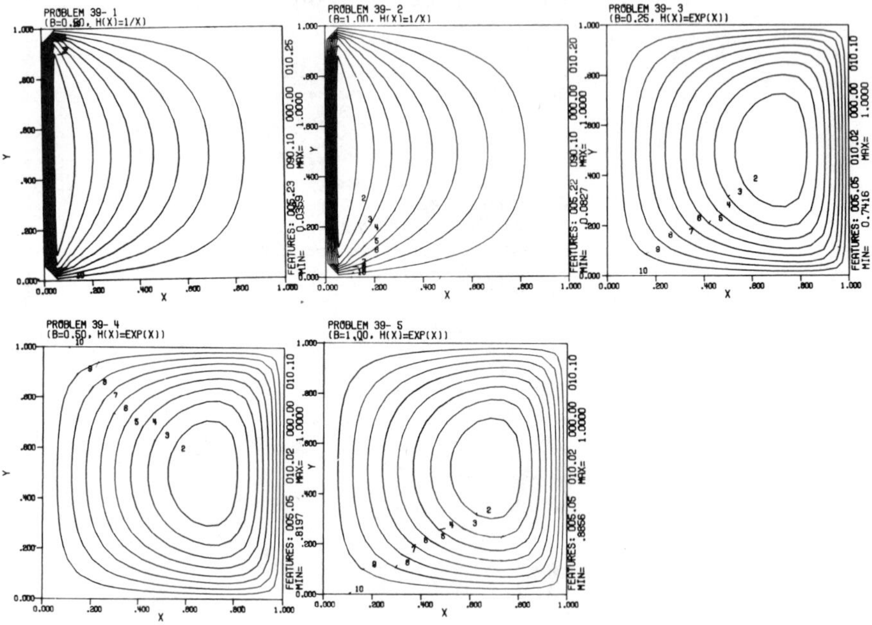

PROB 40 Hadamard's Example [17]

$$u_{xx} + (1+x^2)u_{yy} - y\, u_x = f$$

DOMAIN: Unit square
BC: $u = g$ for $y = 0$ or 1,
$\alpha u + \beta u_x = g$ for $x = 0$ or 1.
TRUE: $\log_{10}[(x+1)/(y+1)] + e^{2(x+y)/(2+x-y)-2}$
OPERATOR: Entire
RIGHT SIDE: Analytic
BOUNDARY CONDITIONS: Mixed
SOLUTION: Analytic
PARAMETER: α and β adjust the contributions to the mixed boundary condition on two sides. Problem 40-2 has $\alpha = 0.15$, $\beta = 0.85$ and 40-3 has $\alpha = 0.85$, $\beta = 0.15$.

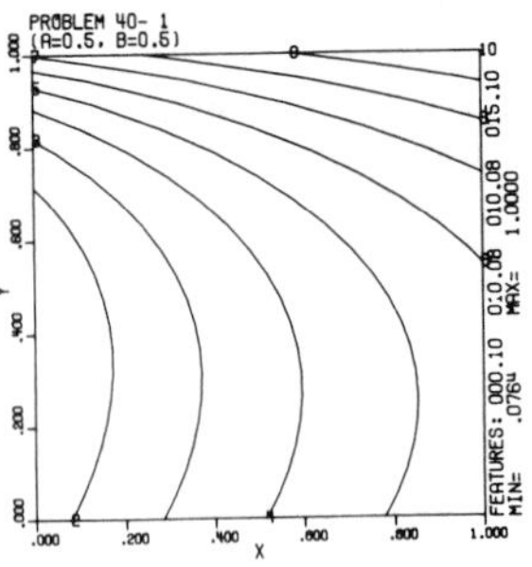

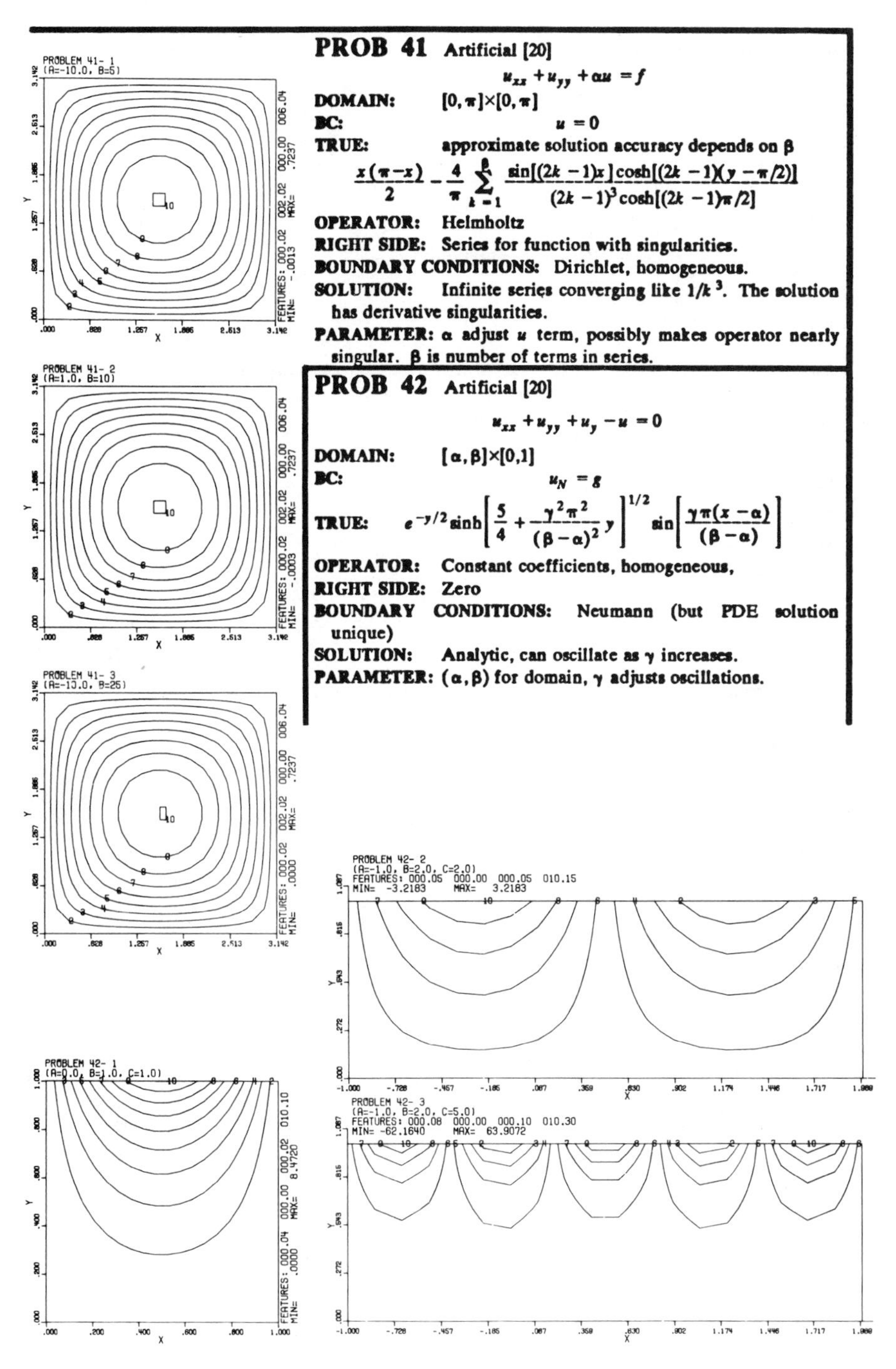

PROB 41 Artificial [20]

$$u_{xx} + u_{yy} + \alpha u = f$$

DOMAIN: $[0,\pi]\times[0,\pi]$

BC: $u = 0$

TRUE: approximate solution accuracy depends on β

$$\frac{x(\pi - x)}{2} - \frac{4}{\pi}\sum_{k=1}^{\beta}\frac{\sin[(2k-1)x]\cosh[(2k-1)(y-\pi/2)]}{(2k-1)^3\cosh[(2k-1)\pi/2]}$$

OPERATOR: Helmholtz

RIGHT SIDE: Series for function with singularities.

BOUNDARY CONDITIONS: Dirichlet, homogeneous.

SOLUTION: Infinite series converging like $1/k^3$. The solution has derivative singularities.

PARAMETER: α adjust u term, possibly makes operator nearly singular. β is number of terms in series.

PROB 42 Artificial [20]

$$u_{xx} + u_{yy} + u_y - u = 0$$

DOMAIN: $[\alpha,\beta]\times[0,1]$

BC: $u_N = g$

TRUE: $$e^{-y/2}\sinh\left[\frac{5}{4} + \frac{\gamma^2\pi^2}{(\beta-\alpha)^2}\, y\right]^{1/2}\sin\left[\frac{\gamma\pi(x-\alpha)}{(\beta-\alpha)}\right]$$

OPERATOR: Constant coefficients, homogeneous,

RIGHT SIDE: Zero

BOUNDARY CONDITIONS: Neumann (but PDE solution unique)

SOLUTION: Analytic, can oscillate as γ increases.

PARAMETER: (α,β) for domain, γ adjusts oscillations.

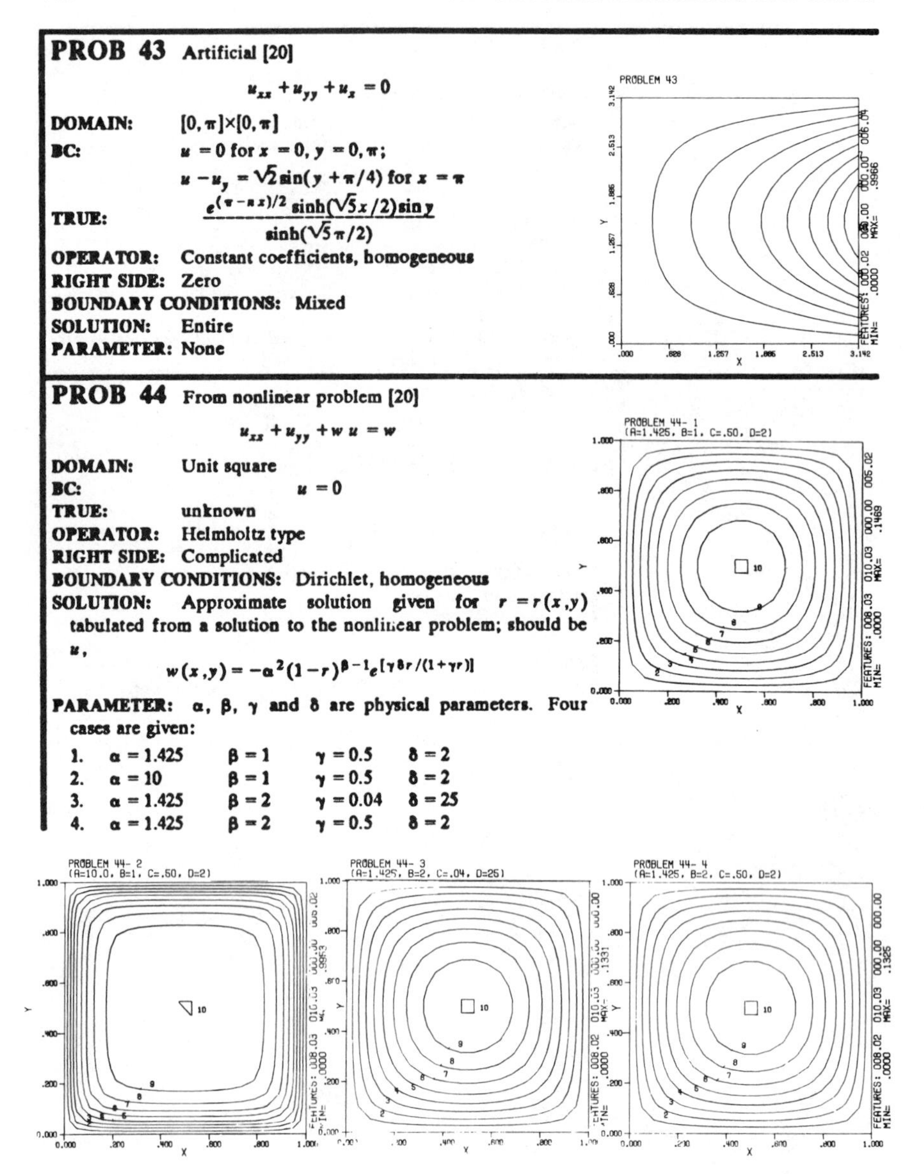

PROB 43 Artificial [20]

$$u_{xx} + u_{yy} + u_x = 0$$

DOMAIN: $[0,\pi]\times[0,\pi]$

BC: $u = 0$ for $x = 0$, $y = 0, \pi$;

$u - u_y = \sqrt{2}\sin(y + \pi/4)$ for $x = \pi$

TRUE: $$\frac{e^{(\pi - x)/2}\sinh(\sqrt{5}x/2)\sin y}{\sinh(\sqrt{5}\pi/2)}$$

OPERATOR: Constant coefficients, homogeneous

RIGHT SIDE: Zero

BOUNDARY CONDITIONS: Mixed

SOLUTION: Entire

PARAMETER: None

PROB 44 From nonlinear problem [20]

$$u_{xx} + u_{yy} + w\,u = w$$

DOMAIN: Unit square

BC: $u = 0$

TRUE: unknown

OPERATOR: Helmholtz type

RIGHT SIDE: Complicated

BOUNDARY CONDITIONS: Dirichlet, homogeneous

SOLUTION: Approximate solution given for $r = r(x,y)$ tabulated from a solution to the nonlinear problem; should be u,

$$w(x,y) = -\alpha^2(1-r)^{\beta-1}e^{[\gamma\delta r/(1+\gamma r)]}$$

PARAMETER: α, β, γ and δ are physical parameters. Four cases are given:

1. $\alpha = 1.425$ $\beta = 1$ $\gamma = 0.5$ $\delta = 2$
2. $\alpha = 10$ $\beta = 1$ $\gamma = 0.5$ $\delta = 2$
3. $\alpha = 1.425$ $\beta = 2$ $\gamma = 0.04$ $\delta = 25$
4. $\alpha = 1.425$ $\beta = 2$ $\gamma = 0.5$ $\delta = 2$

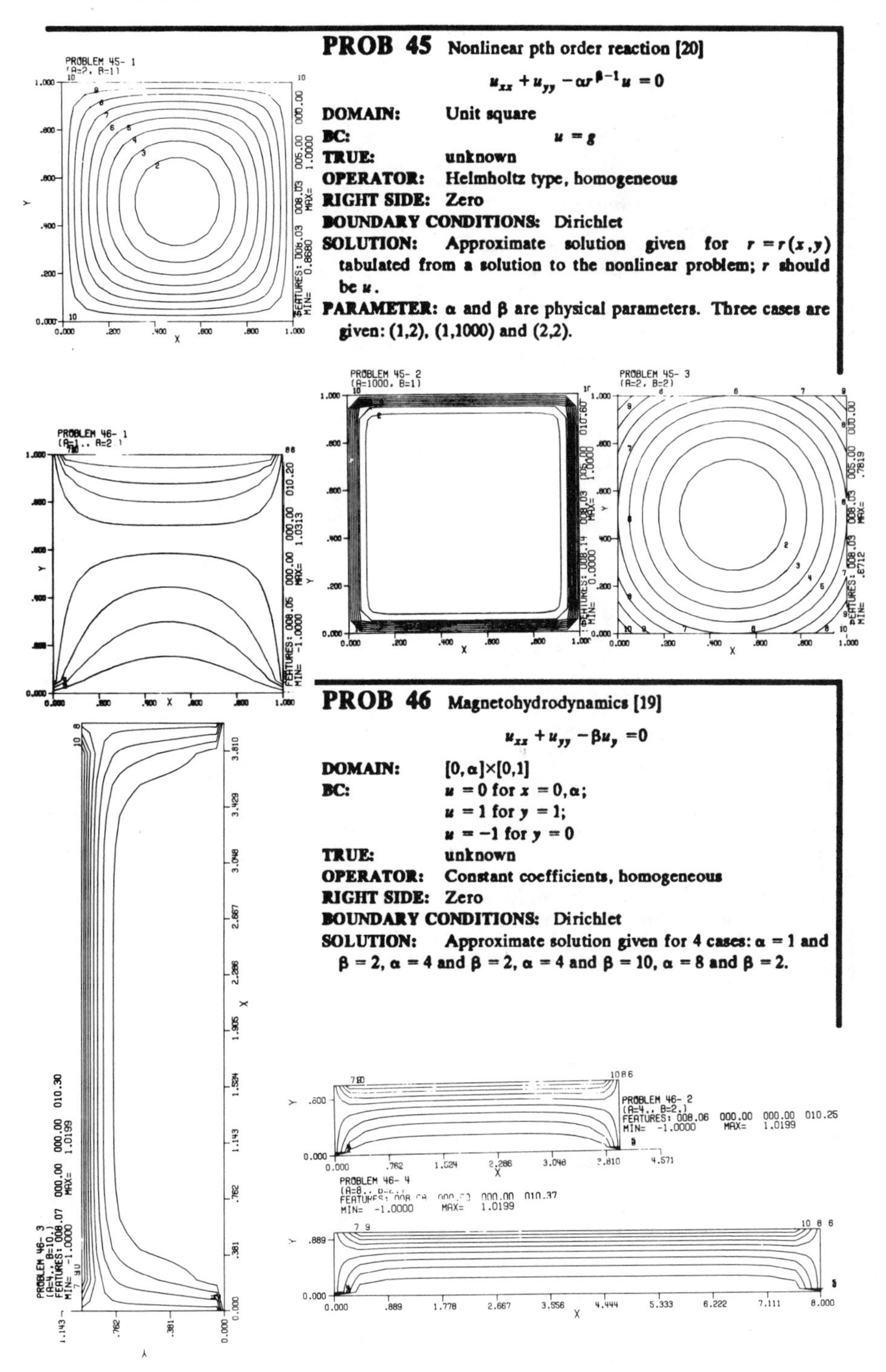

PROB 45 Nonlinear pth order reaction [20]

$$u_{xx} + u_{yy} - \alpha r^{\beta-1} u = 0$$

DOMAIN: Unit square

BC: $u = g$

TRUE: unknown

OPERATOR: Helmholtz type, homogeneous

RIGHT SIDE: Zero

BOUNDARY CONDITIONS: Dirichlet

SOLUTION: Approximate solution given for $r = r(x,y)$ tabulated from a solution to the nonlinear problem; r should be u.

PARAMETER: α and β are physical parameters. Three cases are given: (1,2), (1,1000) and (2,2).

PROB 46 Magnetohydrodynamics [19]

$$u_{xx} + u_{yy} - \beta u_y = 0$$

DOMAIN: $[0,\alpha]\times[0,1]$

BC: $u = 0$ for $x = 0, \alpha$;
$u = 1$ for $y = 1$;
$u = -1$ for $y = 0$

TRUE: unknown

OPERATOR: Constant coefficients, homogeneous

RIGHT SIDE: Zero

BOUNDARY CONDITIONS: Dirichlet

SOLUTION: Approximate solution given for 4 cases: $\alpha = 1$ and $\beta = 2$, $\alpha = 4$ and $\beta = 2$, $\alpha = 4$ and $\beta = 10$, $\alpha = 8$ and $\beta = 2$.

PROB 47 Artificial

$$u_{xx} + u_{yy} = f$$

DOMAIN: Unit square

BC: $u = g$

TRUE: $(xy)^{\alpha/2}$

OPERATOR: Laplace

RIGHT SIDE: Variable singularities

BOUNDARY CONDITIONS: Dirichlet

SOLUTION: Singularity of variable strength.

PARAMETER: α adjusts singularity strength.

PROB 48 Nonlinear diffusion in catalysts [2]

$$u_{xx} + u_{yy} - 1.425 r^{(\gamma - 1)} e^{(\alpha\beta(1-r))/(1+\beta(1-r))} u = 0$$

DOMAIN: Unit square

BC: $u = 1$

TRUE: unknown

OPERATOR: Helmholtz type, homogeneous

RIGHT SIDE: Zero

BOUNDARY CONDITIONS: Dirichlet

SOLUTION: Approximate solution given for $r = r(x,y)$ tabulated from a nonlinear PDE solver; r should be u.

PARAMETER: (α, β, γ) are physical parameters. 5 cases given: (1,0.04,2), (1,0.04,.25), (2,0.04,2), (2,0.04,25) and (2,0.5,2).

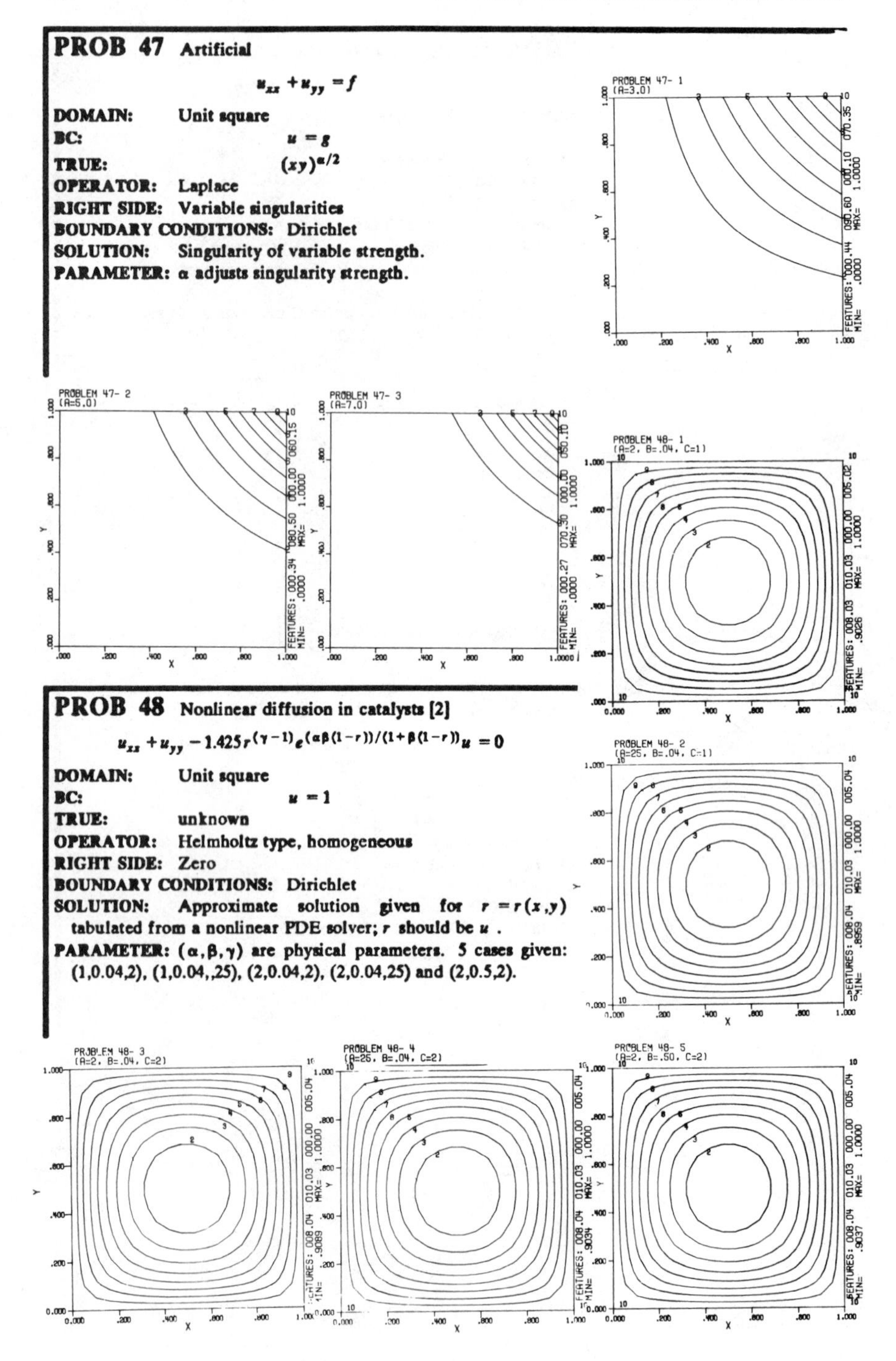

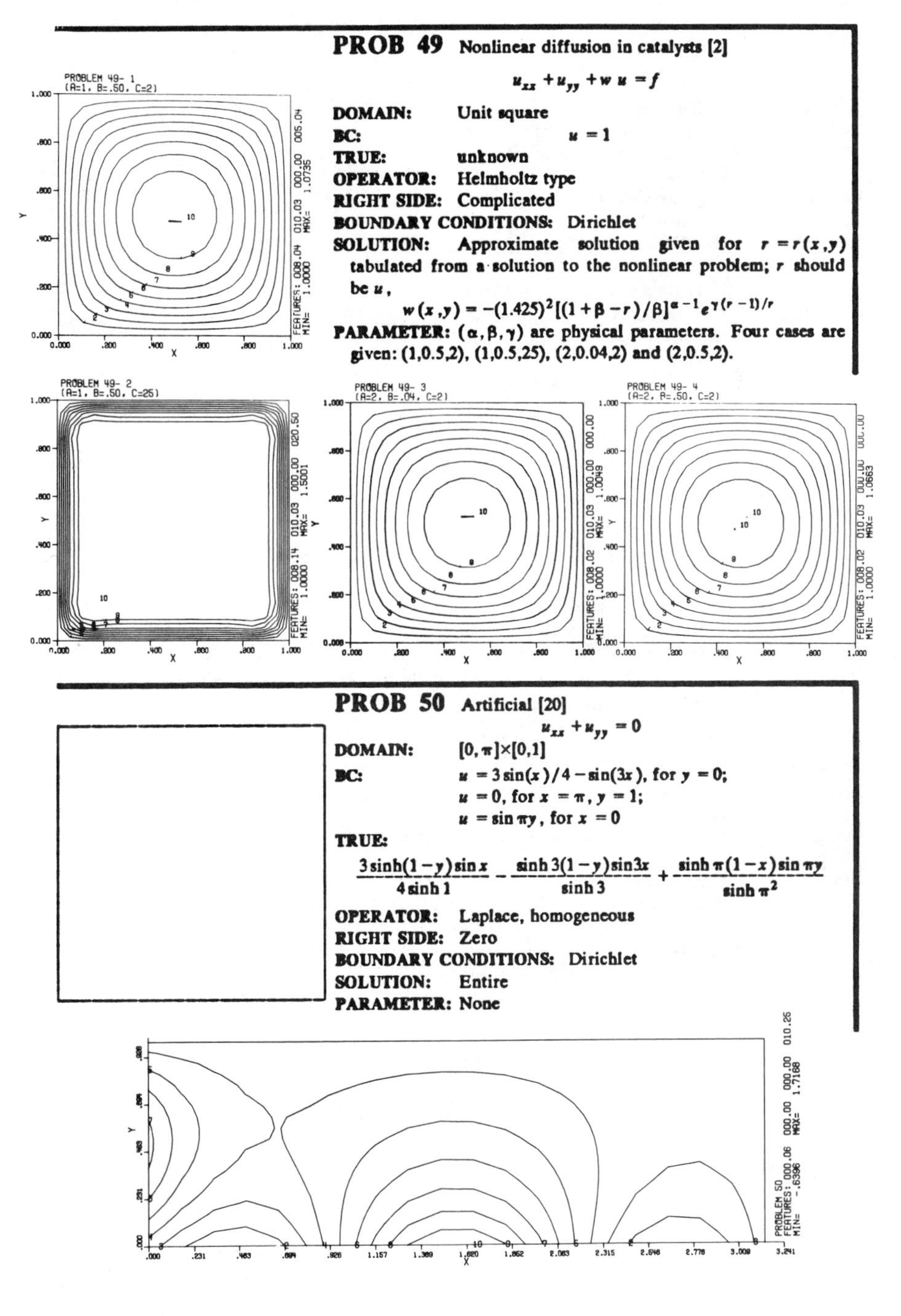

PROB 49 Nonlinear diffusion in catalysts [2]

$$u_{xx} + u_{yy} + w\,u = f$$

DOMAIN: Unit square

BC: $u = 1$

TRUE: unknown

OPERATOR: Helmholtz type

RIGHT SIDE: Complicated

BOUNDARY CONDITIONS: Dirichlet

SOLUTION: Approximate solution given for $r = r(x,y)$ tabulated from a solution to the nonlinear problem; r should be u,

$$w(x,y) = -(1.425)^2[(1+\beta-r)/\beta]^{\alpha-1}e^{\gamma(r-1)/r}$$

PARAMETER: (α,β,γ) are physical parameters. Four cases are given: (1,0.5,2), (1,0.5,25), (2,0.04,2) and (2,0.5,2).

PROB 50 Artificial [20]

$$u_{xx} + u_{yy} = 0$$

DOMAIN: $[0,\pi]\times[0,1]$

BC: $u = 3\sin(x)/4 - \sin(3x)$, for $y = 0$;
$u = 0$, for $x = \pi$, $y = 1$;
$u = \sin\pi y$, for $x = 0$

TRUE:

$$\frac{3\sinh(1-y)\sin x}{4\sinh 1} - \frac{\sinh 3(1-y)\sin 3x}{\sinh 3} + \frac{\sinh\pi(1-x)\sin\pi y}{\sinh\pi^2}$$

OPERATOR: Laplace, homogeneous

RIGHT SIDE: Zero

BOUNDARY CONDITIONS: Dirichlet

SOLUTION: Entire

PARAMETER: None

PROB 51 Fluid Flow [18]

$$u_{xx} + \frac{1}{x}u_x + \frac{1}{x^2}u_{yy} = -10$$

DOMAIN: Unit square

BC: $u = 0$ for $x = 1$; $u_n = 0$ for $x, y = 0$;
$Au_y + Bu = 0$ for $y = 1$

TRUE: unknown

OPERATOR: Singular coefficients

RIGHT SIDE: Constant

BOUNDARY CONDITIONS: Mixed

$$A(x) = \begin{cases} 0 \text{ for } x > \alpha, \\ 1 \text{ for } x \le \alpha, \end{cases} \quad B(x) = \begin{cases} 1 \text{ for } x > \alpha, \\ 0 \text{ for } x \le \alpha. \end{cases}$$

SOLUTION: Has singularity, unusual behavior.

PARAMETER: α adjusts position of change in boundary condition for $y = 1$.

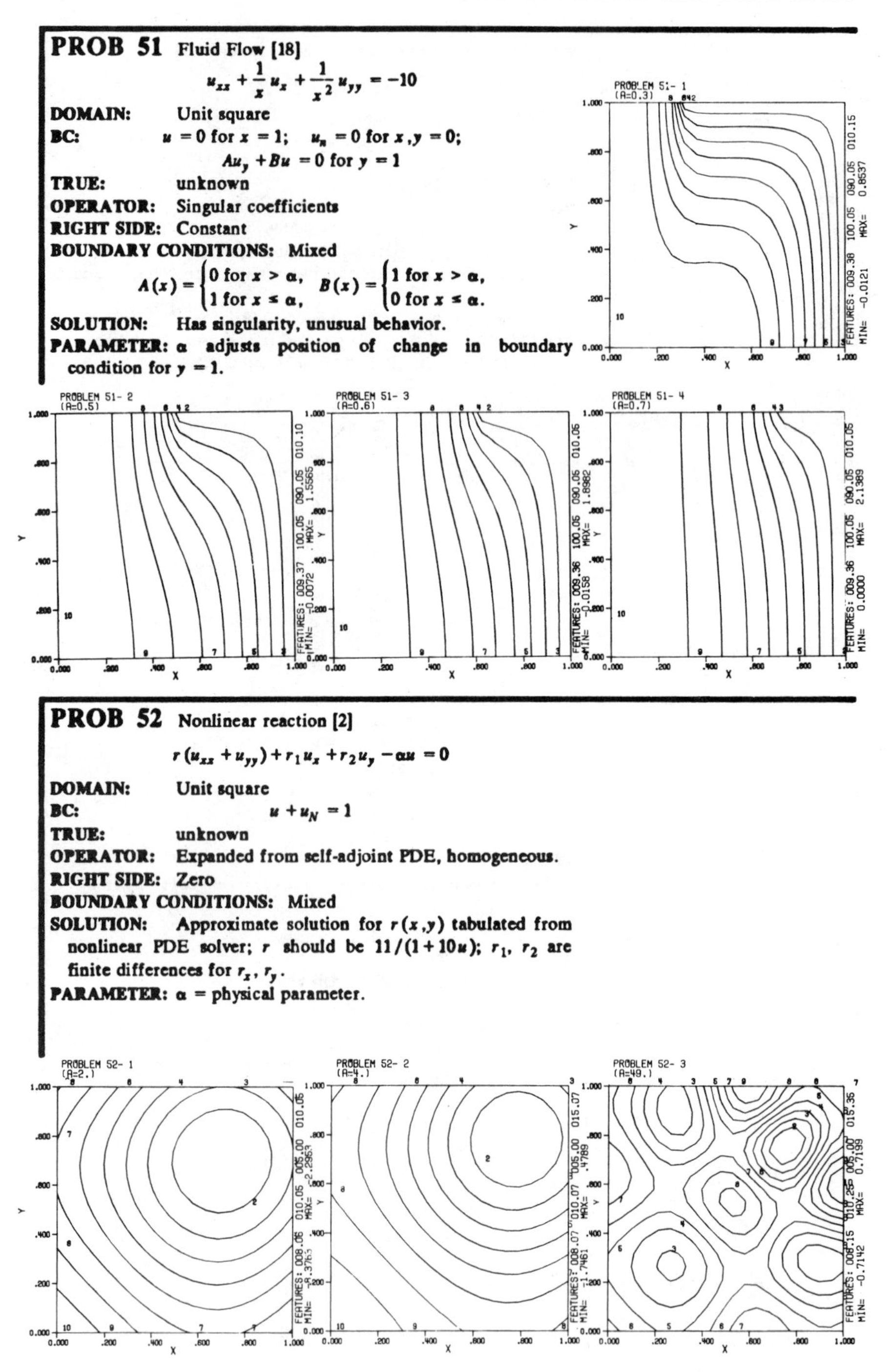

PROB 52 Nonlinear reaction [2]

$$r(u_{xx} + u_{yy}) + r_1 u_x + r_2 u_y - \alpha u = 0$$

DOMAIN: Unit square

BC: $u + u_N = 1$

TRUE: unknown

OPERATOR: Expanded from self-adjoint PDE, homogeneous.

RIGHT SIDE: Zero

BOUNDARY CONDITIONS: Mixed

SOLUTION: Approximate solution for $r(x,y)$ tabulated from nonlinear PDE solver; r should be $11/(1+10u)$; r_1, r_2 are finite differences for r_x, r_y.

PARAMETER: α = physical parameter.

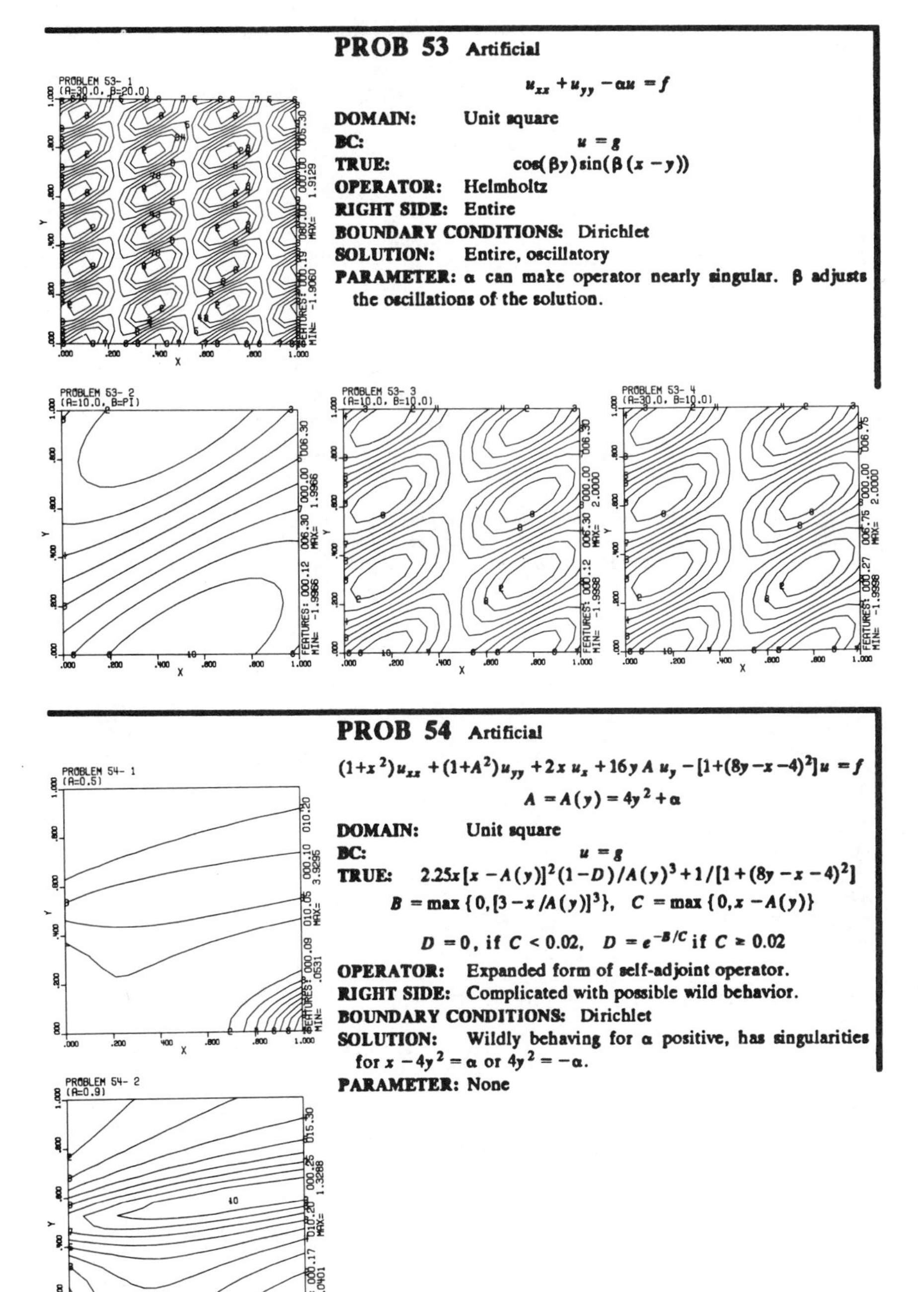

PROB 53 Artificial

$$u_{xx} + u_{yy} - \alpha u = f$$

DOMAIN: Unit square

BC: $u = g$

TRUE: $\cos(\beta y)\sin(\beta(x-y))$

OPERATOR: Helmholtz

RIGHT SIDE: Entire

BOUNDARY CONDITIONS: Dirichlet

SOLUTION: Entire, oscillatory

PARAMETER: α can make operator nearly singular. β adjusts the oscillations of the solution.

PROB 54 Artificial

$$(1+x^2)u_{xx} + (1+A^2)u_{yy} + 2x\, u_x + 16y\, A\, u_y - [1+(8y-x-4)^2]u = f$$

$$A = A(y) = 4y^2 + \alpha$$

DOMAIN: Unit square

BC: $u = g$

TRUE: $2.25x[x - A(y)]^2(1-D)/A(y)^3 + 1/[1+(8y-x-4)^2]$

$$B = \max\{0, [3 - x/A(y)]^3\}, \quad C = \max\{0, x - A(y)\}$$

$$D = 0, \text{ if } C < 0.02, \quad D = e^{-B/C} \text{ if } C \geq 0.02$$

OPERATOR: Expanded form of self-adjoint operator.

RIGHT SIDE: Complicated with possible wild behavior.

BOUNDARY CONDITIONS: Dirichlet

SOLUTION: Wildly behaving for α positive, has singularities for $x - 4y^2 = \alpha$ or $4y^2 = -\alpha$.

PARAMETER: None

PROB 55 Conducting fluid in magnetic field [9]

$$u_{xx} + u_{yy} = 0$$

DOMAIN: $[0,6]\times[0,1]$

BC: $u_x = 0$ for $x = 0$; $u = f$ for $x = 6$;
$Au + Bu_y = g$ for $y = 0$, $Cu + Du_y = h$ for $y = 1$

TRUE: unknown

OPERATOR: Laplace, homogeneous

RIGHT SIDE: Zero

BOUNDARY CONDITIONS: Mixed and complicated:

$$f(y) = e^{\pi(\alpha - y)/2}\sin(\pi y/2)/\alpha^3$$

$$A(x) = \begin{cases} 1 & \text{for } x < \alpha \\ x & \text{for } x \ge \alpha \end{cases} \qquad B(x) = \begin{cases} 0 & \text{for } x < \alpha \\ 1 & \text{for } x \ge \alpha \end{cases}$$

$$C(x) = \begin{cases} 1 & \text{for } x < \alpha \text{ or } k = 1 \\ 0 & \text{for } x \ge \alpha \text{ and } k = 2 \end{cases} \qquad D(x) = \begin{cases} 0 & \text{for } x < \alpha \text{ or } k = 1 \\ 1 & \text{for } x \ge \alpha \text{ and } k = 2 \end{cases}$$

$$G(x) = \begin{cases} 1/\alpha^3 & \text{for } x < \alpha \\ e^{2(\alpha - x)}/\alpha^3 & \text{for } x > \alpha \end{cases} \qquad H(x) = \begin{cases} 0 & \text{for } x < \alpha \text{ or } k = 1 \\ e^{\beta(\alpha - x)}/\alpha^3 & \text{for } x \ge \alpha \text{ and } k = 2 \end{cases}$$

SOLUTION: Has singularities at boundaries, widely varying behavior.

PARAMETER: α, β are physical parameters, c selects different physical models. Four cases are given: 1. $\alpha = 1$, $\beta = 3$, $c = 1$; 2. $\alpha = 1$, $\beta = 3$, $c = 2$; 3. $\alpha = 3$, $\beta = 2$, $c = 1$; 4. $\alpha = 6$, $\beta = 2$, $c = 2$.

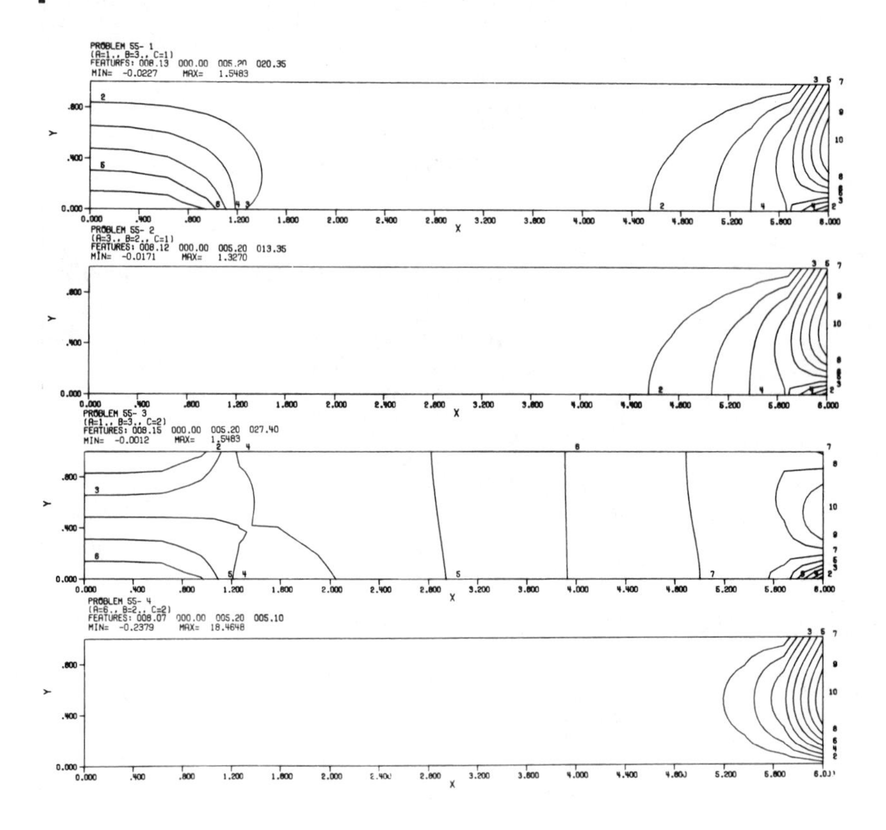

PROB 56 Artificial

$$u_{xx} + \frac{1}{x} u_x + \frac{1}{x^2} u_{yy} = 0$$

DOMAIN: $[0,1]\times[0,\pi]$

BC: $u = g$

TRUE: $\sum_{k=0}^{\alpha} w_k [e^{-z_k x \sin y} \cos(z_k x \cos y) + e^{-z_k x \cos y} \cos(z_k x \sin y)] + \sum_{n=1}^{\beta} a_n x^{2n} \cos(2n y)$

Gauss weights w_k and points z_k depend on α; $a_n = (1, 0.5, 0.25, 1/6, -1/10, -1/15, 1/30, -1/50)$

OPERATOR: Singular coefficients, homogeneous

RIGHT SIDE: Constant

BOUNDARY CONDITIONS: Mixed

SOLUTION: Series expansion approximates electrostatics solution.

PARAMETER: α = order of Gauss quadrature for integral, β = number of terms in expansion.

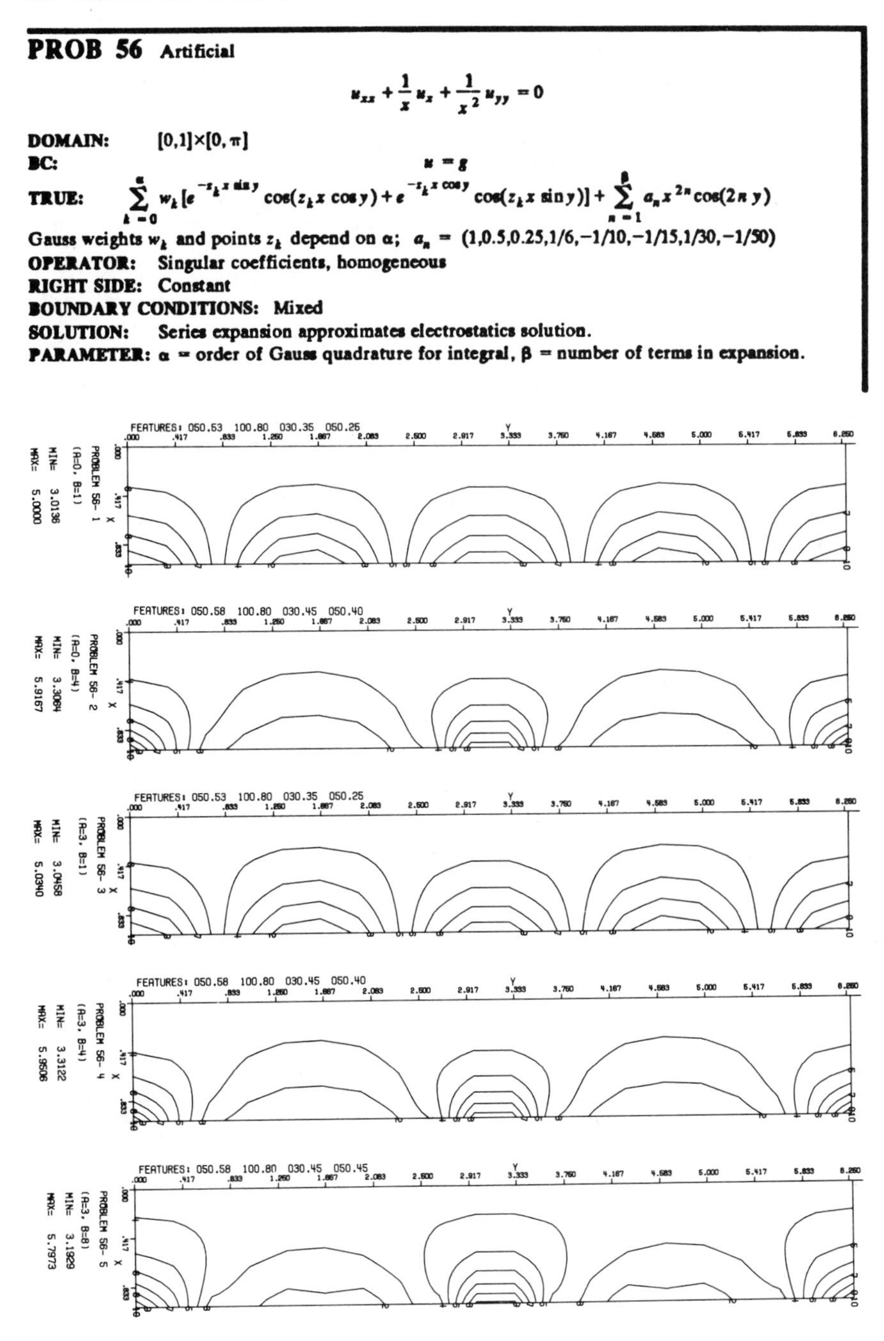

A.6 SOURCES OF PROBLEMS

[1] Ames, W. F., *Nonlinear Partial Differential Equations in Engineering*, Academic Press, 1965.

[2] Aris, R., *The Mathematical Theory of Diffusion and Reaction in Permeable Catalysts*, vols. 1 and 2, Clarendon Press, 1975.

[3] Behnke, H., G. Bertram, L., Collatz, R., Sauer, H. Unger, *Grundzüge der Mathematik* V, Vandenhoeck & Ruprecht, Gottingen, 1968, pp. 353–354.

[4] Berger, M. S., and L. E. Fraenkel, "On the Asymptotic Solution of a Nonlinear Dirichlet Problem", J. Math. Mech. 19 (1970), pp. 553–585.

[5] Collatz, L., *The Numerical Treatment of Differential Equations*, 3rd. Edition, Springer-Verlag, New York, 1966.

[6] Eisenstat, S. C., and M. H. Schultz, "Computation aspects of the finite element method", in *Mathematical Foundations of the Finite Element Method* (A. K. Aziz, ed.), Academic Press, New York, 1972, pp. 505–524.

[7] Eisenstat, S. C., and M. H. Schultz, in *Complexity of Sequential and Parallel Numerical Algorithms* (J. F. Traub, ed.), Academic Press, New York, 1973, pp. 271–282.

[8] Fox, L., *Numerical Solution of Ordinary and Partial Differential Equations*, Pergamon Press, 1962.

[9] Lynch, R. E., P. Gherson, and P. S. Lykoudis, "The use of ELLPACK 77 for solving the laplace equation on a region with interior slits, application to a problem in magnetohydrodynamics", CSD-TR 275, Computer Science Department, Purdue University, 1978.

[10] Greenspan, D., *Lectures on the Numerical Solution of Linear, Singular, and Nonlinear Differential Equations*, Prentice-Hall, Englewood Cliffs, N. J., 1968.

[11] Gupta, O. P., and S. K. Gupta, "Mixed boundary value problems in electrostatics", ZAMM 55 (1975), pp. 715–720.

[12] Houstis, E. N., R. E. Lynch, T. S. Papatheodorou, and J. R. Rice, "Development, evaluation and selection of methods for elliptic partial differential equations", Ann. Assoc. Inter. Calcul. Analog. 11 (1975), pp. 98–103.

[13] Houstis, E. N., R. E. Lynch, T. S. Papatheodorou, and J. R. Rice, "Evaluation of numerical methods for elliptic partial differential equations", J. Comput. Phys. 27 (1978), pp. 323–350.

[14] Houstis, E. N., and T. S. Papatheodorou, "Comparison of fast direct methods for elliptic problems", in *Advances in Computer Methods for Partial Differential Equations*, (R. Vishnevetsky, ed.), IMACS, New Brunswick, N. J., 1977, pp. 46–52.

[15] Kachanov, L. M., *Foundations of the Theory of plasticity*, North-Holland, 1971.

[16] McDonald, B. E., T. P. Coffey, S. Ossockow, and R. N. Sudan, "Preliminary report of numerical solution of type 2 irregularities in the equatorial electrojet", J. Geophy. Res. 79 (1974), 2551–2554.

[17] Mikhlin, S. G., *Linear Equations of Mathematical Physics*, Holt, Rinehart and Winston, 1967.

[18] Soliman, H. M., and A. Feingold, "Analysis of fully developed laminar flow

in longitudinal internally finned tubes", Chem. Eng. J. 14 (1977), pp. 119–128.

[19] Temperley, D. J., "Alternative approaches to the asymptotic solution of $\epsilon\nabla^2 u - u_y = 0$, $0 < \epsilon < 1$, over a rectangle", ZAMM 56 (1976), pp. 461–468.

[20] Weinberger, H. F., *A First Course in Partial Differential Equations*, Blaisdell, 1965.

[21] Trutt, F. C., and E. A. Erdely, "No-load flux distribution in saturated high-speed homopolar inductor alternators", IEEE Trans. on Aerospace 1 (1963), p. 430.

A.7 REFERENCES

Boisvert, R. F., E. N. Houstis, and J. R. Rice, [1979], "A system for performance evaluation of partial differential equations software", IEEE Trans. Software Engrg. 5, pp. 418–425.

Eisenstat, S. C. and M. H. Schultz, [1973], "Computational aspects of the finite element method", in *Complexity of Sequential and Parallel Algorithms* (J.F. Traub, ed.), Academic Press, New York, pp. 271–282.

Houstis, E. N., R. E. Lynch, T. S. Papatheodorou, and J. R. Rice, [1975], "Development, evaluation and selection of methods for elliptic partial differential equations", Ann. Assoc. Calcul. Analog. 11, pp. 98–105.

Houstis, E. N., R. E. Lynch, T. S. Papatheodorou, and J. R. Rice, [1978], "Evaluation of numerical methods for partial differential equations", J. Comput. Phys. 17, pp. 323–350.

Houstis, E. N. and T. S. Papatheodorou, [1977], "Comparison of fast direct methods for elliptic problems", in *Advances in Computer Methods for Partial Differential Equations* II (R. Vichnevetsky, ed.), IMACS, Rutgers University, New Brunswick, N.J., pp. 46–52.

Houstis, E. N. and J. R. Rice, [1980], "An experimental design for the computational evaluation of elliptic partial differential equation solvers", in *The Production and Assessment of Numerical Software*, (M. Delves and M.A. Hennel, eds.), Academic Press, New York, pp. 57–65.

Rice, J. R. [1979], "Methodology for the algorithm selection problem", in Performance Evaluation of Numerical Software (L.D. Fosdick, ed.), North-Holland, Amsterdam, pp. 301–307.

Rice, J. R., E. N. Houstis, and W. R. Dyksen, [1981], "A population of linear, second order elliptic partial differential equations on rectangular domains", Math. Comp. 36, Part 1: pp. 479–484, Part 2: microfiche supplement.

Appendix B

THE PG (PREPROCESSOR GENERATOR) SYSTEM

John F. Brophy
IMSL, Inc.

Libraries and packages of Fortran routines already exist in many areas of scientific computing; however their use demands a familiarity with the details of the routine. The user must allocate storage, determine the proper sequence of routines to call, etc. Therefore, considerable time and effort can be spent dealing with problems of Fortran programming. Such libraries can be made much easier to use if a language is provided to describe the problem to be solved in a natural manner. PG is a program designed to generate preprocessors to convert programs written in such languages to Fortran.

PG is used to write the preprocessor for the ELLPACK language described in this book. To implement a problem oriented language, such as ELLPACK, both a grammar for the language and a driver routine for the preprocessor must be supplied. The driver routine is usually fairly short and is needed to initialize the program, open files, etc. A more detailed description of what is needed is given below. The grammar describes the language to be implemented. PG converts the grammar into a set of Fortran subroutines which are called by the user supplied driver routine.

PG is designed around the template processor described in Appendix C. The PG system contains routines for analyzing and manipulating text strings in the form used by the template processor. The use of these routines is described below.

B.1 INTRODUCTION TO PG GRAMMARS

A preprocessor must perform two basic tasks: parse the input language (e.g. ELLPACK) and generate an output program. Parsing is controlled by grammar **rules** and generating an output program is done by **actions**.

In a PG grammar a line beginning with an asterisk is a comment; blank lines are also considered comments. Either may occur anywhere in the grammar. All lines are assumed to be at most 80 characters long. A statement may be continued to a new line by ending the current line with an ampersand (&).

We start with a number of simple examples from the ELLPACK language and its preprocessor. A very simple PG grammar, consisting of just one rule is

```
PROG     ->     'OPTION.'     NAME     '='     NAME
END.
```

The name of the rule is PROG; rule names must be valid Fortran variable names. To the right of the arrow (->) are **subrules**. Subrules are either character strings to be matched or the name of another rule. PG is **not recursive** and it is illegal for a rule to use itself as a subrule directly or indirectly, as in rule A using rule B using rule A. There are several built-in rules: **NAME** is a built-in rule which matches an alphanumeric string without embedded blanks. The last statement in a PG program must be the built-in rule **END**. Valid input to the preprocessor generated by the above grammar is, for example,

```
OPTION.     WORKSPACE = 5000
```

It is assumed that blanks in the problem oriented language, e.g. ELLPACK, are usually not significant, so blanks are skipped automatically *after* each subrule. The string OPTION. must occur in the first column of the input, but the rest of the line can begin in any column.

The text matched by NAME may not contain any embedded blanks. If it is desired to allow embedded blanks, then the following program can be used

```
PROG     ->     ALFNUM+
END.
```

ALFNUM is a built-in rule which matches any alphanumeric character, (see Table B.1 for a complete list of built-in rules). The '+' is a PG operator and means match one or more occurrences of ALFNUM. Since blanks are skipped after *every* character matched by ALFNUM, embedded blanks are ignored. The '+' may be used with either a rule name, as here, or with a quoted string. The complete set of PG postfix operators is given in Table B.2.

In ELLPACK there are two forms for the OPTION segment, e.g.,

```
OPTION.     WORKSPACE = 5000
```

and

```
OPTION.     MEMORY
```

The PG grammar to define the OPTION segment is

```
PROG     ->     ALFNUM+     '='     ALFNUM+
         ->     ALFNUM+
END.
```

The preprocessor will first try to use the first rule. If it fails, i.e. one of its subrules fails to match the input, then it will try the next **alternative**, marked by -> with no rule name to its left. A rule may have any number of alternatives.

The grammar and its rules define the input language and the **actions** associated with these define the output. The actions are executed when the associated alternative is matched and they correspond to the code generation part of a compiler. Most actions are to set or reset variables in the template processor or to write Fortran statements. For example, the action statement

```
$(TPVNAME) = 'ASTRING'
```

sets the template variable TPVNAME to ASTRING. Template variable names must begin with a letter and contain only letters and digits; there is no limit on their length. A reference to a template variable is indicated by $ preceding the template variable name. The PG grammar used by ELLPACK to set a template variable via the OPTION segment is

```
PROG    ->    'OPTION.'    ALFNUM+    '='    ALFNUM+
                    $($2) = $4
        ->    'OPTION.'  ALFNUM+
                    $($2) = '.TRUE.'
END.
```

Each element in a rule or alternative is numbered in order of its occurrence and **what it matches** is referenced by its number preceded by $. In the first rule above, $1 is 'OPTION'., $2 is whatever is matched by the first ALFNUM+, $3 is '=', and $4 is whatever is matched by the second ALFNUM+.

Thus, if the input is

```
OPTION.    WORKSPACE = 5000
```

then $2 is 'WORKSPACE' and $4 is '5000'. When this line is read by the ELLPACK preprocessor the action $($2) = $4 has the template variable WORKSPACE set to the string '5000'. If the input is

```
OPTION.    MEMORY
```

then the template variable MEMORY is set to .TRUE.. Note that this approach allows one to associate language keywords, (e.g. WORKSPACE and MEMORY in ELLPACK), with template variables.

While all template variables contain text strings, special provision is made for two special classes. A **logical template variable** is one that contains either the text string .TRUE. or .FALSE.. An **integer template variable** is one that contains a character string representing an integer. A logical template variable can be set using the action

```
$L(TPVNAME) = ⟨LOGICAL EXPRESSION⟩
```

and an integer template variable can be set using

```
$I(TPVNAME) = ⟨INTEGER EXPRESSION⟩
```

The $L and $I operators automatically convert the values of the expressions to the proper character string. In addition, both $L(TPVNAME) and $I(TPVNAME) can be used in expressions as if they were Fortran function calls of the corresponding type. In this case, the $L and $I operators convert the character string values of template variables to the proper values to be used in the expression. The use of these features is illustrated by the following grammar.

```
PROG    ->    'OPTION.'    ALFNUM+    '='    ALFNUM+
                    $($2) = $4
                    $I(TOTAL) = $I(TOTAL) + $I($2)
                    $L(TOOBIG) = $I($2)  .GT.  10000
```

Using the same input as before will result in WORKSPACE being set to '5000', adding 5000 to the variable TOTAL and setting TOOBIG to .FALSE..

There is a **string concatenation** operator //. The grammar

```
PROG     ->     'OPTION.'     ALFNUM+     '='     ALFNUM+
                     $(DOUBLE) = $4 // $4
                     $L($2 // SET) = .TRUE.
END.
```

with the same input as above, will set DOUBLE to '50005000' and set WORKSPACESET to .TRUE..

There is an **IF-ELSE-ENDIF** construct in PG, illustrated by the following grammar.

```
PROG  ->  'OPTION.'  ALFNUM+  '='  ALFNUM+
               IF ( $I($2) .LT. IQIVAL($4) )
                   $I($2) = $4
               ENDIF
END.
```

With the same input as above, these actions are interpreted as

IF("integer value of WORKSPACE" .LT. "integer value of '5000' ")
    $I(WORKSPACE) = 5000

Note that $I both converts the value WORKSPACE to an integer (in the IF) and the value of an integer to a template variable value (in the assignment). As explained further below, one cannot use $I($4) in the IF because '5000' is not the name of a template variable; the utility routine IQIVAL produces integer values of literal digit strings. Any number of action statements may occur between the IF and the ELSE and between the ELSE and the ENDIF.

There is also a set of PG **utility routines** to aid in programming the action statements. The logical function

```
LQMTCH ( ⟨STRING1⟩, ⟨STRING2⟩ )
```

returns .TRUE. if ⟨STRING1⟩ the same as ⟨STRING2⟩. The integer function

```
IQIVAL ( ⟨STRING⟩ )
```

returns the numerical value of ⟨STRING⟩. The complete set of these PG functions is given in Table B.4.

The following grammar will set WORKSPACE to the maximum of its previous value and the current value. Note that if the option is not WORKSPACE then no action is taken.

```
PROG  -> 'OPTION'  ALFNUM+  '='  ALFNUM+
               IF (LQMTCH($2, 'WORKSPACE'))
                   $I($2) = MAX0($I($2), IQIVAL($4))
               ENDIF
END.
```

We could *not* use the action

```
$I($2) = MAXO( $I($2), $I($4) )
```

above, because $I applies only to template variables and $4 stands for the string 5000 which is not the name of a template variable. We could, however, have used

```
$(TEMP) = $4
$I($2) = MAXO( $I($2), $I(TEMP) )
```

In addition to the above PG utilities which act like functions there are PG **procedures** which act like subroutine calls. The grammar

```
PROG  ->  'OPTION.'  ALFNUM+  '='  ALFNUM+
               WRITE('        ' // $2 // $3 //$4)
      ->  'OPTION.'  ANY+
               ERROR('ILLEGAL OPTION -' //$2)
END.
```

uses the PG procedures WRITE and ERROR. Given the input

```
OPTION.    LEVEL = 3
```

the statement

```
LEVEL=3
```

is written to the **program** file. If the input is

```
OPTION.    MEMORY
```

then the error message

```
ILLEGAL OPTION - MEMORY
```

is written to the **listing** file and a flag is set indicating that an error has occurred. The complete list of PG procedures is given in Table B.3.

The following grammar illustrates how to allow several options to be separated by dollar signs on the same line and introduces several additional features of PG.

```
PROG      ->   'OPTION.'    OPTEXP ++ '$'
OPTEXP    ->    OPTNAM      '='     ALFNUM+
                   $($1) = $3
                   CLEAR
OPTNAM    ->   ALFNUM+
END.
```

The rule PROG is defined in terms of the subrules OPTEXP and OPTNAM. Given the input

```
OPTION.    LEVEL = 3    $    WORKSPACE = 5000
```

the rule PROG and its actions sets the template variable LEVEL to 3 and sets WORKSPACE to 5000. The **list operator ++** as in

```
⟨SUBRULE1⟩ ++ ⟨SUBRULE2⟩
```

matches a list of ⟨SUBRULE1⟩ type objects separated by objects matching ⟨SUBRULE2⟩. Here we want a list of options separated by dollar signs. The text matched by ALFNUM+ on the right side of the OPTNAM rule is passed back to the OPTEXP rule.

Text would be passed back to the PROG rule if not for the **CLEAR** procedure. If text were always passed back to the PROG rule the system would retain the entire input. This would be very wasteful of space since there is no need to keep the input text once it has been parsed and the relevant actions performed. The CLEAR procedure tells PG that all the text matched so far can be thrown away.

B.2 PG RULE SYNTAX

The basic PG grammar rules are quoted strings and a set of built-in rules. (A quoted string may be thought of as a special kind of built-in rule).

Table B.1 The built-in rules of the PG system. Blanks are automatically skipped after each rule, except for fules marked (no skip).

Rule Name	*Matches*
'TEXT'	quoted text, use " to quote a quote
ALFNUM	a letter or a digit
ANY	anything but an end-of-line
BLANKS	any number of blanks
CHAR	anything but an end-of-line (no skip)
DIGIT	a digit
EOL	end-of-line (no skip)
FEXPR	Fortran expression
FNAME	Fortran-type name
LETTER	a letter
LINE	everything up to an end-of-line
NAME	name (alphanumeric string without embedded blanks)

A basic rule may have a postfix operator applied to it. The postfix operators are given in Table B.2. The postfix operator – is called the **ignore operator**.

Table B.2 The postfix operators in PG. The operators are appended to the rules (denoted by RULE) from Table B.1.

Operator	*Effect*
RULE *	match zero or more occurrences of RULE
RULE +	match one or more occurrences of RULE
RULE ?	match zero or one occurrences of RULE
RULE –	match rule, but ignore matched text. Do not pass the text back up.
RULE ** N	match exactly N (>O) occurrences of RULE
RULE ** O	match rule, but then back up the input so that the text is not considered matched
RULE ?* N	match up to N (>O) occurrences of RULE
RULE ?–	match zero or one occurrence of RULE, but ignore matched text

The **list operator** ++, as in RULE1 ++ RULE2, matches a list of RULE1 type objects separated by RULE2 type objects. For example, FEXPR ++ ',' matches a list of Fortran expressions separated by commas.

B.3 PG ACTIONS

In a PG action a **basic string variable** is either $⟨INTEGER⟩, or a quoted string. If the quoted string contains only letters and digits the quotes may be omitted. A quote may be included in a quoted string by using two quotes. A **string variable** is either a basic string variable, the value of a template variable or a combination of these concatenated using the concatenation operator //. A template variable is referenced by $(⟨SV⟩), where ⟨SV⟩ is a string variable. Recall the definition that $1 refers to the text matched by the first subrule, $2 to the text matched by the second subrule, etc. Examples of string variables are

```
'X+3.6*Y'
'ISN''T'
TEXT
$(TEMPLATE)
$3
ABC // $(VARIABLE) // $2 // $(DEF)
$( ABC // $(XYZ) ) // R12
```

Text matched with an ignore operator attached ('−' or '−?') cannot be accessed with $⟨N⟩ but is counted in the numbering $1, $2, etc.

A PG **action statement** may assign a value to a template variable. The action statement

```
$( ⟨SV1⟩ ) = ⟨SV2⟩
```

assigns the string ⟨SV2⟩ to the template variable ⟨SV1⟩. To append the string ⟨SV2⟩ to ⟨SV1⟩ the ← is used as in the action statement

```
$( ⟨SV1⟩ ) ← ⟨SV2⟩
```

If the value to be assigned to a template variable is either logical ('.TRUE.' or '.FALSE.') or an integer (an optional '+' or '−', followed by a string of digits) then the special assignment statements

```
$L( ⟨SV⟩ ) = LOGICAL-EXPRESSION
$I( ⟨SV⟩ ) = INTEGER-EXPRESSION
```

may be used. The logical and integer expressions follow the usual Fortran rules augmented by the logical and integer valued functions

```
$L( ⟨SV⟩ )
$I( ⟨SV⟩ )
```

with template variable arguments ⟨SV⟩. For example,

```
$I(COUNT) = $I(COUNT) + 1
```

takes the digits in the string named COUNT, converts them to an integer and adds one to the value; the new value is then converted back to a digit string and assigned to COUNT.

There is also a logical function to determine whether a variable has been assigned a value.

```
$D(⟨SV⟩) = .TRUE.     IF ⟨SV⟩ HAS BEEN DEFINED,
         = .FALSE.    OTHERWISE.
```

PG action statements may also involve ordinary Fortran statements with template variables. These statements are local to the set of actions associated with a rule and do not, in general, affect the action statements of other rules (but see the GLOBAL statement below). Fortran variables in these statements have their types determined by the usual Fortran rules. The following example shows how one can access Fortran facilities in a PG action statement using Fortran statements.

```
IWORK = $I(WORKSPACE)/2 + 10
SCALE =  ALOG10( FLOAT(MAX0(IWORK,100)) ) + 1
$I(WORKSPACE) = 10**SCALE + $I(WORKSPACE)
```

The built-in PG procedures are given in Table B.3. They each have one of the forms

```
⟨NAME⟩
⟨NAME⟩ ( ⟨TV⟩ )
⟨NAME⟩ ( ⟨TV1⟩, ⟨TV2⟩ )
```

where ⟨TV⟩, ⟨TV1⟩ and ⟨TV2⟩ are names of template variables.

Conditional execution of action statements is possible using IF-ELSE-ENDIF. It has the form

```
IF ( ⟨LOGICAL-EXPRESSION⟩ )
     ANY NUMBER OF ACTION STATEMENTS
ELSE
     ANY NUMBER OF ACTION STATEMENTS
ENDIF
```

The ELSE clause may be omitted. IF-ELSE-ENDIF statements may be nested.

PG also has a **DO-loop** construct. It has the form

```
DO ⟨I⟩ = ⟨FEXPR1⟩, ⟨FEXPR2⟩, ⟨FEXPR3⟩
     ANY NUMBER OF ACTION STATEMENTS
ENDDO
```

where ⟨I⟩ is a Fortran integer variable and the ⟨FEXPRn⟩ are Fortran integer expressions. The meaning is as in Fortran and the increment expression ⟨FEXPR3⟩ may be omitted; it defaults to one. DO loops may be nested and may include IF statements.

There are several Fortran library routines provided for use in PG action statements. In all cases their arguments are string variables. They are listed in Table B.4. The PG procedures and entities correspond to Fortran subroutines and functions, respectively. For example, corresponding to the PG utility IQIVAL (⟨SV⟩) is the Fortran routine

```
INTEGER FUNCTION IQIVAL(STRING,ILEN)
CHARACTER*1 STRING(ILEN)
```

B.4 SPECIAL SEGMENTS

A PG grammar includes several special **segments**. Segments are declared by their name (starting in column 1 and terminated by a period) in the same way as ELLPACK segments. The first of these is the (optional) **GLOBAL** segment in which Fortran declaration statements may be included. If GLOBAL is used it must

Table B.3 The PG Procedures. A string variable is denoted by ⟨SV⟩.

ERROR(⟨SV⟩)	Prints ⟨SV⟩ as an error message on the listing file and sets an error flag.
WRITE(⟨SV⟩)	Prints ⟨SV⟩ on the normal output file.
FWRITE(⟨SV⟩)	Prints ⟨SV⟩ on the output file, using the Fortran continuation convention if the line is longer than 72 characters.
DONE	Sets a flag indicating normal completion.
CLEAR	Resets PG so that all text matched so far is deleted. This must be done, usually after each input program segment is finished to keep the amount of text stored by PG within set limits.
IGNORE	The text matched by the current rule is cleared. When control returns to the rule for which the current rule is a subrule, it will appear that the subrule succeeded but the corresponding string $⟨N⟩ is null.
REPLACE(⟨SV⟩)	The text matched by the current rule is replaced by ⟨SV⟩. ⟨SV⟩ may contain text matched in the current rule. This procedure allows the text to be rearranged. For the rest of the action in which this procedure appears only $1 is defined and it refers to ⟨SV⟩. When control returns to the calling rule the corresponding $⟨N⟩ has the value ⟨SV⟩.
FAIL	Causes the current rule to fail and control to return to the calling rule without trying any more alternatives of the current rule. This can be used, for example, to write a rule which matches anything but a '$' or an end-of-line marker; as in the example NOTDOL -> '$' FAIL -> ANY

be the first segment in a PG grammar. Each PG grammar rule corresponds to a Fortran subroutine. The declarations in the GLOBAL segment are included in all of these subroutines. Variables which are referenced in actions associated with different rules must be in COMMON blocks. It is also possible to declare the types of variables and arrays. If arrays are declared they should be in COMMON blocks, or else, since they will be declared for each rule, an excessive amount of memory may be needed.

The GLOBAL segment may also be used to set the sizes of various arrays used by PG to other than their default values. The list of these arrays, and the template variables controlling their sizes, is given in Table B.5.

Table B.4. Utility routines for use in actions. The actual arguments to the corresponding Fortran routines have a string variable represented by a character string followed by an integer count of the characters in the string.

Routine Name	*Purpose*
IQNO1 (⟨SV⟩)	Integer function returning -1, 0, or 1 if ⟨SV⟩ is a Fortran expression equal to -1, 0, 1, respectively; otherwise, it returns 2.
IQIVAL (⟨SV⟩)	Integer function returning the value of the string ⟨SV⟩ as an integer.
LQMTCH (⟨SV1⟩, ⟨SV2⟩)	Logical function whose value is .TRUE. if the string variable ⟨SV1⟩ is identical to ⟨SV2⟩.
QQADLS (⟨SV1⟩, ⟨SV2⟩)	Subroutine which adds the Fortran name ⟨SV2⟩ to a comma separated list of names in ⟨SV1⟩, provided ⟨SV2⟩ is not already on the list. This is intended for maintaining a list of variables to be used in Fortran declarations.

Table B.5 Template variables that control the size of PG internal arrays.

Name	*Default*	*Size of*
IQSPMX	1000	CQSTCK (the primary stack)
IQMXIN	2000	CQBFIN (the input buffer)
ICBDIM	1500	input buffer
ICSDIM	1000	character storage array
IHADIM	401	hash table (must be prime)
ISTDIM	6000	integer pointer pool (must be divisible by 3)

An example of a GLOBAL segment is

```
GLOBAL.
      COMMON /ABC/ X, Y, Z
      INTEGER Y
      REAL Z(10)
      $(IQSPMX) = 2000
```

The second special segment is **MAIN**. It is a required segment. It contains actions which are executed in the generated main routine. Usually it includes statements to initialize template variables and a call to a (separately compiled) Fortran routine to open files, etc.

To execute a grammar rule from the actions under MAIN or from a Fortran subprogram one uses

CALL ⟨RULE⟩

The ELLPACK grammar (and the PG sample grammars in this document) use the rule PROG as the base rule, which causes all of the other rules to be executed. In this case exactly one CALL PROG statement is needed to cause an entire program to be parsed.

To open the input file in the separately compiled Fortran routine, set IQINPF in the common block QQCINT (see section B.5) to the Fortran unit number and OPEN the file. Similarly, the listing file unit number is given by IQLSTF and the file for the generated code by IQOUTF. The template processor is invoked by the call

```
CALL TPDRV ( IERR, IN, ILIST, IOUT )
```

where IERR, IN, ILIST, IOUT are the Fortran unit numbers of the error, input, listing and output files, respectively.

The last segment must be **END**. It has no values and no actions.

A complete example of a simple PG grammar is:

```
GLOBAL.
          COMMON /ABC/ X, Y, Z
          INTEGER Y
          REAL Z(10)
          $(IQSPMX) = 2000
MAIN.
          $I(COUNT) = 0
          CALL PRINIT
          CALL PROG
PROG        ->       'OPTION.'   ALFNUM+   '='   ALFNUM+
                        $($2) = $4
                        $I(COUNT) = $I(COUNT) + 1
            ->       'OPTION.'   ALFNUM+
                        $L($2) = .TRUE.
                        $I(COUNT) = $I(COUNT) + 1
END.
```

Note that the GLOBAL, MAIN and END segments appear as special rules.

B.5 PG INTERNALS

In this section we give a description of the data structures used by PG generated programs.

Most PG routines manipulate the **character stack** stored in CQSTCK. Character strings are pushed onto the stack using the routine QQPUSH. Strings are removed from the stack by resetting the pointer IQSP if the rule fails. The action CLEAR resets the pointer to 1. The Fortran data structure for the character stack is:

/QQCSTK/	IQSP	pointer to next free location in CQSTCK
	IQSPMX	dimensioned size of CQSTCK
/QQCCST/	CQSTCK	matched characters

All PG routines get their input through the routine **QQGTCH**. It maintains a circular buffer of the characters read. The data structures for this buffer are

/QQCINP/	IQBFLG	for use by input

	IQNXIN	pointer to next character in CQBFIN to be returned
	IQINB1	pointer to first valid character in CQSTCK
	IQINB2	pointer to last valid character in CQSTCK
	IQMXIN	dimensioned size of CQBFIN
/QQCBIN/	CQBFIN	input characters

The pointer IQINB1 is moved only by the CLEAR action. The characters between IQINB1 and IQNXIN have been tentatively matched; they can be put back into the input queue by adjusting IQNXIN. PG built-in and generated rules save the pointer IQNXIN on entry. If the rule fails then IQNXIN is reset to its previous value.

When IQNXIN reaches IQINB2 the routine **QQINPT** is called to read the next line. QQINPT calls QQREAD to get the next line and prints the line; it skips lines beginning with a comment character and adds the end-of-line marker if the previous line did not end with the continuation character. QQREAD calls **IORDLN** to get the line and strips trailing blanks. All input for PG programs (including input for the template processor) is done through IORDLN. The data structures used by these routines are

/QQCINT/	IQLEVEL	output listing level
	IQINPF	input file unit number
	IQOUTF	output file unit number
	IQLSTF	listing file unit number
	IQCEOL	number of characters used by end-of-line marker
/QQCCHR/	CQCEOL	end-of-line marker
/QQCCIN/	CQCOMT	comment character
	CQCNTN	continuation character
/QQCLOG/	LQERR	.TRUE. if the current rule has failed
	LQEOL	.TRUE. if current character is an end-of-line
	LQFATL	.TRUE. if a fatal error is flagged by the ERROR() action procedure

B.6 INSTALLING PG

The PG distribution tape contains the following files:

1	CSD-TR 405	Users guide for PG
2	PGLIB1.F	Fortran library 1
3	PGLIB2.F	Fortran library 2
4	PGDRIV.F	Driver with OPEN statements for PG
5	PG.F	Fortran routines of PG
6	SYMBOLS	File of symbols used by PG
7	PG.G	Grammar for PG

The files PGLIB1.F and PGLIB2.F together are a library of Fortran routines needed by PG and by the programs generated by PG. They should be compiled and made into a single object module library called PGLIB.

The file PGDRIV.F contains the OPEN statements needed by PG. It may need to be changed on some systems. The file PG.F contains the main PG routines and was itself generated by PG. PGDRIV.F and PG.F should be compiled and link-edited with PGLIB to produce the executable file PG.

The file SYMBOLS is used by PG when it executes. (An OPEN statement in PGDRIV.F refers to it). The file PG.G is the grammar that produces PG.F. Executing PG with PG.G as input should produce PG.F as output.

B.7 ACKNOWLEDGEMENTS

The PG system design and development was supported in part by the National Science Foundation grant MCS 79-26310 and Department of Energy contract DE-ACO-81ER-10997.

B.8 REFERENCES

Brophy, J. [1984], PG — "A preprocessor generator", CSD-TR 472, Computer Science Dept., Purdue University.

Appendix C

THE TEMPLATE PROCESSOR

Calvin Ribbens
John R. Rice
Purdue University

William A. Ward
Mobile College

C.1 INTRODUCTION

The ELLPACK control program is created from a symbol table and program template. The first phase of the preprocessor is to read the ELLPACK program and to derive the values of many variables; these values are placed in the symbol table. The values might be simple numbers or complex Fortran expressions. Once the ELLPACK program is analyzed and all the derived values are put in the symbol table; then these values are substituted into the control program template. The program that does this substitution is the **ELLPACK template processor**. This template processor is a simple macro processor, one designed to manipulate Fortran program text. This macro processor has the basic facilities of a general purpose macro processor and can be used in most macro processor applications.

Several features distinguish this macro processor from others. First, it has a small and syntactically consistent set of commands which are easily learned. Second, it is capable of recursive macro substitution. Finally, it is written in PFORT-portable FORTRAN. Though this last feature somewhat reduces the processor's performance, it is an important advantage because it reduces the human time required to make the processor operational.

The template processor was developed as part of the **TOOLPACK project** [Osterweil, 1982] and was supported by the National Science Foundation under grant MCS-7926310. It was designed and written by John R. Rice and William A. Ward; many improvements were added by Calvin Ribbens. For more information see [Rice and Ward, 1982a], [Rice and Ward, 1982b] and [Ribbens, et al., 1984].

C.2 GENERAL DESCRIPTION

The template processor accepts as input templates or patterns for producing text in some user specified format. This processor is part batch editor, part macro processor, and part interpreter. It is a batch editor in the sense that it capable of manipulating text. It is a macro processor in that it is able to define and redefine macros and substitute their values when references to them occur in the input text. Finally, it is an interpreter because its commands are decoded and executed as they are encountered.

The input consists of commands embedded in lines of text. These commands may take the form of either **substitutions** or **directives**. Two special characters are used to distinguish commands from ordinary text: the **substitution prefix character (CSUB)** and the **directive prefix character (CDIR)**. For the remainder of this appendix the CSUB is assumed to be $ and the CDIR is assumed to be *. The processor operates as follows. Each input line is first scanned for occurrences of the CSUB and if substitutions are required, then they are performed. Then the line is examined to determine if it is a directive line or a text line. If the first nonblank character in the line is the CDIR, then the directive is decoded and executed. Otherwise, the line is assumed to be a text line and is copied to the output file. Note that substitutions may take place in both directive and text lines.

One should view the use of this processor as consisting of two phases. In the first, macro names and their associated values are entered into the symbol table. In the second, text containing directives and references to macros defined in the first phase is processed. In the simplest case, the symbol table might consist of the macros STUDENT and ID along with their respective values 'John Jones' and 22215. The text input might be $STUDENT, ID = $ID and the output would be John Jones, ID = 22215. The processor has three levels of commands:

1. **Directives which build or manipulate the symbol table**:

*APPEND	Append a value to a variable.
*APPEND, *ENDAPP	Append lines of text to a variable.
*DELETE	Delete a variable from the table.
*RESET	Reset a list pointer.
*SET	Assign a value to a variable.
*SET, *ENDSET	Assign lines of text to a variable or perform multiple assignments.

2. **Macro invocations**:

$(NAME) or $NAME	Substitute the value of NAME.
$DEF(NAME)	Substitute '.TRUE.' if NAME is defined and '.FALSE.' otherwise.
*INCLUDE(NAME)	Include lines of text.
$LABEL or $(LABEL)	Substitute a new label.
$LIST(ITEMS)	Substitute the next item from the list ITEMS.
$$	Replace '$$' by '$'. (Embeds a single CSUB in the text.)

3. **Control directives**

*COMMENT, *ENDCOM	Ignore the enclosed comment lines.
*DO, *ENDDO	Specify loops.
*END	End of input.
*IF, *ELSE, *ENDIF	Conditional processing of lines.

C.3 SIMPLE EXAMPLE

The template processor is invoked by a subroutine call of the form

```
      CALL   TPDRV (EUNIT, IUNIT, LUNIT, OUNIT),
```

where EUNIT is the unit number of the error file, IUNIT is the unit number of the input file, LUNIT is the unit number of the listing file, and OUNIT is the unit number of the output file. In the following example, four files are used: The first input file, assigned to unit 4, contains definitions of macros referenced in the second input file, assigned to unit 5. The error and listing files are assigned to 6 and the output file is assigned to unit 7.

Main program:

```
      CALL   TPDRV   (6, 4, 6, 7)
      CALL   TPDRV   (6, 5, 6, 7)
      STOP
      END
```

First input file (on unit 4).

```
*SET
       LASTNAME = 'DOE'
       FIRSTNAME = 'JOHN'
       MONTH = 08
       DAY = 24
       YEAR = 81
       SEMESTER = 'FALL'
*ENDSET
*SET ( NCOURSES = 3 )
*SET ( COURSES = 'BIO 255$$/' )
*APPEND ( COURSES, 'GEO 110$$/' )
*APPEND ( COURSES, 'PSY 201' )
*END
```

TPDRV applied to the first input file merely creates a symbol table for later use.

Second input file (on unit 5).

```
*SET ( NAME = '$$LASTNAME, $$FIRSTNAME' )
*SET ( DATE = '$$MONTH/$$DAY/$$YEAR' )
NAME:          $NAME
DATE:          $DATE
SEMESTER:      $SEMESTER
LIST OF COURSES:
*DO ( I = 1, NCOURSES )
         $I.    $LIST(COURSES)
*ENDDO
*END
```

The processor output on unit 7 is:

```
NAME:          DOE, JOHN
DATE:          08/24/81
SEMESTER:      FALL
LIST OF COURSES:
          1.   BIO 255
          2.   GEO 110
          3.   PSY 201
```

TPDRV also produces a listing of these two files on unit 6.

Note that an **item separator**($/) is embedded in the text by specifying $$/; otherwise, the processor would expect a macro name to follow the CSUB. Also note that the same output results if the two input files are concatenated and read in as one.

C.4 SUBSTITUTION PROCESSING FACILITIES

SIMPLE SUBSTITUTIONS

A reference to the **marco** NAME may be specified by $NAME or by $(NAME). The latter form of substitution is required when the substitution takes place within a block of text and it is necessary to separate the characters of NAME from those in the surrounding text. NAME may be an alphanumeric string of any length but its first character must be alphabetic. A macro is also called a **template variable** and its value is an arbitrary character string. The substitution operator **$** applied to a macro results in the string elements being evaluated according to the rules of the template processor.

Substitutions may be recursive. For example, if we have

Name	Value
A	A(B)(E)
B	BC$(D)
D	D
E	E

then $A results in substituting ABCDE. The depth to which substitutions may be nested depends only on the amount of available workspace. Recursive substitution, along with memory management for the processor and the implementation of do loops and lists, all make use of the same pool of pointers. Therefore, a depth of recursion which is impossible when the symbol table is relatively full might easily be achieved when some space is returned (see *DELETE).

If a single occurrence of $ (the CSUB) is needed in a string, then $$ should be specified. This is similar to the technique used by some FORTRAN compilers to embed quotes in quoted strings.

THE $DEF FUNCTION

The $DEF function is used to determine if a name is defined in the symbol table. $DEF(NAME) returns the string .TRUE. if NAME is defined and returns .FALSE. otherwise. This function is used with the *IF directive discussed below.

THE $LABEL FUNCTION

The special variable LABEL is used to generate Fortran labels or other sequences of increasing integers. Each time LABEL is referenced, its value is incremented by 1. Thus, when $LABEL or $(LABEL) is encountered in the text, the current value of LABEL is substituted, and LABEL is incremented. LABEL must be initialized using a *SET statement and may be assigned a new value at any time. Legal values for LABEL are integers between 1 and 99999 inclusive. If it is set to any other string, an error message is generated the next time LABEL is referenced.

THE $LIST FUNCTION

Lists of items may be created and substituted. The list separator is $/ and thus

```
*SET(XYZ = ' A $$/ B+C $$/ D*E $$/ ')
*SET(THREELINES)
      THIS IS LINE 1
   $$/THIS IS LINE 2
   $$/THIS IS LINE 3
   $$/
*ENDSET
```

creates two lists: XYZ = (A, B+C, D*E) and THREELINES is

```
THIS IS LINE 1
THIS IS LINE 2
THIS IS LINE 3
```

The statement

```
*APPEND (XYZ, ' F-G $$/ HIJ $$/ ')
```

lengthens the list XYZ by adding the two items F−G and HIJ.

The $LIST function is used to substitute items from a list one at a time. Thus the input text

```
X = $LIST(XYZ)
Y = $LIST(XYZ)
Z = $LIST(XYZ) - ($LIST(XYZ))
```

results in the output text

```
X = A
Y = B+C
Z = D*E - (F-G)
```

The next reference to $LIST(XYZ), wherever it might be, produces HIJ. This function treats a variable as a list of items separated by the marker $/. Each list has a pointer which is incremented each time the list is referenced. The pointer may be reset to the start of the list. Thus the input

```
*RESET(XYZ)
K = $LIST(XYZ)
```

results in

```
K=A
```

If the RESET were not present, one would obtain K=HIJ instead. If the pointer reaches the end of the list, then the next reference to the list results in an error message and no substitution is made.

Special Escape Characters

The use of special escape characters has already been illustrated; the complete list of these characters follows:

- $ **Substitution prefix character (CSUB).** It signals that the next name is a variable whose value is to be substituted. The character is doubled to $$ to obtain a $ inside a quoted string.
- * **Directive prefix character (CDIR).** It signals the beginning of a processor directive provided it is the first non-blank character on a line. Otherwise * is an ordinary character.
- $– **End-of-line marker (CEOL).** This special character is placed at the end of every input line. Its effect can br overridden by the continuation marker.
- $+ **Continuation marker (CONC).** This special character at the end of a line of input overrides the end-of-line marker and continues the line with the text from the next line.
- $/ **List item separator (CEOR).** It separates items in a list.

The input text

```
*SET ( AVERYLONGNAME = 'A VERY $+
LONG STRING' )
```

assigns the same value to AVERYLONGNAME as the line

```
*SET ( AVERYLONGNAME = 'A VERY LONG STRING' )
```

Variables may contain end-of-line markers, if LINES has the value

```
T = A $-        A = B $-        B = T $-
```

then *INCLUDE(LINES) results in

```
T = A
A = B
B = T
```

It has already been noted that $$ is used to embed an occurrence of the character $ in a string for later processing. Note that

```
*SET(LINES = ' T=A $$- A=B $$- B=T $$-')
```

or

```
*SET(LINES)
      T=A
      A=B
      B=T
*ENDSET
```

can be used to assign LINES the value used above. If the double $$ is not used in the first case, then the processor attempts to make an immediate substitution for $–; this, of course, fails as there is no variable with – as a name.

Finally note that lists may contain lines and lines may contain lists but the end-of-line marker $- does not indicate an end-of-item marker $/ or vice-versa.

C.5 MACRO PROCESSING DIRECTIVES

This section lists (in alphabetical order) the directives of the template processor which provide the facilities to create the symbol table and to control the substitution phase.

*APPEND AND *ENDAPP

The first form of this construct is

*APPEND (*name*1 , *name*2 or *literal*).

where *literal* is a quoted character string, an integer constant or one of the logical constants .TRUE. or .FALSE. The second form of this statement is

*APPEND (*name*1)
  lines of text
*ENDAPP

This statement is similar in syntax and function to the SET command except that instead of assigning a value to the name, it appends (or, if you prefer, concatenates) the indicated value to the string already associated with *name*1. To emphasize this difference a comma (,) instead of an equal (=) is used to separate the name from the right side. Note that it is much more efficient to append one variable to another by using *APPEND (A, B) than by using *SET (A = 'AB').

*COMMENT AND *ENDCOM

Input files may be documented by bracketing comment lines with the directives *COMMENT and *ENDCOM. All lines between these two directives are ignored.

*DELETE

The statement

*DELETE (*name*)

deletes the variable *name* from the symbol table.

*DO AND *ENDDO

This control statement has a form similar to the Fortran DO-loop:

*DO (*iname* = *i*1, *i*2, *i*3)

The variable *iname* is the loop index and *i*1, *i*2 and *i*3 are the initial, final and incremental values, respectively. The behavior of this construct is very similar to its FORTRAN 66 counterpart, that is

a) The index may not be modified within the range of the do loop.

b) The loop is always performed at least once.

c) *i*1 is the starting value for iname, *i*2 is the termination value, and *i*3, which is optional, is the step size for index. *i*1, *i*2, and *i*3 must be integers or names of template variables containing integer values.

d) The do range is closed by the statement *ENDDO.

e) The depth to which do loops may be nested is restricted only be the amount of available workspace.

To illustrate its use suppose one has a list COEFS with values −12.3, 16.2, −4.9, 8.2,... . Then the three statements

```
*DO (I = 1, 5, 2)
      A($I,1) = B($I) + ($LIST(COEFS))
*ENDDO
```

result in

```
      A(1,1) = B(1) + (-12.3)
      A(3,1) = B(3) + ( 16.2)
      A(5,1) = B(5) + ( -4.9)
```

***END**

This statement terminates the execution of the macro processor.

***IF, *ELSE, AND *ENDIF**

This control statement has the form

*IF (*expression*) directive or text line

or

*IF (*expression*)
    lines in the true range
*ELSE
    lines in the false range
*ENDIF

The *expression* inside the parentheses following the IF must be one of the following:

name (with a logical value)
logical constant (.TRUE., .FALSE.)
*name*1 = *name*2
name = *literal*
name = *integer constant*

For the one-line form, if the value is true, then the line following the closing parenthesis is processed, otherwise it is ignored. For the multi-line form, if the value is true, then the lines in the true range are processed, otherwise, those in the false range are processed. *ELSE's are optional and *IF's may be nested to any depth as illustrated by the following example.

```
*SET  ( TYPEV = 'VALU' )
*SET  ( L1 = 'VALU' )
*SET  ( L2 = 6 )
*SET  ( L3 = .FALSE. )
*IF   ( L1 = TYPEV )
        A = B
        *IF ( L2 = 6 )
                      B = C
                      *ELSE
                      B = D
        *ENDIF
        *IF ( L3 )          C = D
        *IF ( $DEF(L3))           D = E
*ENDIF
```

results in

```
        A = B
                      B = C
        D = E
```

Note that even though L3 is .FALSE., $DEF(L3) is .TRUE. because L3 has been assigned a value.

*INCLUDE

*INCLUDE(*name*) is an alternate form of substitution; it must appear on a line by itself. Its purpose is to allow one to substitute lines of text without performing the substitutions within these lines. There is a switch (see *OPTION below) to turn substitution off and on. If this switch is on, then *INCLUDE(*name*) and $(*name*) are identical. The use of this facility is illustrated by the text

```
*SET(LINSYSCALL)
   *IF(TIMER)
      *INCLUDE(TIME1)
   *ENDIF
   CALL $NAME($MATRIX,$SOLUTION,$RHS,$NUMBEQNS,WORK,IER)
   *IF(TIMER)
      *INCLUDE(TIME2)
   *ENDIF
*ENDSET
```

When this text is processed with the substitution switch off and TIMER = .TRUE., the value of LINSYSCALL has lines of code before and after the subroutine call which time the execution of the subroutine. For example, we might have

```
*SET(TIME2)
   CALL SECOND(TIME2)
   TIME(KTIME) = TIME2-TIME1
   PRINT $TIMELABEL, TIME(KTIME), $NAME
   KTIME = KTIME+1
*ENDSET
```

Later in the processing there will probably be a line *INCLUDE(LINSYSCALL) with the substitution switch turned on. At that point the values for $NAME and $TIMELABEL are substituted in the code to print out the timing information and the name of the subroutine being timed.

***OPTION**

This command has the form

*OPTION (*option name* = *name* or *literal*)

and is used to set internal flags and values which affect the operation of the macro processor. The following table lists the options with their defaults.

Table C.1. The template processor options and defaults values.

Name	*Default*	*Meaning*
CDIR	*	The directive prefix character *.
CEOL	–	The second character of the end-of-line marker.
CEOR	/	The second character of the end-of-item marker.
CONC	+	The second character of the continuation marker.
CSUB	$	The substitution prefix character.
ICPLI	72	The number of characters per line of input.
ICPLO	72	The number of characters per line of output.
IUNITI	5	The input unit number.
IUNITL	6	The listing unit number.
IUNITO	7	The output unit number.
LBREAK	.TRUE.	Try to break lnes longer than ICPLO at a convenient break character.
LCOL1	.TRUE.	Check only column 1 in the input file for occurrences of the CDIR.
LFORT	.TRUE.	Write out long lines by providing Fortran continuation cards.
LISTI	.TRUE.	List input lines as they are read in.
LISTO	.TRUE.	List output lines as they are written out.
LSUB	.TRUE.	Start processing substitutions.
L1TRIP	.TRUE.	Use 1-trip do-loops.

Note that option names which begin with C, I, and L expect a value which consists of a character, an integer, or a logical value, respectively.

***RESET**

*RESET(*name*) resets the list pointer of *name* to the beginning of the list.

***SET AND *ENDSET**

The first form of this statement is

*SET (*name*1 = *name*2 or *literal*).

The value to be assigned on the right of the = may be either a variable name, in which case the value of *name*2 is assigned to *name*1, or a *literal*, in which case the *literal* itself is assigned to *name*1.

The second form of this statement is

*SET (*name*1)
    lines of text *ENDSET

This causes the variable *name*1 to take on as its value the lines of text up to the next matching *ENDSET. End-of-line markers are automatically supplied. These lines of text are processed only for substitution (if the substitution switch is on) and are not examined for directives. In particular, if these lines contain another *SET – *ENDSET pair, the inner *SET will not be processed. For example,

```
*SET( A )
     *SET( B )
          X = Y + Z
     *ENDSET
*ENDSET
```

causes A to take on the value

```
     *SET ( B )$-          X = Y + Z$-       *ENDSET$-'
```

The variable B, however, will not be given a value until $A is encountered later.

The third form of this statement is the multiple assignment and is illustrated by

```
*SET
     NAME1 = VALUE1
     NAME2 = VALUE2
     NAME3 =
             LINE A
             LINE B
*
     NAME4 = VALUE4
*ENDSET
```

The distinguishing feature of this form is that there is no name following *SET. The result of this statement is identical to

```
*SET ( NAME1 = VALUE1 )
*SET ( NAME2 = VALUE2 )
*SET ( NAME3 )
             LINE A
             LINE B
*ENDSET
*SET ( NAME4 = VALUE4 )
```

Note that the * on a line by itself ends a group of text lines in a multiple assignment *SET.

C.6 ANOTHER EXAMPLE

The following example illustrates how the macro processor may be used to generate a program segment. Some background information is needed to understand this example. First, the resulting program calls LINPACK routines [Dongarra et. al., 1979] to factor and possibly solve a system of linear equations. Recall that the LINPACK routine SGECO factors a matrix and produces an estimate of its condition number, SGEFA only factors the matrix, and SGESL takes the factored matrix and solves a linear system given a right side. The double precision versions of these routines are DGECO, DGEFA, and DGESL, respectively. Finally, this example assumes that the template variables CONDNO,

N, SINGLE, and SOLVE have already been set elsewhere to appropriate values.

Input text:

```
*IF (SINGLE)
        *SET ( DECL = 'REAL' )
        *SET ( PREFIX = 'S' )
*ELSE
        *SET ( DECL = 'DOUBLE PRECISION' )
        *SET ( PREFIX = 'D' )
*ENDIF
        $DECL  A($N,$N)
*IF (CONDNO)
        $DECL  RCOND, WORK($N)
*ENDIF
*IF (SOLVE)
        $DECL B($N)
*ENDIF
        INTEGER IPVT($N)
        READ(5,*) A
*IF (CONDNO)
        CALL  $(PREFIX)GECO (A, $N, $N, IPVT, RCOND, WORK)
        WRITE(6,*) RCOND
*ELSE
        CALL  $(PREFIX)GEFA (A, $N, $N, IPVT, INFO)
*ENDIF
*IF (SOLVE)
        READ(5,*) B
        CALL  $(PREFIX)GESL (A, $N, $N, IPVT, B, 0)
        WRITE(6,*) B
*ENDIF
        STOP
        END
```

Output text: assuming SINGLE = .TRUE., CONDNO = .FALSE., SOLVE = .TRUE., and N = 10,

```
        REAL  A(10,10)
        REAL  B(10)
        INTEGER  IPVT(10)
        READ(5,*) A
        CALL  SGEFA (A, 10, 10, IPVT, INFO)
        READ(5,*) B
        CALL  SGESL (A, 10, 10, IPVT, B, 0)
        WRITE(6,*) B
        STOP
        END
```

Output text: assuming SINGLE = .FALSE., CONDNO = .TRUE., SOLVE = .FALSE., and N = 5,

```
        DOUBLE PRECISION  A(5,5)
        DOUBLE PRECISION  RCOND, WORK(5)
        INTEGER  IPVT(5)
        READ(5,*) A
        CALL  DGECO (A, 5, 5, IPVT, RCOND, WORK)
        WRITE(6,*) RCOND
        STOP
        END
```

C.7 PREPARING A TUNED VERSION OF THE TEMPLATE PROCESSOR

The template processor may be tuned by setting the following variables to appropriate values and then applying the basic processor to the template for the template processor. The default value is .FALSE. for all the logical variables listed.

CDC if .TRUE., a Purdue CDC compatible version is produced.

CSTAR1 if .TRUE., Fortran 77 declarations of the form CHARACTER*1 are used instead of integer declarations for Hollerith variables.

DEBUG if .TRUE., MNF (a Fortran compiler) trace statements will be inserted. This should only be used if CDC = .TRUE.

ICBDIM the dimension of the array CBUFFR. This dimension limits the size of text that the APPEND, INCLUDE and SET statements can process. Default = 1500.

ICSDIM the dimension of the array CSTORE. The dimension limits the total amount of information in the symbol table. Default = 20,000

IHADIM the dimension of the array IHASH. This dimension limits the number of names used. This should be a prime number. Default = 809.

ISTDIM the dimension of the array ISTORE. This dimension limits the number of pointers to the main storage area CSTORE. Should be less than ICSDIM. Default = 6000.

NOPACK if .TRUE., all references to the array CSTORE are direct (in-line) instead of being forced through subroutines.

SHORTB if .TRUE. and CDC = .TRUE., short file buffers are used.

STATS if .TRUE., MNF timing statements will be inserted. This should only be used if CDC = .TRUE.

TESTCH if .TRUE., character testing used to check for alphabetic and numeric is performed using in-line IF statements instead of being isolated in separate subroutines.

UNIX if .TRUE., a UNIX compatible version is produced.

C.8 ERROR MESSAGES

Each error message is given preceded by the name of the subroutine which detected the error.

IOREAD – BUFFER SPACE EXCEEDED
Cause: An input line or the multi-line text of an APPEND, INCLUDE, or SET was too long. Solution: Make the text shorter or recompile the processor with CBUFFR dimensioned larger.

MMGET1 – STRING TOO LONG FOR CVALUE(*)
Cause: The buffer passed to MMGET1 was too short. If you are not calling subroutines of the processor yourself, then this argument is probably the internal array CBUFFR. Solution: Recompile the processor with a larger dimension for CBUFFR.

MMHASH – HASH TABLE ARRAY IHASH(*) IS FULL

Cause: The hash table IHASH has been filled because the maximum number of names has been exceeded. Solution: Recompile the processor with a larger dimension for IHASH.

MMNEWI – STORAGE ARRAY ISTORE(*) IS FULL

Cause: All of the pointers used in dynamic memory management have been allocated. This may be due to memory fragmentation or simply to a large number of macro definitions. Solution: Recompile the processor with a larger dimension for ISTORE.

MMPOPC – STRING TOO LONG FOR BUFFER

Cause: The buffer passed to MMPOPC was too short. If you are not calling subroutines of the processor yourself, then this argument is probably the internal array CBUFFR. Solution: Recompile the processor with a larger dimension for CBUFFR.

MMPUT1 – STORAGE ARRAY CSTORE(*) FULL

Cause: The character storage array CSTORE has been filled. Solution: Recompile the processor with a larger dimension for CSTORE.

MMRETI – ATTEMPT TO RETURN INVALID POINTER

Cause: Should not occur under normal circumstances. Solution: Report the problem to the ELLPACK group at Purdue University.

MPITEM – VARIABLE NOT DEFINED

Cause: A reference was made to an undefined macro. Solution: Check to see if the variable name is spelled properly and if it is actually defined.

MPLABL – ILLEGAL LABEL VALUE

Cause: The variable LABEL was assigned a non-integral value or a value that was longer than 5 digits or nonpositive. Solution: Correct LABEL assignment.

MPMAC –VARIABLE NOT DEFINED

Cause: A reference was made to an undefined macro. Solution: Check to see if the variable name is spelled properly and if it is actually defined.

MPPOPN – ILLEGAL VARIABLE NAME

Cause: An illegal variable name was encountered. Solution: Check to make sure the name begins with an alphabetic character and that it contains no special characters.

MPPOPN – MISSING RIGHT PARENTHESIS

Cause: Unbalanced parentheses
Solution: balance parentheses.

TPAPPE – APPEND HAS NO MATCHING ENDAPP

Cause: A multi-line APPEND has no closing ENDAPP. This is only detected when an END statement is encountered, so at this point the APPEND statement has gobbled up the end of your template.

TPCOMM – COMMENT HAS NO MATCHING ENDCOM

Cause: COMMENT has no closing ENDCOM. This is only detected when an END statement is encountered, so at this point the COMMENT statement has gobbled up the end of your template.

TPDO – DO HAS NO MATCHING ENDDO
Cause: A DO statement is not closed by an ENDDO. This is only detected when an END statement is encountered, so at this point the DO statement has gobbled up the end of your template.

TPELSE – IF HAS NO MATCHING ENDIF
Cause: An IF statement is not closed by an ENDIF. This is only detected when an END statement is encountered, so at this point the IF statement has gobbled up the end of your template.

TPENDO – MISPLACED ENDDO
Cause: An ENDDO not preceded by a matching DO has been encountered.

TPENDF – MISPLACED ENDIF
Cause: An ENDIF not preceded by a matching IF has been encountered.

TPEVAL – MISPLACED ENDAPP
Cause: An ENDAPP not preceded by a matching APPEND has been encountered.

TPEVAL – MISPLACED ENDCOM
Cause: An ENDCOM not preceded by a matching COMMENT has been encountered.

TPEVAL – MISPLACED ENDSET
Cause: An ENDSET not preceded by a matching SET has been encountered.

TPEVAL – ILLEGAL OR MISSPELLED DIRECTIVE
Cause: An unknown directive has been encountered.

TPIF –IF HAS NO MATCHING ENDIF
Cause: An IF has no closing ENDIF. This is only detected when an END statement is encountered, so at this point the IF statement has gobbled up the end of your template.

TPINCL – VARIABLE NOT DEFINED
Cause: A reference was made to an undefined macro. Solution: Check to see if the variable name is spelled properly and if it is actually defined.

TPOPT –ILLEGAL OR MISSPELLED OPTION
Cause: An unknown option has been encountered.

TPOPT –OPTION REQUIRES SINGLE CHARACTER
Cause: An incorrect value has been supplied to an option which requires a single character.

TPOPT –OPTION REQUIRES AN INTEGER
Cause: An incorrect value has been supplied to an option which requires a integer value.

TPOPT –OPTION REQUIRES A LOGICAL VALUE
Cause: An incorrect value has been supplied to an option which requires a logical value of .TRUE. or .FALSE.

TPSET –SET HAS NO MATCHING ENDSET
Cause: A multi-line SET has no closing ENDSET. This is only detected when

an END statement is encountered, so at this point the SET statement has gobbled up the end of your template.

TPSYNT – LEFT PARENTHESIS EXPECTED
Cause: A left parenthesis was expected but not found.

TPSYNT – RIGHT PARENTHESIS EXPECTED
Cause: A right parenthesis was expected but not found.

TPSYNT – COMMA EXPECTED
Cause: A comma was expected but not found.

TPSYNT – EQUALS SIGN EXPECTED
Cause: An equals sign was expected but not found.

TPSYNT – VARIABLE EXPECTED
Cause: A special character or constant was encountered where a variable name was expected.

TPSYNT – CONSTANT OR VARIABLE EXPECTED
Cause: A constant or variable name was expected but not found.

TPSYNT – ILLEGAL CHARACTERS AT END OF LINE
Cause: Extra characters were found at the end of the input line.

C.9 REFERENCES

Dongarra, J. J., J. R. Bunch, C. B. Moler and G. W. Stewart [1979], *LINPACK User's Guide*, Soc. Indust. Appl. Math., Philadelphia, PA.

Ribbens, C. J., J. R. Rice and W. A. Ward [1984], "A simple macro processor", ACM Trans. Math. Software 10, Decmeber.

Rice, J. R. and W. A. Ward [1982a], "A simple macro processor", CSD-TR 400, Computer Science Dept., Purdue University.

Rice, J. R. and W. A. Ward [1982b], "A simple macro processor — User's guide", CSD-TR 403, Computer Science Dept., Purdue University.

Appendix D

EXAMPLES OF SOME ELLPACK STATEMENTS

Each ELLPACK program must contain exactly one EQUATION segment, one BOUNDARY segment, and one END segment. Some examples of the first two follow (the first boundary segment form is only valid for rectangular domains).

```
EQUATION.   UXX + UYY + UZZ = F(X,Y,Z)

EQUATION.   UXX + 2.0*UYY + D(X,Y)*UY - U = SIN(X)*EXP(Y)

BOUNDARY.   PERIODIC              ON  X =  0.0
                                  ON  X =  1.0
            U = 0.0               ON  Y = -1.0
            UY + 2.0*U = G(X)     ON  Y =  1.0

BOUNDARY.   UX = 0.0    ON  X=PI, Y=T          FOR T=-1.0 TO 5.0
            U  = 0.0    ON  LINE  PI,5.0 TO 0.0,5.0
            UX = 0.0    ON  LINE 0.0,5.0 TO 0.0,1.0
            SIN(X)*UX + UY = -SINH(Y)  &
                        ON  X=T,  Y=COS(T) FOR T=0.0 TO PI
```

The GRID segment is used to impose a rectangular grid on the domain. Several examples follow (the first form is only valid for rectangular domains).

```
GRID.       11 X POINTS    $    17 Y POINTS

GRID.       11 X POINTS    0.0 TO 1.0
             9 Y POINTS    0.0, .1, .15, .2, .3, .5, .7, .8, 1.0
```

Finally, we list below the output produced by the procedure module LIST MODULES. This gives the names and default parameter values of all problem solving modules (discretization, indexing, solution, triple, output, and procedure) in the current ELLPACK system.

```
 ------------------
 PROCEDURE MODULE
 ------------------

     L I S T   M O D U L E S

     DISCRETIZATION MODULES

     5-POINT STAR
     7-POINT 3D
     COLLOCATION( BCP1    = 0.,     BCP2     = 0.,          &
            DSCARE = .05,  GIVOPT = 1,          &
            USECRN = .FALSE.)
     HERMITE COLLOCATION (BCP1=0.0, BCP2=0.0)
```

```
HODIE (IORDER = 4)
HODIE-ACF (METHOD = -1)
HODIE-HELMHOLTZ ( IORDER = 4, METHOD = 0 )
INTERIOR COLLOCATION
SPLINE GALERKIN(DEGREE=3,NDERV=2)
```

INDEXING MODULES

```
AS IS
HERMITE COLLORDER
INTERIOR COLLORDER
MINIMUM DEGREE
NESTED DISSECTION (NDTYPE = 5)
RED-BLACK (LEVEL = I1LEVL)
REVERSE CUTHILL MCKEE
```

SOLUTION MODULES

```
BAND GE
BAND GE NO PIVOTING
ENVELOPE LDLT
ENVELOPE LDU
JACOBI CG (ITMAX = 100,                                      &
   LEVEL = I1LEVL,IADAPT = 1,ICASE = 1,IDGTS = 0,            &
   ZETA = AMAX1(5.E-6,5.E2*R1EPSM),                          &
   CME = 0.,SME = 0.,FF = .75,OMEGA = 1.,                    &
   SPECR = 0.,BETAB = 0.25)
JACOBI SI  (ITMAX = 100,                                     &
   LEVEL = I1LEVL,IADAPT = 1,ICASE = 1,IDGTS = 0,            &
   ZETA = AMAX1(5.E-6,5.E2*R1EPSM),                          &
   CME = 0.,SME = 0.,FF = .75,OMEGA = 1.,                    &
   SPECR = 0.,BETAB = 0.25)
LINPACK BAND
LINPACK SPD BAND
REDUCED SYSTEM CG (ITMAX = 100,                              &
   LEVEL = I1LEVL,IADAPT = 1,ICASE = 1,IDGTS = 0,            &
   ZETA = AMAX1(5.E-6,5.E2*R1EPSM),                          &
   CME = 0.,SME = 0.,FF = .75,OMEGA = 1.,                    &
   SPECR = 0.,BETAB = 0.25)
REDUCED SYSTEM SI (ITMAX = 100,                              &
   LEVEL = I1LEVL,IADAPT = 1,ICASE = 1,IDGTS = 0,            &
   ZETA = AMAX1(5.E-6,5.E2*R1EPSM),                          &
   CME = 0.,SME = 0.,FF = .75,OMEGA = 1.,                    &
   SPECR = 0.,BETAB = 0.25)
SOR (ITMAX = 100, LEVEL = I1LEVL, IADAPT = 1,                &
  ICASE = 1,    IDGTS = 0,                                 &
  ZETA = AMAX1(5.E-6,5.E2*R1EPSM),                         &
  CME = 0.,     SME = 0.,FF = .75,OMEGA = 1.,              &
  SPECR = 0.,   BETAB = 0.25)
SPARSE GE NO PIVOTING
SPARSE LU UNCOMPRESSED
SYMMETRIC SOR CG (ITMAX = 100,                               &
   LEVEL = I1LEVL,IADAPT = 1,ICASE = 1,IDGTS = 0,            &
   ZETA = AMAX1(5.E-6,5.E2*R1EPSM),                          &
   CME = 0.,SME = 0.,FF = .75,OMEGA = 1.,                    &
   SPECR = 0.,BETAB = 0.25)
SYMMETRIC SOR SI  (ITMAX = 100,                              &
   LEVEL = I1LEVL,IADAPT = 1,ICASE = 1,IDGTS = 0,            &
   ZETA = AMAX1(5.E-6,5.E2*R1EPSM),                          &
   CME = 0.,SME = 0.,FF = .75,OMEGA = 1.,                    &
   SPECR = 0.,BETAB = 0.25)
```

TRIPLE MODULES

```
DYAKANOV-CG ( ITMAX = 100,  DEMAND = 3.0 )
DYAKANOV-CG 4 (ITMAX = 100,  DEMAND = 3.0)
FFT 9-POINT (IORDER = 4)
FISHPAK-HELMHOLTZ
```

```
HODIE 27-POINT 3D
HODIE-FFT (IORDER = 4)
HODIE-FFT 3D (IORDER = 4)
MARCHING ALGORITHM (KGMA = 2)
MULTIGRID MG00 (METHOD=0, UINIT=0, NMIN=2, INEUM=0, &
          ITER=0, IGAMMA=1)
P2C0-TRIANGLES (MEM = 0, NTRI = 0)
SET (U=ZERO)
SET U BY BICUBICS
SET U BY BLENDING
```

OUTPUT MODULES

```
DATA
MAX
NORM
PLOT
PLOT-DOMAIN
RMS
SUMMARY
TABLE
TABLE-BOUNDARY
TABLE-EQUATIONS
TABLE-INDEXES
TABLE-PROBLEM
TABLE-UNKNOWN
TABLE-DOMAIN
```

PROCEDURE MODULES

```
ARC
DISPLAY MATRIX PATTERN ( MATZER = '.', MATNZR = 'X', &
  MATDZR = '0', MATDNZ = 'D', MATBLK = I1NEQN,     &
  MATLNL = 120, EPSMAT = 0.0, MATNBR = 0,          &
  MATNBC = 0,   MATOUT = I1OUTP )
DOMAIN
DOMAIN FILL(NFILL=1,EXTER=.FALSE.)
EIGENVALUES (SCALE = 1.0)
HOLE
LIST MODULES
NON-UNIQUE (X=R1AXGR, Y=R1AYGR, Z=R1AZGR, U=0.0)
PLOT COLLOCATION POINTS (                       &
       BCP1   = 0.,    BCP2   = 0.,     &
       DSCARE = .05,   PTSIZE = 6.,     &
       GIVOPT = 1,     IDPLOT = 0,      &
       USECRN = .FALSE.)
REMOVE (R=ZERO, HXSTEP=-1., HYSTEP=-1., HZSTEP=-1.)
REMOVE BLENDED BC
REMOVE BICUBIC BC
SET UNKNOWNS FOR 5-POINT STAR (UEST=ZERO)
SET UNKNOWNS FOR HODIE-HELMHOLTZ (UEST=ZERO)
```

INDEX